W9-AFO-428

Applied Mathematical Sciences
Volume 110

Applied Mathematical Sciences

1. *John:* Partial Differential Equations, 4th ed.
2. *Sirovich:* Techniques of Asymptotic Analysis.
3. *Hale:* Theory of Functional Differential Equations, 2nd ed.
4. *Percus:* Combinatorial Methods.
5. *von Mises/Friedrichs:* Fluid Dynamics.
6. *Freiberger/Grenander:* A Short Course in Computational Probability and Statistics.
7. *Pipkin:* Lectures on Viscoelasticity Theory.
8. *Giacoglia:* Perturbation Methods in Non-linear Systems.
9. *Friedrichs:* Spectral Theory of Operators in Hilbert Space.
10. *Stroud:* Numerical Quadrature and Solution of Ordinary Differential Equations.
11. *Wolovich:* Linear Multivariable Systems.
12. *Berkovitz:* Optimal Control Theory.
13. *Bluman/Cole:* Similarity Methods for Differential Equations.
14. *Yoshizawa:* Stability Theory and the Existence of Periodic Solution and Almost Periodic Solutions.
15. *Braun:* Differential Equations and Their Applications, 3rd ed.
16. *Lefschetz:* Applications of Algebraic Topology.
17. *Collatz/Wetterling:* Optimization Problems.
18. *Grenander:* Pattern Synthesis: Lectures in Pattern Theory, Vol. I.
19. *Marsden/McCracken:* Hopf Bifurcation and Its Applications.
20. *Driver:* Ordinary and Delay Differential Equations.
21. *Courant/Friedrichs:* Supersonic Flow and Shock Waves.
22. *Rouche/Habets/Laloy:* Stability Theory by Liapunov's Direct Method.
23. *Lamperti:* Stochastic Processes: A Survey of the Mathematical Theory.
24. *Grenander:* Pattern Analysis: Lectures in Pattern Theory, Vol. II.
25. *Davies:* Integral Transforms and Their Applications, 2nd ed.
26. *Kushner/Clark:* Stochastic Approximation Methods for Constrained and Unconstrained Systems.
27. *de Boor:* A Practical Guide to Splines.
28. *Keilson:* Markov Chain Models—Rarity and Exponentiality.
29. *de Veubeke:* A Course in Elasticity.
30. *Shiatycki:* Geometric Quantization and Quantum Mechanics.
31. *Reid:* Sturmian Theory for Ordinary Differential Equations.
32. *Meis/Markowitz:* Numerical Solution of Partial Differential Equations.
33. *Grenander:* Regular Structures: Lectures in Pattern Theory, Vol. III.
34. *Kevorkian/Cole:* Perturbation Methods in Applied Mathematics.
35. *Carr:* Applications of Centre Manifold Theory.
36. *Bengtsson/Ghil/Källén:* Dynamic Meteorology: Data Assimilation Methods.
37. *Saperstone:* Semidynamical Systems in Infinite Dimensional Spaces.
38. *Lichtenberg/Lieberman:* Regular and Chaotic Dynamics, 2nd ed.
39. *Piccini/Stampacchia/Vidossich:* Ordinary Differential Equations in $\mathbf{R}^n$.
40. *Naylor/Sell:* Linear Operator Theory in Engineering and Science.
41. *Sparrow:* The Lorenz Equations: Bifurcations, Chaos, and Strange Attractors.
42. *Guckenheimer/Holmes:* Nonlinear Oscillations, Dynamical Systems and Bifurcations of Vector Fields.
43. *Ockendon/Taylor:* Inviscid Fluid Flows.
44. *Pazy:* Semigroups of Linear Operators and Applications to Partial Differential Equations.
45. *Glashoff/Gustafson:* Linear Operations and Approximation: An Introduction to the Theoretical Analysis and Numerical Treatment of Semi-Infinite Programs.
46. *Wilcox:* Scattering Theory for Diffraction Gratings.
47. *Hale et al:* An Introduction to Infinite Dimensional Dynamical Systems—Geometric Theory.
48. *Murray:* Asymptotic Analysis.
49. *Ladyzhenskaya:* The Boundary-Value Problems of Mathematical Physics.
50. *Wilcox:* Sound Propagation in Stratified Fluids.
51. *Golubitsky/Schaeffer:* Bifurcation and Groups in Bifurcation Theory, Vol. I.
52. *Chipot:* Variational Inequalities and Flow in Porous Media.
53. *Majda:* Compressible Fluid Flow and System of Conservation Laws in Several Space Variables.
54. *Wasow:* Linear Turning Point Theory.
55. *Yosida:* Operational Calculus: A Theory of Hyperfunctions.
56. *Chang/Howes:* Nonlinear Singular Perturbation Phenomena: Theory and Applications.
57. *Reinhardt:* Analysis of Approximation Methods for Differential and Integral Equations.
58. *Dwoyer/Hussaini/Voigt (eds):* Theoretical Approaches to Turbulence.
59. *Sanders/Verhulst:* Averaging Methods in Nonlinear Dynamical Systems.
60. *Ghil/Childress:* Topics in Geophysical Dynamics: Atmospheric Dynamics, Dynamo Theory and Climate Dynamics.

(continued following index)

Odo Diekmann Stephan A. van Gils
Sjoerd M. Verduyn Lunel Hans-Otto Walther

Delay Equations

Functional-, Complex-, and Nonlinear Analysis

With 34 Illustrations

Springer-Verlag
New York Berlin Heidelberg London Paris
Tokyo Hong Kong Barcelona Budapest

Odo Diekmann
Centrum voor Wiskunde en Informatica
1090 GB Amsterdam
The Netherlands
and
Instituut voor Theoretische Biologie
Rijkuniversiteit Leiden
2311 GP Leiden
The Netherlands

Stephan A. van Gils
Faculteit der Toegepaste Wiskunde
Universiteit Twente
7500 AE Enschede
The Netherlands

Sjoerd M. Verduyn Lunel
Faculteit der Wiskunde en Informatica
Universiteit van Amsterdam
1018 TV Amsterdam
The Netherlands

Hans-Otto Walther
Mathematisches Institut
Justus-Liebig-Universität
35392 Giessen
Germany

Mathematics Subject Classification (1991): 76P05, 82C40, 82B40

Library of Congress Cataloging-in-Publication Data
Delay equations: functional-, complex-, and nonlinear analysis / Odo
Diekmann . . . [et al.].
p. cm. – (Applied mathematical sciences; v. 110)
Includes bibliographical references and index.
ISBN 0-387-94416-8
1. Delay differential equations. I. Diekmann, O. II. Series:
Applied mathematical sciences (Springer-Verlag New York Inc.); v.
110.
QA1.A647 vol. 110
[QA371]
510 s–dc20
[515'.35] 94-41858

Printed on acid-free paper.

Production managed by Natalie Johnson; manufacturing supervised by Joseph Quatela.
Photocomposed using the authors' TEX files.
Printed and bound by R.R. Donnelley & Sons, Harrisonburg, VA.
Printed in the United States of America.

9 8 7 6 5 4 3 2 1

ISBN 0-387-94416-8 Springer-Verlag New York Berlin Heidelberg

Preface

The aim of this book is to provide an introduction to the mathematical theory of infinite dimensional dynamical systems by focussing on a relatively simple, yet rich, class of examples, viz. those described by delay differential equations.

It is a textbook giving detailed proofs and many exercises, which is intended both for self-study and for courses at a graduate level. It should also be suitable as a reference for basic results. As the subtitle indicates, the book is about concepts, ideas, results and methods from linear functional analysis, complex function theory, the qualitative theory of dynamical systems and nonlinear analysis.

It gives a motivated introduction to the theory of semigroups of linear operators, emphasizing duality theory and neglecting analytic semigroups (thus it is complementary to an introduction to infinite dimensional dynamical systems focussing on the other relatively simple, yet rich, class of examples, i.e., scalar reaction diffusion equations in one space dimension). It contains an exposition of spectral theory, with special attention to those operators for which all spectral information is contained in an analytic matrix valued function. It introduces the calculus of exponential types of entire functions and exploits this calculus to investigate the behaviour of the resolvent of the generator at infinity, which is a main step to characterise the closure of the span of all eigenvectors and generalized eigenvectors and to investigate the (non-) existence of so-called "small" solutions, which converge to zero faster than any exponential. Essentially, these are Laplace transform methods.

The variation-of-constants formula is the main tool in the development of the local stability and bifurcation theory of equilibrium solutions of nonlinear problems. The center manifold and Hopf bifurcation are treated in detail. Stability of periodic solutions is discussed in terms of Floquet multipliers and Poincaré maps. Subsequently a more global point of view is adopted to study the existence of periodic solutions, in particular so-called slowly oscillating solutions. Here the topological degree and fixed-point theorems are the main tools. A survey of known results on the global dynamics of solutions of delay equations (including some results on chaotic behaviour) completes the book.

From the point of view of applications the most important chapter is perhaps the one on characteristic equations which deals, often by means of examples, with techniques to find the region in parameter space corresponding to the stability of a steady state. At the boundary of that region, bifurcations take place. A formula for the direction of Hopf bifurcation serves as an algorithm to compute this direction in concrete examples. This is often as far as one can get analytically to find out about the possibility of coexistence of local attractors.

After studying this book the reader should have a working knowledge of applied functional analysis and dynamical systems. For purely minded analysts we expect that they become aware of the charm of concrete problems, where often the main difficulty is to find the right mathematical setting. For application oriented readers we expect that they learn to appreciate the extra understanding that mathematical rigour often entails. For people trained in ordinary differential equations the book shows what aspects of operator theory are essential when working in infinite dimensional state spaces. For readers with an operator background it introduces the main ideas concerning the behaviour of dynamical systems. Thus we hope that many different types of readers will find something of value in the book and will, while reading, experience some of the same enjoyment that we had while writing.

It is NOT a handbook for the use of delay equations as mathematical models of physical or biological phenomena. (In fact our opinion is that it is dangerous to model directly in terms of delay equations: careful modelling requires a mechanistic interpretation of the state of a system; of course it is perfectly all right if a delay equation results in the end, possibly after some transformation [255], but one should avoid starting to think in terms of such equations.) Throughout the book, however, it is shown (most of the time by means of exercises) how age dependent population models are covered by exactly the same mathematical theory.

The book was written in many episodes, scattered over a period of approximately six years. Often the obligation to restart to work on it felt like a burden, but when a little later other duties forced us to stop working on it, this felt as an even bigger nuisance. In between, fortunately, it was a pleasure. So now that the project is finished we feel mostly relief but in addition a little bit of excitement since now, finally, the fruits of our efforts are ready for the ultimate test of any book: do you, reader, like it or not?

Amsterdam, December 1994

Odo Diekmann
Stephan van Gils
Sjoerd Verduyn Lunel
Hans-Otto Walther

Contents

Chapter 0

Introduction and preview

0.1 An example of a retarded functional differential equation

Imagine a biological population composed of adult and juvenile individuals. Let $N(t)$ denote the density of adults at time t. Assume that the length of the juvenile period is exactly h units of time for each individual. Assume that adults produce offspring at a per capita rate α and that their probability per unit of time of dying is μ. Assume that a newborn survives the juvenile period with probability ρ and put $r = \alpha\rho$. Then the dynamics of N can be described by the differential equation

$$\dot{N}(t) = -\mu N(t) + rN(t-h) \tag{1.1}$$

which involves a nonlocal term, where N has argument $t - h$, since newborns become adults with some delay. So the rate of change of N involves the current as well as the past values of N. Such equations are called Retarded Functional Differential Equations (RFDE) or, alternatively, Delay Equations.

Equation (1.1) describes the change in N. To fix N, we need an initial condition, say at $t = 0$ (i.e., we start our clock at the time we prescribe the condition).

Example 1.1. The solutions $t \mapsto \sin(\frac{\pi}{2}(t+\frac{1}{2}))$ and $t \mapsto \cos(\frac{\pi}{2}(t+\frac{1}{2}))$ of the equation

$$\dot{x}(t) = -\frac{\pi}{2}x(t-1)$$

coincide at $t = 0$.

It is not enough to specify $N(0)$, since we need to know what to take instead of $rN(t-h)$ for $0 \leq t < h$. So we have to prescribe a function on an interval of length h. The most convenient (though not the most natural from a biological point of view) manner to do this is to prescribe N on the interval $[-h, 0]$ and then to use (1.1) for $t \geq 0$.

So we supplement (1.1) by

$$N(\theta) = \varphi(\theta), \qquad -h \leq \theta \leq 0, \tag{1.2}$$

where φ is a given function. Explicitly we then have for $t \in [0, h]$

$$N(t) = \varphi(0)e^{-\mu t} + r \int_0^t e^{-\mu(t-\tau)}\varphi(\tau - h)\, d\tau. \tag{1.3}$$

Using this expression we can give an expression for N on the interval $[h, 2h]$. Etcetera. Thus the *method of steps* and elementary theory of ordinary differential equations (ODE) provide us with an existence and uniqueness proof.

The key question, of course, is whether the population will ultimately grow without bound or become extinct. In other words, we want to determine the asymptotic behaviour for $t \to \infty$ and how this depends on the parameters r and μ. If we formally substitute $N(t) = N(0)e^{zt}$ into (1.1), we arrive at the *characteristic equation*

$$z = -\mu + re^{-zh}. \tag{1.4}$$

Exercise 1.2.

(i) Show that (1.4) has exactly one real root. Call this root λ_d.

(ii) Show that $\operatorname{Re}\lambda < \lambda_d$ for all other roots λ (the subscript d refers to "dominant" and this qualification should now be clear).

(iii) Show that $\lambda_d > 0$ if $r\mu^{-1} > 1$, whereas $\lambda_d < 0$ if $r\mu^{-1} < 1$.

(iv) Verify that $r\mu^{-1}$ can be interpreted as the expected number of offspring produced by a newborn individual and that, consequently, the result of (iii) is exactly what one expects on the basis of the biological interpretation.

Combining the results of this exercise with our intuition (derived, say, from the theory of ODE) we are led to

Conjecture 1.3. If $\varphi(\theta) \geq 0$ with φ not identically zero, then

$$N(t) \sim e^{\lambda_d t} \quad \text{for } t \to \infty$$

and so the population will grow exponentially when $r > \mu$ and become extinct when $r < \mu$.

In this book we shall introduce techniques and prove general theorems from which the correctness of this conjecture follows. In Chapter I we shall use Laplace transform methods to study *linear* equations, like (1.1), and find that they are quite sufficient for this class of equations. However, if we go beyond and study *nonlinear* equations we need a different perspective as well as other methods.

If competition during the juvenile period influences the probability ρ of survival, we have to replace (1.1) by something else. The equation

$$\dot{N}(t) = -\mu N(t) + f(N(t-h)) \tag{1.5}$$

describes the situation in which competition takes place among individuals in the same age group only. Note that the appropriate initial condition again takes the form (1.2). What we now want is a qualitative theory for equations like (1.5) in much the same spirit as the one for ODE. In this book we shall develop the basic elements of such a theory of dynamical systems in infinite dimensional spaces, using delay equations like (1.1) and (1.5) and similar age-structured population models as our motivating examples.

0.2 Solution operators

In the preceding section we noticed that the information contained in a function φ defined on $[-h, 0]$ is needed and suffices to uniquely fix the future. Hence such a function qualifies as the *state*, at time 0, of the system we describe. Let, for an autonomous (i.e., time translation invariant) system, $S(t)$ denote the operator assigning to the state at some time the state t units of time later. Then necessarily, because of this interpretation,

(i) $S(0) = I$ (the identity operator),

(ii) $S(t+s) = S(t)S(s), \qquad t, s \geq 0,$

where the second property derives from the *uniqueness*. These properties are summarized by saying that the family $\{S(t)\}_{t\geq 0}$ forms a one-parameter semigroup of operators (the adjective "semi" expresses that backward solutions do not necessarily exist or, in other words, that t is restricted to nonnegative values). For a given initial state φ the *orbit* through φ is the subset $\{S(t)\varphi \mid t \geq 0\}$ of the state space. Ever since Poincaré, an important aim in the theory of dynamical systems is to give a qualitative, geometric, description of the collection of all orbits, the so-called phase portrait, possibly restricting attention to the neighbourhood of some steady state (an orbit consisting of one point), periodic solution (a closed orbit) or more complicated invariant set. In bifurcation theory one studies changes in the qualitative properties as parameters involved in the definition of $\{S(t)\}_{t\geq 0}$ vary.

If we want to let these ideas bear on delay equations, we have to specify the state space and to elaborate the definition of $S(t)$.

Exercise 2.1. Let $f : \mathbb{R} \to \mathbb{R}$ be a continuous function and $h > 0$. Use the method of steps to prove that each continuous initial function $\varphi : [-h, 0] \to \mathbb{R}$ extends to a continuous function $N : [-h, \infty) \to \mathbb{R}$ which is differentiable for $t > 0$ and such that (1.2) and, for $t > 0$, (1.5) are satisfied. Show that N is uniquely determined by φ.

This exercise suggests choosing as a state space the Banach space

$$X = C\big([-h, 0], \mathbb{R}\big)$$

of continuous real functions on the interval of length h, provided with the supremum norm. This turns out to be a good choice. Note, however, that also data from larger spaces of measurable and bounded functions on the interval of length h would uniquely define solutions (in a slightly weaker sense) on the domain $[-h, \infty)$, provided we specify precisely the value at $t = 0$. In Chapter II we will find that such a space enters the scene quite naturally even if we start doing analysis in X as defined above.

The state at time t is the "piece" of the extended function N on the interval of length h preceding t, i.e., $\theta \mapsto N(t + \theta; \varphi)$, for $-h \leq \theta \leq 0$.

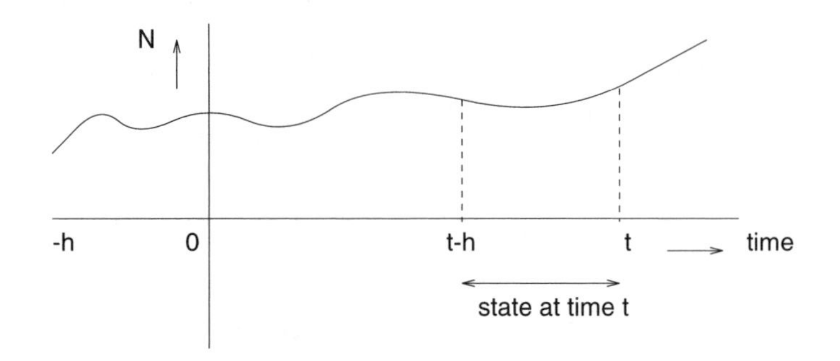

Fig. 0.1. The state at time t.

So we restrict $N(\,\cdot\,;\varphi)$ to the interval $[t - h, t]$ and then, in order to have again an element of X, we perform a shift back to the interval $[-h, 0]$. In still other words and symbols,

$$S(t)\varphi = N_t(\,\cdot\,;\varphi),$$

where we use the notation $x_t(\theta) = x(t + \theta)$. We emphasize that this definition involves two ingredients: we extend φ by solving the RFDE and then we select the relevant piece of information by a shift along the extended function.

Exercise 2.2.

(i) Verify that we obtain, in this manner, a semigroup of operators, starting from equation (1.5).

(ii) Verify that the semigroup $S(t)$ is strongly continuous, which means by definition that for any $\varphi \in X$,

$$\|S(t)\varphi - \varphi\| \to 0 \quad \text{as } t \downarrow 0.$$

Exploit the semigroup property to conclude from this that $t \mapsto S(t)\varphi$ is continuous, i.e., orbits are continuous curves.

The *generator* (in this book denoted by A) of a semigroup of operators is by definition the derivative at $t = 0$. For an infinite dimensional state space this is, as a rule, an unbounded operator. In the theory of partial differential equations one often starts from an abstract differential equation

$$\frac{du}{dt} = Au$$

and then shows that there is a semigroup of solution operators associated with A. Since delay equations are not exactly of this form, the theory for these proceeds somewhat differently. In particular, one first defines the semigroup constructively and only then "computes" the generator.

0.3 Synopsis

In this book we shall study the qualitative properties of the solution operators of delay equations. The emphasis is on autonomous equations, which describe systems for which there are no time-varying inputs. The delays in the equations will always be bounded. See Hino, Murakami and Naito [125] for the theory of equations with unbounded delay.

In the first six chapters we restrict attention to linear equations.

Chapter I deals with the Laplace transform method, a classical approach to represent and analyse solutions of linear autonomous RFDE. First, the one-to-one relation between RFDE and renewal equations (RE) is explained. The latter are Volterra integral equations of convolution type. The convolution character makes them suitable for (inverse) Laplace techniques.

An asymptotic expansion of the solution of the RE is obtained by shifting the path of integration of the inverse Laplace transform to the left. The solutions of the characteristic equation of the RFDE are encountered as poles in this process. Crossing of a finite number of poles means "splitting off" an exponential polynomial (sums of exponentials with polynomial coefficients).

So by this procedure, the solution of an initial-value problem for a RFDE is represented by an exponential polynomial and a remainder term. Several questions arise. What can be said about the convergence of the asymptotic series that is obtained by including more and more poles? If there is no convergence for all initial data, can one determine data for which one has convergence? What can be said about the remaining solutions?

In Chapters II and III, the concept of a solution operator is introduced. As already remarked in Section 0.2, there are two ingredients involved in its construction: extending the function and shifting. In Chapter II the role of the shifting part is analysed in great detail for a linear equation where the extension part is trivial, i.e., the right hand side of the equation is zero. This defines a linear semigroup $\{T_0(t)\}$ on the space X. We compute the adjoint semigroup, $\{T_0^*(t)\}$, on X^*, and its restriction to the largest domain of strong continuity. This is denoted by $\{T_0^{\odot}(t)\}$ ($\odot$ is pronounced as sun) acting on $X^{\odot}$. With assiduity we go through the process of taking the adjoint and restricting to the domain of strong continuity once more, to obtain $\{T_0^{\odot *}(t)\}$ on $X^{\odot *}$ and $\{T_0^{\odot\odot}(t)\}$ on $X^{\odot\odot}$. In the case of delay equations $X^{\odot\odot}$ is isomorphic to X and we call the semigroup $\odot$-reflexive.

It so happens that the space $X^{\odot *}$ is essential for the precise functional analytic formulation of the variation-of-constants formula,which is the most important vehicle for the transfer of results from linear to nonlinear differential equations in local stability and bifurcation theory. It involves, in the case of delay equations, objects which do not "live" in the space X, but in the larger space $X^{\odot *}$. This has made the theory of delay equations somewhat mysterious for people with a functional analytic background. This difficulty was overcome when sun-star calculus was developed in an attempt to unify the theories of delay equations and age-structured population dynamics. One of the reasons to write this book was to demonstrate how sun-star calculus can be used with great advantage when developing the basic theory of RFDE.

In Chapter III it is shown that the general linear autonomous RFDE, with semigroup $\{T(t)\}$, can be viewed as a bounded perturbation of the above-mentioned trivial RFDE. For a bounded perturbation B, mapping X into $X^{\odot *}$, the variation-of-constants equation

$$u(t) = T_0(t-s)u(s) + \int_s^t T_0^{\odot *}(t-\tau)Bu(\tau)\,d\tau, \tag{3.1}$$

which is an abstract integral equation (AIE) involving the notion of a weak* integral, is derived. The linear autonomous RFDE is an example of a bounded perturbation with finite dimensional range and in this case, reduction of the abstract integral equation to a (finite dimensional) renewal equation is possible. Thus we see the connection between the abstract approach and the direct approach of Chapter I.

Chapter IV deals first with the spectral decomposition of the state space for an eventually compact semigroup, such as the one associated with a RFDE. Next it is shown how one can obtain all spectral information from the characteristic matrix (here we need the theory of Jordan chains and the notion of equivalence of operator valued functions). In particular we recover within the abstract framework the decomposition of solutions into an exponential polynomial and a lower order, for $t \to \infty$, remainder as found in Chapter I by the Laplace transform method.

Chapter V is concerned with answers to questions which arose in Chapter I. Results about entire functions, growth properties and exponential type calculus prepare the way to take up the question of convergence of the asymptotic expansions for solutions to linear autonomous RFDE and associated RE. Sufficient conditions for uniform convergence on unbounded intervals are given. The fact that one needs conditions to ensure convergence reflects the possible existence of *small solutions*, i.e., solutions which decay faster than exponentially as $t \to \infty$. These solutions cannot be approximated by the expansions of Chapter I.

The nonexistence of small solutions is characterized in terms of growth properties of the characteristic function whose zeros determine the spectrum of the generator. A closely related problem is: How far is the system of generalized eigenvectors from being complete (in the sense that the closure of the span is the full state space X)? A characterization of completeness and a description of the general situation are given.

Chapter VI on inhomogeneous equations is a link between linear and nonlinear theory. Following the philosophy of Chapter III we are led to consider the AIE

$$u(t) = T(t-s)u(s) + \int_s^t T^{\odot *}(t-\tau)F(\tau)\,d\tau, \qquad t \geq s, \tag{3.2}$$

with F a continuous mapping into $X^{\odot *}$. This is the integrated version of the abstract differential equation

$$\frac{du}{dt}(t) = A^{\odot *}u(t) + F(t). \tag{3.3}$$

Solutions u given by (3.2) are usually called mild solutions of (3.3). In a general context the question in what sense u given by (3.2) satisfies (3.3) is a delicate one. It turns out that this is irrelevant in the context of delay equations. A one-to-one relation between solutions of the inhomogeneous linear RFDE and the corresponding AIE is given explicitly. As an application we derive the Fredholm alternative for periodic inhomogeneities F.

In Chapter VII we collect basic results on existence, uniqueness and smoothness of solutions of nonlinear equations in a parameter dependent context. We derive the principle of linearized stability, which asserts that the stability of a stationary point can be inferred by examining the stability with respect to the linearized equation.

The next three chapters deal with the local theory for nonlinear RFDE. Linearization of a nonlinear RFDE at a stationary point yields the strongly continuous semigroup of solution operators $T(t)$. The spectrum of the generator A is decomposed into the following parts: σ_- in the open left half-plane, σ_0 on the imaginary axis and σ_+ in the open right half-plane. The spectral sets $\sigma_-, \sigma_0, \sigma_+$ define subspaces X_-, X_0, X_+, respectively, which are positively invariant under the semigroup; trajectories in the stable space X_- converge exponentially to 0 as $t \to \infty$. Trajectories in the center space

X_0 and in the unstable space X_+ have extensions in these spaces which are defined on the whole real line. The extensions in X_+ converge to 0 as $t \to -\infty$.

In Chapter VIII we obtain under the hyperbolicity assumption

$$\sigma_0 = \emptyset$$

the existence of the local stable and unstable manifolds. These are tangent to X_- and X_+, respectively, and consist of orbits which behave qualitatively like the orbits of $\{T(t)\}$ in X_- and in X_+, respectively.

Chapter IX contains the construction of local center manifolds, which are tangent to X_0 and positively invariant. Included is a detailed proof that these manifolds are continuously differentiable in case the nonlinearity has this property.

Chapter X applies the previous theory to the phenomenon of Hopf bifurcation, i.e., the appearance of periodic orbits close to a stationary point of a one-parameter family of equations. We derive a formula for the direction of bifurcation which determines whether the bifurcating periodic orbits appear below or above the critical parameter, which is, in many important situations, decisive for their stability character.

The local results of Chapters VIII, IX and X rely on hypotheses about the spectrum or, in other words, about the roots of the characteristic equation associated with the linearized RFDE. There is no general theory which describes how to locate these roots. In Chapter XI we collect a number of techniques, often ad hoc and illustrated in the context of a concrete example, which help to determine the position of the roots relative to the imaginary axis. For most of the examples we also compute the direction of Hopf bifurcation. These are probably the most useful parts of the book for applied mathematicians and scientists that seek to obtain conclusions for concrete models taking the form of a RFDE or something similar.

When we linearize about an orbit that consists of more than one point, we obtain a nonautonomous linear problem. Such problems are touched upon briefly in Chapter XII where we discuss the evolutionary system of linear solution operators

$$U(t,s), \quad t \geq s,$$

which is associated with the initial-value problem for a time-dependent linear RFDE. Then in Chapter XIII we concentrate on the periodic case and the definition of Floquet multipliers. In the ODE context, Floquet theory is used to transform a periodic linear system to an autonomous linear system. We derive an analogous result in the abstract setting.

The results of Chapter XIII are used in Chapter XIV where we consider nonlinear (autonomous) AIE and RFDE in neighbourhoods of periodic orbits, i.e., close to the simplest invariant sets which are not points. The behaviour of orbits close to a periodic orbit is described in terms of the Poincaré map which assigns to points in a transversal hyperplane the intersection of the corresponding trajectory with the same hyperplane at a later time. The Poincaré map has a fixed point at the intersection of the periodic orbit with the hyperplane.

The construction of the Poincaré map requires local smoothness of the semiflow. In the case of RFDE this will, in general, only be satisfied if the period considered is larger than the delay h in the equation.

The linearization of a Poincaré map at its fixed point on the periodic orbit is closely related to the monodromy operators associated with the linear variational equation along the periodic orbit. We establish the precise relations between these maps, their spectra, their generalized eigenspaces and Floquet multipliers under the condition that all spectral points of the monodromy operators except 0 are isolated.

In the final two chapters we no longer concentrate on the neighbourhood of a special solution but instead adopt a more global point of view. As a consequence we need methods from nonlinear analysis beyond the implicit function theorem, in particular the fixed-point index and global bifurcation theorems.

In Chapter XV we study existence of periodic solutions for the nonlinear autonomous RFDE

$$\dot{x}(t) = f(x(t-h)) \tag{3.4}$$

from a global point of view, for nonlinearities which model negative feedback with respect to a stationary state. We consider continuous functions $f : \mathbb{R} \to \mathbb{R}$ which satisfy

$$f(0) = 0 \quad \text{and} \quad \xi f(\xi) < 0 \quad \text{for all } \xi \neq 0.$$

An example for which periodic solutions can explicitly be computed is given.

Scaling the variable t and setting $\alpha = h$ we obtain a one-parameter family of equations

$$\dot{x}(t) = \alpha f(x(t-1)), \quad \alpha > 0, \tag{3.5}$$

on the state space $X = C\big([-1,0],\mathbb{R}\big)$. We prove a global bifurcation theorem due to Nussbaum which asserts, under minimal smoothness and boundedness conditions on f, that an unbounded continuum of nontrivial periodic orbits bifurcates from the zero solution at the critical parameter value

$$\alpha_0 = \frac{\pi}{2f'(0)}.$$

For every $\alpha > \alpha_0$ there exist periodic solutions. The periodic orbits are obtained from fixed points of a return map which is defined by the intersection of trajectories with a cone, analogous to a Poincaré map. We employ a deeper analysis of invariant sets and unstable behaviour of trajectories which in turn permits a straightforward calculation of the index. The methods we introduce play a role also in further studies of the global dynamics generated by nonlinear RFDE.

The final chapter (Chapter XVI) surveys some further results in this direction. We focus on global attractors for RFDE and on existence of chaotic dynamics. In the second section we study the equation

$$\dot{x}(t) = -\mathrm{sign}(x(t-1)). \tag{3.6}$$

This equation with a discontinuous nonlinearity is a limiting case of equations of the form (3.5). It generates a semiflow which can be computed explicitly. In particular, we expose the structure of the global attractor which is expected for the simplest classes of smooth monotone functions f in equation (3.5).

There are eight appendices dealing with

— Functions of bounded variation and integration theory;
— Semigroups of bounded linear operators and their adjoints;
— Holomorphic functions and symbolic calculus;
— Substitution (Nemitskiĭ) operators and their smoothness;
— Banach manifolds;
— Fixed points of parameterized contractions;
— Answers to the exercises on age-dependent population dynamics;
— Hopf bifurcation for ODE.

We have tried to make both the main text and the appendices readable by themselves. As a consequence, some definitions, arguments and results are repeated at different places (this is in particular the case for linear semigroups of operators). We hope that our readers will find this convenient, rather than irritating.

0.4 A few remarks on history

Single differential equations with deviating argument were studied already before the American and French revolutions. Some of them were related to geometric problems. References to these older papers are contained in Erhard Schmitt's systematic exposition of properties of linear RFDE from the beginning of this century [248]. The interest in the field grew rapidly in the second half of the century. Monographs and textbooks that appeared since 1949 include [200, 16, 82, 201, 209, 83, 102, 78]. Many publications on RFDE which appeared in the past two decades have used J.K. Hale's monograph [102] as a basic and unifying reference. The new edition [104] of this book is co-authored by S.M. Verduyn Lunel.

Chapter I

Linear autonomous RFDE

I.1 Prelude: a motivated introduction to functions of bounded variation

Consider a function x defined on some interval of the real axis and taking values in $\mathbb{C}^n$ (we incorporate $\mathbb{R}^n$-valued functions by interpreting $\mathbb{R}^n$ as the subspace of $\mathbb{C}^n$ consisting of elements that have, for each component, a zero imaginary part). We are interested in situations where the variable on the real axis is interpreted as time, and therefore denoted by t, and where the derivative $\dot{x}$ of x at some time t depends linearly on the history of x, that is, on values of x for times less than or equal to t. In this chapter we concentrate on time translation invariant (or, in other words, *autonomous*) problems, where only the values of x matter and not the time t itself. In addition we assume that $\dot{x}$ depends on a finite segment of the history only. More precisely, we postulate the existence of $h > 0$ such that $\dot{x}(t)$ may depend (linearly) on $x(\tau)$ for $t - h \leq \tau \leq t$ but not on $x(\tau)$ for $\tau < t - h$ or for $\tau > t$. Given these restrictions we would like to consider a general equation. However, some more choices have to be made before we can state precisely what is the "general" equation.

Throughout this book x_t denotes the function defined on $[-h, 0]$ by

$$x_t(\theta) = x(t + \theta), \quad \text{for } -h \leq \theta \leq 0. \tag{1.1}$$

Let us assume that x_t is, for all t for which it is defined, a continuous mapping from $[-h, 0]$ into $\mathbb{C}^n$, or, in symbols, $x_t \in \mathcal{C}$, where $\mathcal{C} = C([-h, 0], \mathbb{C}^n)$ equipped with the supremum norm. The class of equations we want to consider in this chapter is

$$\dot{x}(t) = Lx_t, \tag{1.2}$$

where L is a continuous linear mapping from $\mathcal{C}$ into $\mathbb{C}^n$.

To facilitate further analysis it has advantages to represent the mapping L in terms of a function. It is here that functions of bounded variation enter the scene. We first state the main result and only thereafter give the necessary definitions and explanations (for general background information see Appendix I.2).

Theorem 1.1. *(A corollary of the Riesz representation theorem.) Let L be a continuous linear mapping from C into $\mathbb{C}^n$. There exists a unique NBV (normalized bounded variation) function ζ defined on $[0, h]$ with values in $\mathbb{C}^{n\times n}$ such that for all $\varphi \in C$*

$$L\varphi = \int_0^h d\zeta(\theta)\varphi(-\theta), \tag{1.3}$$

where the integral is an n vector whose i^{th} component is equal to

$$\sum_{j=1}^{n} \int_0^h \varphi_j(-\theta) d\zeta_{ij}(\theta).$$

The reason to let the function ζ be defined on $[0, h]$, rather than on $[-h, 0]$, will be explained in Section 2. We write $d\zeta$ in front of φ to indicate that ζ is $n \times n$ matrix-valued and φ is n vector-valued, while taking for granted that in complicated expressions it may be necessary to use brackets to indicate where the factors of the integrand "end". Finally we remark that one can alternatively (but equivalently, see Appendix I.2) represent L in terms of a measure. We choose to work with NBV kernels in order to follow the tradition.

The class of equations (1.2) can now alternatively be written as

$$\dot{x}(t) = \int_0^h d\zeta(\theta)x(t-\theta) \tag{1.4}$$

or, with the notation

$$\langle \zeta, \varphi \rangle_n = \int_0^h d\zeta(\theta)\varphi(-\theta) \tag{1.5}$$

as

$$\dot{x}(t) = \langle \zeta, x_t \rangle_n. \tag{1.6}$$

Here the subscript n is used to indicate that ζ corresponds to a collection of n functionals (the rows of the matrix) and thus defines a mapping L from C into $\mathbb{C}^n$. The rest of this section explains the terminology and the notations above by listing the definitions and some properties of bounded variation functions and Stieltjes integrals. Some more technical background and various other useful results are collected in Appendix I.1.

A *partition* $P = P(t_b, t_e)$ of $[t_b, t_e]$ is a finite ordered set $P = \{\sigma_0, \sigma_1, \ldots, \sigma_N\}$ such that $t_b = \sigma_0 < \sigma_1 < \cdots < \sigma_N = t_e$. The *width* of the partition is

$$\mu(P) = \max_{1 \le j \le N} (\sigma_j - \sigma_{j-1}) . \tag{1.7}$$

Let f be a given function defined on $[a, b]$ with values in $\mathbb{R}$. The *total variation function* $V(f)$ is defined by

$$V(f)(t) = \sup_{P(a,t)} \sum_{j=1}^{N} |f(\sigma_j) - f(\sigma_{j-1})|, \tag{1.8}$$

where the supremum is taken over all partitions of $[a, t]$. Clearly, $V(f)$ is a nonnegative and monotone nondecreasing function of t. The function $V(f)$ may be unbounded. When $V(f)(b)$ is bounded, we say that f is of *bounded variation*, or, in symbols, $f \in \text{BV}$, and we call

$$TV(f) = V(f)(b) \tag{1.9}$$

the *total variation* of f over $[a, b]$. When f is defined on $[a, \infty)$ or on $(-\infty, \infty)$ we use, mutatis mutandae, the same terminology with

$$TV(f) = \lim_{t \to \infty} V(f)(t).$$

To discuss normalization we need to know the following.

Theorem 1.2. *Let $f : [a, b] \to \mathbb{R}$ be of bounded variation then*

$$f(\tau+) = \lim_{\sigma \downarrow \tau} f(\sigma) \quad \text{exists for} \quad \tau \in [a, b)$$

and

$$f(\tau-) = \lim_{\sigma \uparrow \tau} f(\sigma) \quad \text{exists for} \quad \tau \in (a, b]$$

and the set $\{\tau \in (a, b) \mid f(\tau+) \neq f(\tau-)\}$ of discontinuities is at most countable.

We call a BV function $f : [a, b] \to \mathbb{R}$ *normalized* if $f(a) = 0$ and f is continuous from the right on the **open** interval (a, b) [that is, $f(\tau) = f(\tau+)$ at every point $\tau \in (a, b)$]. We shall write $f \in \text{NBV}$ to express that f is a normalized bounded variation function. [Our choice of normalization is motivated by the fact that it yields a simple representation for certain operators $T_0^*(t)$ in Section II.4.]

A complex-valued function f is of (normalized) bounded variation if and only if $\operatorname{Re} f$ and $\operatorname{Im} f$ are of (normalized) bounded variation. An n vector-valued function is of (normalized) bounded variation if and only if all its components are of (normalized) bounded variation.

Let $f : [a, b] \to \mathbb{C}^{n \times n}$ and $\varphi : [a, b] \to \mathbb{C}^n$ be given. For any partition P of $[a, b]$ and any choice of $\tau_j \in [\sigma_{j-1}, \sigma_j]$ we introduce the sum

$$S(f, \varphi, P) = \sum_{j=1}^{N} (f(\sigma_j) - f(\sigma_{j-1}))\varphi(\tau_j). \tag{1.10}$$

Suppose $A \in \mathbb{C}^n$ exists such that

$$\text{for all } \epsilon > 0 \qquad \exists\, \delta = \delta(\epsilon) > 0 \text{ such that } |A - S(f, \varphi, P)| < \epsilon$$

for all partitions P with width $\mu(P) < \delta$ and any choice of "intermediate" points τ_j. We then say that φ is *Riemann-Stieltjes integrable* with respect to f over $[a, b]$ [or, in short, $\varphi \in S(f)$] and we shall write

$$A = \int_a^b df(\tau)\varphi(\tau). \tag{1.11}$$

Consider the case $n = 1$. It is a classical result that $\varphi \in S(f)$ when $f \in \mathrm{BV}$ and $\varphi \in \mathcal{C}$ and that in this case

$$\Big| \int_a^b df(\tau)\varphi(\tau) \Big| \leq TV(f) \sup_{\tau \in [a,b]} |\varphi(\tau)|.$$

In words this says that f defines, via the Riemann-Stieltjes integral, a continuous linear functional on $\mathcal{C}$. The Riesz representation theorem asserts that *all* continuous linear functionals on $\mathcal{C}$ allow such a representation. However, it may occur that two different BV functions represent the same functional. [The easiest example is two BV functions that differ by a constant; note that the sum in (1.10) involves differences of two values of f only.] It is here that normalization is essential: when $f_1, f_2 \in \mathrm{NBV}$ are such that

$$\int_a^b df_1(\tau)\varphi(\tau) = \int_a^b df_2(\tau)\varphi(\tau)$$

for all $\varphi \in \mathcal{C}$, necessarily $f_1 = f_2$. So each continuous linear functional on $\mathcal{C}$ has a *unique* representation in terms of a NBV function and the Riemann-Stieltjes integral. We can reformulate this by saying that NBV, equipped with the total variation norm, is a representation of the dual space of $\mathcal{C}$.

We end with some properties of Riemann-Stieltjes integrals. Here we do not give sharp formulations to avoid further technicalities. Stronger results are presented in Appendix I.1.

Theorem 1.3. (*Integration by parts.*) *Let $f : [a, b] \to \mathbb{C}^{n \times n}$ belong to BV and let $\varphi : [a, b] \to \mathbb{C}^n$ belong to C^1; then*

$$\int_a^b df(\tau)\varphi(\tau) = f(b)\varphi(b) - f(a)\varphi(a) - \int_a^b f(\tau)\varphi'(\tau)\, d\tau. \tag{1.12}$$

Theorem 1.4. *Let $f : [a, b] \to \mathbb{C}^{n \times n}$ be of the form $f(\tau) = \int_a^\tau g(\sigma) d\sigma$ with $g \in L^1([a, b], \mathbb{C}^{n \times n})$; then*

$$\int_a^b df(\tau)\varphi(\tau) = \int_a^b g(\tau)\varphi(\tau) d\tau, \quad \textit{for all } \varphi \in \mathcal{C}.$$

Exercise 1.5. In the spirit of equation (1.2) and Theorem 1.1, determine $\zeta \in \mathrm{NBV}([0,h],\mathbb{C}^{n\times n})$ for each of the following equations:

(a) $\dot{x}(t) = x(t-1), \qquad h=1,\ n=1.$

(b) $\dot{x}(t) = x(t-1) - x(t-2), \quad h=2,\ n=1.$

(c) $\dot{x}_1(t) = -x_2(t-1), \quad \dot{x}_2(t) = 3x_1(t-2), \quad h=2,\ n=2.$

(d) $\dot{x}(t) = \int_{t-1}^{t} x(\tau)dt, \qquad h=1,\ n=1.$

(e) $\dot{x}(t) = \int_{t-1}^{t} x(\tau)dt + x(t-\frac{1}{2}), \quad h=1,\ n=1.$

I.2 Linear autonomous RFDE and renewal equations

Let $\zeta \in \mathrm{NBV} = \mathrm{NBV}\big([0,h],\mathbb{C}^{n\times n}\big)$ be given. As will become clear below, it has advantages to extend the domain of definition of ζ to all of $\mathbb{R}_+$ by putting $\zeta(\tau) = \zeta(h)$ for $\tau \geq h$. Throughout this book we will adopt this convention. We shall study the RFDE

$$\dot{x}(t) = \int_0^h d\zeta(\theta)x(t-\theta). \tag{2.1}$$

To single out a unique solution we have to provide an initial condition at a certain time, which we indicate by t_0. The initial condition should specify the values of x on an interval of length h preceding that time.

Let y satisfy (2.1) for $t \geq t_0$ and the initial condition

$$y(t_0+\theta) = \varphi(\theta), \qquad -h \leq \theta \leq 0,$$

where $\varphi \in \mathcal{C} = C\big([-h,0],\mathbb{C}^n\big)$. Then x, defined for $t \geq 0$ by $x(t) = y(t_0+t)$, satisfies (2.1) for $t \geq 0$ and the initial condition

$$x(\theta) = \varphi(\theta), \qquad -h \leq \theta \leq 0. \tag{2.2}$$

[This is what we mean when we say that (2.1) is time translation invariant.] So we can, without loss of generality, restrict our attention to an initial condition imposed at time zero.

Definition 2.1. A *solution* of the initial-value problem (2.1)–(2.2) on the interval I, where either $I = [-h,t_e)$ with $t_e \in (0,\infty)$ or $I = [-h,\infty)$, is a function $x \in C\big(I,\mathbb{C}^n\big)$ such that

(i) (2.2) holds;

(ii) on $I \cap (0,\infty)$, x is continuously differentiable and (2.1) holds;

(iii) $\lim_{t\downarrow 0} t^{-1}\big(x(t)-\varphi(0)\big)$ exists and equals $\int_0^h d\zeta(\theta)\varphi(-\theta)$.

This section is concerned with the existence, uniqueness and representation of a solution. For $0 \leq t \leq h$ we can combine the two separate pieces of information (2.1) and (2.2) and write

$$\dot{x}(t) = \int_0^t d\zeta(\theta)x(t-\theta) + \int_t^h d\zeta(\theta)\varphi(t-\theta), \tag{2.3a}$$

$$x(0) = \varphi(0). \tag{2.3b}$$

Integrating the first term at the right hand side of (2.3a) by parts we obtain [recall that $\zeta(0) = 0$!]

$$\dot{x}(t) = \int_0^t \zeta(\theta)\dot{x}(t-\theta)d\theta + g(t), \tag{2.4a}$$

$$x(0) = \varphi(0), \tag{2.4b}$$

where

$$g(t) = \zeta(t)\varphi(0) + \int_t^h d\zeta(\theta)\varphi(t-\theta). \tag{2.5}$$

Before we proceed, we introduce some notation and terminology. As usual, we denote the convolution product by $*$. So $k * f \in L^1$ is defined by

$$k * f(t) = \int_0^t k(t-\tau)f(\tau)dt, \tag{2.6}$$

where k is a (possibly $n \times n$ matrix-valued) L^1-function and f is a (possibly n vector- or $n \times n$ matrix-valued) L^1-function. We recall the estimate

$$\|k * f\|_1 \leq \|k\|_1 \, \|f\|_1 \tag{2.7}$$

for later use.

Exercise 2.2. Prove that the convolution product of a bounded function and an L^1-function is continuous.
Hint: Use the fact that the continuous functions are dense in L^1 to prove first that

(i) for $f \in L^1(\mathbb{R}_+)$, the function $t \mapsto \int_0^t f(\sigma)\,d\sigma$ is continuous (in fact, absolutely continuous, see Appendix II.2);

(ii) for $f \in L^1[0,\theta]$, the function $t \mapsto \int_0^\theta |f(t+\sigma) - f(\sigma)|\,d\sigma$ is continuous or, in words, translation is continuous in L^1. Here f is defined to be zero outside its original domain of definition $[0,\theta]$.

Exercise 2.3. Prove that the convolution product of two (N)BV functions belongs to (N)BV $\cap\, C$.

Remark. See the paper [194] by Mikusin'ski and Ryll-Nardzewski for further results on the smoothness of the convolution product, given certain assumptions on the two factors.

Equations of the form

$$y = \zeta * y + f, \tag{2.8}$$

where the *kernel* ζ and the *forcing function* f are given and y is the unknown, are called (linear) *renewal equations* or, alternatively, Volterra convolution integral equations (of the second kind). We shall use the abbreviation RE for renewal equation.

Accordingly, (2.4a) is a RE for $\dot{x}$ with kernel ζ and forcing function g defined in terms of ζ and φ by (2.5).

Lemma 2.4. *The function g defined by* (2.5) *is continuous.*

Proof. A simple manipulation yields

$$\begin{aligned} g(t_1) - g(t_2) &= (\zeta(t_1) - \zeta(t_2))\varphi(0) - \int_{t_2}^{t_1} d\zeta(\theta)\varphi(t_1 - \theta) \\ &\quad + \int_{t_2}^{h} d\zeta(\theta)[\varphi(t_1 - \theta) - \varphi(t_2 - \theta)] \\ &= \int_{t_2}^{t_1} d\zeta(\theta)[\varphi(0) - \varphi(t_1 - \theta)] \\ &\quad + \int_{t_2}^{h} d\zeta(\theta)[\varphi(t_1 - \theta) - \varphi(t_2 - \theta)]. \end{aligned}$$

Now choose $|t_1 - t_2|$ so small that $|\varphi(\sigma) - \varphi(\sigma + \tau)| < \frac{1}{2}(TV(\zeta))^{-1}\epsilon$ for $\sigma \in [-h, 0]$ and $|\tau| \leq |t_1 - t_2|$ with $\sigma + \tau \in [-h, 0]$ (this is possible due to the uniform continuity of φ). Then $|g(t_1) - g(t_2)| < \frac{\epsilon}{2} + \frac{\epsilon}{2} = \epsilon$. □

A solution of (2.4a) is now, by definition, a continuous function $\dot{x}$ defined for $0 \leq t < t_e$ with $0 < t_e \leq \infty$. Given $\dot{x}$, we can define x for $t > 0$ by

$$x(t) = \varphi(0) + \int_0^t \dot{x}(\tau)d\tau \tag{2.9}$$

and for $-h \leq \theta \leq 0$ by

$$x(\theta) = \varphi(\theta).$$

Our manipulations have shown that there is a one-to-one correspondence between solutions of the RFDE (2.1) with initial condition (2.2) and solutions of the RE (2.4a) satisfying (2.4b) and with g given by (2.5). It should now be clear that we defined ζ on $[0,h]$ (rather than introducing the kernel on [-h,0]) in order to facilitate the reformulation as a renewal equation.

By integration we can rewrite (2.4) as a renewal equation for x:

$$\begin{aligned} x(t)-\varphi(0) &= \int_0^t \int_0^s \zeta(\theta)\dot{x}(s-\theta)d\theta ds + \int_0^t g(s)ds \\ &= \int_0^t \zeta(\theta)\int_\theta^t \dot{x}(s-\theta)dsd\theta + \int_0^t g(s)ds \\ &= \int_0^t \zeta(\theta)[x(t-\theta)-x(0)]d\theta + \int_0^t g(s)ds \end{aligned}$$

and so

$$x = \zeta * x + f \tag{2.10}$$

with

$$\begin{aligned} f(t) &= \varphi(0) + \int_0^t g(s)\,ds - \int_0^t \zeta(\theta)\,d\theta\varphi(0) \\ &= \varphi(0) + \int_0^t \Big(\int_s^h d\zeta(\theta)\varphi(s-\theta)\Big)\,ds. \end{aligned} \tag{2.11}$$

Exercise 2.5. Verify that

$$f(t) = \varphi(0) + \int_0^h [\zeta(t+\sigma)-\zeta(\sigma)]\varphi(-\sigma)\,d\sigma. \tag{2.12}$$

Hint: First write

$$\int_s^h d\zeta(\theta)\varphi(s-\theta) = \int_0^h d_\sigma\zeta(\sigma+s)\varphi(-\sigma),$$

where the notation d_σ is used to make it unambiguously clear that σ is the integration variable; next use Fubini's theorem (Appendix I.2) and Theorem 1.4.

Lemma 2.6. *The function f defined in terms of ζ and φ by (2.11) is Lipschitz continuous on $[0,h]$ and constant for $t \geq h$.*

Proof.

$$\frac{df}{dt}(t) = \int_t^h d\zeta(\theta)\varphi(t-\theta)$$

and this is a bounded function of t which vanishes for $t \geq h$. □

Our manipulations show that any solution of (2.4a) and (2.4b), with g given by (2.5), yields a solution of (2.10), with f given by (2.11). To prove the converse statement we would have to show that any solution of (2.10) with f given by (2.11) is differentiable. We can, however, avoid that exercise by proving that both (2.4) and (2.10) admit one and only one solution. So we now concentrate on proving existence and uniqueness of solutions of renewal equations. We shall do so in an L^1-context first and only later discuss regularity under various assumptions on kernel and forcing function.

The *resolvent* R of a renewal equation with kernel ζ is defined as the matrix-valued solution of

$$R = R * \zeta + \zeta. \tag{2.13}$$

As we shall show below we can equivalently introduce R as the (matrix-valued) solution of

$$R = \zeta * R + \zeta \tag{2.14}$$

since it will turn out that

$$\zeta * R = R * \zeta. \tag{2.15}$$

The key property of the resolvent concerns the representation of the solution of a RE with kernel ζ and arbitrary forcing function.

Theorem 2.7. *Let $y = \zeta * y + g$; then*

$$y = g + R * g. \tag{2.16}$$

Proof. Suppose $y = \zeta * y + g$; then

$$\begin{aligned} R * y &= R * \zeta * y + R * g \\ &= (R - \zeta) * y + R * g. \end{aligned}$$

Hence $\zeta * y = R * g$ and substituting this into the equation we arrive at (2.16). □

We now discuss the existence and the uniqueness of the solution of (2.13), under the assumption that $\zeta \in L^1([0,\infty), \mathbb{C}^{n\times n})$, which suffices for our purposes (the proofs will make clear that one can derive results under weaker conditions; see Gripenberg et al. [97]).

Lemma 2.8. *For t_e sufficiently small, equation (2.14) has at most one solution in $L^1([0,t_e], \mathbb{C}^{n\times n})$.*

Proof. Consider the mapping $\mathcal{L}$ from $L^1([0,t_e], \mathbb{C}^{n\times n})$ into itself defined by

$$\mathcal{L}\psi = \psi * \zeta + \zeta.$$

Then

$$\mathcal{L}\psi_1 - \mathcal{L}\psi_2 = (\psi_1 - \psi_2) * \zeta$$

and, consequently,

$$\|\mathcal{L}\psi_1 - \mathcal{L}\psi_2\| \leq \|\zeta\| \, \|\psi_1 - \psi_2\|$$

where the norm is the $L^1\big([0,t_e],\mathbb{C}^{n\times n}\big)$ norm. Consequently, $\|\zeta\| < 1$ for t_e sufficiently small, and for such t_e, the map $\mathcal{L}$ has, by the contraction mapping theorem, a *unique* fixed point. □

Theorem 2.9. *There exists a unique solution R of* (2.14) *and*

$$R = \zeta + \zeta * \zeta + \zeta * \zeta * \zeta + \cdots \tag{2.17}$$

with convergence in an exponentially weighted L^1-space. As a consequence, $\zeta * R = R * \zeta$.

Proof. The idea is to use a contraction mapping argument just as in the proof of Lemma 2.8, but now for functions defined on $[0,\infty)$ while using a suitable weight function to achieve contractivity. Note that for any $\gamma > 0$, equation (2.13) is equivalent to

$$R(t)e^{-\gamma t} = \int_0^t R(t-\tau)e^{-\gamma(t-\tau)}\zeta(\tau)e^{-\gamma\tau}\,d\tau + \zeta(t)e^{-\gamma t}.$$

Choose γ such that

$$\int_0^\infty |\zeta(\tau)|\, e^{-\gamma\tau}\,d\tau < 1$$

and define $\mathcal{L}$ on $L^1\big([0,\infty),\mathbb{C}^{n\times n}\big)$ by

$$(\mathcal{L}\psi)(t) = \int_0^t \psi(t-\tau)\zeta(\tau)e^{-\gamma\tau}\,d\tau + \zeta(t)e^{-\gamma t};$$

then $\mathcal{L}$ is a contraction mapping on all of L^1 which, consequently, has a unique fixed point obtained by successive approximations in the form of the series expansion

$$\zeta^\gamma + \zeta^\gamma * \zeta^\gamma + \cdots$$

where

$$\zeta^\gamma(t) = e^{-\gamma t}\zeta(t).$$

Multiplying by $e^{\gamma t}$ we arrive at (2.17) and the identity $\zeta * R = R * \zeta$ then follows at once. This argument yields existence and the representation (2.17), but the uniqueness assertion is weaker than the result of Lemma 2.8. Clearly, however, any solution on $[0,\infty)$ yields, by restriction, a solution on $[0,t_e]$. □

The smoothness results presented in Exercises 2.2 and 2.3 immediately lead to the following two theorems:

Theorem 2.10. *For a continuous function g, the solution $y = g + R * g$ of the renewal equation $y = \zeta * y + g$ is continuous.*

Theorem 2.11. *For $\zeta \in BV$ the resolvent R is of bounded variation on bounded intervals and R inherits the normalization of ζ except for the constancy for $t \geq h$. Likewise, the solution of the renewal equation belongs locally to* BV *when both kernel and forcing function are BV.*

We now summarize the conclusions obtained so far in this section in the following theorem.

Theorem 2.12. *Let $\zeta \in NBV$ and $\varphi \in \mathcal{C} = C\big([-h,0],\mathbb{C}^n\big)$ be given. Define g and f in terms of ζ and φ by, respectively,* (2.5),

$$g(t) = \zeta(t)\varphi(0) + \int_t^h d\zeta(\theta)\varphi(t-\theta),$$

and (2.12),

$$f(t) = \varphi(0) + \int_0^h \big[\zeta(t+\sigma) - \zeta(\sigma)\big]\varphi(-\sigma)\,d\sigma.$$

The RFDE (2.1),

$$\dot{x}(t) = \int_0^h d\zeta(\theta)x(t-\theta),$$

provided with the initial condition (2.2),

$$x(\theta) = \varphi(\theta), \qquad -h \leq \theta \leq 0,$$

admits a unique solution on $[-h,\infty)$. For $t \geq 0$ this solution coincides with the unique solution of the renewal equation (2.10),

$$x = \zeta * x + f,$$

whereas the derivative $\dot{x}$ coincides with the unique solution of the renewal equation (2.4a),

$$\dot{x} = \zeta * \dot{x} + g.$$

So with R defined as the resolvent of ζ we have the following alternative representations:

$$x = f + R * f \tag{2.18}$$

and

$$x(t) = \varphi(0) + \int_0^t y(\tau)\,d\tau, \qquad y = g + R * g. \tag{2.19}$$

Proof. The one-to-one correspondence of solutions of (2.1)–(2.2) and of (2.4a) and (2.4b) follows directly from the manipulations that lead from one formulation to the other. Now (2.4a) admits a unique solution which is, moreover, represented by $g+R*g$. If we integrate (2.4a) and use (2.4b), we arrive at (2.10). So x is a solution of (2.10). But (2.10) admits at most one solution and we conclude that, for $t \geq 0$, (2.10) gives a complete characterization of the solution of the RFDE and that, moreover, the representation (2.18) holds. □

So far we have restricted ourselves to continuous initial functions φ. When relaxing the smoothness condition on φ, we have to realize that the value of $x(0)$ needs to be prescribed in a precise manner. However, in prescribing the history, we may use functions whose point values are only almost everywhere defined, since the history is only used to calculate the right hand side of the differential equation and "innocent" indeterminacies will disappear upon integration. This phenomenon is clearly reflected in formula (2.12) for the forcing function of the renewal equation for x. If we rewrite (2.12) in the form

$$f(t) = \alpha + \int_0^h [\zeta(t+\sigma) - \zeta(\sigma)]\varphi(-\sigma)\, d\sigma, \tag{2.20}$$

we see that initial data in the form of the couple (α, φ), with $\alpha \in \mathbb{C}^n$ and $\varphi \in L^\infty([-h,0], \mathbb{C}^n)$ still yield a continuous f and hence a continuous solution x of the renewal equation (2.10). Of course x is not a solution of the RFDE (2.1) in the sense of our Definition 2.1, but our manipulations above should provide sufficient motivation to call x at least a "generalized" solution. We shall see in Chapters II and III that formulations in terms of semigroups of operators also lead us to consider initial data in the form of couples (α, φ) with $\alpha \in \mathbb{C}^n$ and $\varphi \in L^\infty$.

A special class of such initial data consists of couples $(\alpha, 0)$. By choosing $\alpha = e_i$, $i = 1, \ldots, n$, where e_i is the ith unit vector in $\mathbb{C}^n$, and combining the n solutions into one matrix-valued solution, we get what is traditionally called the *fundamental matrix solution* Q. In other words, Q is the solution of the (matrix) renewal equation

$$Q = \zeta * Q + I, \tag{2.21}$$

where I is the constant function with as its only value the $n \times n$ identity matrix. Hence $Q = I + R * I$ or

$$Q(t) = I + \int_0^t R(\tau)\, d\tau \tag{2.22}$$

and

$$\dot{Q} = R. \tag{2.23}$$

We conclude that the fundamental matrix solution and the resolvent contain exactly the same information concerning the kernel ζ.

Our strategy in this section has been to rewrite the initial-value problem for a linear autonomous RFDE as a renewal equation and then to use the resolvent to give a representation of the solution. This approach is very effective when dealing with existence and uniqueness questions and it can be used to answer questions about the smoothness of solutions. Next comes the problem of determining the asymptotic behaviour of the solution for $t \to \infty$. To solve that problem we still use the renewal equation, but we use a new tool to represent its solution. In the next section we turn our attention to that tool.

Exercise 2.13. Show that there can be at most one solution of the simultaneous equations $R = R * \zeta + \zeta$ and $R = \zeta * R + \zeta$.

Exercise 2.14. Let R be the resolvent of the kernel ζ. Verify that R^T is the resolvent of the kernel ζ^T. (Here M^T denotes the transpose of the matrix M.)

Exercise 2.15. Use Theorem 2.12 to prove that the RFDE

$$\dot{x}(t) = \int_0^h d\zeta(\tau)x(t-\tau), \qquad t \geq 0,$$

is equivalent with the integral equation

$$x(t) - \int_{-h}^t \zeta(t-\tau)x(\tau)\,d\tau = \text{constant}, \qquad t \geq 0.$$

I.3 Solving renewal equations by Laplace transformation

The *Laplace transform* of a measurable and locally integrable function g with domain of definition $\mathbb{R}_+$ is defined by

$$\bar{g}(z) = \lim_{T\to\infty} \int_0^T e^{-zt} g(t)\,dt \tag{3.1}$$

for those values of the complex variable z for which the limit exists. We shall restrict ourselves to those properties of the Laplace transform that we need in the course of our analysis of renewal equations. If g is a σ_0-exponentially bounded function, i.e., there exists $C > 0$ such that for almost all $t \in \mathbb{R}_+$

$$|g(t)| \leq Ce^{\sigma_0 t}, \tag{3.2}$$

then $\overline{g}$ is defined in the half-plane $\operatorname{Re} z > \sigma_0$ since the integral converges absolutely for those values of z. Moreover, $\overline{g}$ is an analytic function of z and bounded in this half-plane. In fact

$$|\overline{g}(z)| \leq \frac{C}{\operatorname{Re} z - \sigma_0}. \tag{3.3}$$

The key property of the Laplace transform, which makes it such a useful tool for analysing renewal equations, is that it converts convolution products into algebraic products:

Lemma 3.1. *Let both f and g be σ_0-exponentially bounded; then*

$$\overline{f * g}(z) = \overline{f}(z)\overline{g}(z) \quad \textit{for } \operatorname{Re} z > \sigma_0. \tag{3.4}$$

We leave the proof as an exercise to the reader. Note that the result holds when g takes values in $\mathbb{C}^n$ ($\mathbb{R}^n$) and f in $\mathbb{C}^{n\times n}$ ($\mathbb{R}^{n\times n}$) as well as when both g and f take values in $\mathbb{C}^{n\times n}$ ($\mathbb{R}^{n\times n}$).

Throughout the rest of this section, ζ denotes an element of NBV = NBV$\big([0,h],\mathbb{C}^{n\times n}\big)$ and $f : \mathbb{R}_+ \to \mathbb{C}^n$ is assumed to be continuous, of bounded variation and constant for $t \geq h$. The renewal equation

$$x = \zeta * x + f \tag{3.5}$$

has a unique solution, which is continuous, of bounded variation and at most of exponential growth for $t \to \infty$ [recall Theorem 2.9 and the representation (2.18)]. Laplace transformation yields the algebraic equation

$$\overline{x} = \overline{\zeta}\,\overline{x} + \overline{f} \tag{3.6}$$

which we can solve to obtain an explicit expression for $\overline{x}$, viz.

$$\overline{x} = (I - \overline{\zeta})^{-1}\overline{f}. \tag{3.7}$$

The idea is now to obtain an explicit representation of x itself by using the inverse transformation, which is presented in our next result from the standard Laplace transform literature (see Widder [307, 7.3-5] or Doetsch [71, 24.4]).

Theorem 3.2. *Let g be a σ_0-exponentially bounded function that is of bounded variation on bounded intervals. Then for $\gamma > \sigma_0$ and $t > 0$ we have the inversion formula*

$$\frac{g(t+) + g(t-)}{2} = \lim_{\omega\to\infty} \frac{1}{2\pi i} \int_{\gamma - i\omega}^{\gamma + i\omega} e^{zt}\overline{g}(z)\, dz, \tag{3.8}$$

whereas for $t = 0$ we have

$$\frac{g(0+)}{2} = \lim_{\omega \to \infty} \frac{1}{2\pi i} \int_{\gamma - i\omega}^{\gamma + i\omega} \overline{g}(z)\, dz. \tag{3.9}$$

To facilitate the formulation of results like (3.8) and (3.9) in the following, we introduce some

Notation 3.3. $L(\gamma)$ denotes the line $\{z \mid \mathrm{Re}\, z = \gamma\}$ parallel to the imaginary axis in the complex plane. By $\int_{L(\gamma)} \cdots dz$ we denote the so-called principal value integral $\lim_{\omega \to \infty} \int_{\gamma - i\omega}^{\gamma + i\omega} \cdots dz$.

For completeness we also state a uniqueness result.

Theorem 3.4. *Let both f and g be σ_0-exponentially bounded functions on $\mathbb{R}_+$. If $\overline{f}(z) = \overline{g}(z)$ for $\mathrm{Re}\, z > \sigma_0$, then $f(t) = g(t)$ for almost all $t \geq 0$.*

From formula (3.8) it follows that the value of the complex line integral is independent of the choice of $\gamma > \sigma_0$. We shall now prove this directly in order to demonstrate the use of certain complex integration techniques that will be used repeatedly in the sequel. Define $\Gamma_N(\gamma, \gamma')$ to be the closed oriented contour in the complex plane consisting of four straight line segments through the vertices $\gamma - iN, \gamma' - iN, \gamma' + iN$ and $\gamma + iN$.

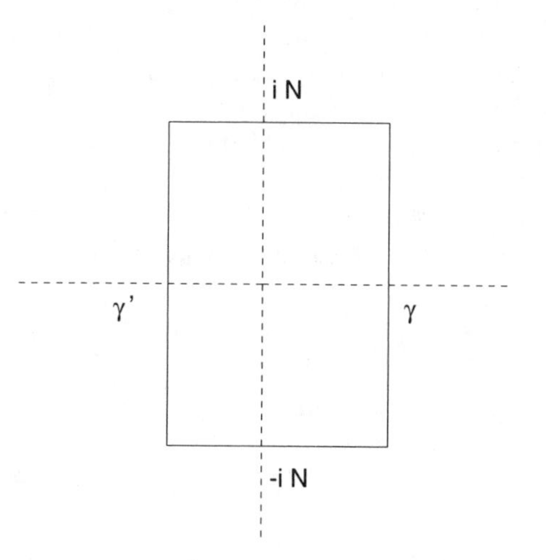

Fig. I.1. The contour $\Gamma_N(\gamma, \gamma')$.

Since $\overline{g}$ is analytic in the half-plane $\mathrm{Re}\, z > \sigma_0$, the Cauchy Theorem tells us that

$$\frac{1}{2\pi i}\int_{\Gamma_N(\gamma,\gamma')} e^{zt}\overline{g}(z)\,dz = 0 \tag{3.10}$$

for $\gamma' > \gamma > \sigma_0$. By taking the limit $N \to \infty$ we can draw the conclusion that

$$\frac{1}{2\pi i}\int_{L(\gamma)} e^{zt}\overline{g}(z)\,dz = \frac{1}{2\pi i}\int_{L(\gamma')} e^{zt}\overline{g}(z)\,dz, \tag{3.11}$$

provided we show that

$$\lim_{N\to\pm\infty}\Big|\int_{\gamma+iN}^{\gamma'+iN} e^{zt}\overline{g}(z)dz\Big| = 0. \tag{3.12}$$

In order to prove (3.12) we need

Lemma 3.5. (Riemann-Lebesgue.) *If f belongs to $L^1(\mathbb{R}_+)$, then*

$$\lim_{\omega\to\pm\infty}\Big|\int_0^\infty e^{i\omega t} f(t)\,dt\Big| = 0.$$

A proof can be found in [123, Thm. 21.39].

The Riemann-Lebesgue lemma tells us that $\lim_{N\to\pm\infty}\overline{g}(\theta+iN) = 0$ for each $\theta\in[\gamma,\gamma']$. Lebesgue's dominated convergence theorem then leads to the conclusion that (3.12) is indeed correct.

We may now apply Theorem 3.2 with $\overline{x}$ given explicitly by (3.7). However, as we will see below, it has advantages to first rewrite the right hand side of (3.7) in a form which allows for an analytic continuation to a meromorphic function on the *whole* complex plane.

For a given function f with values in $\mathbb{C}^n$, defined and of bounded variation on $\mathbb{R}_+$ and constant for $t \geq h$ [so when $f(0) = 0$ and f is continuous from the right on $(0,h)$ we have $f \in \mathrm{NBV}([0,h],\mathbb{C}^n)$, but we do not require such a normalization here], integration by parts leads to the identity

$$\overline{f}(z) = \int_0^\infty e^{-zt} f(t)\,dt = \frac{1}{z}\big[f(0) + \int_0^h e^{-zt}\,df(t)\big] \tag{3.13}$$

for z with $\operatorname{Re} z > 0$. Likewise we have for $\zeta \in \mathrm{NBV}$ that

$$\overline{\zeta}(z) = \frac{1}{z}\int_0^h e^{-zt}\,d\zeta(t). \tag{3.14}$$

Combining the two identities we obtain

$$(I - \overline{\zeta}(z))^{-1}\overline{f}(z) = \Delta(z)^{-1}\big(f(0) + \int_0^h e^{-zt}\,df(t)\big), \tag{3.15}$$

where the so-called *characteristic matrix* $\Delta(z)$ is defined by

(3.16)
$$\Delta(z) = zI - \int_0^h e^{-zt}\, d\zeta(t).$$

Since both Δ and $z \mapsto f(0) + \int_0^h e^{-zt}\, df(t)$ are entire functions, the right hand side of (3.15) is a meromorphic function with, possibly, poles at the roots of the *characteristic equation*

(3.17)
$$\det \Delta(z) = 0.$$

Exercise 3.6. Show that $z \mapsto \Delta(z)^{-1}$ is the Laplace transform of the fundamental matrix solution Q [defined in (2.22)]. Next derive the identity $\overline{R}(z) = z\Delta(z)^{-1} - I$ for the Laplace transform of the resolvent.

We are now ready to state a first representation result.

Theorem 3.7. *For $\zeta \in NBV$ and $f : \mathbb{R}_+ \to \mathbb{C}^n$ continuous, of bounded variation and constant for $t \geq h$ the solution x of the renewal equation* (3.5)

$$x = \zeta * x + f$$

admits for $t > 0$ the representation

(3.18)
$$x(t) = \frac{1}{2\pi i} \int_{L(\gamma)} e^{zt} \Delta(z)^{-1} \big(f(0) + \int_0^h e^{-z\theta}\, df(\theta) \big)\, dz$$

for $\gamma > \sup\{\operatorname{Re} z : \det \Delta(z) = 0\}$.

"Proof". Note first of all that $\Delta(z)$ is nonsingular for $\operatorname{Re} z$ sufficiently large [indeed, one can conclude from (3.16) that $zI - \Delta(z) \to 0$ for $\operatorname{Re} z \to \infty$; for a precise proof see Theorem 4.4 (i)]. Since x is continuous, of bounded variation on bounded intervals, and exponentially bounded, we obtain (3.18) for γ sufficiently large by combining (3.8) with (3.7) and (3.15). Next we can shift the contour to the left until we encounter the first singularity, provided we verify that the appropriate analogue of (3.12) holds. This last step requires a further analysis of the characteristic matrix $\Delta(z)$, which we perform in the next section (see in particular Lemma 4.5). □

Exercise 3.8. Prove that the RFDE (2.1) has a solution of the form $x(t) = e^{\lambda t} v$ for some $v \in \mathbb{C}^n$ iff $\det \Delta(z)$ is zero at $z = \lambda$ and that necessarily $\Delta(\lambda) v = 0$.

I.4 Estimates for det $\Delta(z)$ and related quantities

In order to obtain information from the representation (3.18) by shifting the line $L(\gamma)$ we need first to derive some results about the zeros of det $\Delta(z)$ and about the behaviour of the integrand for $\operatorname{Im} z \to \infty$ and $\operatorname{Re} z$ bounded.

Theorem 4.1. *For given $\zeta \in NBV$ there exists positive constants C_0 and C such that*

$$|\det \Delta(z)| \geq \frac{1}{2}|z|^n \tag{4.1}$$

for those values of $z \in \mathbb{C}$ for which

$$|z| \geq C_0|e^{-zh}| \quad \text{and} \quad |z| > C. \tag{4.2}$$

The proof is based on the following two lemma's.

Lemma 4.2. *For any $\eta \in NBV([0,h],\mathbb{C})$ and any $C_0 > 0$ the inequality*

$$\Big|\int_0^h e^{-zt}\, d\eta(t)\Big| \leq \frac{1}{C_0} TV(\eta)|z| \tag{4.3}$$

holds for $z \in \mathbb{C}$ satisfying (4.2) with $C \geq C_0$.

Proof. For $0 \leq t \leq h$

$$|e^{-zt}| = e^{-t\operatorname{Re} z} \leq \max\{1, e^{-h\operatorname{Re} z}\}.$$

Hence

$$\Big|\int_0^h e^{-zt}\, d\eta(t)\Big| \leq TV(\eta)\max\{1, e^{-h\operatorname{Re} z}\} \leq \frac{1}{C_0} TV(\eta)|z|$$

provided (4.2) is satisfied with $C \geq C_0$. □

Lemma 4.3. *For* det $\Delta(z)$ *we have the representation*

$$\det \Delta(z) = z^n - \sum_{j=1}^{n}\Big(\prod_{k=1}^{j}\int_0^h e^{-zt}\, d\eta_{jk}(t)\Big)z^{n-j}, \tag{4.4}$$

where η_{jk} is a linear combination of elements ζ_{il} of ζ.

Proof. $\Delta(z) = zI - \int_0^h e^{-zt}\, d\zeta(t)$ and, consequently, we can write det $\Delta(z)$ as a nth order polynomial in z with coefficients as in (4.4). □

Proof of Theorem 4.1. From (4.4) we deduce that

$$|\det \Delta(z)| \geq \big||z|^n - \sum_{j=1}^{n} \prod_{k=1}^{j} \Big| \int_0^h e^{-zt}\, d\eta_{jk}(t)\Big| |z|^{n-j}\big|.$$

Next, Lemma 4.2 yields, with $q(C,\eta) := \sum_{j=1}^{n} C^{-j} \prod_{k=1}^{j} C_0^{-1} TV(\eta_{jk})$, that

$$|\det \Delta(z)| \geq |z|^n (1 - q(C,\eta))$$

for $z \in \mathbb{C}$ satisfying (4.2) and C large enough to have $q(C,\eta) < 1$. We arrive at (4.1) by choosing C such that $q(C,\eta) < \frac{1}{2}$. □

As easy corollaries of Theorem 4.1 we have the following collection of propositions (recall that the zeros of an analytic function are isolated and cannot have a finite accumulation point).

Theorem 4.4. *Concerning the zeros of the entire function* $\det \Delta(z)$, *the following results hold:*

(i) $\sup\{\operatorname{Re} z \mid \det \Delta(z) = 0\} < \infty$, *i.e.,* $\det \Delta(z)$ *has a zero free right half-plane.*
(ii) $\#\{z \mid \det \Delta(z) = 0,\ \gamma_- \leq \operatorname{Re} z \leq \gamma_+\} < \infty$ *for all finite values of* γ_- *and* γ_+ *with* $\gamma_- < \gamma_+$, *i.e.,* $\det \Delta(z)$ *has at most finitely many zeros in a given vertical strip.*
(iii) *For zeros of* $\det \Delta(z)$ *in the left half-plane, necessarily*

$$|\operatorname{Im} z| \leq C e^{-h \operatorname{Re} z},$$

where C *is the constant introduced in Theorem* 4.1.

We now turn our attention to the integrand in (3.18). Define

$$F(z) = \Delta(z)^{-1}\big(f(0) + \int_0^h e^{-z\theta}\, df(\theta)\big) \tag{4.5}$$

with f and ζ as in Theorem 3.7.

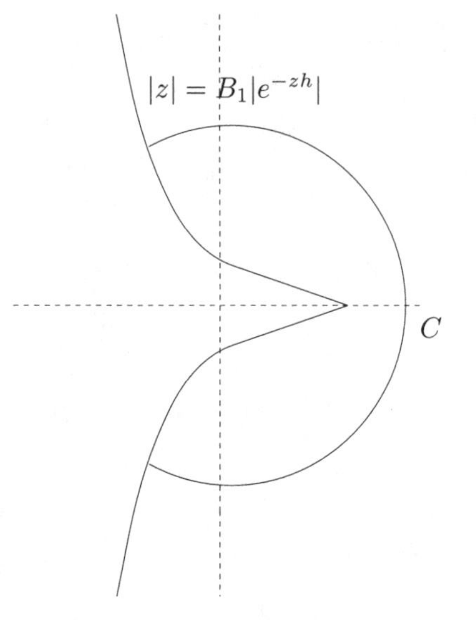

Fig. I.2. Three curves in the complex plane which together divide the plane into three connected components, one of which (the "right" one) is free of zeros of det $\Delta(z)$.

Lemma 4.5.

$$\lim_{\omega\to\pm\infty} e^{(s+i\omega)t}F(s+i\omega) = 0$$

uniformly for $\gamma_- \leq s \leq \gamma_+$ and compact t-sets.

Proof. From Theorem 4.1 we know that, given γ_- and γ_+, for ω sufficiently large, $|\det \Delta(z)| \geq \frac{1}{2}|z|^n$, where $z = s+i\omega$ and $\gamma_- \leq s \leq \gamma_+$. Let adj $\Delta(z)$ denote the matrix of cofactors of $\Delta(z)$, i.e., the elements of adj $\Delta(z)$ are the $(n-1)\times(n-1)$ subdeterminants of $\Delta(z)$. Lemma 4.2 tells us that

$$|(\text{adj } \Delta(z))_{ij}| \leq K|z|^{n-1}$$

for some constant $K_0 > 0$ and $z = s+i\omega$, $\gamma_- \leq s \leq \gamma_+$ and ω large enough. Since

$$\Delta(z)^{-1} = \frac{1}{\det \Delta(z)} \text{adj } \Delta(z),$$

we obtain that

$$|(\Delta(z)^{-1})_{ij}| \leq \frac{2K_0}{|z|}$$

for z as before. Both e^{zt} and $\int_0^h e^{-z\theta}\, df(\theta)$ are uniformly bounded for $z = s+i\omega$, $\gamma_- \leq s \leq \gamma_+$ and t in a given compact set. Hence there exists a constant K_1 such that

$$|e^{zt}F(z)| \leq \frac{K_1}{|z|}$$

for these values of z and t, with ω sufficiently large. □

Now that we have Lemma 4.5 available, the "proof" of Theorem 3.7 becomes a true proof. Note, however, that Lemma 4.5 does not restrict $\operatorname{Re} z$ to the zero free right half-plane. So we can continue to shift the line $L(\gamma)$ to the left while keeping track of the residues corresponding to the singularities of $\Delta(z)^{-1}$ that we pass. This is exactly what we are going to do in the next section.

I.5 Asymptotic behaviour for $t \to \infty$

The starting point for an analysis of the asymptotic behaviour of the solution of the RFDE (or the equivalent RE) is the representation (3.18). The singularities of the integrand $e^{zt}F(z)$ are poles of finite order at those values of z for which $\det \Delta(z) = 0$. Our first result gives some information on the residue in a given pole.

Lemma 5.1. *Let λ be a zero of $\det \Delta(z)$ of order m. Then*

$$\operatorname*{Res}_{z=\lambda} e^{zt}F(z) = p(t)e^{\lambda t}, \tag{5.1}$$

where p is a $\mathbb{C}^n$-valued polynomial in t of degree less than or equal to $m-1$.

Proof. In a neighbourhood of $z = \lambda$ we have the series expansions

$$\Delta(z)^{-1} = \frac{1}{\det \Delta(z)} \operatorname{adj} \Delta(z) = \sum_{k=-m}^{\infty} (z-\lambda)^k A_k,$$

$$f(0) + \int_0^h e^{-z\theta}\, df(\theta) = \sum_{k=0}^{\infty} (z-\lambda)^k v_k,$$

$$e^{zt} = e^{\lambda t} e^{(z-\lambda)t} = e^{\lambda t} \sum_{k=0}^{\infty} \frac{t^k}{k!} (z-\lambda)^k.$$

Since the residue in $z = \lambda$ equals the coefficient of the $(z-\lambda)^{-1}$-term of the Laurent expansion of $e^{zt}F(z)$ in a neighbourhood of $z = \lambda$, a multiplication of the above series expansions yields the desired result. □

Exercise 5.2. With the assumptions and notation of Lemma 5.1, let $y(t) = p(t)e^{\lambda t}$. Show that y satisfies both the RFDE (2.1) and the translation invariant integral equation

$$y(t) = \int_0^{\infty} \zeta(\theta) y(t-\theta)\, d\theta \tag{5.2}$$

(which is called the *limiting* equation of the renewal equation).
Hint: Write $y(t) = \frac{1}{2\pi i} \oint e^{zt}F(z)\, dz$ and verify by substitution that

$$\dot{y}(t) - \int_0^h d\zeta(\theta)y(t-\theta) \quad \text{as well as} \quad y(t) - \int_0^\infty \zeta(\theta)y(t-\theta)\,d\theta$$

are given by a Cauchy integral with a regular integrand and, hence, are zero.

Now choose any $\gamma \in \mathbb{R}$ such that $\det \Delta(z) \neq 0$ on the line $\operatorname{Re} z = \gamma$. The Cauchy theorem, the last lemma and Lemma 4.5 together imply that the following generalization of (3.18) holds:

$$x(t) = \sum_{j=1}^{l} p_j(t)e^{\lambda_j t} + \frac{1}{2\pi i}\int_{L(\gamma)} e^{zt}F(z)\,dz, \tag{5.3}$$

where $\lambda_1, \ldots, \lambda_l$ are the finitely many zeros of $\det \Delta(z)$ to the right of $L(\gamma)$ (cf. Theorem 4.4). It remains to verify that the integral term is, asymptotically for $t \to \infty$, of higher order.

Lemma 5.3.

$$\frac{1}{2\pi i}\int_{L(\gamma)} e^{zt}F(z)\,dz = o(e^{\gamma t}) \quad \textit{for } t \to \infty. \tag{5.4}$$

Proof. We have to prove that

$$\Big(\lim_{N\to\infty} \frac{1}{2\pi}\int_{-N}^{N} e^{it\omega}F(\gamma+i\omega)\,d\omega\Big) \to 0 \quad \text{for } t \to \infty.$$

In the proof of Lemma 4.5 we showed that

$$|F(\gamma+i\omega)| \leq \frac{K_1}{|\gamma+i\omega|}, \quad \text{for } |\omega| \text{ large},$$

but this does not, unfortunately, guarantee that $F(\gamma+i\omega)$ is an L^1-function of ω (in other words, the integral above does not necessarily converge absolutely) and, hence, we cannot apply the Riemann-Lebesgue Lemma 3.5 directly. For every fixed N, however, the Riemann-Lebesgue Lemma tells us that

$$\lim_{t\to\infty}\int_{-N}^{N} e^{it\omega}F(\gamma+i\omega)\,d\omega = 0.$$

So if we prove that the limits $t \to \infty$ and $N \to \infty$ are interchangeable, we obtain the wanted conclusion. For this it suffices to show that the convergence for $N \to \infty$ is uniform for $t \geq T_0$ for some fixed value T_0.

First note that for $\eta \in \mathrm{NBV}\big([0,h],\mathbb{C}\big)$ the integral

$$\int_0^h e^{-z\theta}d\eta(\theta)$$

is bounded for $z = \gamma + i\omega$, $-\infty < \omega < \infty$. Since $\Delta(z) = zI - \int_0^h e^{-z\theta} d\zeta(\theta)$, it follows that for $z = \gamma + i\omega$

$$\det \Delta(z) = z^n + O(|\omega|^{n-1}) \quad \text{as } |\omega| \to \infty$$

and

$$\operatorname{adj} \Delta(z) = z^{n-1} I + O(|\omega|^{n-2}) \quad \text{as } |\omega| \to \infty.$$

Hence, for such z

$$F(z) = \frac{1}{z}\big(f(0) + \int_0^h e^{-z\theta}\, df(\theta)\big) + O(|\omega|^{-2}) \quad \text{as } |\omega| \to \infty.$$

For the $O(|\omega|^{-2})$ term we can estimate $|e^{it\omega}|$ by one and uniformity in t is automatic (or, in other words, the contribution of this term to the integral converges absolutely). So we can concentrate on the first term. We shall use partial integration

$$\begin{aligned}
&\int_{-N}^{N} \frac{1}{\gamma + i\omega}\big(f(0)e^{it\omega} + \int_0^h e^{i(t-\theta)\omega - \gamma\theta}\, df(\theta)\big)\, d\omega \\
&= \frac{1}{\gamma + i\omega}\big(\frac{1}{it} f(0)e^{it\omega} + \int_0^h \frac{1}{i(t-\theta)} e^{i(t-\theta)\omega - \gamma\theta}\, df(\theta)\big)\Big|_{-N}^{N} \\
&\quad + \int_{-N}^{N} \frac{1}{(\gamma + i\omega)^2}\big(\frac{1}{t} f(0)e^{it\omega} + \int_0^h \frac{1}{t-\theta} e^{i(t-\theta)\omega - \gamma\theta}\, df(\theta)\big)\, d\omega.
\end{aligned}$$

For $t \geq T_0$ there exist constants K_0, K_1 and K_2 such that

$$\Big|\frac{1}{it} f(0)e^{\pm itN}\Big| \leq \frac{K_0}{T_0}$$

and

$$\Big|\int_0^h \frac{1}{i(t-\theta)} e^{\pm i(t-\theta)N - \gamma\theta}\, df(\theta)\Big| \leq \frac{K_1}{T_0 - h},$$

so the boundary terms converge to zero as $N \to \infty$, uniformly for $t \geq T_0$. The remaining integral converges absolutely and estimates like

$$\Big|\int_N^\infty \frac{1}{(\gamma + i\omega)^2} \frac{1}{t} f(0)e^{it\omega}\, d\omega\Big| \leq \frac{K_2}{T_0} \int_N^\infty \frac{d\omega}{\gamma^2 + \omega^2}$$

easily show that the convergence for $N \to \infty$ is uniform in t for $t \geq T_0$. □

We are now ready to formulate the main result.

Theorem 5.4. *Let x be a solution of the RFDE* (2.1) *corresponding to some initial function φ. For any $\gamma \in \mathbb{R}$ such that* $\det \Delta(z) \neq 0$ *on the line* $\operatorname{Re} z = \gamma$ *we have the asymptotic expansion*

$$x(t) = \sum_{j=1}^{l} p_j(t) e^{\lambda_j t} + o(e^{\gamma t}) \quad \textit{for } t \to \infty, \tag{5.5}$$

where $\lambda_1, \ldots, \lambda_l$ are the finitely many zeros of $\det \Delta(z)$ *with real part exceeding γ and where $p_j(t)$ is a $\mathbb{C}^n$-valued polynomial in t of degree less than or equal to $m_j - 1$ with m_j the multiplicity of λ_j as a zero of* $\det \Delta(z)$.

Corollary 5.5. *If* $\det \Delta(z)$ *has no zeros in the right half-plane* $\{z \mid \operatorname{Re} z \geq 0\}$, *all solutions of the RFDE* (2.1) *converge to zero exponentially as $t \to \infty$.*

Exercise 5.6. Prove that the RFDE (2.1) has a solution of the form $x(t) = (tv + w)e^{\lambda t}$ with $v, w \in \mathbb{C}^n$ provided $\Delta(\lambda)v = 0$ and $\Delta'(\lambda)v + \Delta(\lambda)w = 0$. Convince yourself that one can equivalently formulate the condition on v and w as

$$\Delta(z)(v + (z - \lambda)w) = O((z - \lambda)^2) \quad \text{for } z \to \lambda.$$

(See Exercise IV.5.11 for the appropriate generalization.)

Exercise 5.7. Extend the assertions of Exercise 5.2 to the setting of Theorem 5.4, either by using linearity and the result of Exercise 5.2 or by a slight modification of the proof of Exercise 5.2.

Theorem 5.4 tells us that, in order to describe the asymptotic behaviour of solutions, we have to look for the right most zeros of $\det \Delta(z)$. It does not, however, tell us exactly how the polynomials $p_j(t)$ depend on properties of $\Delta(z)$ in the spirit of Exercises 3.8 and 5.6 and on the initial condition φ. In principle, one can deduce this dependence by extending the analysis of the proof of Lemma 5.1, whereas taking into account how f depends on φ [see (2.13)]. Essentially this is what we shall do in Chapter IV in a more general functional analytic setting.

What happens when we let $\gamma \to -\infty$ in the series expansion (5.5)? Do we get a convergent infinite series and if so, is it a faithful representation of $x(t)$ or does $x(t)$ contain a component which goes to zero (as $t \to \infty$) faster than any exponential? A solution which does go to zero faster than any exponential is called a *small solution.* How do we recognise from the NBV kernel ζ [or from the characteristic matrix $\Delta(z)$, for that matter] whether or not small solutions exist? And how do we recognise whether the infinite expansion in polynomial-exponential functions is convergent and represents the solution for arbitrary initial data or, otherwise, how can we characterise those initial data for which it does? These are questions we shall address in Chapter V, using Laplace transformation and exponential type calculus for entire functions of order one as our main tool. But first, in the next two chapters, we shall introduce a more geometric way of looking at RFDE. In particular we shall be concerned with solution operators acting on the space of initial functions, rather than with $\mathbb{C}^n$-valued solutions. As a consequence, the next two chapters are functional analytic in spirit. Chapter IV, on spectral theory, brings complex analysis, in particular residue calculus for isolated poles, back to the fore. In a sense Chapter IV repeats much of the analysis of this chapter, but in a slightly different language and setting while clearly, we hope, explaining the relationship between the geometric point of view of Chapters II and III and the analysis of $\mathbb{C}^n$-valued solutions as presented in this chapter. Chapter V then shows how one can obtain much stronger results by exploiting more sophisticated results from complex analysis.

I.6 Comments

The material of this chapter is a mixture of classical results that can be found in textbooks such as Bellman and Cooke [16], Hale [102, 104] and Miller [195]. For a recent comprehensive survey of the theory of (resolvents of) Volterra integral equations we refer to Gripenberg et al. [97]. For the classical theory of functions of bounded variation, we refer to the work by Natanson [203].

Chapter II
The shift semigroup

II.1 Introduction

A retarded functional differential equation (RFDE) consists of a rule to extend the "initial" condition. Starting from a given function defined on $[-h, 0]$ one obtains, by solving the equation, a function defined on $[-h, \omega)$, with $\omega = \infty$ under mild conditions on the right hand side (such that blow up of solutions in finite time can be excluded).

The dynamical system point of view is to consider the given function on $[-h, 0]$ as the initial state and the piece of the extended function with the domain of definition $[t-h, t]$ as the state at time t. Thus, given the rule of infinitesimal change embodied in the differential equation, the state at some arbitrary time consists of the information needed to uniquely fix the future.

In Sections 3 and 5 of Chapter I the Laplace transform was used to analyse linear systems, and the idea of "state" and state transformation played hardly any role. For nonlinear systems one cannot use Laplace transformation and then it becomes most convenient, if not necessary, to consider the possible states as elements of a state space and to describe the qualitative and geometric aspects of the time evolution in terms of properties of operators mapping this state space into itself. Even in the linear case this may yield additional insight. So in this chapter we shall reconsider linear systems, but in a new perspective.

The state space shall be some space of functions defined on $[-h, 0]$. To obtain a mapping, indexed by t, which maps such a space into itself, one has to shift the piece of extended function between $t-h$ and t back to the interval $[-h, 0]$. In this slightly more abstract way of thinking about delay equations, there are two basic ingredients for the construction of solution operators: extending the function and shifting it back.

The first ingredient is specific for a particular equation, but the second is general, i.e., the same for all delay equations. Our approach will be to study the second ingredient first in considerable detail for the special case in which the extension rule is as simple as possible and then to consider other extension rules as "perturbations". This may seem long winded. Indeed, we will invest quite some energy, time and space in the formal analysis of a trivial equation. It will turn out, however, that the functional analytic framework which we develop is very well suited for the study of general RFDEs. It gives, in particular, a rigorous basis for the variation-of-constants formula which will make our subsequent analysis rather easy. Thus our investment is refunded with considerable interest.

In this chapter we shall introduce strongly continuous semigroups and their adjoints by means of a concrete example. In Appendix II we present a more systematic development, including proofs of some of the key results. Depending on their background, readers may wish to skip the appendix completely, read it first in detail, or go back and forth. In order to make the appendix and this chapter independently readable, we introduce and explain various concepts at both places.

II.2 The prototype problem

Consider the scalar equation

$$\dot{x} = 0$$

as a RFDE. As our state space we choose $X = C\big([-h, 0], \mathbb{C}\big)$ provided with the usual supremum norm. So we study the initial-value problem

$$\begin{aligned} \dot{x}(t) &= 0 \quad \text{for } t \geq 0, \\ x(\theta) &= \varphi(\theta) \quad \text{for } -h \leq \theta \leq 0 \end{aligned} \tag{2.1}$$

with φ a given element of X. Clearly the solution is

$$x(t) = \begin{cases} \varphi(t), & -h \leq t \leq 0, \\ \varphi(0), & t \geq 0. \end{cases} \tag{2.2}$$

In order to distinguish the time at which we inspect the state from the variable passing through the interval $[-h, 0]$ we shall, as usual in the theory of delay equations (see Hale [102]), write throughout this book

$$x_t(\theta) := x(t + \theta) \quad \text{for } t \geq 0 \text{ and } -h \leq \theta \leq 0. \tag{2.3}$$

With this notation, $x_t \in X$ is the state at time t.

Of course the solution x depends on the initial condition φ and whenever appropriate we shall incorporate this into the notation by writing $x = x(\,\cdot\,; \varphi)$. For each $t \geq 0$

$$(T_0(t)\varphi)(\theta) = \begin{cases} \varphi(t + \theta), & \text{if } -h \leq t + \theta \leq 0 \\ \varphi(0), & \text{if } t + \theta \geq 0 \end{cases} \tag{2.4}$$

defines a bounded linear operator $T_0(t) : X \to X$. The operator $T_0(t)$ maps the initial state φ at time zero onto the state x_t at time t.

A family $T = \{T(t)\}_{t\geq 0}$ of bounded linear operators on a Banach space X such that

(i) $T(0) = I$ (the identity),

(ii) $T(t)T(s) = T(t+s)$ for $t, s \geq 0$,

(iii) for any $\varphi \in X$, $\|T(t)\varphi - \varphi\| \to 0$ as $t \downarrow 0$

is called a *strongly continuous semigroup of operators* or, in short, a $\mathcal{C}_0$-*semigroup*. One can associate with such a semigroup the abstract differential equation

$$\frac{d}{dt}\big(T(t)\varphi\big) = A\big(T(t)\varphi\big), \tag{2.5}$$

where, by definition,

$$A\varphi = \lim_{t\downarrow 0} \frac{1}{t}\big(T(t)\varphi - \varphi\big) \tag{2.6}$$

with

$$\mathcal{D}(A) = \{\varphi \mid \lim_{t\downarrow 0} \frac{1}{t}\big(T(t)\varphi - \varphi\big) \text{ exists}\}. \tag{2.7}$$

The linear operator A, which is in general unbounded, is called the *infinitesimal generator* of the semigroup T and is a closed densely defined operator. We also note that for any $\mathcal{C}_0$-semigroup $T(t)$ there exist constants $M, \omega > 0$ such that $\|T(t)\| \leq Me^{\omega t}$ for $t \geq 0$. See Appendix II for a more detailed presentation of these and other basic results concerning semigroups of operators.

Clearly T_0 defined by (2.4) is a $\mathcal{C}_0$-semigroup. But what is its generator A_0 and what is the relation between (2.5) and (2.1)?

Lemma 2.1. *The generator of* T_0 *is given by*

$$\mathcal{D}(A_0) = \{\varphi \mid \dot{\varphi} \in C([-h,0],\mathbb{C}),\ \dot{\varphi}(0) = 0\}, \quad A_0\varphi = \dot{\varphi}.$$

Proof. Let $\varphi \in \mathcal{D}(A_0)$. Put $\psi = A_0\varphi$. From (2.4) and

$$0 = \lim_{t\downarrow 0} \|\frac{1}{t}\big(T_0(t)\varphi - \varphi\big) - \psi\| = \lim_{t\downarrow 0} \sup_{-h\leq\theta\leq 0} \big|\frac{1}{t}\big((T_0(t)\varphi(\theta) - \varphi(\theta)\big) - \psi(\theta)\big|$$

it follows that φ is right-differentiable on $[-h, 0)$ with right derivative ψ and, moreover, that necessarily $\psi(0) = 0$. The continuity of ψ implies that actually φ is differentiable and $\dot{\varphi} = \psi$. (Indeed,

$$\Big|\frac{\varphi(\theta - t) - \varphi(\theta)}{-t} - \psi(\theta)\Big| \leq \Big|\frac{\varphi(s+t) - \varphi(s)}{t} - \psi(s)\Big| + |\psi(s) - \psi(s+t)|,$$

where $s = \theta - t$; both terms on the right hand side converge to zero as $t \downarrow 0$, uniformly in s.)

Conversely, suppose that $\varphi \in C^1$ and $\dot{\varphi}(0) = 0$. Define $\varphi(t) = \varphi(0)$ for $t \geq 0$; then

$$\left|\frac{1}{t}\left(\varphi(t+\theta) - \varphi(\theta)\right) - \dot{\varphi}(\theta)\right| = \left|\frac{1}{t}\int_0^t \left(\dot{\varphi}(\theta+\sigma) - \dot{\varphi}(\theta)\right) d\sigma\right|$$

converges, as $t \downarrow 0$, to zero uniformly for $-h \leq \theta \leq 0$. □

So we see that in our particular case the abstract problem (2.5) is, on the one hand, a restricted version of (2.1), since φ has to be differentiable and $\dot{\varphi}(0) = 0$, but that, on the other hand, (2.5) incorporates the translation (shifting) explicitly since, in terms of $u(t,\theta) = \big(T_0(t)\varphi\big)(\theta)$ it takes the form of the partial differential equation (PDE)

$$\frac{\partial u}{\partial t} = \frac{\partial u}{\partial \theta}, \tag{2.8}$$

which describes translation with speed one. For $t + \theta > 0$ this equation automatically holds by the very definition of T_0. If we formulate the abstract equation (2.5), we require it to hold for $t+\theta \leq 0$ as well and as a consequence φ has to satisfy additional conditions.

It is a general fact that differentiation is the generator of translation. In the present context we have, in addition to translation, a rule for the extension of the initial function beyond its original domain of definition. It appears that this rule is incorporated in $\mathcal{D}(A_0)$ in the form of the condition $\dot{\varphi}(0) = 0$. If we change the rule we will change the domain of definition of the generator. This will give rise to unpleasant technical complications if we want to relate solutions of various equations to each other by means of a variation-of-constants formula. It is a source of much trouble and confusion in the study of delay equations.

As we will show in the following, one can avoid the technical complications by using duality theory of semigroups. The main idea here is to embed X into a larger space $X^{\odot *}$ which is defined in terms of a combination of properties of the space X and the prototype semigroup T_0. Basically, $X^{\odot *}$ is like a second dual space but with canonical restrictions in terms of the semigroup built into the construction.

On $X^{\odot *}$ one has a semigroup $T_0^{\odot *}$ which reduces to T_0 when restricted to X. It will turn out that one has a notion of generator on the space $X^{\odot *}$ as well, for which translation and extension are *both* described by the *action* of the operator, whereas the domain is determined by the translation only. As a consequence the domain is independent of the specific rule for extension. This is the main advantage of introducing $X^{\odot *}$.

The price one has to pay is that the semigroup on $X^{\odot *}$ is not strongly continuous, i.e., property (iii) does not hold. Therefore we do not abandon X altogether. We keep thinking of X as our basic state space but exploit the larger space $X^{\odot *}$ whenever it is convenient for our purposes. In other words, we run with the hare and hunt with the hounds!

Exercise 2.2. This is the first of a series of exercises, all dealing with age-dependent population dynamics and renewal equations. Set $X = L^1([0,h],\mathbb{C})$ and

$$(T_0(t)\varphi)(a) = \begin{cases} \varphi(a-t), & \text{if } a \geq t, \\ 0, & \text{if } a < t. \end{cases} \tag{2.9}$$

We say that φ is absolutely continuous, and we write $\varphi \in \mathrm{AC}$ when φ is the integral of an L^1-function (see Appendix II.2 for a different characterization).

(i) Show that T_0 is a strongly continuous semigroup generated by A_0 with

$$\mathcal{D}(A_0) = \{\varphi \in \mathrm{AC} \mid \varphi(0) = 0\}, \quad A_0\varphi = -\dot{\varphi}. \tag{2.10}$$

Hint: Use $(zI - A_0)^{-1} = \int_0^\infty e^{-zt} T_0(t)\,dt$ for $\operatorname{Re}\lambda$ sufficiently large (see Appendix II, Proposition 1.11).

(ii) Verify that in this case the abstract differential equation (2.5) can be, when we put $u(t,a) = (T_0(t)\varphi)(a)$, written as

$$\begin{aligned} &\frac{\partial u}{\partial t} + \frac{\partial u}{\partial a} = 0, \qquad 0 \leq a \leq h,\ t > 0, \\ &u(t,0) = 0, \qquad t > 0. \end{aligned} \tag{2.11}$$

The first order partial differential equation (PDE) is a bookkeeping equation (balance law) for an age-structured population (with predestinated death at age h; of course the interpretation requires that we restrict our attention to real-valued functions). The boundary condition $u(t,0) = 0$ tells us that the population does not reproduce [note that newborns have age zero by definition and that, consequently, $u(t,0)$ equals the population birth rate]. We refer to Metz and Diekmann [193] for background information on modelling aspects.

II.3 The dual space

Let X be a Banach space. The space of continuous linear (real or complex valued, depending on the context, i.e., on whether X is a real or a complex Banach space) functionals on X is called the *dual space* of X and denoted by X^*. As usual we shall denote elements of X^* by x^* and write $\langle x^*, x\rangle$ instead of $x^*(x)$. Equipped with the norm

$$\|x^*\| = \sup_{\|x\| \leq 1} \left|\langle x^*, x\rangle\right|$$

the space X^* becomes a Banach space. It then follows that

$$\|x\| = \sup_{\|x^*\| \leq 1} \left|\langle x^*, x\rangle\right|.$$

Apart from the norm topology on X^* we shall also work with the so-called *weak* topology*. By definition this is the weakest topology such that all functionals

$$x^* \mapsto \langle x^*, x \rangle \quad \text{for } x \in X$$

are continuous. Most important for us is that a sequence $x_n^* \in X^*$ converges in the weak* topology to $x^* \in X^*$ if and only if for all $x \in X$ the sequence $\langle x_n^*, x \rangle$ converges to $\langle x^*, x \rangle$ as $n \to \infty$.

When X is a space of functions it is convenient (and quite often possible) to represent X^* by a space of functions or measures and to give a "concrete" definition of the pairing $\langle \cdot , \cdot \rangle$. When $X = C([-h, 0], \mathbb{C})$, we can, as explained in Section I.1, identify X^* with the space of normalized bounded variation functions, the pairing being given by the Riemann-Stieltjes integral. Alternatively but equivalently we can think of X^* as the space of finite complex Borel measures; see Appendix I.

For reasons which were indicated in Section I.2 and which will become even more clear in Section III.5 we shall take as the essential domain of definition of the NBV functions not the interval $[-h, 0]$ but the mirror image under time reversal $[0, h]$. In fact we shall extend the domain of definition of $f \in \mathrm{NBV}$ to all of $\mathbb{R}$ by putting $f(\theta) = 0$ for $\theta \leq 0$ and $f(\theta) = f(h)$ for $\theta \geq h$. Recalling the convention introduced in Section I.1 we write

$$\langle f, \varphi \rangle = \int_0^h df(\theta)\varphi(-\theta)$$

and sometimes, even though φ may not be defined for $\theta < -h$,

$$\langle f, \varphi \rangle = \int_0^\infty df(\theta)\varphi(-\theta).$$

(Note that the values of φ for $\theta < -h$ do not matter since f is constant for $\theta \geq h$.)

II.4 The adjoint shift semigroup

Given a bounded linear operator $L : X \to X$, its *adjoint* $L^* : X^* \to X^*$ is defined by

$$\langle x^*, Lx \rangle = \langle L^* x^*, x \rangle,$$

for every $x \in X$ and $x^* \in X^*$. This relation uniquely defines a bounded linear operator and

$$\|L^*\| = \|L\|$$

(recall that the norm of a bounded linear operator $K : Y \to Z$, where Y and Z are Banach spaces, is defined by $\|K\| = \sup_{\|y\| \leq 1} \|Ky\|$).

Let T be a $\mathcal{C}_0$-semigroup on X. We shall write $T^*(t)$ to denote $\big(T(t)\big)^*$. Let T^* denote the family of adjoint operators on X^*, i.e., $T^* = \{T^*(t)\}_{t\geq 0}$. Then $T^*(0) = I$ and T^* is a semigroup, i.e.,

$$T^*(t)T^*(s) = T^*(t+s), \quad t, s \geq 0,$$

but T^* is not necessarily strongly continuous. However, given $x^* \in X^*$, the mapping $t \mapsto \langle T^*(t)x^*, x\rangle$ is continuous for all $x \in X$ or, in other words, orbits are continuous in the weak* topology. We shall call T^* the *adjoint semigroup*.

Let T_0 be the shift semigroup defined by (2.2). Then

$$\begin{aligned}\langle f, T_0(t)\varphi\rangle &= \int_0^t df(\theta)\varphi(0) + \int_t^\infty df(\theta)\varphi(t-\theta)\\ &= f(t)\varphi(0) + \int_0^\infty d_\sigma f(t+\sigma)\varphi(-\sigma).\end{aligned}$$

We conclude that

$$\big(T_0^*(t)f\big)(\theta) = f(t+\theta) \quad \text{for } \theta > 0, \tag{4.1}$$

since our convention that elements of X^* are zero for $\theta \leq 0$ and continuous from the right on $(0, h)$, then implies that $T_0^*(t)f$ has a jump of magnitude $f(t)$ at $\theta = 0$ and this jump contributes the term $f(t)\varphi(0)$. So T_0^* is a shift semigroup too. If we choose, for instance, $f(\theta) = 0$ for $\theta < h$ and $f(h) \neq 0$, then

$$\|T_0^*(t)f - f\| = 2|f(h)|, \quad t > 0,$$

which illustrates that T_0^* is not strongly continuous.

II.5 The adjoint generator and the sun subspace

We now discuss two related questions: can one define a generator for T^* and does there exist a maximal invariant subspace on which T^* is strongly continuous? The general theory is presented in Butzer and Beerens [23], Hille and Phillips [124], Clément et al. [45] and van Neerven [204]. See Appendix II for an introduction. We shall summarize the main results without proof.

The *adjoint* A^* of a densely defined unbounded operator A is defined by $x^* \in \mathcal{D}(A^*)$ if and only if $y^* \in X^*$ exists such that

$$\langle x^*, Ax\rangle = \langle y^*, x\rangle$$

for all $x \in \mathcal{D}(A)$, and in that case,

$$A^*x^* = y^*.$$

Now let A be the generator of a $\mathcal{C}_0$-semigroup T on X; then the adjoint A^* is the generator of the adjoint semigroup T^* in the weak* sense, i.e.,

$$\frac{1}{t}\langle T^*(t)x^* - x^*, x\rangle \quad \text{converges for all } x \in X \text{ as } t \downarrow 0$$

if and only if $x^* \in \mathcal{D}(A^*)$, and in that case, the limit equals $\langle A^*x^*, x\rangle$.

The domain $\mathcal{D}(A^*)$ is weak* dense but not necessarily norm dense. In fact one can show (see Appendix II) that the norm closure

$$X^{\odot} := \overline{\mathcal{D}(A^*)} \tag{5.1}$$

is precisely the subspace on which T^* is strongly continuous, i.e.,

$$X^{\odot} = \{x^* \in X^* \mid \lim_{t\downarrow 0} \|T^*(t)x^* - x^*\| = 0\}. \tag{5.2}$$

The pronunciation of $\odot$ is "sun".

The restriction of T^* to the invariant subspace $X^{\odot}$ is strongly continuous. We denote the restriction by $T^{\odot}$. According to the general theory (Appendix II) the generator $A^{\odot}$ of $T^{\odot}$ is the part of A^* in $X^{\odot}$, i.e.,

$$\mathcal{D}(A^{\odot}) = \{x^{\odot} \in \mathcal{D}(A^*) \mid A^*x^{\odot} \in X^{\odot}\} \quad \text{and} \quad A^{\odot}x^{\odot} = A^*x^{\odot}.$$

We are now in exactly the same position as when we started: on some Banach space $X^{\odot}$ we have a $\mathcal{C}_0$-semigroup $T^{\odot}$ generated by $A^{\odot}$. So we can introduce the dual space $X^{\odot *}$ and the semigroup of adjoint operators $T^{\odot *}$ which is strongly continuous on $X^{\odot\odot} := \overline{\mathcal{D}(A^{\odot *})}$.

The pairing between elements of X and $X^{\odot}$ can be used to define an embedding (note that $X^{\odot}$ is weak* dense in X^*) j of X into $X^{\odot *}$. Of course

$$T^{\odot *}(t)jx = j(T(t)x)$$

and therefore $j(X) \subset X^{\odot\odot}$. Whenever $j(X) = X^{\odot\odot}$ we call X *$\odot$-reflexive* with respect to T. Recently, de Pagter [231] improved a result of Phillips by showing that $j(X) = X^{\odot\odot}$ if and only if the resolvent $(zI - A)^{-1}$ is weakly compact. It is sometimes convenient to present the interrelationship of the various spaces schematically in the following "duality" diagram:

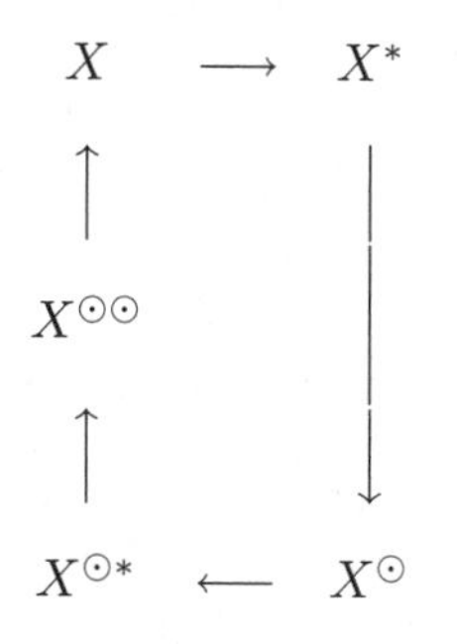

In the case of $\odot$-reflexivity the diagram simplifies to

$$\begin{array}{ccc} X & \longrightarrow & X^* \\ \uparrow & & \downarrow \\ X^{\odot *} & \longleftarrow & X^{\odot} \end{array}$$

Note that such a diagram has nothing to do with the commutative diagrams of algebra.

We now return to our particular example where T_0 denotes the shift semigroup given by (2.4).

Theorem 5.1. *$f \in \mathcal{D}(A_0^*)$ if and only if for $\theta > 0$*

$$f(\theta) = f(0+) + \int_0^\theta g(\sigma)\, d\sigma, \tag{5.3}$$

where $g \in NBV$ with $g(h) = 0$; for such f we have $A_0^* f = g$.

Proof. Let $f \in \mathcal{D}(A_0^*)$ and $A_0^* f = g$. For all $\varphi \in C^1$ with $\dot{\varphi}(0) = 0$, i.e., $\varphi \in \mathcal{D}(A_0)$, we have, using partial integration,

$$\begin{aligned} \int_0^\infty df(\theta)\dot{\varphi}(-\theta) = \langle f, A_0\varphi\rangle = \langle g, \varphi\rangle \\ &= \int_0^\infty dg(\theta)\varphi(-\theta) \\ &= \int_0^h g(\theta)\dot{\varphi}(-\theta)\, d\theta + g(h)\varphi(-h). \end{aligned}$$

Since this has to be an identity for functions φ which are constant, necessarily $g(h) = 0$. For any s, t with $0 < s < t < h$ we may choose a sequence $\varphi_n \in \mathcal{D}(A_0)$ such that $\dot{\varphi}_n(\theta)$ converges pointwise in θ and monotone increasing in n as $n \to \infty$ to the characteristic function of the interval $[-t, -s]$. Applying Lebesgue's monotone convergence theorem (see Theorem 2.1 of Appendix I) we find that

$$f(t) - f(s) = \int_s^t df(\theta) = \int_s^t g(\theta)\, d\theta.$$

Letting $s \downarrow 0$ we obtain

$$f(t) = f(0+) + \int_0^t g(\theta)\, d\theta.$$

Conversely, let f have this form with $g \in$ NBV. Then f is constant for $t \geq h$ if and only if $g(h) = 0$. Moreover, for all $\varphi \in C^1$ with $\dot{\varphi}(0) = 0$ we have

$$\begin{aligned} \langle f, A_0\varphi\rangle &= \int_0^\infty df(\theta)\dot{\varphi}(-\theta) = \int_0^\infty g(\theta)\dot{\varphi}(-\theta)\, d\theta \\ &= \int_0^\infty dg(\theta)\varphi(-\theta) = \langle g, \varphi\rangle \end{aligned}$$

from which we conclude that $f \in \mathcal{D}(A_0^*)$ and $A_0^* f = g$. □

Our next step is to characterise the closure of $\mathcal{D}(A_0^*)$. To this end the following two observations are useful:

(i) the total variation norm of $f(\theta) = f(0+) + \int_0^\theta g(\sigma)\, d\sigma$ equals $|f(0+)| + \|g\|_{L^1}$;

(ii) bounded variation functions are dense in L^1.

So when taking the closure, the form (5.3) is retained, but g is allowed to be in L^1. In words: the closure of $\mathcal{D}(A_0^*)$ consists precisely of the functions that are absolutely continuous on $(0, h]$ but are allowed to have a jump at zero.

Theorem 5.2. *For the shift semigroup we have*

$$X^\odot = \Big\{ f \in NBV \mid f(t) = c + \int_0^t g(\theta)\, d\theta \text{ for } t > 0, \text{ where } c \in \mathbb{C} \text{ and } g \in L^1 \text{ with } g(\theta) = 0, \text{ for (almost all) } \theta \geq h \Big\}.$$

Elements of $X^\odot$ are completely specified by $c \in \mathbb{C}$ and $g \in L^1$. [Note that $f(0+) = c$.] More precisely, the space $X^\odot$ is isometrically isomorphic to $\mathbb{C} \times L^1([0,h], \mathbb{C})$ equipped with the norm

$$\|(c, g)\| = |c| + \|g\|_{L^1}. \tag{5.4}$$

We shall frequently work with this representation of $X^\odot$ while adopting the convention that the L^1-function is extended to (h, ∞) by zero. In these coordinates we have

$$T_0^\odot(t)(c, g) = \Big(c + \int_0^t g(\sigma)\, d\sigma, g(t + \cdot) \Big). \tag{5.5}$$

Working with two "concrete" representations of an abstractly defined space introduces the danger of confusion. To oppose that danger we recapitulate the situation. Whenever we have a $\mathcal{C}_0$-semigroup on a Banach space X we can define the subspace $X^\odot$ of the dual space X^*. When we identify X^* with a function space Y we automatically obtain an identification of $X^\odot$ with a subspace, call it Z_1, of Y. Now let Z_2 be another function space which is isometrically isomorphic with Z_1, equipped with the topology inherited from Y. Then we can, alternatively, represent $X^\odot$ by Z_2. (The point of doing this is that Z_2 may form a more convenient starting point for finding a suitable representation of $X^{\odot *}$.) A careful description of this situation requires symbols for the isometric isomorphisms between X^* and Y and Z_1 and Z_2. Using these symbols one can then describe how operators defined on X^* act on Z_1 or on Z_2. The price is a rather heavy notation which makes statements hard to read. So we have chosen not to follow the safe road. We shall write $X^{\odot} = Z_1$ as well as $X^{\odot} = Z_2$ and we shall indicate operators on X^* and their "concrete" representations on Y, Z_1 or Z_2 by one and the same symbol.

Either from Theorem 5.1 or from (5.5) and known results about the generator of translation in L^1 (see Exercise 2.2, Appendix II or [23, 1.3.13], we obtain

Theorem 5.3. *For the shift semigroup we have*

$$\begin{aligned}\mathcal{D}(A_0^{\odot}) = \{f \mid f(t) = c + \int_0^t g(\theta)\, d\theta \text{ for } t > 0, \text{ where } c \in \mathbb{C} \\ \text{and } g \in \mathrm{AC}(0,h) \text{ with } g(\theta) = 0 \text{ for } \theta \geq h\}\end{aligned}$$

and $A_0^{\odot} f = g$. *Alternatively, we write*

$$\mathcal{D}(A_0^{\odot}) = \{(c,g) \mid c \in \mathbb{C} \text{ and } g \in \mathrm{AC}(0,h) \text{ with } g(\theta) = 0 \text{ for } \theta \geq h\}$$

and

$$A_0^{\odot}(c,g) = (g(0+), \dot{g}).$$

We shall represent $X^{\odot *}$ by $\mathbb{C} \times L^{\infty}([-h,0],\mathbb{C})$.

Exercise 5.4. Verify that the norm on $X^{\odot}$ given in (5.4) leads to the norm

$$\|(\alpha,\varphi)\| = \sup\{|\alpha|, \|\varphi\|_{\infty}\}$$

on $X^{\odot *} = \mathbb{C} \times L^{\infty}([-h,0],\mathbb{C})$, the pairing between $X^{\odot *}$ and $X^{\odot}$ being given by

$$\langle(\alpha,\varphi),(c,g)\rangle = \alpha c + \int_0^h \varphi(-\theta) g(\theta)\, d\theta. \tag{5.6}$$

From the explicit formula (5.5) for $T_0^{\odot}$ we deduce

$$T_0^{\odot *}(t)(\alpha,\varphi) = (\alpha, \varphi_t^{\alpha}), \tag{5.7}$$

where by definition

$$\varphi_t^{\alpha}(\theta) = \begin{cases} \varphi(t+\theta), & \text{if } t+\theta \leq 0, \\ \alpha, & \text{if } t+\theta > 0. \end{cases}$$

So φ_t^{α} is obtained by extending φ with the value α for $\theta > 0$ and then shifting it over t (if we want to be completely precise, we have to talk about equivalence classes and representatives).

Theorem 5.5. *For the shift semigroup we have*

$$\mathcal{D}\big(A_0^{\odot *}\big) = \big\{(\alpha,\varphi) \mid \varphi \in \mathrm{Lip}(\alpha)\big\} \text{ and } A_0^{\odot *}(\alpha,\varphi) = (0,\dot{\varphi}),$$

where $\mathrm{Lip}(\alpha)$ *denotes the subset of* $L^\infty\big([-h,0],\mathbb{C}\big)$ *whose elements contain a Lipschitz continuous function which assumes the value* α *at* $\theta = 0$.

Proof. Let $(\alpha,\varphi) \in \mathcal{D}\big(A_0^{\odot *}\big)$ and $A_0^{\odot *}(\alpha,\varphi) = (\beta,\psi)$. Then for all $g \in \mathrm{AC}$ with $g(\theta) = 0$ for $\theta \geq h$ we know, using Theorem 5.1, that

$$\begin{aligned}
\alpha g(0) + \int_0^h \varphi(-\theta)\dot{g}(\theta)\,d\theta &= \langle A_0^{\odot *}(\alpha,\varphi),(c,g)\rangle \\
&= \beta c + \int_0^h \psi(-\theta) g(\theta)\,d\theta \\
&= \beta c + \alpha g(0) + \int_0^h \big(\alpha + \int_0^{-\theta} \psi(\sigma)\,d\sigma\big)\dot{g}(\theta)\,d\theta.
\end{aligned}$$

So necessarily $\beta = 0$ (since c is arbitrary). Since $\psi \in L^\infty\big([-h,0],\mathbb{C}\big)$ it follows that $\varphi \in \mathrm{Lip}(\alpha)$.

Conversely, let $\varphi \in \mathrm{Lip}(\alpha)$. Then $\varphi(-\theta) = \alpha + \int_0^{-\theta} \psi(\sigma)\,d\sigma$ for some $\psi \in L^\infty\big([-h,0],\mathbb{C}\big)$ (See Appendix II.2: first note that a Lipschitz function is absolutely continuous and so can be written as the integral of an L^1-function; next use the almost everywhere differentiability together with the bound

$$\Big|\frac{\varphi(\theta_1) - \varphi(\theta_2)}{\theta_1 - \theta_2}\Big| \leq K$$

to conclude that the L^1-function is essentially bounded.) Exactly as above we obtain the identity

$$\alpha g(0) + \int_0^h \varphi(-\theta)\dot{g}(\theta)\,d\theta = \int_0^h \psi(-\theta) g(\theta)\,d\theta$$

for any $g \in \mathrm{AC}$ with $g(\theta) = 0$ for $\theta \geq h$. We conclude that $(\alpha,\varphi) \in \mathcal{D}\big(A_0^{\odot *}\big)$ and $A_0^{\odot *}(\alpha,\varphi) = (0,\psi)$. □

Taking the closure of $\mathcal{D}\big(A_0^{\odot *}\big)$ we lose the Lipschitz condition, but the continuity remains. So

$$X^{\odot\odot} = \overline{\mathcal{D}\big(A_0^{\odot *}\big)} = \big\{(\alpha,\varphi) \mid \varphi \in C(\alpha)\big\}, \tag{5.8}$$

where $C(\alpha)$ denotes the closed subspace of $L^\infty\big([-h,0],\mathbb{C}\big)$ whose elements contain a continuous function with the value α at zero. The embedding j of X into $X^{\odot *}$ assigns to φ the couple $(\varphi(0),\varphi)$. Hence $X^{\odot\odot} = j(X)$ and X is $\odot$-reflexive with respect to the shift semigroup T_0. Alternatively, one can base this conclusion on the compactness of $(\lambda I - A_0)^{-1}$ and the general result of Phillips-de Pagter quoted earlier. But in the present situation the direct verification, carried out above, is just as easy as to verify the compactness of $(\lambda I - A_0)^{-1}$.

In the following we shall omit the embedding operator j in our notation and identify X and $X^{\odot\odot}$. In other words, we shall go back and forth between

$$\varphi \in X = C([-h,0],\mathbb{C})$$

and

$$(\varphi(0),\varphi) \in X^{\odot\odot} = \{(\alpha,\varphi) \in \mathbb{C} \times L^\infty([-h,0],\mathbb{C}) \mid \varphi \in C(\alpha)\}.$$

This should not create any confusion.

We are now full circle. Starting from X and T_0 we arrived at $X^{\odot\odot}$ and $T_0^{\odot\odot}$ which are just the same objects in different colors. So what did we gain?

We introduced a space $X^{\odot *}$ in which X lies embedded and a natural extension of T_0 to this larger space. Thus we enlarged our "vocabulary" and, as we will see further on, this turns out to be extremely helpful. Moreover, since our construction involves dual spaces and adjoint operators we save some labour at a later stage of the analysis when we deal with Fredholm alternatives, projection operators, etc.

In the space $X^{\odot *}$ the initial function is defined in the L^∞-sense on $[-h,0]$, but the extension is based on an additional precisely defined point value at $\theta = 0$. The $\mathbb{C}$ component singles out the information on which the extension is based. The differential equation has a zero in this component (see Theorem 5.5) since we extend by a constant. If we change the rule for extension, we shall change the differential equation in this component only. On the space $X^{\odot *}$ this is easily done. Subsequently, we may restrict our attention to X and take for the generator the part in X.

Exercise 5.6. (A continuation of Exercise 2.2) When $X = L^1([0,h],\mathbb{C})$ we can take the representation

$$X^* = \{f \in L^\infty \mid \text{ essential support } f \subset [0,h]\}.$$

(Note: Here there is no need to make a reflection and to represent X^* by functions defined on $[-h,0]$; the point is that "age" is, although strongly correlated with "time", not the same as "time"; see Section XII.2 for more information on the "backward" character of the adjoint problem.) The adjoint of T_0 on X defined by [cf. (2.9)]

$$(T_0(t)\varphi)(a) = \begin{cases} \varphi(a-t), & a \geq t, \\ 0, & a < t \end{cases}$$

is then given by

$$(T_0^*(t)f)(a) = f(a+t).$$

Prove that

$$\mathcal{D}(A_0^*) = \{f \in \text{Lip} \mid f(a) = 0 \quad \text{for } a \geq h\}, \qquad A_0^* f = \dot{f},$$

and

$$X^{\odot} = C_0([0,h),\mathbb{C}) = \{f \in C(\mathbb{R}_+,\mathbb{C}) \mid f(a) = 0 \quad \text{for } a \geq h\}.$$

Next verify that

$$\mathcal{D}(A_0^{\odot}) = \{f \in C^1(\mathbb{R}_+,\mathbb{C}) \mid f(a) = 0, \quad a \geq h\}, \qquad A_0^{\odot} f = \dot{f}.$$

Representing $X^{\odot *}$ by NBV$([0,h),\mathbb{C})$, where the normalization includes [as indicated by the notation $[0,h)$] that the function is continuous at $a = h$, prove that

$$\mathcal{D}(A_0^{\odot *}) = \{\varphi \mid \varphi(a) = \int_0^a \psi(\alpha)\,d\alpha \text{ for } a \geq 0, \text{ for some } \psi \in \text{NBV}\},$$
$$A_0^{\odot *}\varphi = -\psi,$$

and

$$(T_0^{\odot *}(t)\varphi)(a) = \begin{cases} \varphi(a-t), & a \geq t, \\ \varphi(0) = 0, & a < t. \end{cases}$$

Finally, verify that the embedding $j : X \to X^{\odot *}$ is given by

$$j(\varphi)(a) = \int_0^a \varphi(\alpha)\,d\alpha$$

and that X is $\odot$-reflexive with respect to T_0.

Exercise 5.7. Verify that $T_0^{\odot *}(t)X^{\odot *} \subset X$ and $T_0^*(t)X^* \subset X^{\odot}$ for $t \geq h$, both for RFDE and for the setting of the preceding exercise.

II.6 The prototype system

If we consider the system of equations

$$\dot{x}(t) = 0 \quad \text{for } x(t) \in \mathbb{C}^n$$

as a RFDE with state space $X = C\big([-h,0],\mathbb{C}^n\big)$, we have, since there is no coupling, just n identical copies of the scalar case. So all of the results above carry over to the case of a system immediately: simply replace everywhere $\mathbb{C}$ by $\mathbb{C}^n$. Of course, coupling may arise as soon as we change the rule for extension and then the $\mathbb{C}^n$ structure becomes important. So in the next chapter we deal with systems from the very beginning.

Exercise 6.1. Let $X = Y \times Z$, where Y and Z are Banach spaces, and define for $x = (y,z)$

$$\|x\| = \|y\| + \|z\|.$$

Let $S_1(t)$ be a $\odot$-reflexive $\mathcal{C}_0$-semigroup on Y and $S_2(t)$ a $\odot$-reflexive $\mathcal{C}_0$-semigroup on Z. Show that, for $x = (y,z)$,

$$T(t)x = (S_1(t)y, S_2(t)z)$$

defines a $\odot$-reflexive $\mathcal{C}_0$-semigroup on X.
Hint: Represent X^* by $Y^* \times Z^*$.

II.7 Comments

Adjoint semigroups were first studied by Phillips [239] (also see Hille and Phillips [124]). A very readable exposition is given in the book of Butzer and Berens [23]. Many new (as well as old) aspects of the theory are developed in the recent lecture notes by van Neerven [204]. In Appendix II we collect those results and proofs which are most relevant in the context of delay equations. The idea that one can use perturbation theory for adjoint semigroups to study delay equations is first presented in Diekmann [65], building on the work by Clément, Diekmann, Gyllenberg, Heijmans and Thieme [44, 45, 46, 47]. The symbol $\odot$ appears in Hille and Phillips [124] without a hint concerning the pronunciation. As it is the original Chinese character for the sun, it seems logical to call it by that name.

Chapter III

Linear RFDE as bounded perturbations

III.1 The basic idea, followed by a digression on weak* integration

Consider the linear autonomous RFDE

$$(1.1) \qquad \dot{x}(t) = \int_0^h d\zeta(\theta)x(t-\theta), \qquad t \geq 0 \quad \text{and } x(t) \in \mathbb{C}^n,$$

where ζ denotes a $n \times n$ matrix-valued function whose entries belong to NBV. Alternatively we can write (see I.1.5)

$$(1.2) \qquad \dot{x}(t) = \langle \zeta, x_t \rangle_n.$$

How can we represent the equation in our abstract framework? Keeping in mind our discussion about the two components "shifting" and "extending" it is tempting to try

$$(1.3) \qquad \frac{d}{dt}x_t = A_0^{\odot *}x_t + (\langle \zeta, x_t \rangle_n, 0)$$

or, in other symbols,

$$(1.4) \qquad \frac{d}{dt}x_t = (A_0^{\odot *} + B)x_t,$$

where $B : X \to X^{\odot *}$ is defined by

$$(1.5) \qquad B\varphi = (\langle \zeta, \varphi \rangle_n, 0).$$

Note that $\mathcal{R}(A_0^{\odot *}) \cap \mathcal{R}(B) = \{(0,0)\}$; this reflects the independence of the "shift" and the "extend" components.

The operator B describing the rule for the extension is not only linear but also bounded. This property we owe to the fact that the range space $X^{\odot *}$ is large enough. Note, moreover, that even though we still conceive of x_t as an element of X, the differential equation is an identity for elements of $X^{\odot *}$!

The program now is as follows. First we formally integrate (1.4) to obtain the variation-of-constants equation

$$x_t = T_0(t)\varphi + \int_0^t T_0^{\odot *}(t-\tau)Bx_\tau d\tau. \tag{1.6}$$

We then show that for each initial condition $\varphi \in X$ this integral equation has a unique solution $x_t = x_t(\varphi) \in X$. If we pay attention to the first component (i.e., the value at $\theta = 0$) only, the equation reads

$$x(t) = \varphi(0) + \int_0^t \langle \zeta, x_\tau \rangle_n d\tau \tag{1.7}$$

and so the continuity of $t \mapsto x_t$ implies that $x(t)$ is continuously differentiable on $\mathbb{R}_+$ and satisfies (1.2). So, indeed, we solved (1.2) provided with the initial condition $x_0 = \varphi$.

We shall show in Section 4 that one can recover the renewal equation (I.2.4a) by applying B to (1.6) and exploiting the fact that B has finite dimensional range.

Defining $T(t)\varphi = x_t(\varphi)$ we obtain a $\mathcal{C}_0$-semigroup on X. It turns out (see Section 2) that, as a consequence of the boundedness of B, the spaces $X^{\odot}$ and $X^{\odot\odot}$ are the same for $T(t)$ and for $T_0(t)$, a property which can be expressed by saying that the duality structure does not change under bounded perturbation.

In conclusion of this introductory section we want to lay the basis for the precise interpretation of the integral in formula (1.6) above.

Let $I \subset \mathbb{R}$ be an interval and let $q : I \to X^*$ be such that

$$\langle q(\,\cdot\,), x \rangle \in L^1\big([a,b], \mathbb{C}\big), \qquad \text{for all } x \in X.$$

We claim that there exists $Q \in X^*$ for which

$$\langle Q, x \rangle = \int_I \langle q(\sigma), x \rangle \, d\sigma, \qquad \text{for all } x \in X$$

and we shall call Q the *weak* integral* of q over the interval I. Moreover, we shall simply write

$$Q = \int_I q(\sigma)\, d\sigma$$

whenever it is clear from the context that the weak* integral is meant. [For the arguments which substantiate the claim, see Appendix II.3.13.]

The following sequence of exercises explains how this definition applies to (1.6) and, moreover, introduces some useful manipulations and identities.

Exercise 1.1. Let K be a strongly continuous family of bounded linear operators on a Banach space X, i.e., for all $x \in X$ the function $\sigma \to K(\sigma)x$ is continuous from $[a,b]$ to X. Let $h : [a,b] \to X^*$ be norm continuous. Interpret

$$\int_a^b K(\sigma)^* h(\sigma)\, d\sigma$$

as a weak* integral (note that under these assumptions one can work with the Riemann integral since all integrands are continuous functions).

Exercise 1.2. Let $f : \mathbb{R}_+ \to X^{\odot *}$ be a norm continuous function. Convince yourself that

$$\int_r^s T_0^{\odot *}(t-\tau) f(\tau)\, d\tau, \quad 0 \le r \le s \le t < \infty,$$

is well defined as a weak* integral (with values in $X^{\odot *}$).

The next two exercises will play a role in Section 3.

Exercise 1.3. Show that for $g \in L^1([a,b], \mathbb{C})$ and fixed $x^* \in X^*$ one can define an element $\int_a^b K(\sigma)^* g(\sigma) x^* d\sigma$ of X^* by requiring that

$$\langle \int_a^b K(\sigma)^* g(\sigma) x^* d\sigma, x \rangle = \int_a^b g(\sigma) \langle x^*, K(\sigma) x \rangle d\sigma$$

for all $x \in X$. Here K is as in Exercise 1.1.

Exercise 1.4. Let $\eta \in L^1$ and $x^{\odot *} \in X^{\odot *}$ be given. Convince yourself that

$$\int_r^s T_0^{\odot *}(t-\tau) \eta(\tau) x^{\odot *}\, d\tau, \qquad 0 \le r \le s \le t < \infty,$$

is well defined as a weak* integral (with values in $X^{\odot *}$).

The last two exercises are needed when we want to show that the solution of (1.6) has the semigroup property.

Exercise 1.5. Let $L : X \to X$ be a bounded linear operator. Verify that

$$L^* \int_a^b K(\sigma)^* h(\sigma)\, d\sigma = \int_a^b L^* K(\sigma)^* h(\sigma)\, d\sigma.$$

(Cautionary remark: It is not allowed to replace L^* by an arbitrary bounded linear operator on X^*.)

Exercise 1.6. Let $f : \mathbb{R}_+ \to X^{\odot *}$ be norm continuous. Show that

$$T_0^{\odot *}(h) \int_r^s T_0^{\odot *}(t-\tau) f(\tau)\, d\tau = \int_r^s T_0^{\odot *}(t+h-\tau) f(\tau)\, d\tau.$$

III.2 Bounded perturbations in the sun-reflexive case

Throughout this section T_0 denotes a $\mathcal{C}_0$-semigroup on a Banach space X which is assumed to be $\odot$-reflexive with respect to T_0. The generator of T_0 is denoted by A_0. Of course the shift semigroup on $X = C([-h,0],\mathbb{C}^n)$ is our motivating example, but the results have more general applicability.

We start with some technical results.

Lemma 2.1. *Let* $f : \mathbb{R}_+ \to X^{\odot *}$ *be norm continuous. Denote the subset* $\{(t,s,r) \mid 0 \le r \le s \le t < \infty\}$ *of* $\mathbb{R}^3$ *by* Ω. *Define* $w : \Omega \to X^{\odot *}$ *as the following weak* integral:*

$$w(t,s,r) = \int_r^s T_0^{\odot *}(t-\tau) f(\tau)\, d\tau. \tag{2.1}$$

Then w *is norm continuous and takes values in* X.

Proof. Equivalently, we can write

$$w(t,s,r) = \int_{t-s}^{t-r} T_0^{\odot *}(\sigma) f(t-\sigma)\, d\sigma.$$

Hence

$$\begin{aligned} w(t_2,s_2,r_2) - w(t_1,s_1,r_1) &= \int_{I_2 \backslash I_1} T_0^{\odot *}(\sigma) f(t_2-\sigma)\, d\sigma \\ &\quad + \int_{I_1 \cap I_2} T_0^{\odot *}(\sigma)\{f(t_2-\sigma) - f(t_1-\sigma)\} d\sigma \\ &\quad - \int_{I_1 \backslash I_2} T_0^{\odot *}(\sigma) f(t_1-\sigma)\, d\sigma, \end{aligned}$$

where $I_i = [t_i - s_i, t_i - r_i]$, $i = 1,2$. The norms of the first and the last term at the right hand side are less than

$$\Big(\sup_{\sigma \in I_1 \cup I_2} ||T_0^{\odot *}(\sigma)||\Big) \big(\max\{\text{meas}(I_2\backslash I_1), \text{meas}(I_1 \backslash I_2)\}\big) \Big(\sup_{r \le \sigma \le s} ||f(\sigma)||\Big)$$

from which we conclude that they tend to zero when

$$|(t_2,s_2,r_2) - (t_1,s_1,r_1)| \to 0.$$

The norm of the middle term is bounded by

$$\Big(\sup_{\sigma \in I_1 \cap I_2} ||T_0^{\odot *}(\sigma)||\Big) \Big(\sup_{\tau \in I_1 \cap I_2} ||f(t_2-\tau) - f(t_1-\tau)||\Big)$$

and, by invoking the norm continuity of f, we arrive at the same conclusion. Hence w is norm continuous.

Now recall that we have identified X with its embedding $j(X)$ in $X^{\odot *}$ and that, by our assumption of $\odot$-reflexivity,

$$X = X^{\odot\odot} = \{x^{\odot *} \mid ||T_0^{\odot *}(h)x^{\odot *} - x^{\odot *}|| \to 0 \text{ as } h \downarrow 0\}.$$

So if we show that

$$||T_0^{\odot *}(h)w(t,s,r) - w(t,s,r)|| \to 0 \quad \text{as } h \downarrow 0,$$

we can conclude that $w(t,s,r) \in X$.

Finally, observe that (see Exercise 1.6)

$$T_0^{\odot *}(h)w(t,s,r) = w(t+h,s,r)$$

and note that the norm continuity of w is proved above. □

Exercise 2.2. When f is not norm continuous but of the special form $f(t) = \eta(t)x^{\odot *}$ for some $\eta \in L^1$, formula (2.1) still makes sense (Exercise 1.4). Modify the proof to arrive at the same conclusion for functions f of this form. (Hint: Translation is continuous in L^1; see, for instance, Exercise I.2.2, Appendix II.2 or Butzer and Berens [23].)

Lemma 2.3.. *Let $f : \mathbb{R}_+ \to X^{\odot *}$ be norm continuous. Define $v : \mathbb{R}_+ \to X^{\odot *}$ as the weak* integral*

$$v(t) = \int_0^t T_0^{\odot *}(t-\tau)f(\tau)\,d\tau. \tag{2.2}$$

Then v is norm continuous, takes values in X and

$$||v(t)|| \le M\frac{e^{\omega t}-1}{\omega} \sup_{0\le\tau\le t} ||f(\tau)||, \tag{2.3}$$

where M and ω are such that $||T_0(t)|| \le Me^{\omega t}$. Moreover

$$\frac{1}{t}v(t) \xrightarrow{*} f(0) \quad \text{as } t \downarrow 0,$$

where $\xrightarrow{*}$ indicates weak convergence.*

Proof. Since $v(t) = w(t,t,0)$, the first two assertions follow from Lemma 2.1. The estimate (2.3) is a straightforward corollary of $||T_0^*(t)|| = ||T_0(t)|| \le Me^{\omega t}$. Finally,

$$\frac{1}{t}\langle v(t), x^{\odot}\rangle = \frac{1}{t}\int_0^t \langle f(\tau), T_0^{\odot}(t-\tau)x^{\odot}\rangle d\tau \to \langle f(0), x^{\odot}\rangle, \quad \text{as } t \downarrow 0$$

for arbitrary $x^{\odot} \in X^{\odot}$. □

On the level of the generator, we now introduce a perturbation in the form of a bounded linear operator $B : X \to X^{\odot *}$. We want to construct a perturbed semigroup $T(t)$ by solving the variation-of-constants equation

$$T(t)\varphi = T_0(t)\varphi + \int_0^t T_0^{\odot *}(t-\tau)BT(\tau)\varphi d\tau \tag{2.4}$$

by successive approximations. The proof that this method works will appear to be almost identical to the proof in case of a (truly) bounded operator from X into X as given in, for instance, Pazy [233, Sect. 3.1] or Davis [59, Sect.3.1].

Theorem 2.4. *Let $B : X \to X^{\odot *}$ be a bounded operator. There exists a unique C_0-semigroup $\{T(t)\}_{t\geq 0}$ such that (2.4) holds.*

Proof. Define inductively

$$T_k(t)\varphi = \int_0^t T_0^{\odot *}(t-\tau)BT_{k-1}(\tau)\varphi d\tau \quad \text{for } k \geq 1. \tag{2.5}$$

Then Lemma 2.3 implies that $T_k(t)$ maps X into X and that $t \mapsto T_k(t)\varphi$ is continuous. We claim that the estimate $\|T_0(t)\| \leq Me^{\omega t}$ implies that

$$\|T_k(t)\| \leq Me^{\omega t}\frac{M^k\|B\|^k t^k}{k!}.$$

The proof is by induction and follows directly from the definition of T_k.

Next we define

$$T(t) = \sum_{k=0}^{\infty} T_k(t) \tag{2.6}$$

and note that our estimate implies that the series converges in the operator norm, uniformly for t in bounded intervals. It follows that $t \mapsto T(t)\varphi$ is continuous for every $\varphi \in X$. The definitions (2.5) and (2.6) imply that $T(t)\varphi$ satisfies (2.4).

To prove uniqueness let $\{S(t)\}_{t\geq 0}$ be a family of bounded linear operators on X for which $t \mapsto S(t)\varphi$ is continuous for every $\varphi \in X$ and the equation

$$S(t)\varphi = T_0(t)\varphi + \int_0^t T_0^{\odot *}(t-\tau)BS(\tau)\varphi d\tau$$

holds. Subtracting this equation from (2.4) we find

$$T(t)\varphi - S(t)\varphi = \int_0^t T_0^{\odot *}(t-\tau)B\big(T(\tau)\varphi - S(\tau)\varphi\big)d\tau$$

which yields the estimate

$$\|T(t)\varphi - S(t)\varphi\| \leq M\|B\|e^{\omega t}\int_0^t e^{-\omega\tau}\|T(\tau)\varphi - S(\tau)\varphi\|d\tau.$$

According to Gronwall's inequality (see Hale [101, I.6.6]) this implies that $\|T(t)\varphi - S(t)\varphi\|$ is identically zero.

Finally, we exploit the uniqueness to prove that T is a semigroup. Starting from equation (2.4) we manipulate as follows:

$$\begin{aligned} T(t+s)\varphi &= T_0(t+s)\varphi + \int_0^{t+s} T_0^{\odot *}(t+s-\tau)BT(\tau)\varphi d\tau \\ &= T_0(t+s)\varphi + \int_0^{s} T_0^{\odot *}(t+s-\tau)BT(\tau)\varphi d\tau \\ &\quad + \int_0^{t} T_0^{\odot *}(t-\sigma)BT(\sigma+s)\varphi d\sigma \\ &= T_0(t)\{T_0(s)\varphi + \int_0^{s} T_0^{\odot *}(s-\tau)BT(\tau)\varphi d\tau\} \\ &\quad + \int_0^{t} T_0^{\odot *}(t-\sigma)BT(\sigma+s)\varphi d\sigma \\ &= T_0(t)T(s)\varphi + \int_0^{t} T_0^{\odot *}(t-\sigma)BT(\sigma+s)\varphi d\sigma. \end{aligned}$$

This is exactly the same equation as (2.4) but now with initial data $T(s)\varphi$. So the solution is $T(t)T(s)\varphi$. But our manipulations have shown that we can also express the solution in the form $T(t+s)\varphi$ and therefore uniqueness implies that necessarily the semigroup property $T(t+s)\varphi = T(t)T(s)\varphi$ holds for all $\varphi \in X$. Finally, by putting $t = 0$ in (2.4), we find that $T(0)\varphi = T_0(0)\varphi = \varphi$, i.e., $T(0) = I$. □

Exercise 2.5. Let $\{T_0(t)\}$ be eventually compact, i.e., $T_0(t)$ is compact for $t \geq t_0$ for some t_0 and let $B : X \to X^{\odot *}$ be bounded. Prove that $\{T(t)\}$ is eventually compact.
Hint: Use Gronwall's inequality.

Our next task is to determine the generator of T, which we shall call A. As it turns out, the simplest approach follows a roundabout way: we shall first determine $A^*, A^{\odot}$ and $A^{\odot *}$. In order to do this we shall first verify that $X^{\odot}$ is the space of strong continuity of $T^*(t)$ as well.

It is convenient to provide the difference of the perturbed and the unperturbed semigroup with a name, so we define

$$U(t)\varphi = \int_0^t T_0^{\odot *}(t-\tau)BT(\tau)\varphi d\tau \tag{2.7}$$

and note that, according to Lemma 2.3, $\|U(t)\| \to 0$ as $t \downarrow 0$. It follows that $\|U^*(t)\| \to 0$ as well, and as a consequence, $t \mapsto T^*(t)x^*$ is norm continuous at $t = 0$ if and only if $t \mapsto T_0^*(t)x^*$ has this property. We conclude:

Lemma 2.6. *The subspace $X^\odot$, at first defined in terms of the unperturbed semigroup T_0^*, is as well the subspace of strong continuity for the perturbed adjoint semigroup T^*. In particular, $X^\odot$ is invariant under T^*.*

Next we analyse the behaviour for t tending to zero; but we need to straighten out an aspect of our notation first. The adjoint of B maps $X^{\odot**}$ into X^*. The Banach space $X^\odot$ can be considered as a subspace of its second dual space $X^{\odot**}$ and, consequently, we can restrict the adjoint of B to $X^\odot$. We shall not introduce a new symbol for this restriction but simply write $B^* : X^\odot \to X^*$.

Lemma 2.7. *For arbitrary $x^\odot \in X^\odot$ and $x \in X$ one has that*

$$\frac{1}{t}\langle U^*(t)x^\odot, x\rangle \to \langle B^*x^\odot, x\rangle \quad \text{as } t \downarrow 0.$$

Proof.

$$\begin{aligned}\frac{1}{t}\langle U^*(t)x^\odot, x\rangle &= \frac{1}{t}\langle x^\odot, U(t)x\rangle = \frac{1}{t}\int_0^t \langle T_0^\odot(t-\tau)x^\odot, BT(\tau)x\rangle d\tau \\ &= \frac{1}{t}\int_0^t \langle B^*T_0^\odot(t-\tau)x^\odot, T(\tau)x\rangle d\tau \to \langle B^*x^\odot, x\rangle\end{aligned}$$

as $t \downarrow 0$. □

Now recall that, according to the general theory,

$$\mathcal{D}(A^*) = \left\{x^* \mid \frac{1}{t}(T^*(t)x^* - x^*) \text{ converges weak}^* \text{ as } t \downarrow 0\right\}$$

and that for $x^* \in \mathcal{D}(A^*)$

$$\frac{1}{t}(T^*(t)x^* - x^*) \xrightarrow{*} A^*x^* \quad \text{as } t \downarrow 0.$$

Lemma 2.6 implies that $\mathcal{D}(A^*) \subset X^\odot$ and Lemma 2.7 tells us that for all $x^\odot \in X^\odot$ the expression $\frac{1}{t}U^*(t)x^\odot$ converges weak* to $B^*x^\odot$. By definition, $\frac{1}{t}(T_0^*(t)x^\odot - x^\odot)$ converges weak* if and only if $x^\odot \in \mathcal{D}(A_0^*)$. Combining this information we obtain:

Corollary 2.8. $\mathcal{D}(A^*) = \mathcal{D}(A_0^*)$ *and* $A^* = A_0^* + B^*$.

Let, as usual, $T^\odot$ denote the restriction of T^* to the invariant subspace $X^\odot$. The generator of $T^\odot$ is the part of $A_0^* + B^*$ in $X^\odot$:

Corollary 2.9. *The C_0-semigroup $T^{\odot}$ is generated by the operator $A^{\odot}$ defined by*

$$\mathcal{D}(A^{\odot}) = \{x^{\odot} \in \mathcal{D}(A_0^*) : (A_0^* + B^*)x^{\odot} \in X^{\odot}\}, \qquad A^{\odot} = A_0^* + B^*.$$

Both T_0^* and T^* leave $X^{\odot}$ invariant and so the same must be true for their difference U^*. Let $U^{\odot}$ denote the restriction, then $\|U^{\odot}(t)\| \to 0$ as $t \downarrow 0$ and likewise $\|U^{\odot *}(t)\| \to 0$. It follows that the space $X^{\odot\odot}$ is the same for $T_0^{\odot *}$ and $T^{\odot *}$ and, in particular, we find

Theorem 2.10. *The space X is $\odot$-reflexive with respect to the perturbed semigroup.*

As a reformulation of Lemma 2.7 we have

Lemma 2.11. *For arbitrary $x \in X$ and $x^{\odot} \in X^{\odot}$*

$$\frac{1}{t}\langle x^{\odot}, U(t)x\rangle \to \langle Bx, x^{\odot}\rangle, \qquad \text{as } t \downarrow 0.$$

Reasoning exactly as above we deduce

Corollary 2.12. $\mathcal{D}(A^{\odot *}) = \mathcal{D}(A_0^{\odot *})$ *and* $A^{\odot *} = A_0^{\odot *} + B$.

Corollary 2.13. *The C_0-semigroup T is generated by the operator A defined by*

$$\mathcal{D}(A) = \{x \in \mathcal{D}(A_0^{\odot *}) : A_0^{\odot *}x + Bx \in X\}, \qquad A = A_0^{\odot *} + B.$$

Exercise 2.14. In our formulation of the corollary above we have suppressed the embedding operator j. Show that the more elaborate formulation is

$$\mathcal{D}(A) = \{x \in j^{-1}(\mathcal{D}(A_0^{\odot *})) : A_0^{\odot *}jx + Bx \in j(X)\},$$
$$Ax = j^{-1}(A_0^{\odot *}jx + Bx).$$

In the next section we shall elaborate the special case that B has finite dimensional range, which applies to RFDE, in more detail. But before doing so, we collect a number of somewhat technical variations on our main theme in an interlude. This material may be skipped at first and then only consulted when it is used later. Another short cut is to read only equation (2.12), Theorem 2.16, Corollary 2.18 and Lemma 2.23.

Interlude: (Integrated) Variants of the variation-of-constants equation

Before looking at integrated semigroups we collect a number of identities which are derived from an interchange of (i) the role of $\{T_0(t)\}$ and $\{T(t)\}$ and/or (ii) the role of X and $X^{\odot}$. From a "pure" mathematical point of view it is arbitrary which of the two semigroups one considers to be a perturbation of the other. Therefore, one expects that the identity

$$T_0(t)\varphi = T(t)\varphi - \int_0^t T^{\odot *}(t-\tau)BT_0(\tau)\varphi d\tau \tag{2.8}$$

holds, or, written in another way,

$$T(t)\varphi = T_0(t)\varphi + \int_0^t T^{\odot *}(t-\tau)BT_0(\tau)\varphi d\tau. \tag{2.9}$$

There are many ways to check that this is indeed the case. For instance, one can use the expansion (2.5) – (2.6) or one can consider (2.8) as an equation, with $T(t)$ given and $T_0(t)$ unknown, and then compute the generator of the solution, which will turn out to be A_0. The unique correspondence between generator and $\mathcal{C}_0$-semigroup then yields the conclusion that the solution is indeed $T_0(t)$.

The assumption of $\odot$-reflexivity yields a symmetric duality diagram

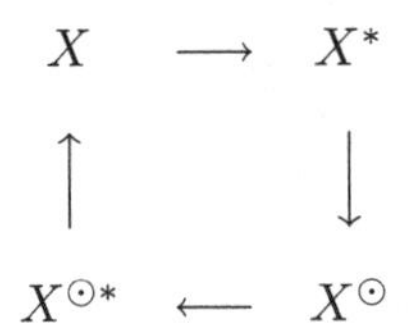

and one may just as well use $T_0^{\odot}, T_0^*$ and B^* in order to constructively define $T^{\odot}$. If one takes adjoints, the order of the factors reverses. So starting from (2.4) we obtain

$$T^{\odot}(t)x^{\odot} = T_0^{\odot}(t)x^{\odot} + \int_0^t T^*(t-\tau)B^*T_0^{\odot}(\tau)x^{\odot}d\tau, \tag{2.10}$$

whereas (2.9) yields

$$T^{\odot}(t)x^{\odot} = T_0^{\odot}(t)x^{\odot} + \int_0^t T_0^*(t-\tau)B^*T^{\odot}(\tau)x^{\odot}d\tau. \tag{2.11}$$

Thus we derived four formulations of the variation-of-constants equation, viz. (2.4), (2.9), (2.10) and (2.11), which all carry exactly the same information.

In general one cannot write down an equation for the weak* continuous semigroups T^* and $T^{\odot *}$, respectively acting on the "big" spaces X^* and

$X^{\odot*}$. However, for the so-called *integrated semigroup* $W(t) : X^{\odot*} \to X$, defined by

$$W(t)x^{\odot*} = \int_0^t T^{\odot*}(\tau)x^{\odot*}d\tau, \tag{2.12}$$

we have a variation-of-constants equation which does make sense on the "big" space.

Exercise 2.15. Use Lemma 2.1 to show that $W(t)$, at first defined as a mapping from $X^{\odot*}$ into $X^{\odot*}$, takes values in X. Note that without the assumption of $\odot$-reflexivity we can only conclude that $W(t)$ takes values in $X^{\odot\odot}$.

Theorem 2.16. *For every $x^{\odot*} \in X^{\odot*}$ and every $t \geq 0$*

$$W(t)x^{\odot*} = W_0(t)x^{\odot*} + \int_0^t T_0^{\odot*}(t-\tau)BW(\tau)x^{\odot*}d\tau. \tag{2.13}$$

Note: Here and in the following we refrain from repeating a definition like (2.12) with an index 0 to introduce the analogous object corresponding to the unperturbed semigroup. Similarly we shall not repeat identities like (2.14) below with an index 0. The proof of Theorem 2.16 is somewhat technical. Readers wanting to skip the rest of this interlude are advised to just have a brief look at Corollary 2.18.

For the proof we need another auxiliary result.

Lemma 2.17. *For all $x^* \in X^*, x^{\odot*} \in X^{\odot*}$ and $t \geq 0$ the following identity holds:*

$$\langle x^*, \int_0^t T^{\odot*}(\tau)x^{\odot*}d\tau\rangle = \langle x^{\odot*}, \int_0^t T^*(\tau)x^*d\tau\rangle. \tag{2.14}$$

Proof. First we take $x^* = x^{\odot} \in X^{\odot}$. Then the identity follows at once from the definition of the weak* integral and the fact that $T^{\odot*}(t)$ is the adjoint $T^{\odot}(t)$.

Next we approximate arbitrary x^* by $1/h \int_0^h T^*(\sigma)x^*d\sigma \in X^{\odot}$. On the left hand side we have

$$\begin{aligned}
&\frac{1}{h}\langle \int_0^h T^*(\sigma)x^*d\sigma, \int_0^t T^{\odot*}(\tau)x^{\odot*}d\tau\rangle \\
&\quad = \frac{1}{h}\int_0^h \langle x^*, T(\sigma)\int_0^t T^{\odot*}(\tau)x^{\odot*}d\tau\rangle d\sigma \\
&\quad \to \langle x^*, \int_0^t T^{\odot*}(\tau)x^{\odot*}d\tau\rangle \quad \text{as } h \downarrow 0.
\end{aligned}$$

On the right hand side,

$$\begin{aligned}
&\frac{1}{h}\langle x^{\odot *}, \int_0^t T^*(\tau)\int_0^h T^*(\sigma)x^* d\sigma\, d\tau\rangle \\
&\quad = \frac{1}{h}\langle x^{\odot *}, \int_0^h \int_\sigma^{t+\sigma} T^*(\theta)x^* d\theta\, d\sigma\rangle \\
&\quad = \frac{1}{h}\int_0^h \langle x^{\odot *}, \int_\sigma^{t+\sigma} T^*(\theta)x^* d\theta\rangle\, d\sigma \to \langle x^{\odot *}, \int_0^t T^*(\theta)x^* d\theta\rangle
\end{aligned}$$

as $h \downarrow 0$. □

Corollary 2.18. *$W(t)^*$ maps X^* into $X^\odot$ and $W(t)^* = W^\odot(t)$, where $W^\odot(t) : X^* \to X^\odot$ is defined by*

$$W^\odot(t)x^* = \int_0^t T^*(\tau)x^* d\tau. \tag{2.15}$$

Corollary 2.19. *Let $\{x_n^{\odot *}\}$ be a sequence in $X^{\odot *}$ converging weak* to $x_\infty^{\odot *}$. Then $\{W(t)x_n^{\odot *}\}$ converges weakly in X to $W(t)x_\infty^{\odot *}$.*

Proof of Theorem 2.16. From the variation-of-constants formula (2.4) it follows that for every $\varphi \in X$ and $x^\odot \in X^\odot$

$$\begin{aligned}
\langle x^\odot, \int_0^t T(\tau)\varphi d\tau\rangle &= \langle x^\odot, \int_0^t T_0(\tau)\varphi d\tau\rangle \\
&\quad + \int_0^t \int_0^\tau \langle B^* T_0^\odot(\sigma)x^\odot, T(\tau-\sigma)\varphi\rangle d\sigma d\tau.
\end{aligned}$$

By changing the order of integration in the last term we find that

$$\langle x^\odot, W(t)\varphi\rangle = \langle x^\odot, W_0(t)\varphi\rangle + \int_0^t \langle B^* T_0^\odot(\sigma)x^\odot, W(t-\sigma)\varphi\rangle d\sigma.$$

Let $x^{\odot *} \in X^{\odot *}$ and let $\{x_n\}$ be a sequence in X converging to $x^{\odot *}$ in the weak* sense [for instance, take $x_n = n(nI - A^{\odot *})^{-1}x^{\odot *}$]. The last identity above with $\varphi = x_n$, Corollary 2.19, and the dominated convergence theorem together imply that by letting $n \to \infty$ we obtain

$$\langle x^\odot, W(t)x^{\odot *}\rangle = \langle x^\odot, W_0(t)x^{\odot *}\rangle + \int_0^t \langle B^* T_0^\odot(\sigma)x^\odot, W(t-\sigma)x^{\odot *}\rangle d\sigma$$

and from this, (2.13) immediately follows. □

Exercise 2.20. Formulate variants of (2.13) corresponding to (2.9), (2.10) and (2.11).

Exercise 2.21. For given $f \in L^1$ define $F(t) = \int_0^t f(\tau)\, d\tau$. Show, by integration by parts, that

$$\int_0^t W(t-\tau)f(\tau)x^{\odot *}\,d\tau = \int_0^t T^{\odot *}(t-\tau)F(\tau)x^{\odot *}\,d\tau.$$

Exercise 2.22. Verify that $W(t)x^{\odot *}$, considered as an $X^{\odot *}$-valued function, is weak* differentiable with derivative $T^{\odot *}(t)x^{\odot *}$.

We conclude this interlude with a result that we shall need in the proof of Proposition VII.5.4, but which is very natural by itself. If we consider (informally) the inhomogeneous differential equation

$$\frac{du}{dt} = A_0^{\odot *}u + Bu + f$$

for some given function $f : [0, s] \to X^{\odot *}$, and with initial condition $u(0) = \varphi$, it is clear that u is given explicitly by

$$u(t) = T(t)\varphi + \int_0^t T^{\odot *}(t-\tau)f(\tau)\,d\tau \tag{2.16}$$

whereas, on the other hand, u should also satisfy the integral equation

$$u(t) = T_0(t)\varphi + \int_0^t T_0^{\odot *}(t-\tau)\{Bu(\tau) + f(\tau)\}\,d\tau. \tag{2.17}$$

The next lemma states that our expectations are warranted.

Lemma 2.23. *Let $f : [0, s] \to X^{\odot *}$ be continuous. Let $B : X \to X^{\odot *}$ be a bounded linear operator and let $T(t)$ be the C_0-semigroup on X generated by the part of $A_0^{\odot *} + B$ in X. The function u defined by* (2.16) *is the unique solution of* (2.17).

Proof. Uniqueness follows from Gronwall's inequality exactly as in the proof of Theorem 2.5. Substituting (2.16) into (2.17) and using the relation

$$T(t) = T_0(t) + \int_0^t T_0^{\odot *}(\tau)BT(t-\tau)\,d\tau,$$

we find that we have to check equality of

$$\int_0^t T^{\odot *}(t-\tau)f(\tau)\,d\tau$$

and

$$\int_0^t T_0^{\odot *}(t-\tau)f(\tau)\,d\tau + \int_0^t T_0^{\odot *}(\tau)B\int_0^{t-\tau} T^{\odot *}(t-\tau-\sigma)f(\sigma)\,d\sigma\,d\tau.$$

For $t = 0$, equality clearly holds. In order to manipulate with well-defined quantities, we integrate both functions from zero to t. We claim this yields for the last term of the second expression

$$\int_0^t \left(W(t-\sigma) - W_0(t-\sigma)\right) f(\sigma)\, d\sigma$$

which is, indeed, exactly the difference of the integral of the first expression and the integral of the first term of the second expression. So it only remains to substantiate our claim. In order to do so we pair with an arbitrary element $x^\odot$ of $X^\odot$ and manipulate as follows:

$$\begin{aligned}
&\langle \int_0^t \int_0^\tau T_0^{\odot *}(\sigma) B \int_0^{\tau-\sigma} T^{\odot *}(\tau-\sigma-\eta) f(\eta)\, d\eta\, d\sigma\, d\tau, x^\odot \rangle \\
&= \int_0^t \int_0^\tau \langle \int_0^{\tau-\sigma} T^{\odot *}(\tau-\sigma-\eta) f(\eta)\, d\eta, B^* T_0^\odot(\sigma) x^\odot \rangle\, d\sigma\, d\tau \\
&= \int_0^t \int_\sigma^t \langle \int_0^{\tau-\sigma} T^{\odot *}(\tau-\sigma-\eta) f(\eta)\, d\eta, B^* T_0^\odot(\sigma) x^\odot \rangle\, d\tau\, d\sigma \\
&= \int_0^t \langle \int_\sigma^t \int_0^{\tau-\sigma} T^{\odot *}(\tau-\sigma-\eta) f(\eta)\, d\eta\, d\tau, B^* T_0^\odot(\sigma) x^\odot \rangle\, d\sigma \\
&= \int_0^t \langle \int_0^{t-\sigma} \int_{\eta+\sigma}^t T^{\odot *}(\tau-\sigma-\eta) f(\eta)\, d\tau\, d\eta, B^* T_0^\odot(\sigma) x^\odot \rangle\, d\sigma \\
&= \int_0^t \langle \int_0^{t-\sigma} W(t-\eta-\sigma) f(\eta)\, d\eta, B^* T_0^\odot(\sigma) x^\odot \rangle\, d\sigma \\
&= \langle \int_0^t T_0^{\odot *}(\sigma) B \int_0^{t-\sigma} W(t-\eta-\sigma) f(\eta)\, d\eta\, d\sigma, x^\odot \rangle \\
&= \langle \int_0^t \int_0^{t-\eta} T_0^{\odot *}(\sigma) B W(t-\eta-\sigma)\, d\sigma\, f(\eta)\, d\eta, x^\odot \rangle \\
&= \langle \int_0^t \left(W(t-\eta) - W_0(t-\eta)\right) f(\eta)\, d\eta, x^\odot \rangle,
\end{aligned}$$

where, in the last step, we have used the identity (2.13). □

Exercise 2.24. Let f be of the special form $f(t) = \eta(t) x^{\odot *}$, where $x^{\odot *}$ is a given element of $X^{\odot *}$ and η is a scalar L^1-function. Prove the analogue of Lemma 2.23. Hint: Use Exercise 1.4.

III.3 Perturbations with finite dimensional range

As formula (1.5) shows, the perturbation operator B has finite dimensional range in the case of delay equations. In this section we shall elaborate the consequences of this property. In particular, we shall reduce the abstract variation-of-constants formula to a (finite dimensional) renewal equation. It will turn out that the renewal equation contains all the relevant information, even when the initial condition is taken from the "big" space.

Assume that the range of B is contained in the n-dimensional subspace of $X^{\odot *}$ spanned by the linearly independent $r_1^{\odot *}, \dots, r_n^{\odot *} \in X^{\odot *}$. Then there exist $r_1^*, \dots, r_n^* \in X^*$ such that for all $\varphi \in X$

$$B\varphi = \sum_{j=1}^{n} \langle r_j^*, \varphi \rangle r_j^{\odot *}. \tag{3.1}$$

The variation-of-constants equation (2.4) written as

$$T(t)\varphi = T_0(t)\varphi + \sum_{j=1}^{n} \int_0^t T_0^{\odot *}(t-\tau) y_j(\tau) r_j^{\odot *} \, d\tau, \tag{3.2}$$

where the components of the $\mathbb{C}^n$-valued function y are defined by

$$y_j(t) = \langle r_j^*, T(t)\varphi \rangle, \tag{3.3}$$

is an explicit expression for $T(t)\varphi$ once y is known!

In an attempt to derive an equation for y we apply r_i^* to both sides of (3.2). Unfortunately, we cannot immediately interchange the order of applying r_i^* and performing the integration since, in general,

$$\langle r_i^*, T_0^{\odot *}(t) r_j^{\odot *} \rangle$$

is not defined. However, as we shall show below, the function

$$t \mapsto \langle r_i^*, T_0^{\odot *}(t) r_j^{\odot *} \rangle$$

exists in the sense of L^∞_{loc} and one can indeed interchange the order to arrive at a renewal equation for y. But we need to make technical preparations.

We recall from (2.12) the definition

$$W(t)x^{\odot *} = \int_0^t T^{\odot *}(\tau) x^{\odot *} \, d\tau.$$

Lemma 3.1. *For any $x^* \in X^*$ and $x^{\odot *} \in X^{\odot *}$ the function*

$$t \mapsto \langle x^*, W(t)x^{\odot *} \rangle$$

is locally Lipschitz continuous and consequently has a well-defined derivative in the sense of L^∞_{loc}.

Proof. Let M and ω be such that $||T(t)|| \le Me^{\omega t}$. Then

$$\begin{aligned} \left| \langle x^*, W(t_1)x^{\odot *} - W(t_2)x^{\odot *} \rangle \right| &\le M \frac{|e^{\omega t_2} - e^{\omega t_1}|}{\omega} ||x^*|| \, ||x^{\odot *}|| \\ &\le M e^{\omega s} ||x^*|| \, ||x^{\odot *}|| \, |t_2 - t_1| \end{aligned}$$

for some s between t_1 and t_2. □

Corollary 3.2. *There exists an $n \times n$ matrix-valued function R with entries belonging to L^∞_{loc} such that*

$$\int_0^t R_{ij}(\tau)\,d\tau = \langle r_i^*, W(t) r_j^{\odot *}\rangle, \quad t \geq 0. \tag{3.4}$$

Nota bene: We refrain from stating the corresponding results with index 0.

We recall from Exercises 1.4 and 2.2 that for any $\eta \in L^1$ the weak* integral

$$\int_0^t T^{\odot *}(t-\tau)\eta(\tau)x^{\odot *}\,d\tau$$

defines a continuous function of t taking values in X.

Lemma 3.3. *For any $\eta \in L^1$ the following identity holds:*

$$\langle r_i^*, \int_0^t T_0^{\odot *}(t-\tau)\eta(\tau)r_j^{\odot *}\,d\tau\rangle = \int_0^t R_{0ij}(t-\tau)\eta(\tau)\,d\tau. \tag{3.5}$$

Proof. Clearly the identity holds for $t = 0$. Integrating the left hand side we obtain

$$\begin{aligned} &\langle r_i^*, \int_0^t \int_0^s T_0^{\odot *}(s-\tau)\eta(\tau)r_j^{\odot *}\,d\tau\,ds\rangle \\ &= \langle r_i^*, \int_0^t \int_0^{t-\tau} T_0^{\odot *}(\sigma)r_j^{\odot *}\,d\sigma\,\eta(\tau)\,d\tau\rangle \\ &= \int_0^t \langle r_i^*, W_0(t-\tau)r_j^{\odot *}\rangle\eta(\tau)\,d\tau \\ &= \int_0^t \int_0^{t-\tau} R_{0ij}(\sigma)\,d\sigma\,\eta(\tau)\,d\tau. \end{aligned}$$

Since integration of the right hand side of (3.5) yields exactly the same result and the functions which are integrated are continuous, the integrands have to be identical for all values of t. □

Corollary 3.4. *For given $x^{\odot *} \in X^{\odot *}$ define the n vector-valued function y with L^∞_{loc} elements by*

$$\int_0^t y_i(\tau)\,d\tau = \langle r_i^*, W(t)x^{\odot *}\rangle. \tag{3.6}$$

Then

$$y = y_0 + R_0 * y, \tag{3.7}$$

where $y_{0i}(t) = \frac{d}{dt}\langle r_i^*, W_0(t)x^{\odot *}\rangle$ and

$$R_{0ij}(t) = \frac{d}{dt}\langle r_i^*, W_0(t) r_j^{\odot *}\rangle.$$

Proof. Applying r_i^* to the integrated variation-of-constants formula (2.13) and using Lemma 3.3 we obtain the renewal equation

$$Y = Y_0 + R_0 * Y,$$

where

$$Y_i(t) := \langle r_i^*, W(t) x^{\odot *}\rangle.$$

Lemma 3.1 yields the representation

$$Y(t) = \int_0^t y(\tau)\, d\tau$$

for $y \in L^\infty_{loc}$. The renewal equation (3.7) is obtained by differentiation of the renewal equation for Y. □

Corollary 3.5. *The kernel R is the resolvent of the kernel R_0, i.e.,*

$$R = R_0 + R_0 * R. \tag{3.8}$$

Proof. If we choose $x^{\odot *} = r_j^{\odot *}$ and compare (3.6) with (3.4), we see that (3.8) is a special case of (the matrix version of) (3.7). □

Theorem 3.6. *The C_0-semigroup $T^{\odot *}$ has the following representation*

$$T^{\odot *}(t) x^{\odot *} = T_0^{\odot *}(t) x^{\odot *} + \sum_{j=1}^{n} \int_0^t T_0^{\odot *}(t-\tau) y_j(\tau) r_j^{\odot *}\, d\tau. \tag{3.9}$$

Proof. The identity

$$\int_0^t T_0^{\odot *}(t-\tau) Y_j(\tau) r_j^{\odot *}\, d\tau = \int_0^t W_0(t-\tau) y_j(\tau) r_j^{\odot *}\, d\tau$$

follows by partial integration (Exercise 2.20). As a consequence, the integrated variation-of-constants formula (2.13) takes the form

$$W(t) x^{\odot *} = W_0(t) x^{\odot *} + \sum_{j=1}^{n} \int_0^t W_0(t-\tau) y_j(\tau) r_j^{\odot *}\, d\tau.$$

We can now apply weak* differentiation (i.e., form the pairing with an arbitrary element of $X^\odot$ and differentiate with respect to t) to arrive at (3.9) (cf. Exercise 2.21). □

Corollary 3.7. *For all* $x^{\odot *} \in X^{\odot *}$ *and* $t \geq 0$.

$$T^{\odot *}(t)x^{\odot *} - T_0^{\odot *}(t)x^{\odot *} \in X.$$

Proof. Recall the observation preceding Lemma 3.3. □

Thus we managed to express the semigroup of solution operators of the perturbed problem in terms of known quantities (i.e., quantities with index 0) and the solution of the renewal equation (3.7). This means that the renewal equation, which is in a sense a projection of the abstract equation on the range of B, contains all relevant information.

In general, both the forcing function y_0 and the solution y belong to L^∞_{loc}. But when we restrict to initial conditions in X, both y_0 and y can be represented by continuous functions.

Exercise 3.8. Show that for $x^{\odot *} = \varphi \in X$, (3.6) reduces to (3.3).

To conclude this section we show that $T^*(t)$ allows a representation in terms of solutions of the renewal equation with transposed kernel R_0^T.

Theorem 3.9. *The* $\mathcal{C}_0$*-semigroup* T^* *has the following representation*

$$T^*(t)x^* = T_0^*(t)x^* + \sum_{j=1}^{n} \int_0^t T_0^*(t-\tau)z_j(\tau)r_j^* d\tau, \tag{3.10}$$

where z *is the solution of*

$$z = z_0 + R_0^T * z \tag{3.11}$$

with z_0 *defined by*

$$\int_0^t z_{0j}(\tau)\, d\tau = \langle r_j^{\odot *}, W_0^{\odot}(t)x^* \rangle, \tag{3.12}$$

where

$$W_0^{\odot}(t)x^* = \int_0^t T_0^*(\tau)x^* d\tau. \tag{3.13}$$

Proof. From (3.1) we infer that $B^* : X^{\odot} \to X^*$ has the representation

$$B^* x^{\odot} = \sum_{j=1}^{n} \langle r_j^{\odot *}, x^{\odot} \rangle r_j^*. \tag{3.14}$$

Applying $r_j^{\odot *}$ to the adjoint integrated variation-of-constants formula [cf. Exercise 2.19 and recall definition (2.15)]

$$W^{\odot}(t)x^* = W_0^{\odot}(t)x^* + \int_0^t T_0^*(t-\tau)B^*W^{\odot}(\tau)x^*d\tau, \tag{3.15}$$

we obtain, using the appropriate variant of Lemma 3.3 and the identity

$$\int_0^t R_{0ij}(\tau)\,d\tau = \langle r_j^{\odot *}, W_0^{\odot}(t)r_i^*\rangle \tag{3.16}$$

[which follows from the definition (3.4) and Corollary 2.18],

$$Z_j(t) = Z_{0j}(t) + \sum_{i=1}^{n}\int_0^t R_{0ij}(t-\tau)Z_i(\tau)d\tau,$$

where, by definition,

$$Z_j(t) = \langle r_j^{\odot *}, W^{\odot}(t)x^*\rangle.$$

Defining z by

$$\int_0^t z(\tau)\,d\tau = Z(t)$$

and differentiating we obtain (3.11). The rest of the proof parallels the proof of Theorem 3.6. □

Exercise 3.10. Show that $T^*(t) - T_0^*(t)$ maps X^* into $X^{\odot}$.

III.4 Back to RFDE

Let $X = C\big([-h,0],\mathbb{C}^n\big)$ and

$$\big(T_0(t)\varphi\big)(\theta) = \begin{cases} \varphi(t+\theta), & -h \le t+\theta \le 0, \\ \varphi(0), & t+\theta \ge 0. \end{cases} \tag{4.1}$$

Then $X^* = \mathrm{NBV}\big([0,h],\mathbb{C}^n\big)$ and

$$X^{\odot *} = \mathbb{C}^n \times L^{\infty},$$

where $L^{\infty} = L^{\infty}\big([-h,0],\mathbb{C}^n\big)$. Consider a matrix-valued NBV function ζ. The ith row of ζ defines an element r_i^* of X^*. Let e_i denote the ith unit vector in $\mathbb{C}^n$; then $r_i^{\odot *} = (e_i, 0)$ belongs to $X^{\odot *}$. The operator $B : X \to X^{\odot *}$ defined by [cf. (1.5)]

$$B\varphi = \big(\langle \zeta, \varphi\rangle_n, 0\big) \tag{4.2}$$

is exactly of the form (3.1), with r_i^* and $r_i^{\odot *}$ as defined above (note that the matrix structure of ζ derives from the fact that we do not only have n elements r_i^* of X^* to specify B but that, in addition, each of these is itself a $\mathbb{C}^n$-valued function).

Theorem 4.1. *Let, with $\{T_0(t)\}$ and B as defined above, $\{T(t)\}$ be the semigroup defined by the abstract integral equation* (2.4)

$$T(t)\varphi = T_0(t)\varphi + \int_0^t T_0^{\odot *}(t-\tau)BT(\tau)\varphi \, d\tau.$$

Let $x(\cdot\,;\varphi)$ be the solution of the RFDE

$$\dot{x}(t) = \int_0^h d\zeta(\theta)x(t-\theta), \qquad t \geq 0, \tag{4.3}$$

with initial condition

$$x(\theta) = \varphi(\theta), \qquad -h \leq \theta \leq 0.$$

Then

$$T(t)\varphi = x_t(\cdot\,;\varphi). \tag{4.4}$$

The proof of this theorem is based on an identification of the renewal equation derived in the last section and the one derived in Section I.2. To deduce this identification we need some preparations, in which we elaborate the ingredients of the renewal equation of the last section in the present specific situation.

Our first aim is to determine the kernel R_0. The following corollary of the characterisations derived in Section II.5 is useful.

Lemma 4.2. *For $f \in X^{\odot}$, $\langle r_i^{\odot *}, f\rangle = f_i(0+)$.*

"Proof". See Theorem II.5.2 and the discussion following that theorem. □

Lemma 4.3. *For $\eta \in L^1_{loc}$*

$$\int_0^t T_0^{\odot *}(t-\tau)\eta(\tau)r_i^{\odot *} d\tau = e_i \int_0^{\max\{0,t+\,\cdot\,\}} \eta(\sigma)\, d\sigma. \tag{4.5}$$

Note that here the left hand side is considered as an element of $X = C([-h,0],\mathbb{C}^n)$. A more precise formulation would put j^{-1} in front of the left hand side, where j is the embedding of X into $X^{\odot *}$ given by $j(\varphi) = (\varphi(0),\varphi)$; cf. Section II.5.

Proof of Lemma 4.3.

$$\begin{aligned}
\langle \int_0^t T_0^{\odot *}(t-\tau)\eta(\tau)r_i^{\odot *}\,d\tau, x^{\odot}\rangle &= \int_0^t \eta(\tau)\langle r_i^{\odot *}, T_0^{\odot}(t-\tau)x^{\odot}\rangle d\tau \\
\text{(cf. II.4.1 and Lemma 4.2)} \qquad &= \int_0^t \eta(\tau)x_i^{\odot}(t-\tau)\,d\tau \\
&= \int_0^t \int_0^{t-\tau} \eta(\sigma)\,d\sigma\,dx_i^{\odot}(\tau) \\
&= \int_0^h \int_0^{\max\{0,t-\tau\}} \eta(\sigma)\,d\sigma\,dx_i^{\odot}(\tau) \\
&= \langle x^{\odot}, \varphi\rangle
\end{aligned}$$

with $\varphi(\theta) = e_i \int_0^{\max\{0,t+\theta\}} \eta(\sigma)\,d\sigma$ for $-h \leq \theta \leq 0$. □

Corollary 4.4. $W_0(t)r_i^{\odot *} = e_i \max\{0, t + \cdot\}$.

Corollary 4.5. $R_0 = \zeta$.

Proof. By definition,

$$\int_0^t R_{0ij}(\tau)\,d\tau = \langle r_i^*, W_0(t)r_j^{\odot *}\rangle.$$

So in this special case

$$\int_0^t R_{0ij}(\tau)\,d\tau = \int_0^h \max\{0, t-\theta\}d\zeta_{ij}(\theta) = \int_0^t \zeta_{ij}(\sigma)\,d\sigma.$$

□

Next we calculate an expression for the forcing function y_0 in the renewal equation, or, rather, for its integral.

Lemma 4.6. *For* $x^{\odot *} = (\alpha, \varphi) \in \mathbb{C}^n \times L^\infty$

$$\int_0^t y_0(\tau)\,d\tau = \int_0^h d\zeta(\theta) \int_{-\theta}^{t-\theta} \varphi^\alpha(\sigma)\,d\sigma, \tag{4.6}$$

where, by definition (cf. II.5.7*),*

$$\varphi^\alpha(\theta) = \begin{cases} \varphi(\theta), & -h \leq \theta \leq 0, \\ \alpha, & \theta > 0. \end{cases} \tag{4.7}$$

When $x^{\odot *} \in X$, *i.e.,* φ *is continuous and* $\alpha = \varphi(0)$, *this implies that*

$$y_0(t) = \zeta(t)\varphi(0) + \int_t^h d\zeta(\theta)\varphi(t-\theta). \tag{4.8}$$

Proof. Recalling (II.5.7) we note that

$$T_0^{\odot *}(t)x^{\odot *} = (\alpha, \varphi_t^{\alpha})$$

and therefore

$$\big(W_0(t)x^{\odot *}\big)(\theta) = \int_0^t \varphi_\tau^{\alpha}(\theta)\, d\tau = \int_\theta^{t+\theta} \varphi^{\alpha}(\sigma)\, d\sigma.$$

By definition,

$$\int_0^t y_0(\tau)\, d\tau = \langle r^*, W_0(t)x^{\odot *}\rangle_n$$

and, since $r^* = \zeta$, (4.6) follows. When φ^α is continuous we can differentiate (4.6) to obtain

$$\begin{aligned} y_0(t) &= \int_0^h d\zeta(\theta)\varphi^{\alpha}(t-\theta) = \int_0^t d\zeta(\theta)\varphi(0) + \int_t^h d\zeta(\theta)\varphi(t-\theta) \\ &= \zeta(t)\varphi(0) + \int_t^h d\zeta(\theta)\varphi(t-\theta). \end{aligned}$$

□

Thus we have identified in this special case the renewal equation (3.7) with the renewal equation (I.2.4a), obtained by manipulating the functional differential equation.

Combining the general representation (3.9) with Lemma 4.3 we see that (recall Corollary 3.7)

$$T^{\odot *}(t)x^{\odot *} - T_0^{\odot *}(t)x^{\odot *} = \int_0^{\max\{0,t+\,\cdot\,\}} y(\tau)\, d\tau.$$

So if we denote the $\mathbb{C}^n$ component of $T^{\odot *}(t)x^{\odot *}$ by $x(t)$, we have

$$x(t) = \alpha + \int_0^t y(\tau)\, d\tau, \quad t \geq 0,$$

and for the L^∞-component,

$$\big(T^{\odot *}(t)x^{\odot *}\big)_{(2)}(\theta) = \begin{cases} \varphi(t+\theta), & t+\theta \leq 0, \\ \alpha + \int_0^{t+\theta} y(\tau)\, d\tau = x(t+\theta), & t+\theta \geq 0 \end{cases}$$

which shows that $y = \dot{x}$ and that the action of the semigroup $T^{\odot *}(t)$ corresponds indeed to translation along the solution of the RFDE as described in Section II.1! In particular, we have, by restriction of the initial condition to elements of X, obtained a proof of Theorem 4.1.

Corollary 4.7. *The semigroup $T(t)$ corresponding to the RFDE* (4.3) *is compact for $t \geq h$.*

Proof. See Exercise 2.5. An alternative direct proof goes as follows. Let Ω be an arbitrary bounded subset of X. Since $\|T(t)\| \leq Me^{\omega t}$ we know that $\cup_{0\leq\tau\leq t} T(\tau)\Omega$ is bounded. Hence $\dot{x}(\tau) = \langle\zeta, x_\tau\rangle_n$ is bounded on $[0,t]$ uniformly for $\varphi \in \Omega$. The Arzela-Ascoli theorem then implies that $\{x : [0,t] \to \mathbb{C}^n | \varphi \in \Omega\}$ is precompact in $C([0,t], \mathbb{C}^n)$. Hence for $t \geq h$ we have that $\{x_t = T(t)\varphi | \varphi \in \Omega\}$ is a precompact subset of X. □

Recalling the proof of Corollary 3.5 we note that

$$T^{\odot *}(t) r^{\odot *} = \big(Q(t), Q(t+\cdot)\big), \tag{4.9}$$

where

$$Q(t) = \begin{cases} 0, & \text{for } t < 0, \\ I + \int_0^t R(\sigma)\, d\sigma, & \text{for } t \geq 0 \end{cases} \tag{4.10}$$

[or, in words, Q is the fundamental matrix solution (cf. I.2.22); one should read (4.9) as follows: if we give $r^{\odot *}$ index j, we should take the jth column of Q at the right hand side]. This observation will be used when studying inhomogeneous linear systems in Chapter VI.

Next, let us have a closer look at the generators. Since $(A_0^{\odot *} + B)\varphi = (\langle\zeta, \varphi\rangle_n, \dot{\varphi})$ it follows at once that the part A of $A_0^{\odot *} + B$ in $X = \mathcal{C}$ is given by

$$A\varphi = \dot{\varphi} \quad \text{with} \quad \mathcal{D}(A) = \{\varphi \in C^1 \mid \dot{\varphi}(0) = \langle\zeta, \varphi\rangle_n\}. \tag{4.11}$$

As an immediate consequence of Theorem 4.1 we obtain:

Corollary 4.8. *The semigroup $T(t)$ corresponding to the RFDE* (4.7) *maps X into $\mathcal{D}(A)$ for $t \geq h$.*

Exercise 4.9. Show that $T(t)$ maps X into $\mathcal{D}(A^m)$ for $t \geq mh$ and $m = 1, 2, \ldots$.

Note that in taking the part in X, all information about the particular equation is shifted from the action into the domain! Indeed, the action is the same for all equations.

Similarly, $(A_0^* + B^*)f = \dot{f} + f(0+)\zeta$ and so

$$\begin{aligned} \mathcal{D}(A^{\odot}) &= \{f \in AC(0,h] \mid \dot{f} + f(0+)\zeta \in \mathrm{AC}(0,h]\} \\ &= \{(c, g) \mid g + c\zeta \in \mathrm{AC}(0,h]\} \end{aligned} \tag{4.12}$$

with

$$A^{\odot} f = \dot{f} + f(0+)\zeta.$$

Exercise 4.10. Show that (4.6) can be rewritten as

$$\int_0^t y_0(\tau)\, d\tau = \int_{-h}^0 (\zeta(t-\tau) - \zeta(-\tau))\varphi(\tau)\, d\tau + \int_0^t \zeta(\tau)\, d\tau\, \alpha$$

and call this function $Y_0(t)$. Verify that

$$Y(t) = x(t) - \alpha$$

satisfies the renewal equation

$$Y = Y_0 + \zeta * Y.$$

Exercise 4.11. This is a continuation of Exercise II.5.6, to which we refer for definitions of $X, T_0(t)$, etc. (But now we work with $\mathbb{C}^n$-valued functions, i.e., systems of equations.) Consider $B : X \to X^{\odot *}$ defined by

$$B\varphi = \sum_{i=1}^{n} \langle \beta_i, \varphi \rangle \delta_i,$$

where $\beta_i \in X^* \simeq L^\infty([0,h], \mathbb{C}^n)$ and $\delta_i \in X^{\odot *} \simeq \mathrm{NBV}([0,h), \mathbb{C}^n)$ is defined by

$$\delta_i(a) = e_i \quad \text{for } a > 0$$

[recall that the normalization includes that $\delta_i(0) = 0$ and that, consequently, δ_i corresponds to the measure concentrated in $a = 0$ at the ith component].

Show that $R_0 = \beta$, where β denotes the matrix with β_i as the ith row, and that

$$\Big(\int_0^t T_0^{\odot *}(t-\tau)\delta_i \eta(\tau)\, d\tau\Big)(a) = \begin{cases} e_i \eta(t-a), & a < t, \\ 0, & a \geq t. \end{cases}$$

Next prove that for $\varphi \in X$

$$(T(t)\varphi)(a) = \begin{cases} \varphi(a-t), & a \geq t, \\ y(t-a), & a < t, \end{cases}$$

where y is the solution of the renewal equation

$$y = \beta * y + y_0$$

with

$$y_0(t) = \int_0^t \beta(\alpha)\varphi(\alpha - t)\, d\alpha.$$

Derive this renewal equation directly from the first order PDE

$$\frac{\partial u}{\partial t} + \frac{\partial u}{\partial a} = 0, \quad 0 \leq a \leq h, \ t > 0,$$

provided with the initial condition

$$u(0,a) = \varphi(a)$$

and the boundary condition

$$u(t,0) = \langle \beta, u(t,\cdot) \rangle_n$$

which describes the population birth rate (cf. Exercise II.2.2; here $\langle \beta, \varphi \rangle_n$ is the n vector with components $\langle \beta_i, \varphi \rangle$) in terms of the population composition and the functionals β. So, for $n = 1$, one can interpret β as the age-specific per capita birth rate, i.e., the age-specific probability per unit of time of producing offspring. Hint: Define $y(t) = u(t,0)$ and use integration along characteristics.

The difference with RFDE is that (apart from L^1 versus C) here we prescribe the value of the function itself at the "instream" boundary point $a = 0$ rather than the value of the time derivative.

See Webb [304] and Inaba [133] for further information on functional analytic aspects of age-dependent population growth.

The renewal equation for the birth rate was first derived by Sharpe and Lotka [249] from first principles (note that the present context explains the name "renewal"; originally the name arose in a similar economic context). The $\odot *$-framework allows us to show the direct link between the renewal equation and the abstract variation-of-constants formula. The renewal equation is by far the easiest tool to prove existence and uniqueness of solutions, but the abstract version of it enables us, in later chapters, to prove qualitative results, like linearized stability, Hopf bifurcation, etc., by standard arguments.

Exercise 4.12. Determine A and $A^{\odot}$ for the setting of age-dependent population growth described in the foregoing exercise.

Exercise 4.13. When death occurs before reaching the "ultimate" age h we have to work with the PDE

$$\frac{\partial m}{\partial t} + \frac{\partial m}{\partial a} = -\mu m,$$

where $\mu = \mu(a)$ is the age-specific per capita death rate (the probability per unit of time of dying). Define the survival function $\mathcal{F}$ by

$$\mathcal{F}(a) = \exp(-\int_0^a \mu(\alpha)\, d\alpha).$$

Verify that the transformation

$$m(t,a) = \mathcal{F}(a)u(t,a)$$

leads to the equation

$$\frac{\partial u}{\partial t} + \frac{\partial u}{\partial a} = 0,$$

whereas the boundary condition

$$m(t,0) = \int_0^h \beta(\alpha)m(t,\alpha)\, d\alpha$$

transforms into

$$u(t,0) = \int_0^h \beta(\alpha)\mathcal{F}(\alpha)u(t,\alpha)\, d\alpha.$$

For any monotone decreasing continuous function $\mathcal{F}$ with $\mathcal{F}(h) > 0$ the multiplication mapping $\varphi \mapsto \mathcal{F}\varphi$ is an isomorphism on L^1.

III.5 Interpretation of the adjoint semigroup

In this section we give Theorem 3.9 concrete form in the special case of RFDE. This will lead to a new interpretation of the action of the adjoint semigroup.

Lemma 5.1. *Let* $x^* = f \in NBV([0,h],\mathbb{C}^n)$. *Then the forcing function* z_0 *defined by* (3.12) *equals* f, *i.e.,* $z_0 = f$.

Proof. Recall from (II.4.1) that $\big(T_0^*(t)f\big)(\theta) = f(t+\theta)$. It follows that

$$\big(W_0^{\odot}(t)f\big)(\theta) = \int_0^t f(\tau+\theta)\,d\tau.$$

The definition (3.12) and Lemma 4.1 then show that

$$\int_0^t z_0(\tau)\,d\tau = \int_0^t f(\tau)\,d\tau.$$

□

In words, Lemma 5.1 tells us that the forcing function in the renewal equation and the initial state for the adjoint semigroup are identical!

Theorem 5.2.

$$\big(T^*(t)f\big)(\theta) = f(t+\theta) + \int_0^t \zeta^T(t+\theta-\tau)z(\tau)\,d\tau, \tag{5.1}$$

where z *is the solution of the renewal equation*

$$z = f + \zeta^T * z. \tag{5.2}$$

Proof. For RFDE we have

$$\int_0^t T_0^*(t-\tau)z_j(\tau)r_j^*\,d\tau = \int_0^t \zeta_j(t-\tau+\cdot)z_j(\tau)\,d\tau,$$

where ζ_j denotes the jth row of ζ. Hence, (3.10) reads in the present case

$$\big(T^*(t)f\big)_i(\theta) = f_i(t+\theta) + \sum_{j=1}^n \int_0^t \zeta_{ji}(t-\tau+\theta)z_j(\tau)\,d\tau, \quad i = 1,2,\ldots,n.$$

□

Remark **5.3**. In the proof above we have used that for RFDE

$$\left(\int_0^t T_0^*(t-\tau)x^*\eta(\tau)\,d\tau\right)(\theta) = \int_0^t x^*(t-\tau+\theta)\eta(\tau)\,d\tau \tag{5.3}$$

for $\eta \in L^1_{loc}(\mathbb{R}_+, \mathbb{C})$; or, in other words, that one can replace the weak* integral by a family, parameterized by θ, of integrals of $\mathbb{C}^n$-valued functions. The verification of this identity involves Fubini's theorem.

Exercise 5.4. Prove (5.3).

Exercise 5.5. Show that $z(t)$ is the jump of $T^*(t)f$ at $\theta = 0$.

At this point we make a digression. The state of a dynamical system is the information which suffices to uniquely determine the future development. As we have seen, the initial function concept qualifies as such for delay equations. However, this is not the only possibility. One may as well put the relevant information in the forcing function of a renewal equation. In that case the state at a later time should be a forcing function again. If we start from the equation

$$z = f + \zeta * z,$$

the state at time t, which we denote by $S(t)f$, should be the forcing function in the renewal equation which is satisfied by the translate z_t of z, i.e.,

$$z_t = S(t)f + \zeta * z_t.$$

A straightforward calculation will show that $S(t)f$ thus defined is given by the right hand side of (5.1)!

So there are (at least) two ways to keep books of the information which fixes the future evolution of a delay system: initial functions and forcing functions. Each way induces a natural semigroup construction. Taking adjoints involves two actions: (i) the structure of $\mathbb{C}^n$ requires changing over to the transposed matrix and (ii) the structure of time dependence requires changing over from the initial function point of view to the forcing function point of view or vice versa.

Exercise 5.6. (A continuation of Exercise 4.11.) Prove that for the "age-dependent population dynamics" semigroup too the forcing function z_0 of the (transposed) renewal equation and the initial state for the adjoint semigroup are one and the same thing, i.e., prove that for $x^* = f \in L^\infty$ and z_0 defined by

$$\int_0^t z_0(\tau)\,d\tau = \langle \delta, W_0^{\odot}(t)f\rangle$$

we have that $z_0(t) = f(t)$. Next show that

$$(T^*(t)f)(a) = f(t+a) + \int_0^t \beta^T(t+a-\tau)z(\tau)\,d\tau,$$

where z satisfies

$$z = f + \beta^T * z.$$

Observe that the only difference with the RFDE case [see (5.1) and (5.2)] lies in the function space! In particular, the forcing functions have to be zero for $a \geq h$, rather than constant as in the RFDE case.

Exercise 5.7. Consider a RFDE with $\zeta(h) = 0$. Verify that, for the adjoint semigroup, the space of (NBV-forcing) functions which are zero at h (and hence for $\theta \geq h$) is invariant.

III.6 Equivalent description of the dynamics

Now that we realise that there is not just one canonical way of doing the bookkeeping of states and state transformations, we wonder how we can understand and express the interrelationship. Let, as before, $T(t)$ denote the semigroup acting on initial conditions of RFDE (Section 4) and $S(t)$ the semigroup acting on forcing functions of RE according to

$$x_t = S(t)f + \zeta * x_t. \tag{6.1}$$

How are $T(t)$ and $S(t)$ interrelated? An answer to this question of course requires a precise specification of the space of forcing functions on which $S(t)$ acts, but we postpone such a specification and at first proceed somewhat formally.

Motivated by our calculations in Section I.2 we introduce a mapping F as follows (cf. I.2.12):

$$(F\varphi)(t) = \varphi(0) + \int_0^h [\zeta(t+\sigma) - \zeta(\sigma)]\varphi(-\sigma)\,d\sigma. \tag{6.2}$$

Then F maps an initial condition onto the corresponding forcing function for the renewal equation satisfied by the solution x of the RFDE. So if we apply F to x_s, we obtain the forcing function in the renewal equation satisfied by x_s which is, by definition, $S(s)F\varphi$. Since, on the other hand, $x_s = T(s)\varphi$, we have proved:

Lemma 6.1. $FT(s) = S(s)F$.

In general, F is not invertible (for instance, when ζ is constant for $t \geq \frac{1}{2}h$ and φ has its support in $[-h, -\frac{1}{2}h]$ then $F\varphi = 0$; less trivial examples of null spaces of F are given in Section V.4). Hence the intertwining relation of Lemma 6.1 does not establish the equivalence of $T(s)$ and $S(s)$. We have to look for another mapping between initial functions and forcing functions.

For a given forcing function f the RE

$$x = f + \zeta * x \tag{6.3}$$

has a unique solution x and, in particular, the restriction of x to the interval $[0, h]$ is uniquely determined by f. Conversely, if we consider x on the interval $[0, h]$ as given, then we can define f on $[0, h]$ by $f = x - \zeta * x$ and extend f to $[h, \infty)$ by requiring that $f(t) = f(h)$ for $t \geq h$. To meet the requirement that an initial function is defined on $[-h, 0]$, rather than on $[0, h]$, we have to incorporate the appropriate shift in the mapping.

Definition 6.2. $(Gf)(\theta) = x(h + \theta, f)$, $-h \leq \theta \leq 0$, where $x(\cdot\,, f)$ is the solution of the RE (6.3).

Theorem 6.3. *Let $\widetilde{C}$ denote the space of continuous functions on $\mathbb{R}_+$ with values in $\mathbb{C}^n$ which are constant on $[h, \infty)$, equipped with the supremum norm. Then $G : \widetilde{C} \to C$ is a one-to-one bounded linear surjection whose bounded inverse is given by*

$$(G^{-1}\varphi)(t) = \begin{cases} \varphi_h(t) - \zeta * \varphi_h(t), & 0 \leq t \leq h, \\ \varphi_h(h) - \zeta * \varphi_h(h), & t \geq h. \end{cases}$$

Moreover,

$$T(s)G = GS(s) \tag{6.4}$$

and consequently the semigroups $T(s)$ and $S(s)$ are isomorphic.

Proof. The representation $x = f + R * f$ shows at once that G is linear and bounded. Injectivity follows from the uniqueness of solutions of the RE. Likewise the expression for the inverse of G follows from uniqueness. To conclude that G^{-1} is bounded, one can either use this explicit expression or invoke the closed graph theorem. □

Exercise 6.4. Show that $GF = T(h)$ and $FG = S(h)$.

Exercise 6.5. Determine the generator A_S of $S(t)$ as a semigroup on $\widetilde{C}$.

Exercise 6.6. Let now the generator of $T(t)$ be denoted by A_T. Show that F maps $\mathcal{D}(A_T)$ into $\mathcal{D}(A_S)$ and that for all $\varphi \in \mathcal{D}(A_T)$

$$FA_T\varphi = A_S F\varphi.$$

Exercise 6.7. Show that G maps $\mathcal{D}(A_S)$ onto $\mathcal{D}(A_T)$ and that for all $f \in \mathcal{D}(A_S)$

$$GA_S f = A_T G f.$$

Exercise 6.8. Define $K : L^1([0,h],\mathbb{C}^n) \to L^1([-h,0],\mathbb{C}^n)$ by

$$(Kg)(\theta) = g(h+\theta) + R^T * g(h+\theta)$$

or, in other "words", by

$$(Kg)(\theta) = y(h+\theta, g), \qquad -h \le \varphi \le 0,$$

where y is the solution of the RE

$$y = g + \zeta^T * y.$$

Verify that

$$\int_0^h g(\theta)(Gf)(-\theta)\,d\theta = \int_0^h (Kg)(-\theta)f(\theta)\,d\theta.$$

Exercise 6.9. Verify that, with

$$\mathcal{C}^* = \mathrm{NBV}([0,h],\mathbb{C}^n) \quad \text{and} \quad \widetilde{\mathcal{C}}^* = \mathrm{NBV}([-h,0],\mathbb{C}^n),$$

the action of the adjoint G^* of G corresponds to solving the RE with transposed kernel on the interval $[0,h]$ and shifting the solution back to $[-h,0]$.
Hint: The foregoing exercise shows that this is true for absolutely continuous functions.

III.7 Complexification

So far, we have worked with complex-valued functions and, as a consequence, with complex Banach spaces. In later chapters, when dealing with nonlinear problems, we shall start out with real-valued functions and real Banach spaces. When discussing, say, behaviour near an equilibrium as in Chapters VIII and IX, we shall study the linearized system by means of the spectral theory of Chapter IV, which requires a complex Banach space as the underlying structure. Therefore we have to pay attention to the complexification procedure, by which a real Banach space is embedded into a complex one and by which linear operators on the real space are extended to operators on the complex space. This procedure is less trivial than it may seem at first thought, especially since our duality framework forces us to simultaneously complexify several spaces, while preserving certain relations, notably "being the dual space" and "being a subspace". Here we have, once more, to make a choice in the rigour of our notation: shall we introduce symbols for all natural isometric isomorphisms involved and use them consistently, or shall we "identify" spaces that are related to each other by a natural isometry? We shall do the latter, while presenting at the end a series of exercises dealing with the first possibility.

Our presentation in this section has been inspired by Ruston [246] as far as the general theory is concerned. A special feature of our discussion will be the complexification of the variation-of-constants equation and our main conclusion can be summarized by saying that the complex variation-of-constants equation is simply twice the real one.

Definition 7.1. Let X be a real vector space. The Cartesian product $X \times X$ can be given the structure of a complex vector space by defining the following operations:

(i) *addition:* $(x, y) + (u, v) = (x + u, y + v)$;
(ii) *multiplication by a complex number:*

$$(\alpha + i\beta)(x, y) = (\alpha x - \beta y, \alpha y + \beta x).$$

The complex vector space obtained in this manner is denoted by $X_{\mathbb{C}}$.

By identifying $x \in X$ with $(x, 0) \in X_{\mathbb{C}}$, we can embed X into $X_{\mathbb{C}}$. Since $(x, y) = (x, 0) + i(y, 0)$ we can then use the alternative notation $x + iy$ to denote (x, y).

Now suppose that X is actually a normed space. Can we lift the norm of X to $X_{\mathbb{C}}$? Is there a canonical way to do this?

A first and somewhat disappointing observation is that the candidate

$$\|x + iy\| = \sqrt{\|x\|^2 + \|y\|^2} \tag{7.1}$$

does not necessarily do the job. Indeed, $\|(\alpha + i\beta)(x + iy)\| = |\alpha + i\beta| \|x + iy\|$ if and only if

$$\sqrt{\|\alpha x - \beta y\|^2 + \|\alpha y + \beta x\|^2} = \sqrt{\alpha^2 + \beta^2}\sqrt{\|x\|^2 + \|y\|^2} \tag{7.2}$$

and this relation does not hold in general (it does when, for instance, X is an inner product space).

Exercise 7.2. Find an example for which (7.2) does not hold.

Exercise 7.3. Show that the inequality $\|x + iy\| \leq \|x\| + \|y\|$ has to hold whenever we require the (special case of the triangle) inequality $\|x + iy\| \leq \|x\| + \|iy\|$ and the homogeneity of the norm with respect to multiplication by a complex number.

Exercise 7.4. Show that the "quasi-orthogonality" condition $\|x + iy\| \geq \|x\|$ and the homogeneity of the norm with respect to multiplication by a complex number together imply the estimate $\|x + iy\| \geq \max\{\|x\|, \|y\|\}$.

Definition 7.5. Let X be a real normed space. A complex norm $\|\cdot\|$ on $X_{\mathbb{C}}$ is called *admissible* if and only if

$$\max\{\|x\|, \|y\|\} \leq \|x + iy\| \leq \|x\| + \|y\| \tag{7.3}$$

for all $x, y \in X$. The vector space $X_{\mathbb{C}}$ together with an admissible norm is called a *complexification* of X.

Exercise 7.6. Verify that all admissible norms on $X_{\mathbb{C}}$ are equivalent.

Exercise 7.7. Show that $\|x + i0\| = \|x\| = \|0 + ix\|$ for any admissible norm.

The following are examples of recipes to construct an admissible norm on $X_{\mathbb{C}}$ from the given norm on X:

$$\|x + iy\| = \sup_{-\pi \le \varphi \le +\pi} \|x \cos\varphi + y \sin\varphi\|, \tag{7.4}$$

$$\|x + iy\| = \inf \sum_{r=1}^{n} |\alpha_r + i\beta_r| \|x_r\|, \tag{7.5}$$

where the infimum is taken over all finite families of triples (α_r, β_r, x_r), $r = 1, 2, \ldots, n$, with α_r and β_r real numbers and x_r a set of points of X such that $\sum_{r=1}^{n} \alpha_r x_r = x$ and $\sum_{r=1}^{n} \beta_r x_r = y$.

$$\|x + iy\| = \sup_{-\pi \le \varphi \le +\pi} \sqrt{\|x \cos\varphi - y \sin\varphi\|^2 + \|y \cos\varphi + x \sin\varphi\|^2}. \tag{7.6}$$

One can easily prove that (7.4) is the least admissible norm on $X_{\mathbb{C}}$ and (7.5) the greatest (cf. Ruston [246]). The norm (7.6) reduces to (7.1) in certain situations, e.g., when X is an inner product space. We conclude that there is not a unique way to provide $X_{\mathbb{C}}$ with a norm, but rather a multitude of admissible ways.

Exercise 7.8. Let X be a real normed space and let $X_{\mathbb{C}}$ be a complexification of X. Prove that $X_{\mathbb{C}}$ is complete if and only if X is complete.

Exercise 7.9. Write down isomorphisms between the following spaces:

(i) $(C(I, X))_{\mathbb{C}}$ and $C(I, X_{\mathbb{C}})$;

(ii) $\mathbb{C}^n$ and $(\mathbb{R}^n)_{\mathbb{C}}$;

(iii) $(C([-h, 0], \mathbb{R}^n))_{\mathbb{C}}$ and $C([-h, 0], \mathbb{C}^n)$.

Next we address the following question: are the operations of "taking the dual space" and "complexifying a space" interchangeable? It turns out that the answer is yes, provided we are careful enough about the choice of norms and careless enough about an isometry.

Let $x^* + iy^* \in (X^*)_{\mathbb{C}}$ and $x + iy \in X_{\mathbb{C}}$. We define a pairing between two such elements by

$$\langle x^* + iy^*, x + iy \rangle = \langle x^*, x \rangle - \langle y^*, y \rangle + i\big(\langle x^*, y \rangle + \langle y^*, x \rangle\big) \tag{7.7}$$

In this manner $x^* + iy^*$ defines a bounded complex linear functional on $X_{\mathbb{C}}$, hence an element, say z^*, of $X_{\mathbb{C}}$. We claim that

$$\|z^*\| \leq \|x^*\| + \|y^*\| \tag{7.8}$$

where on the left hand side the norm is from $(X_{\mathbb{C}})^*$ and on the right hand side from X^*. To substantiate this claim we rewrite (7.7) in the form

$$\langle z^*, x + iy\rangle = \langle x^*, x\rangle + i\langle x^*, y\rangle + i\big(\langle y^*, x\rangle + i\langle y^*, y\rangle\big) \tag{7.9}$$

and invoke the following lemma:

Lemma 7.10.

$$\big|\langle x^*, x\rangle + i\langle x^*, y\rangle\big| \leq \|x^*\|\,\|x + iy\|.$$

The proof of the lemma is based on the observation that (7.4) is the *least* admissible norm and on

Exercise 7.11.

(i) Let a and b be real numbers. Prove that

$$\sup_{-\pi \leq \varphi \leq +\pi} |a\cos\varphi + b\sin\varphi| = |a + ib|.$$

(ii) Show that

$$\sup_{-\pi \leq \varphi \leq +\pi} |\langle x^*, x\cos\varphi + y\sin\varphi\rangle| = |\langle x^*, x\rangle + i\langle x^*, y\rangle|.$$

(iii) Show that

$$\sup_{-\pi \leq \varphi \leq +\pi} \|x\cos\varphi + y\sin\varphi\| = \sup_{\|x^*\| \leq 1} |\langle x^*, x\rangle + i\langle x^*, y\rangle|.$$

So with every element $x^* + iy^* \in (X^*)_{\mathbb{C}}$ we can associate $z^* \in (X_{\mathbb{C}})^*$ such that (7.8) holds. Conversely, let $z^* \in (X_{\mathbb{C}})^*$ be given. Define $x^*, y^* \in X^*$ by

$$\begin{aligned} \langle x^*, x\rangle &= \operatorname{Re}\langle z^*, x + i0\rangle, \\ \langle y^*, x\rangle &= \operatorname{Im}\langle z^*, x + i0\rangle; \end{aligned} \tag{7.10}$$

then clearly $\|x^*\| \leq \|z^*\|$ and $\|y^*\| \leq \|z^*\|$ whereas z^* is exactly the element of $(X_{\mathbb{C}})^*$ associated with $x^* + iy^* \in (X^*)_{\mathbb{C}}$. Combining the inequalities into

$$\max\big\{\|x^*\|, \|y^*\|\big\} \leq \|z^*\| \leq \|x^*\| + \|y^*\|, \tag{7.11}$$

we see that the norm

$$\|x^* + iy^*\| = \sup_{\|x+iy\| \leq 1} \big|\langle x^* + iy^*, x + iy\rangle\big| \tag{7.12}$$

is an admissible norm on $(X^*)_{\mathbb{C}}$. We conclude that we can represent $(X_{\mathbb{C}})^*$ by $(X^*)_{\mathbb{C}}$, with the pairing defined by (7.7), provided we choose the admissible norm (7.12). So once we fix an admissible norm on $X_{\mathbb{C}}$, an admissible norm on $(X^*)_{\mathbb{C}}$ is induced and the recipe to obtain the norm on $(X^*)_{\mathbb{C}}$ from the norm on X^* may very well be different from the recipe to obtain the norm on $X_{\mathbb{C}}$ from the norm on X. The next theorem summarizes the conclusion.

Theorem 7.12. *The spaces $(X_{\mathbb{C}})^*$ and $(X^*)_{\mathbb{C}}$ can be identified, i.e., they are isometrically isomorphic provided we give, given an admissible norm on $X_{\mathbb{C}}$, the appropriate admissible norm to $(X^*)_{\mathbb{C}}$.*

Exercise 7.13. Prove that $X_{\mathbb{C}}$ is reflexive if and only if X is reflexive.

Exercise 7.14. Prove that one may as well first choose an admissible norm on $(X^*)_{\mathbb{C}}$ and then define an admissible norm on $X_{\mathbb{C}}$ by

$$\|x+iy\| = \sup_{\|x^*+iy^*\|\leq 1} |\langle x^*+iy^*, x+iy\rangle|.$$

Throughout the following we shall use $(X^*)_{\mathbb{C}}$ to represent $(X_{\mathbb{C}})^*$, with pairing (7.7) and norming (7.12). We now turn our attention to the complexification of linear operators and their adjoints.

Definition 7.15. Let $L : X \to Z$ be a linear operator with domain of definition $\mathcal{D}(L) \subset X$. The *complexification* $L_{\mathbb{C}}$ of L maps $X_{\mathbb{C}}$ into $Z_{\mathbb{C}}$ and is defined by

$$\mathcal{D}(L_{\mathbb{C}}) = \{x+iy \in X_{\mathbb{C}} \mid x \in \mathcal{D}(L) \text{ and } y \in \mathcal{D}(L)\},$$
$$L_{\mathbb{C}}(x+iy) = Lx + iLy.$$

As an immediate consequence of our representation of $(X_{\mathbb{C}})^*$ we have

Lemma 7.16. $(L_{\mathbb{C}})^* = (L^*)_{\mathbb{C}}$, *whenever the adjoints are defined, i.e., whenever $\mathcal{D}(L)$ is dense.*

It is instructive to write out the proof in detail oneself. If one does not manage, one can check the elaboration that we present now.

Proof. (i) Let $z^* \in \mathcal{D}((L^*)_{\mathbb{C}})$. This means that $x^*, y^* \in \mathcal{D}(L^*)$ exist such that $z^* = x^* + iy^*$ and then $(L^*)_{\mathbb{C}} z^* = L^*x^* + iL^*y^*$. Now consider $x, y \in \mathcal{D}(L)$. Then

$$\begin{aligned}\langle z^*, L_{\mathbb{C}}(x+iy)\rangle &= \langle z^*, Lx+iLy\rangle = \langle x^*+iy^*, Lx+iLy\rangle \\ &= \langle x^*, Lx\rangle - \langle y^*, Ly\rangle + i(\langle y^*, Lx\rangle + \langle x^*, Ly\rangle) \\ &= \langle L^*x^* + iL^*y^*, x+iy\rangle\end{aligned}$$

and we conclude that $z^* \in \mathcal{D}\big((L_{\mathbb{C}})^*\big)$ and

$$\big(L_{\mathbb{C}}\big)^* z^* = L^*x^* + iL^*y^* = \big(L^*\big)_{\mathbb{C}} z^*.$$

(ii) Conversely, assume that $z^* = x^* + iy^* \in \mathcal{D}\big(\big(L_{\mathbb{C}}\big)^*\big)$ and

$$\big(L_{\mathbb{C}}\big)^* z^* = \xi^* + i\eta^*.$$

Then for all $x, y \in \mathcal{D}(L)$ we have

$$\begin{aligned}\langle x^*, Lx\rangle - \langle y^*, Ly\rangle + i\big(\langle x^*, Ly\rangle + \langle y^*, Lx\rangle\big) &= \langle x^* + iy^*, Lx + iLy\rangle \\ = \langle z^*, Lx + iLy\rangle = \langle \big(L_{\mathbb{C}}\big)^* z^*, x + iy\rangle &= \langle \xi^* + i\eta^*, x + iy\rangle \\ = \langle \xi^*, x\rangle - \langle \eta^*, y\rangle + i\big(\langle \xi^*, y\rangle + \langle \eta^*, x\rangle\big).\end{aligned}$$

So, in particular,

$$\langle x^*, Lx\rangle - \langle y^*, Ly\rangle = \langle \xi^*, x\rangle - \langle \eta^*, y\rangle.$$

Taking $y = 0$ we find that necessarily $x^* \in \mathcal{D}(L^*)$ and $\xi^* = L^*x^*$. Similarly, by taking $x = 0$ we find $y^* \in \mathcal{D}(L^*)$ and $\eta^* = L^*y^*$. So

$$\begin{aligned}\big(L_{\mathbb{C}}\big)^* z^* &= \xi^* + i\eta^* = L^*x^* + iL^*y^* \\ &= \big(L^*\big)_{\mathbb{C}}(x^* + iy^*) = \big(L^*\big)_{\mathbb{C}} z^*.\end{aligned}$$

□

Corollary 7.17. *Let $T(t)$ be a $\mathcal{C}_0$-semigroup on X with generator A. Then*

(i) *$A_{\mathbb{C}}$ generates $(T(t))_{\mathbb{C}}$,*

(ii) *$(A^*)_{\mathbb{C}}$ is the weak* generator of $(T^*(t))_{\mathbb{C}}$,*

(iii) $\overline{\mathcal{D}\big((A^*)_{\mathbb{C}}\big)} = (X^{\odot})_{\mathbb{C}}$,

(iv) *$X_{\mathbb{C}}$ is $\odot$-reflexive with respect to $(T(t))_{\mathbb{C}}$ if and only if X is $\odot$-reflexive with respect to $T(t)$.*

Note that in (iii) we suppress another natural isomorphism in our notation: that between the complexification of a subspace and the corresponding subspace of the complexification. Note that the idea behind (iv) is that we represent $((X^{\odot})_{\mathbb{C}})^*$ by $(X^{\odot *})_{\mathbb{C}}$ and then use the embedding of X into $X^{\odot *}$ to induce an embedding of $X_{\mathbb{C}}$ into $(X^{\odot *})_{\mathbb{C}}$. We can represent this somewhat symbolically in the "complexified" duality diagram

$$\begin{array}{ccc} X_{\mathbb{C}} & \longrightarrow & (X^*)_{\mathbb{C}} \\ \uparrow & & \Big| \\ (X^{\odot\odot})_{\mathbb{C}} & & \Big| \\ \uparrow & & \downarrow \\ (X^{\odot *})_{\mathbb{C}} & \longleftarrow & (X^{\odot})_{\mathbb{C}} \end{array}$$

Next we consider the variation-of-constants formula. Let $T_0(t)$ be a $\mathcal{C}_0$-semigroup on X with generator A_0. Assume that X is $\odot$-reflexive with respect to $T_0(t)$. Let there be given a bounded linear operator $B : X \to X^{\odot *}$. By definition, $B_{\mathbb{C}} : X_{\mathbb{C}} \to (X^{\odot *})_{\mathbb{C}}$. The complexified variation-of-constants equation is defined to be

$$(T(t))_{\mathbb{C}} = (T_0(t))_{\mathbb{C}} + \int_0^t (T_0^{\odot *}(t-\tau))_{\mathbb{C}} B_{\mathbb{C}} (T(\tau))_{\mathbb{C}}\, d\tau, \tag{7.13}$$

where the weak* integral is based on the identification of $(X^{\odot *})_{\mathbb{C}}$ with $((X^{\odot})_{\mathbb{C}})^*$. Note that the two components are *uncoupled*. So, in effect, this complexified equation is simply twice the real equation.

$$T(t) = T_0(t) + \int_0^t T_0^{\odot *}(t-\tau) B T(\tau)\, d\tau \tag{7.14}$$

and, in particular, $(T(t))_{\mathbb{C}}$, defined as the complexification of $T(t)$, is the solution.

The following series of exercises is intended for readers who do not like our identifications of spaces and who prefer to see the isometries written.

Exercise 7.18. Let $k : (X_{\mathbb{C}})^* \to (X^*)_{\mathbb{C}}$ be the isometry of Theorem 7.12. Show that for any linear mapping $L : X \to X$ we have

$$k(L_{\mathbb{C}})^* = (L^*)_{\mathbb{C}} k.$$

Exercise 7.19. Let $T_0(t)$ be a $\mathcal{C}_0$-semigroup on X. Show that k maps $(X_{\mathbb{C}})^{\odot}$ onto $(X^{\odot})_{\mathbb{C}}$.

Exercise 7.20. Let $l : ((X^{\odot})_{\mathbb{C}})^* \to (X^{\odot *})_{\mathbb{C}}$ be the analogue of k for the space $X^{\odot}$. Let $k^* : ((X^{\odot})_{\mathbb{C}})^* \to (X_{\mathbb{C}})^{\odot *}$ be the adjoint of k, considered as a mapping from $(X_{\mathbb{C}})^{\odot}$ to $(X^{\odot})_{\mathbb{C}}$. Define $m : (X^{\odot *})_{\mathbb{C}} \to (X_{\mathbb{C}})^{\odot *}$ by $m = k^* l^{-1}$. Show that

$$\begin{aligned}\langle m(x^{\odot *} + i y^{\odot *}), z^{\odot}\rangle &= \langle x^{\odot *}, x^{\odot}\rangle - \langle y^{\odot *}, y^{\odot}\rangle \\ &\quad + i(\langle x^{\odot *}, y^{\odot}\rangle + \langle y^{\odot *}, x^{\odot}\rangle),\end{aligned}$$

where $x^{\odot} + i y^{\odot} = k z^{\odot}$.

Exercise 7.21. Let the real Banach space X be $\odot$-reflexive with respect to the $\mathcal{C}_0$-semigroup $\{T_0(t)\}$. Given $B : X \to X^{\odot *}$, define $\widetilde{B} : X_{\mathbb{C}} \to (X_{\mathbb{C}})^{\odot *}$ by $\widetilde{B} = m B_{\mathbb{C}}$. We can now consider $\widetilde{B}$ as a perturbation operator in the framework of the "standard" duality diagram

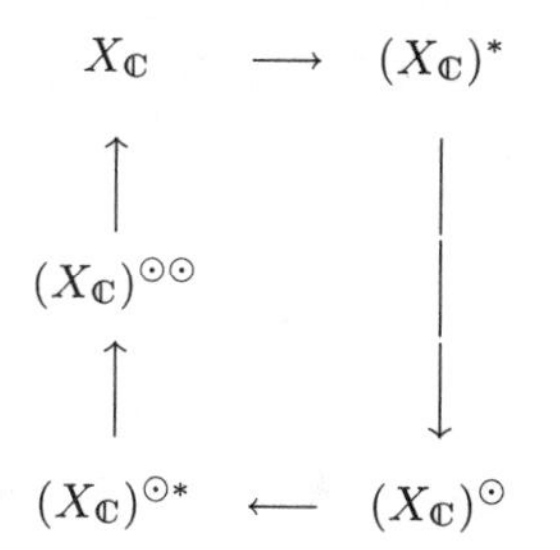

More precisely, we can consider the "standard" variation-of-constants equation

$$S(t) = (T_0(t))_{\mathbb{C}} + \int_0^t ((T_0(t-\tau))_{\mathbb{C}})^{\odot *} \widetilde{B} S(\tau)\, d\tau \tag{7.15}$$

and use the results of Section 3. Verify that the solution $S(t)$ is obtained by complexification of the solution $T(t)$ of the real equation (7.14), i.e., verify that $S(t) = (T(t))_{\mathbb{C}}$.

We conclude this section with some concepts and results which will turn out to be useful in the context of spectral theory. The *conjugation* map $C : X_{\mathbb{C}} \to X_{\mathbb{C}}$ is defined by

$$C(x+iy) = x - iy. \tag{7.16}$$

Whenever convenient, we shall also use the notation $C\varphi = \overline{\varphi}$. Note that $C^2 = I$.

Definition 7.22. Let $L : X_{\mathbb{C}} \to X_{\mathbb{C}}$ be a linear operator with domain $\mathcal{D}(L)$. Then $\overline{L} : X_{\mathbb{C}} \to X_{\mathbb{C}}$ is the linear operator defined by
(i) $\mathcal{D}(\overline{L}) = C\mathcal{D}(L)$;
(ii) $\overline{L} = CLC$.

Note that this makes sense since C maps $C\mathcal{D}(L)$ into $\mathcal{D}(L)$. Also note that $\overline{\alpha L} = \overline{\alpha}\overline{L}$ for $\alpha \in \mathbb{C}$.

Lemma 7.23. *$L = \overline{L}$ iff L is the complexification of a map $K : X \to X$.*

Proof. If $L = K_{\mathbb{C}}$ then

$$\begin{aligned} \overline{L}(x+iy) &= CL(x-iy) = C(Kx - iKy) \\ &= Kx + iKy = L(x+iy). \end{aligned}$$

Next suppose $L = \overline{L}$. Define $\mathcal{D}(K) = \{x \mid x + i0 \in \mathcal{D}(L)\}$. We claim that L maps $\mathcal{D}(K) + i0$ into $X + i0$. Indeed, suppose that $L(x+i0) = \alpha + i\beta$; then $\overline{L}(x+i0) = CL(x+i0) = \alpha - i\beta$ and this equals $\alpha + i\beta$ iff $\beta = 0$. Hence L induces a map $K : \mathcal{D}(K) \to X$ via $L(x+i0) = Kx + i0$. Subsequently, we note that, by the complex linearity of L,

$$L(0+ix) = L(e^{i\frac{\pi}{2}}(x+i0)) = e^{i\frac{\pi}{2}}L(x+i0)$$
$$= iL(x+i0) = 0+iKx.$$

Taken together, this implies that

$$L(x+iy) = Kx + iKy$$

or, in words, L is the complexification of K. □

Exercise 7.24. For a given operator $L : X_{\mathbb{C}} \to X_{\mathbb{C}}$ such that $C\mathcal{D}(L) = \mathcal{D}(L)$, let $\operatorname{Re} L, \operatorname{Im} L : X_{\mathbb{C}} \to X_{\mathbb{C}}$ with domain $\mathcal{D}(L)$ be defined by

$$\operatorname{Re} L = \frac{1}{2}(L+\overline{L}),$$
$$\operatorname{Im} L = \frac{1}{2i}(L-\overline{L}).$$

Show that

$$L(x+iy) = L_1 x - L_2 y + i(L_1 x + L_2 y),$$

where L_1 and L_2 are the "reelifications" of $\operatorname{Re} L$ and $\operatorname{Im} L$, respectively, with $\mathcal{D}(L_i) = \{x \mid x+i0 \in \mathcal{D}(L)\}$.

Exercise 7.25. Let $L = K_{\mathbb{C}}$. Prove that L is a projection iff K is a projection.

Exercise 7.26. Prove that L is invertible iff $\overline{L}$ is invertible and $(\overline{L})^{-1} = \overline{L^{-1}}$.

Exercise 7.27. Let $\phi \in X_{\mathbb{C}}$ be such that $\overline{\phi} \neq \phi$. Define

$$Y = \{z\phi + \overline{z}\overline{\phi} \mid z \in \mathbb{C}\}.$$

Prove that $Y \subset X + i0$ and that Y, considered as a subspace of X, is two dimensional.

Exercise 7.28. Let $\phi \in X_{\mathbb{C}}$ and $\psi \in X^*_{\mathbb{C}}$ be such that $\langle \psi, \phi \rangle = 1$ and $\overline{\phi} \neq \phi$. Let P denote the one-dimensional projection on $X_{\mathbb{C}}$ defined by $P\varphi = \langle \psi, \varphi \rangle \phi$. Show that $\overline{P}\varphi = \langle \overline{\psi}, \varphi \rangle \overline{\phi}$ and that $P + \overline{P}$ is the complexification of a two-dimensional projection on X.

Exercise 7.29. Let $\phi_1, \ldots, \phi_k \in X_{\mathbb{C}}$ and $\psi_1, \ldots, \psi_m \in X_{\mathbb{C}}$ be such that $\overline{\phi}_j \neq \phi_j$ whereas $\overline{\psi}_j = \psi_j$. Assume that $\phi_1, \ldots, \phi_k, \overline{\phi}_1, \ldots, \overline{\phi}_k, \psi_1, \ldots, \psi_m$ are linearly independent. Define

$$Y = \{\sum_{j=1}^{k}(z_j\phi_j + \overline{z}_j\overline{\phi}_j) + \sum_{j=1}^{m} x_j\psi_j \mid z_j \in \mathbb{C}, \quad x_j \in \mathbb{R}\}.$$

Prove that $Y \subset X + i0$ and that Y, considered as a subspace of X, has dimension $2k + m$.

III.8 Remarks about the non-sun-reflexive case

Let $T_0(t)$ be a $\mathcal{C}_0$-semigroup on the Banach space X and let A_0 denote its generator. Assume that $X^{\odot\odot}$ is "bigger" than (the embedding into $X^{\odot *}$ of) X. If we now introduce a perturbation operator $B : X \to X^{\odot *}$, a first question to ask is whether B admits an extension to a continuous linear operator $\widetilde{B} : X^{\odot\odot} \to X^{\odot *}$. For if it does, we can consider the variation-of-constants equation

$$S(t) = T_0^{\odot\odot}(t) + \int_0^t T_0^{\odot *}(t-\tau)\widetilde{B}S(\tau)\,d\tau \tag{8.1}$$

and prove, exactly as in Section 2, that a perturbed $\mathcal{C}_0$-semigroup $S(t)$ exists on $X^{\odot\odot}$. Subsequently, we may then ask whether or not $S(t)$ leaves X invariant. If it does, we are essentially in the same situation as before, despite the lack of $\odot$-reflexivity. The first part of this section deals with the program sketched above: extending B and analyzing whether X is invariant under $S(t)$. The main application we have in mind is RFDE with infinite delay. The second part, which is of relevance for structured population models, concentrates at the right hand side of the duality diagram by considering perturbed semigroups on $X^{\odot}$ and X^* when there is, by lack of $\odot$-reflexivity, no perturbed semigroup on X.

When $X^{\odot\odot} = X$ we have automatically a duality pairing between $X^{\odot\odot}$ and X^* since we have one between X and X^*. Below we show that even when $X^{\odot\odot} \neq X$ we can define a pairing between $X^{\odot\odot}$ and X^* which now, however, involves the behaviour of the (unperturbed) semigroup for $t \downarrow 0$ (or, equivalently, the behaviour of the resolvent for $z \to \infty$). This pairing is the key tool for carrying out the extension of B.

Theorem 8.1. *For any $x^{\odot\odot} \in X^{\odot\odot}$ and $x^* \in X^*$ the limits*

$$\lim_{t\downarrow 0} \frac{1}{t}\langle x^{\odot\odot}, \int_0^t T_0^*(\tau)x^*\,d\tau\rangle$$

and

$$\lim_{z\to\infty} z\langle x^{\odot\odot}, (zI - A_0^*)^{-1}x^*\rangle$$

exist and are equal to each other. Let $[x^{\odot\odot}, x^*]$ denote the common limit; then, with M such that $\|T_0(t)\| \le Me^{\omega t}$,

$$\big|[x^{\odot\odot}, x^*]\big| \le M\|x^{\odot\odot}\|\,\|x^*\|.$$

Moreover,

$$[x^{\odot\odot}, x^{\odot}] = \langle x^{\odot\odot}, x^{\odot}\rangle \quad \text{for all } x^{\odot} \in X^{\odot}$$

and

$$[x, x^*] = \langle x^*, x\rangle \quad \text{for all } x \in X.$$

Proof. First recall that for all $x^{\odot} \in X^{\odot}$

$$\lim_{t\downarrow 0} \| \frac{1}{t} \int_0^t T_0^{\odot}(\tau)x^{\odot}\, d\tau - x^{\odot} \| = 0$$

and

$$\lim_{z\to\infty} \|z(zI - A_0^{\odot})^{-1}x^{\odot} - x^{\odot}\| = 0.$$

Take any $x^{\odot\odot} \in \mathcal{D}(A_0^{\odot *})$ and fix $\mu > \omega$; then

$$\begin{aligned}
&\frac{1}{t}\langle x^{\odot\odot}, \int_0^t T_0^*(\tau)x^*\, d\tau\rangle \\
&\quad = \frac{1}{t}\langle (\mu I - A_0^{\odot *})x^{\odot\odot}, (\mu I - A_0^{\odot})^{-1} \int_0^t T_0^*(\tau)x^*\, d\tau\rangle \\
&\quad = \frac{1}{t}\langle (\mu I - A_0^{\odot *})x^{\odot\odot}, \int_0^t T_0^{\odot}(\tau)(\mu I - A_0^*)^{-1}x^*\, d\tau\rangle \\
&\quad \to \langle (\mu I - A_0^{\odot *})x^{\odot\odot}, (\mu I - A_0^*)^{-1}x^*\rangle \quad \text{as } t \downarrow 0.
\end{aligned}$$

Similarly

$$\begin{aligned}
&z\langle x^{\odot\odot}, (zI - A_0^*)^{-1}x^*\rangle \\
&\quad = z\langle (\mu I - A_0^{\odot *})x^{\odot\odot}, (\mu I - A_0^{\odot})^{-1}(zI - A_0^*)^{-1}x^*\rangle \\
&\quad = z\langle (\mu I - A_0^{\odot *})x^{\odot\odot}, (zI - A_0^{\odot})^{-1}(\mu I - A_0^*)^{-1}x^*\rangle \\
&\quad \to \langle (\mu I - A_0^{\odot *})x^{\odot\odot}, (\mu I - A_0^*)^{-1}x^*\rangle \quad \text{as } z \to \infty.
\end{aligned}$$

The estimates

$$z\|\, (zI - A_0^*)^{-1})\| \le \frac{Mz}{z - \omega}$$

and

$$\frac{1}{t}\| \int_0^t T_0^*(\tau)\, d\tau\| \le \frac{M(e^{\omega t} - 1)}{\omega t}$$

imply, together with the fact that $\mathcal{D}(A_0^{\odot *})$ is dense in $X^{\odot\odot}$, that the limits exist and equal each other for arbitrary $x^{\odot\odot} \in X^{\odot\odot}$ (the proof uses the triangle inequality repeatedly). The same estimates yield the inequality

$$\big|[x^{\odot\odot}, x^*]\big| \le M\|x^{\odot\odot}\|\, \|x^*\|.$$

For $x^* = x^{\odot} \in X^{\odot}$ we have that

$$\begin{aligned}
\lim_{t\downarrow 0} \frac{1}{t}\langle x^{\odot\odot}, \int_0^t T_0^{\odot}(\tau)x^{\odot}\, d\tau\rangle &= \lim_{z\to\infty} z\langle x^{\odot\odot}, (zI - A_0^{\odot})^{-1}x^{\odot}\rangle \\
&= \langle x^{\odot\odot}, x^{\odot}\rangle
\end{aligned}$$

and hence $[x^{\odot\odot}, x^{\odot}] = \langle x^{\odot\odot}, x^{\odot}\rangle$. Similarly, if $x^{\odot\odot} = x \in X$, we can write

$$\frac{1}{t}\langle x^{\odot\odot}, \int_0^t T_0^*(\tau)x\,d\tau\rangle = \frac{1}{t}\langle x^*, \int_0^t T_0(\tau)x\,d\tau\rangle$$

and

$$z\langle x^{\odot\odot}, (zI - A_0^*)^{-1}x^*\rangle = z\langle x^*, (zI - A_0)^{-1}x\rangle$$

to obtain $\langle x^*, x\rangle$ in the limit $t \downarrow 0$, respectively $z \to \infty$. □

Exercise 8.2. Prove that

(i) $[T_0^{\odot\odot}(t)x^{\odot\odot}, x^*] = [x^{\odot\odot}, T_0^*(t)x^*]$;

(ii) $\langle A_0^{\odot *}x^{\odot\odot}, x^{\odot}\rangle = [x^{\odot\odot}, A_0^* x^{\odot}]$ for $x^{\odot\odot} \in \mathcal{D}(A_0^{\odot *})$ and $x^{\odot} \in \mathcal{D}(A_0^*)$;

(iii) $[x^{\odot\odot}, \int_0^t T_0^*(t-\tau)f(\tau)\,d\tau] = \int_0^t [T_0^{\odot\odot}(t-\tau)x^{\odot\odot}, f(\tau)]\,d\tau$ for any norm continuous function $f : \mathbb{R}_+ \to X^*$.

Let $B : X \to X^{\odot *}$ be a bounded linear operator. Its adjoint B^* maps $X^{\odot **}$ into X^*. Let C denote the restriction of B^* to (the embedding into $X^{\odot **}$ of) $X^{\odot}$, so $C : X^{\odot} \to X^*$. The adjoint C^* of C maps X^{**} into $X^{\odot *}$. As Theorem 8.1 tells us, there exists an embedding of $X^{\odot\odot}$ into X^{**}. By restricting C^* to this embedding, we induce a bounded linear mapping $\widetilde{B} : X^{\odot\odot} \to X^{\odot *}$. We conclude that the answer to the first question of the beginning of this section is yes!

Exercise 8.3. Fill in the details of our statement that equation (8.1) defines a $\mathcal{C}_0$-semigroup $S(t)$ on $X^{\odot\odot}$.

Theorem 8.4. *The following conditions are equivalent:*

(i) *X is invariant under $S(t)$;*

(ii) *$(zI - A_0^{\odot *})^{-1}B$ maps X into X for all $z \in \rho(A_0)$.*

Proof. (ii) ⇒ (i): Let G denote the generator of $S(t)$. Taking the Laplace transform of (8.1) we obtain

$$(zI - G)^{-1} = (zI - A_0^{\odot\odot})^{-1} + (zI - A_0^{\odot *})^{-1}\widetilde{B}(zI - G)^{-1} \tag{8.2}$$

from which it follows that

$$\begin{aligned}(zI - G)^{-1} &= (I - (zI - A_0^{\odot *})^{-1}\widetilde{B})^{-1}(zI - A_0^{\odot\odot})^{-1} \\ &= \sum_{k=0}^{\infty}\big((zI - A_0^{\odot *})^{-1}\widetilde{B}\big)^k(zI - A_0^{\odot\odot})^{-1}.\end{aligned}$$

Since $T_0^{\odot\odot}(t)$ leaves X invariant, the same must be true for the operator $(zI - A_0^{\odot\odot})^{-1}$. So this series expansion and our assumption (ii) together imply that $(zI - G)^{-1}$ maps X into X. Hence $G(zI - G)^{-1}$ maps X into X. From general semigroup theory (Appendix II) we know that

$$S(t)x = \lim_{z\to\infty} \sum_{k=0}^{\infty} \frac{t^k}{k!} \left[zG(zI-G)^{-1} \right]^k x.$$

Hence, since X is a closed subspace of $X^{\odot\odot}$, $S(t)x \in X$ or, in other words, (i) holds.
(i) $\Rightarrow$ (ii): Since both $S(t)$ and $T_0^{\odot\odot}(t)$ map X into X, it follows that $(zI-G)^{-1} - (zI-A_0^{\odot\odot})^{-1}$ maps X into X. Let $T(t)$ denote the restriction of $S(t)$ to X and let A be its generator. Then A is the part of G in X. For $x \in \mathcal{D}(A)$ we can deduce from (8.2) that

$$(zI - A_0^{\odot *})^{-1} Bx = \left((zI-G)^{-1} - (zI - A_0^{\odot\odot})^{-1} \right)(zI - A)x.$$

Since the right hand side belongs to X, so does the left hand side. Since $\mathcal{D}(A)$ is dense in X we conclude that (ii) holds. $\square$

Exercise 8.5. Let $X = C_0(\mathbb{R}_-, \mathbb{C})$. Define $T_0(t)$ by

$$(T_0(t)\varphi)(\theta) = \begin{cases} \varphi(t+\theta), & \text{for } \theta \le -t, \\ \varphi(0), & \text{for } -t \le \theta \le 0. \end{cases}$$

Show that $X^{\odot *} = \mathbb{C} \times L^\infty(\mathbb{R}_-, \mathbb{C})$ and that

$$X^{\odot\odot} = \{(\alpha, \varphi) \mid \varphi \in \mathrm{BUC}(\mathbb{R}_-, \mathbb{C}) \text{ and } \varphi(0) = \alpha\} \simeq \mathrm{BUC}(\mathbb{R}_-, \mathbb{C}),$$

where BUC denotes the Banach space of bounded, uniformly continuous functions equipped with the supremum norm.

Exercise 8.6. In the setting of the foregoing exercise, let $B : X \to X^{\odot *}$ be defined by

$$B\varphi = \left(\int_0^\infty h(\tau)\varphi(-\tau)\, d\tau, 0 \right) \quad \text{for some } h \in L^1(\mathbb{R}_+, \mathbb{C}).$$

Show that the extension $\widetilde{B}$ is given by the same expression. Moreover, convince yourself that the semigroup $S(t)$ defined by (8.1) corresponds to translation along the solution of the infinite delay equation

$$\dot{x}(t) = \int_0^\infty h(\tau) x(t-\tau)\, d\tau, \quad t \ge 0,$$
$$x(\theta) = \varphi(\theta), \quad -\infty < \theta \le 0.$$

(See the book [125] by Hino, Murakami and Naito for much more general results. Section 5.5 of that book gives a systematic exposition of the extension of the ideas and results of this chapter to the case of infinite delay equations.)

We conclude this section with a trivial, yet useful, remark. Suppose that a perturbation is given in the form of a bounded linear operator $C : X^{\odot} \to X^*$. Then we can consider the variation-of-constants equation

$$T^{\odot}(t) = T_0^{\odot}(t) + \int_0^t T_0^*(t-\tau)CT^{\odot}(\tau)d\tau \tag{8.3}$$

and construct a $\mathcal{C}_0$-semigroup $T^{\odot}(t)$ on $X^{\odot}$. It turns out that one can extend this semigroup to the "big" space X^* by the intertwining formula

$$T^{\times}(t) = (zI - A^{\times})T^{\odot}(t)(zI - A^{\times})^{-1} \tag{8.4}$$

where $\mathcal{D}(A^{\times}) = \mathcal{D}(A_0^*)$ and $A^{\times} = A_0^* + C$. Note that this formula only makes sense if $\mathcal{D}(A_0^*)$ is invariant under $T^{\odot}(t)$. So we have to prove that this is indeed the case. In order to do so we need a new concept.

For any semigroup $\{S(t)\}_{t\geq 0}$ of bounded linear operators on a Banach space Z we define the *Favard class* of S by

$$\text{Fav}\,(S) = \left\{ z \in Z \mid \limsup_{t\downarrow 0} \frac{1}{t} \|S(t)z - z\| < \infty \right\}.$$

Note that the semigroup property guarantees that Fav (S) is invariant under $S(t)$. From general semigroup theory (Theorem 3.19 of Appendix II) we know

Lemma 8.7. *Let $T_0(t)$ be a $\mathcal{C}_0$-semigroup on X. Then* Fav $(T_0^*) = \mathcal{D}(A_0^*)$.

We are now ready to formulate the result which guarantees that (8.4) makes sense.

Theorem 8.8. Fav $(T^{\odot})$ = Fav $(T_0^{\odot}) = \mathcal{D}(A_0^*)$. *As a consequence $\mathcal{D}(A_0^*)$ is invariant under $T^{\odot}(t)$.*

Proof. Recalling Lemma 2.3 we conclude from (8.3) that $\|T^{\odot}(t) - T_0^{\odot}(t)\| = O(t)$ for $t \downarrow 0$. Hence $T^{\odot}(t)$ and $T_0^{\odot}(t)$ do have the same Favard class, and the result follows from Lemma 8.7. □

Exercise 8.9. Prove that the infinitesimal generator $A^{\odot}$ of $T^{\odot}$ is the part of $A^{\times}$ in $X^{\odot}$.

Exercise 8.10. Let $B : X \to X^{\odot *}$ be given and let $C : X^{\odot} \to X^*$ be the restriction of the adjoint of B. Let $S(t)$ be the semigroup on $X^{\odot\odot}$ defined by (8.1) and $T^{\times}(t)$ the semigroup on X^* defined by (8.4) and (8.3). Show that the duality relation

$$[S(t)x^{\odot\odot}, x^*] = [x^{\odot\odot}, T^{\times}(t)x^*]$$

holds.

Exercise 8.11. With the notation of the last exercise, assume that X is invariant under $S(t)$. Let $T(t)$ denote the restriction of $S(t)$ to X. Prove that $T^{\times}(t) = T^*(t)$.

Exercise 8.12. (Not so easy.) Let $T_0(t)$ and $T(t)$ be two $\mathcal{C}_0$-semigroups on X such that $\|T_0(t) - T(t)\| = O(t)$ as $t \downarrow 0$. Prove that a bounded linear operator $B : X \to X^{\odot *}$ exists such that for all $x \in X$

$$T(t)x = T_0(t)x + \int_0^t T_0^{\odot *}(t-\tau)BT(\tau)x\,d\tau.$$

(See Diekmann, Gyllenberg and Heijmans [67].)

Exercise 8.13. Consider age-dependent population dynamics without any upper bound on the ages and show that our results apply
Hint: Let X^* be the space of bounded Borel measures on $\mathbb{R}_+$ and $X^{\odot}$ the space of absolutely continuous measures, i.e., measures obtained by integrating an element of $L^1(\mathbb{R})$.

III.9 Comments

The material of this chapter was developed by Clément, Diekmann, Gyllenberg, Heijmans and Thieme [44, 45, 46, 47] and Diekmann [65]. Duality theory for RFDE was started by Henry [120]. The interpretation of the action of the adjoint semigroup was first presented for equations with infinite delay by Burns and Herdman [22] (see [97] for a systematic exposition).

The idea that one can associate with a Volterra integral equation a dynamical system acting on a space of forcing functions goes back to Miller and Sell [197] and Miller [196]. For the formulation of the duality principle in the case of finite delay it is essential to restrict to forcing functions with *compact support*. This was noted by Diekmann [64] (also see [63]).

The structural operators F and G from Section 6 were introduced and studied by Delfour and Manitius [60, 61] (also see [157]).

For other recent developments in semigroup theory, such as integrated semigroups, we refer to [8, 58, 62, 96, 148, 171, 206, 273, 274]. For an integrated semigroup approach to delay equations see Adimy [2].

Chapter IV

Spectral theory

IV.1 Introduction

In Chapter I we studied the large time behaviour of solutions of linear retarded functional differential equations. This study was based on the observation that for positive time, the solution $x(\,\cdot\,;\varphi)$ of the RFDE

$$\dot{x}(t) = \int_0^h d\zeta(\theta)x(t-\theta), \qquad x_0 = \varphi, \tag{1.1}$$

satisfies a renewal equation (I.2.10)

$$x = \zeta * x + f.$$

The forcing function f is locally of bounded variation and given by (cf. I.2.12)

$$f(t) = \varphi(0) + \int_0^h [\zeta(t+\sigma) - \zeta(\sigma)]\varphi(-\sigma)\,d\sigma.$$

The renewal equation behaves nicely under the Laplace transform, and from the inverse Laplace transform we derived the following representation (cf. I.5.3)

$$x(t) = \sum_{j=1}^{l} p_j(t)e^{\lambda_j t} + \frac{1}{2\pi i}\int_{L(\gamma)} e^{zt}\Delta(z)^{-1}\big(f(0) + \int_0^h e^{-z\theta}df(\theta)\big)\,dz. \tag{1.2}$$

In this chapter, we shall rephrase this result in operator language using the spectral mapping theorem which relates the spectrum of the unknown semigroup $T(t)$ to that of the known generator A.

The goal is to obtain explicit formulas for the coefficients of the polynomials p_j, as linear operators acting on the initial condition φ, in terms of the spectral data of the generator. We shall see that all information is contained in the characteristic matrix Δ given by (I.3.16) and that the polynomials p_j can be built from the Jordan chains of $\Delta(z)$ at $z = \lambda_j$.

IV.2 Spectral decomposition for eventually compact semigroups

Throughout this chapter, $X = (X, \|\cdot\|)$ will denote a complex Banach space.

Let $L : \mathcal{D}(L) \to X$ be a linear operator with domain $\mathcal{D}(L)$ in X. A complex number λ belongs to the *resolvent set* $\rho(L)$ of L if and only if the *resolvent* $(zI - L)^{-1}$ exists and is bounded, i.e.,

(i) $\lambda I - L$ is one-to-one (injective);

(ii) $\mathcal{R}(\lambda I - L) = X$;

(iii) $(\lambda I - L)^{-1}$ is bounded.

Note that for closed operators, (iii) is superfluous, since it is a direct consequence of the other assumptions by the closed graph theorem. The *spectrum* $\sigma(L)$ is by definition the complement of $\rho(L)$ in $\mathbb{C}$.

The *point spectrum* $\sigma_p(L)$ is the set of those $\lambda \in \mathbb{C}$ for which $\lambda I - L$ is not one-to-one, i.e., $L\varphi = \lambda\varphi$ for some $\varphi \neq 0$. One then calls λ an *eigenvalue* and φ an *eigenvector* corresponding to λ.

The null space $\mathcal{N}(\lambda I - L)$ is called the *eigenspace* and its dimension the *geometric multiplicity* of λ. The *generalized eigenspace* $\mathcal{M}_\lambda = \mathcal{M}_\lambda(L)$ is the smallest closed linear subspace that contains all $\mathcal{N}((\lambda I - L)^j)$ for $j = 1, 2, \ldots$ and its dimension $M(L; \lambda)$ is called the *algebraic multiplicity* of λ. If, in addition, λ is an isolated point in $\sigma(L)$ and $M(L; \lambda)$ is finite, then λ is called an *eigenvalue of finite type.* When $M(L; \lambda) = 1$ we say that λ is a *simple* eigenvalue. A class of operators for which the eigenvalues are of finite type is formed by the compact operators.

Definition 2.1. A $\mathcal{C}_0$-semigroup $T(t)$ is called *eventually compact* if there exists a $t_0 \geq 0$ such that for all $t \geq t_0$, $T(t)$ is a compact operator.

We begin this section with the spectral mapping principle for eventually compact semigroups. It may be convenient for the reader to read Section 4 of Appendix II first.

Theorem 2.2. *Let $A : \mathcal{D}(A) \to X$ be the generator of a $\mathcal{C}_0$-semigroup $T(t)$. If $T(t)$ is eventually compact, then*

$$\sigma(T(t)) \subset \big(\sigma_p(T(t)) \cup \{0\}\big) = \big(e^{t\sigma_p(A)} \cup \{0\}\big). \tag{2.1}$$

Proof. The inclusion (2.1) holds for $t = 0$. Since $T(t)$ is eventually compact, there exists $t_0 \geq 0$ such that $T(t)$ is a compact operator for $t \geq t_0$. For any $t_1 > 0$, there exists an $n \in \mathbb{N}$ such that $nt_1 \geq t_0$. So the operator $T(t_1)^n = T(nt_1)$ is compact and, consequently,

$$\sigma(T(t_1)^n) \subset \big(\sigma_p(T(t_1)^n) \cup \{0\}\big).$$

From the polynomial spectral mapping theorem, see Exercise II.4.8 of Appendix II, we conclude

$$\sigma(T(t_1)) \subset \sigma_p(T(t_1)) \cup \{0\}.$$

The second relation follows from Theorem 4.17 of Appendix II. □

Let $r_\sigma(T(t))$ denote the spectral radius of a $\mathcal{C}_0$-semigroup $T(t)$ and let ω_0 denote the growth bound, i.e.,

$$\omega_0 = \lim_{t\to\infty} \frac{1}{t} \log \|T(t)\|$$

(cf. Appendix II.4). We repeat Proposition 4.10 of Appendix II here as

Lemma 2.3. $r_\sigma(T(t)) = e^{\omega_0 t}$.

Define the *spectral bound* of an (unbounded) operator A to be

$$s(A) = \sup\{\operatorname{Re}\lambda \mid \lambda \in \sigma(A)\}.$$

From Exercise 4.14 of Appendix II, one can conclude that, in general, $\omega_0 \neq s(A)$. But from the spectral mapping principle for the point spectrum we have a positive result for eventually compact semigroups.

Theorem 2.4. *Let $A : \mathcal{D}(A) \to X$ be the generator of a $\mathcal{C}_0$-semigroup $T(t)$. If $T(t)$ is eventually compact, then the growth bound of $T(t)$ equals the spectral bound of A:*

$$\omega_0 = s(A).$$

Proof. Since for z with $\operatorname{Re} z > \omega_0$,

$$(zI - A)^{-1}x = \int_0^\infty e^{-zt}T(t)x\,dt, \quad x \in X,$$

we find that $s(A) \leq \omega_0$. To prove the opposite inequality, we use that $T(t)$ is an eventually compact semigroup. From Theorem 2.2 it follows that for $t_0 \geq 0$ fixed, the spectral radius of $T(t_0)$ satisfies

$$r_\sigma(T(t_0)) = e^{s(A)t_0}.$$

Therefore, Lemma 2.3 yields the result. □

Theorem 2.4 will be especially useful in combination with the following standard result about isolated spectral points (see [89] or [272]). A point $z = \lambda$ is called a *pole* of $(zI - A)^{-1}$ of order m when the Laurent expansion of $(zI - A)^{-1}$ is given by $\sum_{j=-m}^{\infty} B_j(z-\lambda)^j$ with $B_{-m} \neq 0$.

Theorem 2.5. *Let $A : \mathcal{D}(A) \to X$ be a closed operator. If $z = \lambda$ is a pole of $(zI - A)^{-1}$ of order m, then*

(i) $\mathcal{M}_\lambda(A) = \mathcal{N}\big((\lambda I - A)^m\big)$ *and m is the smallest integer with this property;*

(ii) $\mathcal{R}\big((\lambda I - A)^m\big) = \mathcal{R}\big((\lambda I - A)^{m+1}\big)$ *and this space is closed;*

(iii) $X = \mathcal{M}_\lambda(A) \oplus \mathcal{R}_\lambda(A)$, *where* $\mathcal{R}_\lambda(A) = \mathcal{R}\big((\lambda I - A)^m\big)$;

(iv) $\sigma(A|_{\mathcal{M}_\lambda}) = \{\lambda\}$;

(v) *the spectral projection P_λ onto $\mathcal{M}_\lambda$ along $\mathcal{R}_\lambda$ can be represented by a Dunford integral*

$$P_\lambda = \frac{1}{2\pi i}\int_{\Gamma_\lambda} (zI - A)^{-1}\, dz = \operatorname*{Res}_{z=\lambda}\,(zI - A)^{-1}, \tag{2.2}$$

where Γ_λ is a small circle such that λ is the only singularity of $(zI - A)^{-1}$ inside Γ_λ;

(vi) *if $m = 1$ and $\dim \mathcal{N}\big(\lambda I - A\big) = 1$, then*

$$P_\lambda \psi = \langle \phi^*, \psi\rangle \phi,$$

where ϕ and ϕ^* are eigenvectors of A and A^*, respectively, corresponding to the eigenvalue λ, normalized such that $\langle \phi^*, \phi\rangle = 1$.

Exercise 2.6. Prove Theorem 2.5(vi).
Hint: Use that the $\mathcal{R}(P_\lambda)$ can be characterized as follows:

$$\mathcal{R}(P_\lambda) = \{\varphi \in X \mid \langle \psi, \varphi\rangle = 0, \quad \text{for all } \psi \in \mathcal{R}((\lambda I - A^*)^m)\}.$$

Exercise 2.7. Let A be the generator of a $\mathcal{C}_0$-semigroup $T(t)$. Show that $\mathcal{M}_\lambda(A)$ and $\mathcal{R}_\lambda(A)$ are $T(t)$-invariant and that the restricted semigroup $T(t)|_{\mathcal{M}_\lambda(A)}$ has a bounded generator.
Hint: Prove that, for $z \neq \lambda$, zero does not belong to the spectrum of the resolvent $((zI - A|_{\mathcal{M}_\lambda})^{-1}$. (See [202, Thm A.III.3.3].)

So, if $A : \mathcal{D}(A) \to X$ is the generator of a $\mathcal{C}_0$-semigroup $T(t)$ such that

$$\Lambda = \Lambda(\beta) = \{\lambda \in \sigma(A) \mid \operatorname{Re}\lambda > \beta\}$$

is a finite set of poles of $(zI - A)^{-1}$, we have the decomposition

$$X = \mathcal{M}_\Lambda \oplus \mathcal{R}_\Lambda,$$

where

$$\mathcal{M}_\Lambda = \bigoplus_{\lambda\in\Lambda} \mathcal{M}_\lambda(A), \qquad \mathcal{R}_\Lambda = \bigcap_{\lambda\in\Lambda} \mathcal{R}_\lambda(A)$$

are $T(t)$-invariant subspaces; the spectral projection $P_\Lambda : X \to X$ onto $\mathcal{M}_\Lambda$ is given by

$$P_\Lambda = \sum_{\lambda \in \Lambda} \operatorname*{Res}_{z=\lambda} (zI - A)^{-1}.$$

Furthermore, the restriction of $T(t)$ to $\mathcal{M}_\Lambda$ has a bounded generator and hence defines a *group* on $\mathcal{M}_\Lambda$. (In other words, within the positively invariant subspace $\mathcal{M}_\Lambda$ we have unique backward extensions for all time.)

To provide the exponential estimates on the complementary space $\mathcal{R}_\Lambda$ we have to make additional assumptions.

Exercise 2.8. Let $T(t)$ be an eventually compact $\mathcal{C}_0$-semigroup with generator A.

(i) Prove that for every $\beta \in \mathbb{R}$,

$$\Lambda(\beta) = \{\lambda \in \sigma(A) \mid \operatorname{Re} \lambda > \beta\}$$

is a finite set of isolated eigenvalues of A.

(ii) Show the following identity for the resolvent of A (cf. Lemma 4.15 of Appendix II):

$$(zI - A)^{-1} = \int_0^{t_0} e^{-zt} T(t)\, dt + e^{-zt_0} T(t_0)(zI - A)^{-1}$$

and conclude, by taking t_0 sufficiently large, that the spectral projections P_λ, $\lambda \in \sigma(A)$, are compact and, consequently, have finite dimensional range.

(iii) Prove that for every $\beta \in \mathbb{R}$, $\Lambda(\beta)$ is a finite set of eigenvalues of finite type of A.

Hint: See [202, Thm B.IV.2.1].

We are now ready to state the main results of this section, which give a decomposition of the state space related, via the spectrum, to a decomposition of $\mathbb{C}$ in some right and left half-planes, together with the corresponding exponential estimates.

Theorem 2.9. *Let $A : \mathcal{D}(A) \to X$ be the generator of an eventually compact $\mathcal{C}_0$-semigroup $T(t)$ and let*

$$\Lambda = \Lambda(\beta) = \{\lambda \in \sigma(A) \mid \operatorname{Re} \lambda > \beta\}.$$

Then

$$X = \mathcal{M}_\Lambda \oplus \mathcal{R}_\Lambda$$

and, for any β, there exist positive constants K and δ such that

$$\begin{aligned} \|T(t)P_\Lambda\| &\le K e^{(\beta+\delta)t} \|P_\Lambda\|, \quad t \le 0, \\ \|T(t)(I - P_\Lambda)\| &\le K e^{(\beta+\delta)t} \|I - P_\Lambda\|, \quad t \ge 0, \end{aligned}$$

where P_Λ denotes the spectral projection associated with Λ.

Proof. It remains to prove the exponential estimates. Choose $\delta > 0$ such that $\beta + \delta < \inf\{\operatorname{Re}\lambda \mid \lambda \in \sigma(A),\ \operatorname{Re}\lambda > \beta\}$. Since Λ is a finite set, it follows from Exercise 2.7 that the restriction of $T(t)$ to $\mathcal{M}_\Lambda$ has a bounded generator with spectrum in the right half-plane $\operatorname{Re} z > \beta + \delta$ only. The first exponential estimate follows immediately.

The restriction of $T(t)$ to $\mathcal{R}_\Lambda$ defines a $\mathcal{C}_0$-semigroup with the part of A in $\mathcal{R}_\Lambda$ as its generator and the second exponential estimate therefore follows from Theorem 2.4. □

Corollary 2.10. *Let $A : \mathcal{D}(A) \to X$ be the generator of an eventually compact $\mathcal{C}_0$-semigroup $T(t)$. If $\sigma(A) \subset \{z \in \mathbb{C} \mid \operatorname{Re} z < 0\}$, then there exist positive constants K and δ such that*

$$\|T(t)\varphi\| \le Ke^{-\delta t}\|\varphi\|, \quad t \ge 0,\ \varphi \in X.$$

We now show how these results can easily be "lifted" to the "bigger" space $X^{\odot *}$. Let P be any projection on X. The decomposition

$$X = \mathcal{R}(P) \oplus \mathcal{R}(I - P)$$

induces a decomposition

$$X^* = \mathcal{R}(P^*) \oplus \mathcal{R}(I - P^*)$$

and one may identify $\mathcal{R}(P^*) = \big(\mathcal{R}(P)\big)^*$.

So when $z = \lambda$ is a pole of $(zI - A)^{-1}$ we have a decomposition

$$X^* = \mathcal{M}_\lambda^* \oplus \mathcal{R}_\lambda^*,$$

where $\mathcal{M}_\lambda^* = \mathcal{R}(P_\lambda^*)$. Note that the Dunford representation for the spectral projection shows that this is exactly the decomposition which one obtains when applying Theorem 2.5 directly to A^* [note in particular that λ is a pole of order m of $(zI - A^*)^{-1}$ when it is a pole of order m of $(zI - A)^{-1}$].

Theorem 2.11. *Let $A : \mathcal{D}(A) \to X$ be the generator of an eventually compact $\mathcal{C}_0$-semigroup $T(t)$ and let*

$$\Lambda = \Lambda(\beta) = \{\lambda \in \sigma(A) \mid \operatorname{Re}\lambda > \beta\}.$$

Then

$$X^* = \mathcal{M}_\Lambda^* \oplus \mathcal{R}_\Lambda^*$$

and, for any β, there exist positive constants K and δ such that

$$\|T^*(t)P_\Lambda^*\| \le Ke^{(\beta+\delta)t}\|P_\Lambda^*\|, \quad t \le 0,$$
$$\|T^*(t)(I - P_\Lambda^*)\| \le Ke^{(\beta+\delta)t}\|I - P_\Lambda^*\|, \quad t \ge 0,$$

where P_Λ denotes the spectral projection associated with Λ.

Proof. First of all, we recall that for every $L : X \to X$ we have $\|L\| = \|L^*\|$. Rewriting the first exponential estimate in Theorem 2.9 in the form

$$\|P_\Lambda T(t)\big|_{\mathcal{M}_\Lambda}\| \leq K e^{(\beta+\delta)t} \|P_\Lambda\|, \quad t \leq 0,$$

we obtain, by taking adjoints, that

$$\|T^*(t)\big|_{\mathcal{M}^*_\Lambda}\| \leq K e^{(\beta+\delta)t} \|P^*_\Lambda\|, \quad t \leq 0,$$

which is identical to the first exponential estimate of the present theorem since $\mathcal{M}^*_\Lambda$ is invariant under $T^*(t)$. The second estimate is proved similarly. □

Theorem 2.12. *Let $A : \mathcal{D}(A) \to X$ be the generator of an eventually compact $\mathcal{C}_0$-semigroup $T(t)$. If*

$$\Lambda = \Lambda(\beta) = \{\lambda \in \sigma(A) \mid \operatorname{Re}\lambda > \beta\},$$

then

$$\Lambda = \{\lambda \in \sigma(A^{\odot *}) \mid \operatorname{Re}\lambda > \beta\}$$

and

$$X^{\odot *} = \mathcal{M}^{\odot *}_\Lambda \oplus \mathcal{R}^{\odot *}_\Lambda .$$

Furthermore there exist positive constants K and δ such that

$$\|T^{\odot *}(t) P^{\odot *}_\Lambda\| \leq K e^{(\beta+\delta)t} \|P^{\odot *}_\Lambda\|, \quad t \leq 0,$$
$$\|T^{\odot *}(t)(I - P^{\odot *}_\Lambda)\| \leq K e^{(\beta+\delta)t} \|I - P^{\odot *}_\Lambda\|, \quad t \geq 0,$$

where P_Λ denotes the spectral projection associated with Λ.

When X is $\odot$-reflexive with respect to $\{T(t)\}$, $\mathcal{M}^{\odot *}_\Lambda$ is a subspace of X.

Proof. It suffices to observe that the norms do not change when taking the restrictions of $P^*_\Lambda T^*(t)|_{\mathcal{M}^*_\Lambda}$ and $(I - P^*_\Lambda)T^*(t)$ to $X^\odot$. On $X^\odot$ we again have the situation of an eventually compact $\mathcal{C}_0$-semigroup $T^\odot(t)$ and hence we can apply Theorem 2.11 to $T^\odot(t)$ and $A^\odot$. Finally, we note that in the $\odot$-reflexive case, $\mathcal{D}(A^{\odot *})$, and hence $\mathcal{M}^{\odot *}_\Lambda$, is contained in (the embedding of) X. □

In the second part of this section we analyse the action of a semigroup on a generalized eigenspace corresponding to an eigenvalue λ of finite type [i.e., λ is an isolated point in $\sigma(A)$ and $\mathcal{M}_\lambda$ has finite dimension].

Let $\lambda \in \sigma_p(A)$ be an eigenvalue of finite type of the generator A and let $\{\varphi_1, \ldots, \varphi_{m_\lambda}\}$ be a basis of eigenvectors and generalized eigenvectors of A at λ. Define the row m_λ-vector $\Phi = (\varphi_1, \ldots, \varphi_{m_\lambda})$. Since $\mathcal{M}_\lambda$ is invariant under A, there exists a $m_\lambda \times m_\lambda$ matrix M such that

$$A\Phi = \Phi M. \tag{2.3}$$

Exercise 2.13. Show that the matrix M in (2.3) has λ as its only eigenvalue.

Lemma 2.14. *The action of $T(t)$ on Φ is given by $T(t)\Phi = \Phi e^{tM}$.*

Proof. For $j = 1, \ldots, m_\lambda$ we have $\varphi_j \in \mathcal{D}(A)$. So $T(t)\varphi_j$ is differentiable and $\frac{d}{dt}T(t)\varphi_j = T(t)A\varphi_j$. This shows

$$\frac{d}{dt}T(t)\Phi = T(t)A\Phi = T(t)\Phi M.$$

□

One can reinterpret this as follows. Let $\varphi \in \mathcal{M}_\lambda$ be arbitrary. Define $y(t)$ to be the m_λ-column vector of coordinates of $T(t)\varphi$ with respect to the basis Φ [in other words, $T(t)\varphi = \Phi y(t)$]. Then $y(t) = e^{tM}y(0)$ and

$$\dot{y}(t) = My(t). \tag{2.4}$$

In other words, the restriction of $T(t)$ to the finite dimensional invariant subspace $\mathcal{M}_\lambda$ is generated by the ordinary differential equation (2.4).

Note, in particular, that for $\varphi \in \mathcal{M}_\lambda$ the definition of $T(t)\varphi$ for $t < 0$ is given by

$$T(t)\varphi = \Phi e^{tM} y(0) \quad \text{for } t < 0. \tag{2.5}$$

Let $\Lambda = \{\lambda_1, \ldots, \lambda_m\}$ be a finite set of eigenvalues of A. Let Φ_{λ_j} be a basis of eigenvectors and generalized eigenvectors for A at λ_j, $j = 1, \ldots, m$. Define M_{λ_j} to be the $m_{\lambda_j} \times m_{\lambda_j}$-matrix such that

$$A\Phi_{\lambda_j} = \Phi_{\lambda_j} M_{\lambda_j}.$$

Put $\Phi_\Lambda = (\Phi_{\lambda_1}, \ldots, \Phi_{\lambda_m})$ and $M_\Lambda = \operatorname{diag}(M_{\lambda_1}, \ldots, M_{\lambda_m})$.

Repeating the process as described earlier in Lemma 2.14 we obtain the following result.

Theorem 2.15. *Let $\Lambda = \{\lambda_1, \ldots, \lambda_m\}$ be a finite set of eigenvalues of finite type of A. Let $\mathcal{M}_\Lambda$ be the $m_{\lambda_1} + \cdots + m_{\lambda_m}$ dimensional subspace spanned by Φ_Λ. Then*

$$T(t)\Phi_\Lambda = \Phi_\Lambda e^{tM_\Lambda}$$

and this relation defines the action of $T(t)$ on $\mathcal{M}_\Lambda$ for $t \in \mathbb{R}$.

A combination of the previous results now yields the following result for eventually compact $\mathcal{C}_0$-semigroups.

Theorem 2.16. *Let $A : \mathcal{D}(A) \to X$ be the generator of an eventually compact $\mathcal{C}_0$-semigroup $T(t)$ and let*

$$\Lambda = \{\lambda \in \sigma(A) \mid \operatorname{Re}\lambda > \beta\} = \{\lambda_1, \ldots, \lambda_m\}.$$

If $\mathcal{M}_\Lambda$ denotes the $m_{\lambda_1} + \cdots + m_{\lambda_m}$ dimensional subspace spanned by Φ_Λ, then

$$T(t)\Phi_\Lambda = \Phi_\Lambda e^{tM_\Lambda}$$

and this relation defines the action of $T(t)$ on $\mathcal{M}_\Lambda$ for $t \in \mathbb{R}$. There exists a $T(t)$-invariant complementary subspace $\mathcal{R}_\Lambda$ such that

$$X = \mathcal{M}_\Lambda \oplus \mathcal{R}_\Lambda,$$

and for $\delta > 0$ sufficiently small there exists a positive constant K such that

$$\begin{aligned} \|T(t)P_\Lambda\| &\leq Ke^{(\beta+\delta)t}\|P_\Lambda\|, \quad t \leq 0, \\ \|T(t)(I - P_\Lambda)\| &\leq Ke^{(\beta+\delta)t}\|I - P_\Lambda\|, \quad t \geq 0, \end{aligned}$$

where P_Λ denotes the spectral projection associated with Λ.

A representation for Φ requires a concrete representation of A. In the next section, when we specialize to delay equations, we shall give such a representation.

We conclude this section with some observations which are useful whenever $X = Y_{\mathbb{C}}$, i.e., X is the complexification of some real Banach space Y. Note that in applications this is usually the case and that this situation arises in particular whenever we obtain linear problems by linearization of nonlinear problems. For relevant definitions and notation we refer back to Section III.7.

Let A be a linear operator on Y, possibly unbounded. By the spectrum, eigenvalues, etc., of A we shall mean the spectrum, eigenvalues, etc., of $A_{\mathbb{C}}$. Since (see Definition III.7.22)

$$\overline{zI - A_{\mathbb{C}}} = \bar{z}I - A_{\mathbb{C}},$$

we have (cf. Exercise III.7.26)

Lemma 2.17. *$\lambda \in \sigma(A)$ if and only if $\overline{\lambda} \in \sigma(A)$.*

Moreover, a short look at the Dunford integral in Theorem 2.5(v) should convince the reader that the following result is correct.

Theorem 2.18. *Let λ be a pole of $(zI - A_{\mathbb{C}})^{-1}$; then*

(2.6) $$P_{\overline{\lambda}} = \overline{P_\lambda},$$

where P_λ and $P_{\overline{\lambda}}$ are the projection operators associated with λ and $\overline{\lambda}$, respectively.

Combining (2.6) with Lemma III.7.23 and Exercise III.7.25 we conclude

Corollary 2.19. *Let λ with $\operatorname{Im}\lambda \neq 0$ be a pole of $(zI - A_{\mathbb{C}})^{-1}$; then $P_\lambda + P_{\overline{\lambda}}$ is the complexification of a projection operator P_λ^Y on Y.*

Exercise 2.20.

(i) Let ϕ be an eigenvector of $A_{\mathbb{C}}$ corresponding to the eigenvalue λ. Show that $\overline{\phi}$ is an eigenvector of $A_{\mathbb{C}}$ corresponding to the eigenvalue $\overline{\lambda}$.

(ii) Assume that $P_\lambda\psi = \langle\phi^*, \psi\rangle\phi$ for some $\phi^* \in X^*$ with $\langle\phi^*, \phi\rangle = 1$ and ϕ as above. Show that

$$P_{\overline{\lambda}}\psi = \langle\overline{\phi^*}, \psi\rangle\overline{\phi}$$

and that

$$P_\lambda^Y y + i0 = \langle\phi^*, y + i0\rangle\phi + \langle\overline{\phi^*}, y + i0\rangle\overline{\phi}.$$

(In Chapter X we shall take the freedom of omitting "$+i0$" in such expressions.)

(iii) Verify that P_λ^Y has a two-dimensional range.
Hint: See Exercises III.7.27 and III.7.28.

Exercise 2.21. Extend Theorem 2.18 and Corollary 2.19 to more general spectral subsets.

Exercise 2.22. Let λ with $\operatorname{Im}\lambda = 0$ be a pole of $(zI - A_{\mathbb{C}})^{-1}$ on the real axis. Prove that P_λ is the complexification of a projection operator P_λ^Y on Y.

IV.3 Delay equations

Since the semigroup associated with RFDE (1.1) is eventually compact we can apply Theorem 2.16. In this section we shall derive an explicit representation for the resolvent of the infinitesimal generator of the semigroup associated with RFDE (1.1) and use this expression to compute the spectral projections.

Theorem 3.1. *Let $A : \mathcal{D}(A) \to \mathcal{C}$ denote the generator of the $\mathcal{C}_0$-semigroup associated with RFDE* (1.1). *Then the resolvent of A has the following explicit representation:*

$$\begin{aligned}((zI - A)^{-1}\varphi)(\theta) &= e^{z\theta}\{\Delta(z)^{-1}[\varphi(0)\\ &\quad + \int_0^h d\zeta(\tau)\int_0^\tau e^{-z\sigma}\varphi(\sigma-\tau)\,d\sigma] + \int_\theta^0 e^{-z\sigma}\varphi(\sigma)\,d\sigma\}.\end{aligned} \tag{3.1}$$

Proof. Let $\psi = (zI - A)^{-1}\varphi$. From the definition of A it follows that

$$(zI - A)\psi = \varphi$$

if and only if ψ satisfies conditions (see Section III.4)

(i) $z\psi - \dot{\psi} = \varphi$;

(ii) $z\psi(0) - \int_0^h d\zeta(\theta)\psi(-\theta) = \varphi(0)$;

(iii) $\dot{\psi} \in \mathcal{C}$.

Define

$$\psi(\theta) = e^{z\theta}\psi(0) + \int_\theta^0 e^{z(\theta-s)}\varphi(s)\,ds,$$

where $-h \leq \theta \leq 0$. Then ψ satisfies the conditions (i) and (iii). Moreover, condition (ii) becomes

$$\Delta(z)\psi(0) = \varphi(0) + \int_0^h d\zeta(\tau) \int_0^\tau e^{-z\sigma}\varphi(\sigma - \tau)\,d\sigma.$$

This proves representation (3.1). □

Exercise 3.2. Use the identity

$$(zI - A)^{-1} = \int_0^\infty e^{-zt}T(t)\,dt$$

(see Proposition 1.11 of Appendix II) to give an alternative derivation of the representation (3.1).
Hint: Use (I.3.15) and (I.2.11).

Corollary 3.3. *The spectrum of A consists of point spectrum only and is given by*

$$\sigma(A) = \sigma_p(A) = \{\lambda \in \mathbb{C} \mid \det \Delta(\lambda) = 0\}.$$

Proof. Because of Theorem 3.1 we know that

$$\{z \in \mathbb{C} \mid \det \Delta(z) \neq 0\} \subset \rho(A).$$

To prove the reverse inclusion, choose $\lambda \in \mathbb{C}$ such that $\det \Delta(\lambda) = 0$ and define

$$\varphi(\theta) = e^{\lambda\theta}\varphi^0 \quad \text{for } -h \leq \theta \leq 0,$$

where $\varphi^0 \neq 0$ is an element of the null space of $\Delta(\lambda)$. Then

$$\dot{\varphi}(0) = \lambda\varphi^0 = \int_0^h e^{-\lambda\theta} d\zeta(\theta)\varphi^0 = \langle \zeta, \varphi \rangle_n,$$

so $\varphi \in \mathcal{D}(A)$ and, moreover,

$$A\varphi = \dot{\varphi} = \lambda\varphi.$$

We conclude that $\lambda \in \sigma_p(A)$. □

Let Φ be a basis for the generalized eigenspace of A at some eigenvalue λ, as in (2.3).

Lemma 3.4. *The basis Φ is given by $\Phi(\theta) = \Phi(0)e^{\theta M}$ for $-h \le \theta \le 0$.*

Proof. The result follows immediately from the concrete form of A. □

A combination of Lemma 2.14 and Lemma 3.4 yields the following result:

Proposition 3.5. $\big(T(t)\Phi\big)(\theta) = \Phi(0)e^{(t+\theta)M}$ *for* $-h \le \theta \le 0$.

So, in order to obtain a complete description of the subspace $\mathcal{M}_\lambda$ and the behaviour of solutions in that subspace, we have to compute the row m_λ-vector $\Phi(0) = \big(\varphi_1(0), \dots, \varphi_{m_\lambda}(0)\big)$, where each $\varphi_i(0)$ is an n-vector, and the $m_\lambda \times m_\lambda$ matrix M. In this section we present the results for RFDE in a special and relatively simple case. Note first of all that an approach to compute the row vector $\Phi(0)$ is contained in Theorem 2.5 (vi). It states that the spectral projection onto $\mathcal{M}_\lambda$ is given by the residue of the resolvent of A at $z = \lambda$. So, $\Phi(0)$ can be found by providing a basis for the range of $\operatorname{Res}_{z=\lambda}(zI - A)^{-1}$. The next section is devoted to a systematic procedure to construct such a basis and in Section 5 we present the general results for delay equations. It will turn out that all essential information is contained in the $n \times n$-matrix function $\Delta(z)$.

In the next part of this section we shall restrict ourselves to the case that

$$\Delta(z)^{-1} = \frac{H(z)}{z - \lambda} \quad \text{with } H \text{ analytic at } z = \lambda. \tag{3.2}$$

(In the next section we shall see that a sufficient, but not necessary, condition for this to be true is that λ is a simple zero of $\det \Delta$.)

Using representation (3.1) for $(zI - A)^{-1}$, it follows that

$$\begin{aligned} P_\lambda\varphi &= \operatorname*{Res}_{z=\lambda}(zI - A)^{-1}\varphi \\ &= e^{\lambda\cdot} H(\lambda)\big[\varphi(0) + \int_0^h d\zeta(\theta) \int_0^\theta e^{-\lambda\sigma}\varphi(\sigma - \theta)\,d\sigma\big]. \end{aligned} \tag{3.3}$$

Lemma 3.6. *Under assumption* (3.2) *we have that the generalized eigenspace at λ, $\mathcal{M}_\lambda = \mathcal{R}(P_\lambda)$, is given by*

$$\mathcal{M}_\lambda = \{\theta \mapsto e^{\lambda\theta}v \mid v \in \mathcal{N}(\Delta(\lambda))\}$$

and equals the eigenspace at λ (i.e., the geometric and the algebraic multiplicity of λ are equal).

Proof. First note that

$$\mathcal{R}\big(H(\lambda)\big) = \mathcal{N}\big(\Delta(\lambda)\big).$$

Indeed, from (3.2) it follows that $\Delta(z)H(z) = H(z)\Delta(z) = (z-\lambda)I$. So $\Delta(\lambda)H(\lambda) = 0$. Assume that $v \in \mathbb{C}^n$ is such that

$$\Delta(\lambda)v = 0$$

and define

$$w = \lim_{z\to\lambda} \frac{1}{z-\lambda}\Delta(z)v = \Delta'(\lambda)v.$$

Then

$$H(\lambda)w = \lim_{z\to\lambda} \frac{1}{z-\lambda}H(z)\Delta(z)v = \lim_{z\to\lambda}\frac{z-\lambda}{z-\lambda}v = v.$$

So $H(\lambda)w = v$. The proof of the lemma is now easily completed using representation (3.3) for P_λ. □

The characterisation of $\mathcal{M}_\lambda$ for eigenvalues of A such that (3.2) holds implies that in this case

$$\Phi(0) = (v_1, \ldots, v_m),$$

where $\{v_j\}_{j=1}^m$ is a basis for $\mathcal{N}\big(\Delta(\lambda)\big)$ and $M = \lambda I_m$. In Section 5 we present a systematic procedure to compute $\Phi(0)$ for arbitrary eigenvalues of A (see Theorem 5.3 and Exercise 5.8).

Exercise 3.7. Use representation (3.3) for P_λ to prove that the adjoint projection P_λ^* is given by

$$(P_\lambda^* f)(t) = \int_0^h df(\tau)e^{-\lambda\tau}H(\lambda)[I + \int_0^t \int_\sigma^h d\zeta(\theta)e^{\lambda(\sigma-\theta)}\, d\sigma].$$

Use this representation to prove that the eigenvectors of A^* at λ are given by

$$(I + \int_0^t \int_\sigma^h d\zeta(\theta)^T e^{\lambda(\sigma-\theta)}\, d\sigma)u, \qquad t > 0,$$

with the vector u such that $\Delta(\lambda)^T u = 0$.

Exercise 3.8. Let $\varphi(\theta) = e^{\lambda\theta}v$ for some $v \in \mathcal{N}\big(\Delta(\lambda)\big)$. Check that $P_\lambda\varphi = \varphi$.

Exercise 3.9. Verify that $P_\lambda^2 = P_\lambda$.
Hint: Use that $v = H(\lambda)\Delta'(\lambda)v$ whenever $v \in \mathcal{N}\big(\Delta(\lambda)\big)$; cf. the proof of Lemma 3.6.

Exercise 3.10. Assume, in addition to (3.2), that $\dim \mathcal{N}\big(\Delta(\lambda)\big) = 1$. Show that $P_\lambda\psi = \langle\phi^*, \psi\rangle\phi$, where $A\phi = \lambda\phi$, $A^*\phi^* = \lambda\phi^*$ and $\langle\phi^*, \phi\rangle = 1$ [cf. Theorem 2.5(vi)].

Exercise 3.11. Derive an explicit representation for $(zI - A_0^{\odot *} - B)^{-1}$ in the case of RFDE. In addition, give a characterisation of $\mathcal{N}((zI - A_0^{\odot *} - B))$ for z in the spectrum.
Hint: If you do not manage, see Section 5.

Exercise 3.12. Prove, for RFDE, that z is an algebraically simple eigenvalue of A if and only if

(i) $\dim \mathcal{N}(\Delta(\lambda)) = 1$;
(ii) $u\Delta'(\lambda)v \neq 0$ for $u \neq 0$ and $v \neq 0$ such that $\Delta(\lambda)^T u = 0$ and $\Delta(\lambda)v = 0$.

Show that in this case the projection operator is given by

$$P_\lambda \varphi = \frac{u[\varphi(0) + \int_0^h d\zeta(\theta) \int_0^\theta e^{-\lambda\sigma} \varphi(\sigma - \theta)\, d\sigma]}{u\Delta'(\lambda)v} v e^{\lambda \cdot}.$$

Exercise 3.13. Consider the RFDE

$$\dot{x}(t) = x(t) - x(t - \frac{1}{2}).$$

Verify that

$$\Delta(z) = z - 1 + e^{-\frac{1}{2}z}$$

and show that $\lambda = 0$ is an algebraically simple eigenvalue of the infinitesimal generator with associated projection operator

$$P_0 \varphi = 2(\varphi(0) - \int_0^{\frac{1}{2}} \varphi(-\theta)\, d\theta)\mathbf{1},$$

where $\mathbf{1}$ denotes the constant function with value 1.

Exercise 3.14. Suppose that

$$\Delta(z)^{-1} = \frac{K(z)}{(z - \lambda)^2}$$

with K analytic at $z = \lambda$. Show that for regular functions F we have

$$\operatorname*{Res}_{z=\lambda} \Delta(z)^{-1} F(z) = K'(\lambda)F(\lambda) + K(\lambda)F'(\lambda).$$

The last exercises of this section are relevant for Sections IX.10 and XI.2. Also see Example IV.5.13.

Exercise 3.15. Consider the equation

$$\dot{x}(t) = x(t) - x(t - 1).$$

(i) Verify that $\Delta(z) = z - 1 + e^{-z}$.
(ii) Show that the corresponding infinitesimal generator has eigenvalue zero with associated projection operator

$$P_0\varphi = (2\int_0^1 (1-\theta)\varphi(-\theta)\,d\theta + \frac{2}{3}(\varphi(0) - \int_0^1 \varphi(-\theta)\,d\theta))\mathbf{1}$$
$$+ 2(\varphi(0) - \int_0^1 \varphi(-\theta)\,d\theta)I,$$

where $I(\theta) = \theta$, and conclude that $\mathbf{1}$ is an eigenvector, I a generalized eigenvector and that the algebraic multiplicity is two.

(iii) Verify that the matrix M introduced in (2.3) is in this case given by

$$M = \begin{pmatrix} 0 & 1 \\ 0 & 0 \end{pmatrix}$$

and that, consequently,

$$e^{tM} = \begin{pmatrix} 1 & t \\ 0 & 1 \end{pmatrix}.$$

(iv) Check the assertion of Proposition 3.5 for this particular case.

Exercise 3.16. Do Exercise IX.9.4.

IV.4 Characteristic matrices, equivalence and Jordan chains

In this section we shall study the structure of the matrix M introduced in equation (2.3). The final result will provide an algorithm for the construction of a canonical basis Φ such that M has Jordan canonical form with respect to this basis.

Let $A : \mathcal{D}(A) \to X$ be a closed unbounded operator. If λ is an eigenvalue of finite type, the operator $L = A\,|_{\mathcal{M}_\lambda}$ is a bounded operator from a finite dimensional space into itself. So the situation is reduced to the finite dimensional case, which we shall, therefore, discuss first.

Let X be a finite dimensional space, say $\dim X = m$. Let $L : X \to X$ be a bounded linear operator. If A denotes the matrix representation of L with respect to some basis, then the eigenvalues of L are precisely given by the roots of the characteristic polynomial

$$C(z) = \det\big(zI - A\big). \tag{4.1}$$

Over the scalar field $\mathbb{C}$ the characteristic polynomial can be factorized into a product of m linear factors

$$C(z) = \prod_{j=1}^{m}(z - \lambda_j),$$

where $\lambda_j \in \sigma(L)$. Define the *algebraic multiplicity* of λ_j to be the number of times the factor $(z - \lambda_j)$ appears, or, in other words, the order of λ_j as a zero of the determinant. The dimension of $\mathcal{N}(\lambda_j I - L)$ is called the *geometric multiplicity* of λ_j. The characteristic polynomial is an annihilating polynomial of L, i.e., $C(L) = 0$. The *minimal polynomial* M of L is defined to be an annihilating polynomial of L that divides any other annihilating polynomial. Necessarily, M is of the form

$$M(z) = \prod_{j=1}^{l} (z - \lambda_j)^{k_j}, \tag{4.2}$$

where $\sigma(L) = \{\lambda_1, \ldots, \lambda_l\}$, and for $j = 1, \ldots, l$, the number k_j is positive and less than or equal to the algebraic multiplicity of λ_j. Define

$$\mathcal{M}_j = \mathcal{N}\big((\lambda_j I - L)^{k_j}\big).$$

This is a L-invariant subspace, i.e., $L\mathcal{M}_j \subseteq \mathcal{M}_j$, and we can define the part of L in $\mathcal{M}_j$, i.e., $L_j = L\,|_{\mathcal{M}_j} \colon \mathcal{M}_j \to \mathcal{M}_j$.

Define

$$Q_j(z) = \frac{M(z)}{(z - \lambda_j)^{k_j}}, \qquad j = 1, \ldots, l;$$

then Q_j are relatively prime and there exist polynomials R_j, $j = 1, \ldots, l$, such that

$$Q_1(z)R_1(z) + \cdots + Q_n(z)R_n(z) = 1.$$

(See [235, Chapter 6].) Set $F_j(z) = Q_j(z)R_j(z)$ and define $P_j : X \to X$ by

$$P_j = F_j(L), \qquad j = 1, \ldots, l. \tag{4.3}$$

Exercise 4.1. The linear operators P_j defined in (4.3) satisfy
(i) $P_m P_k = P_k P_m = 0$ for $k, m = 1, \ldots, l$ and $m \neq k$;
(ii) $P_j^2 = P_j$ and $P_j \neq 0$ for $j = 1, \ldots, l$;
(iii) $P_1 + \cdots + P_l = I$;
(iv) $\mathcal{R}(P_j) = \mathcal{M}_j$.
Hint: Use that $P_j = F_j(L)$, where F_j is a polynomial.

Exercise 4.1 is concerned with a special case of the operational calculus (see Appendix III) and yields $X = \mathcal{M}_1 \oplus \cdots \oplus \mathcal{M}_l$. The operator L decomposes accordingly

$$L = \bigoplus_{j=1}^{l} L_j.$$

This decomposition is unique (up to the order of summands). The action of L can be broken down to the study of the action of L_j. To continue the decomposition one studies the structure of the subspaces $\mathcal{M}_j$ more closely.

Let $\lambda \in \sigma(L)$. The subspace $\mathcal{M}_\lambda = \mathcal{N}\big((\lambda I - A)^{k_\lambda}\big)$ is called the *generalized eigenspace* associated to the eigenvalue λ. A vector x is called a *generalized eigenvector of order* r if

$$(\lambda I - L)^r x = 0 \qquad \text{while} \quad (\lambda I - L)^{r-1} x \neq 0.$$

Suppose x_{r-1} is a generalized eigenvector of order r; then there are vectors $(x_{r-2}, \dots, x_1, x_0)$ for which $x_0 \neq 0$ and

$$\begin{aligned} Lx_0 &= \lambda x_0, \\ Lx_1 &= \lambda x_1 + x_0, \\ &\vdots \\ Lx_{r-1} &= \lambda x_{r-1} + x_{r-2} \end{aligned} \tag{4.4}$$

and hence $x_j \in \mathcal{N}\big((\lambda I - L)^{j+1}\big)$. Such a sequence is called a *Jordan chain.* The chain can also be conceived as being associated with x_0: the ordered set $(x_0, x_1, \dots, x_{r-1})$ of vectors in X is a Jordan chain for L at λ if and only if $x_0 \neq 0$ and

$$(zI - L)[x_0 + (z-\lambda)x_1 + \cdots + (z-\lambda)^{r-1}x_{r-1}] = O((z-\lambda)^r) \tag{4.5}$$

for $|z - \lambda| \to 0$. Obviously, the length of the Jordan chain is less than or equal to k_λ and a Jordan chain consists of linearly independent elements.

Exercise 4.2. Prove the last statement that a Jordan chain consists of linearly independent elements.

The *rank* of x_0 is the maximal length of a Jordan chain starting with x_0. Define the Jordan subspace associated with $x_0 \in \mathcal{N}(\lambda I - L)$ by

$$\mathcal{J}_\lambda(x_0) = \operatorname{span}\{x_0, x_1, \dots, x_{r-1}\},$$

where $(x_0, x_1, \dots, x_{r-1})$ is a Jordan chain of maximal length.

Exercise 4.3. Prove that a Jordan subspace is L-invariant and contains only one independent eigenvector.

Let $\mathcal{K} \subset X$ be a L-invariant subspace. If there exists, for some $y \in \mathcal{K}$, a basis for $\mathcal{K}$ of the form

$$\{y, Ly, L^2y, \dots, L^{k-1}y\}, \qquad k = \dim \mathcal{K},$$

then $\mathcal{K}$ is said to be a *cyclic subspace* generated by y.

Exercise 4.4. Prove that a Jordan subspace is cyclic.

Let $\{x_0, \dots, x_{r-1}\}$ be a Jordan chain of length r and let $\mathcal{J}_\lambda(x_0)$ be the Jordan subspace. By Exercise 4.4 there exists a $y \in \mathcal{J}_\lambda(x_0)$ such that $\{y, Ly, \dots, L^{r-1}y\}$ is a basis for $\mathcal{J}_\lambda(x_0)$.

Exercise 4.5. Prove that the vectors

$$\{(L-\lambda I)^{r-1}y, (L-\lambda I)^{r-2}y, \dots, y\} \tag{4.6}$$

form a basis for $\mathcal{J}_\lambda(x_0)$ as well.
Hint: Look at the minimal polynomial for $L\mid_{\mathcal{J}_\lambda(x_0)}$.

Exercise 4.6. Show that with respect to the basis given in (4.6) the restriction $L\mid_{\mathcal{J}_\lambda(x_0)}$ has the following $r\times r$ matrix representation

$$L\mid_{\mathcal{J}_\lambda(x_0)} = \begin{pmatrix} \lambda & 1 & & 0 \\ & \ddots & \ddots & \\ 0 & & \lambda & 1 \\ & & & \lambda \end{pmatrix}. \tag{4.7}$$

The matrix in (4.7) is referred to as a Jordan block of order r corresponding to the eigenvalue λ.

One can organize the Jordan chains according to the procedure described by Gohberg and Sigal [90]. Choose an eigenvector, say $x_{1,0}$, with maximal rank, say r_1. Next, choose a Jordan chain

$$(x_{1,0}, \dots, x_{1,r_1-1})$$

of length r_1 and let N_1 be the complement in $\mathcal{N}(\lambda I - L)$ of the subspace spanned by $x_{1,0}$. In N_1 we choose an eigenvector $x_{2,0}$ of maximal rank, say r_2, and let

$$(x_{2,0}, \dots, x_{2,r_2-1})$$

be a corresponding Jordan chain of length r_2. We continue in the same manner: let N_2 be the complement in N_1 of the subspace spanned by $x_{2,0}$ and replace N_1 by N_2 in the above-described procedure.

In this way, we obtain a basis $\{x_{1,0}, \dots, x_{p,0}\}$ of $\mathcal{N}(\lambda I - L)$ and a corresponding *canonical system* of Jordan chains

$$x_{1,0}, \dots, x_{1,r_1-1}, \dots, x_{p,0}, \dots, x_{p,r_p-1}. \tag{4.8}$$

Because of the construction, the canonical system of Jordan chains is a basis for the generalized eigenspace and

$$\mathcal{M}_\lambda = \bigoplus_{j=1}^{p} \mathcal{J}(x_{j,0}).$$

With respect to this basis, the operator $L\mid_{\mathcal{M}_\lambda}$ has Jordan normal form $J(\lambda)$ with λ on the main diagonal. We shall call such a basis a *canonical basis of eigenvectors and generalized eigenvectors for L at λ*. The partitioning of the basis (4.8) corresponds to the partitioning of the Jordan matrix in Jordan blocks. The numbers $r_1 \geq \cdots \geq r_p$ (which do not depend on the particular choice of the basis) are the sizes of the Jordan blocks, and they are called the *partial multiplicities* of the eigenvalue λ. The largest partial multiplicity is equal to the smallest number q_λ such that

$$\mathcal{M}_\lambda = \mathcal{N}\big((\lambda I - L)^{q_\lambda}\big)$$

and is called the *ascent* of λ. The matrix A is similar to the block diagonal Jordan matrix with the blocks constructed by the procedure described above. Hence the characteristic polynomial can be written in the form

$$C(z) = \det(zI - A) = \prod_{j=1}^{k} \det\big(zI - J(\lambda_j)\big) = \prod_{j=1}^{k} (z - \lambda_j)^{\sum_{i=1}^{p} r_i}.$$

We conclude that the sum of the partial multiplicities equals the algebraic multiplicity of an eigenvalue. It follows that the dimension of the generalized eigenspace $\mathcal{M}_\lambda$ is equal to the algebraic multiplicity, as we defined it in terms of the multiplicity of λ as a zero of $C(z) = \det(zI - A)$. Frequently the algebraic multiplicity is defined as $\dim \mathcal{M}_\lambda$. The result derived above shows that the two definitions are equivalent.

In infinite dimensions there is, in general, no characteristic equation and the theory as described above does not carry over. However, for the generator A of the semigroup associated with RFDE (1.1), we have seen that the spectrum of A is given by the zeros of $\det \Delta$, where Δ is given by (I.3.16). In the second part of this section, we shall see that we can develop a theory as above for a class of unbounded operators, which includes the generators associated with delay equations.

Suppose $K : \Omega \to \mathcal{L}(X, Y)$ and $M : \Omega \to \mathcal{L}(X', Y')$ are operator valued functions, holomorphic on an open set Ω in the complex plane $\mathbb{C}$. The operator functions K and M are called *equivalent* on Ω if there exist holomorphic operator functions $E : \Omega \to \mathcal{L}(X', X)$ and $F : \Omega \to \mathcal{L}(Y, Y')$, whose values are bijective operators, such that

$$M(z) = F(z)K(z)E(z), \quad z \in \Omega. \tag{4.9}$$

Again let $K : \Omega \to \mathcal{L}(X, Y)$ be a holomorphic operator function. A point λ is called a *characteristic value* of K if there exists a vector $x_0 \in X$, $x_0 \neq 0$, such that

$$K(\lambda)x_0 = 0.$$

An ordered set $(x_0, x_1, \ldots, x_{k-1})$ of vectors in X is called a *Jordan chain* for K at λ if $x_0 \neq 0$ and

$$K(z)[x_0 + (z - \lambda)x_1 + \cdots + (z - \lambda)^{k-1}x_{k-1}] = O((z - \lambda)^k) \tag{4.10}$$

for $|z - \lambda| \to 0$. The number k is called the *length* of the chain and the maximal length of a chain starting with x_0 is called the *rank* of x_0. The holomorphic function

$$\sum_{l=0}^{k-1}(z-\lambda)^l x_l$$

in (4.10) is called a *root function* of K corresponding to λ.

Lemma 4.7. *If two holomorphic operator functions K and M are equivalent, then there is a one-to-one correspondence between their Jordan chains.*

Proof. Since the equivalence relation (4.9) implies that the null spaces $\mathcal{N}(K(\lambda))$ and $\mathcal{N}(M(\lambda))$ are isomorphic, it suffices to show that there is a one-to-one correspondence between the Jordan chains of length k, $k \geq 1$, of K and M at λ. If $(x_0, \ldots, x_{k-1})$ is a Jordan chain for K at λ of length k, then the equivalence relation implies that

$$M(z)E(z)^{-1}\sum_{l=0}^{k-1}(z-\lambda)^l x_l = O((z-\lambda)^k) \tag{4.11}$$

for $|z-\lambda| \to 0$. If $\sum_{l=0}^{k-1}(z-\lambda)^l y_l$ denotes the Taylor expansion of order k around $z=\lambda$ for the holomorphic function

$$E(z)^{-1}\sum_{l=0}^{k-1}(z-\lambda)^l x_l,$$

then

$$M(z)\sum_{l=0}^{k-1}(z-\lambda)^l y_l = O((z-\lambda)^k) \quad \text{for } |z-\lambda| \to 0$$

and $(y_0, \ldots, y_{k-1})$ is a Jordan chain for M at λ of length k. So, we proved that a Jordan chain for K at λ of length k induces a Jordan chain for M at λ of length k. Since the role of K and M can be interchanged, the proof is complete. □

Our aim is to find, for certain holomorphic operator functions, an equivalence with holomorphic functions which are essentially *matrix* valued. The main motivation for this is that, as we now show first, the Jordan chains can be constructed by a systematic procedure in the matrix case.

Throughout this section $\Delta : \Omega \to \mathcal{L}(\mathbb{C}^n)$ will denote a holomorphic *matrix*-valued function. If the determinant of Δ is not identically zero, then we denote by $m(\lambda, \Delta)$ the order of λ as a zero of $\det \Delta$ and by $k(\lambda, \Delta)$ the order of λ as a pole of the matrix-valued function Δ^{-1}.

Exercise 4.8. Let λ be an isolated characteristic value of Δ. Prove that the Jordan chains for Δ at λ have finite rank.
Hint: Consider the expression for $\sum_{l=0}^{k-1}(z-\lambda)^l y_l$ and conclude that the rank of a Jordan chain is less than or equal to the multiplicity of λ as a zero of $\det \Delta$.

Let λ be an isolated characteristic value of Δ. We can organize the chains according to the procedure described in Section 1 [with the null space of $\Delta(\lambda)$ replacing the eigenspace of L]. In this way, we obtain a basis $\{x_{1,0},\ldots,x_{p,0}\}$ of $\mathcal{N}(\Delta(\lambda))$ and a corresponding *canonical system* of Jordan chains

$$x_{1,0},\ldots,x_{1,r_1-1},\ldots,x_{p,0},\ldots,x_{p,r_p-1}$$

for Δ at λ.

It is easy to see that the rank of any eigenvector $x_0 \in \mathcal{N}(\Delta(\lambda))$ corresponding to the characteristic value λ is equal to one of the r_j for $1 \le j \le p$. Thus the integers $r_1,\ldots,r_p$ do not depend on the particular choices made in the procedure described above and are called the *zero multiplicities* of Δ at λ. Their sum $r_1+\cdots+r_p$ is called the *algebraic multiplicity* of Δ at λ and will be denoted by $\mathrm{M}(\Delta(\lambda))$.

In the linear case $\Delta(z) = zI - A$, a Jordan chain $(x_0,\ldots,x_{k-1})$ for Δ at λ satisfies

$$\begin{aligned}(A-\lambda)x_0 &= 0,\\ (A-\lambda)x_1 &= x_0,\\ &\vdots\\ (A-\lambda)x_{k-1} &= x_{k-2}\end{aligned}$$

and hence

$$\{x_{i,l} \mid i=1,2,\ldots,p;\ l=0,1,\ldots,r_{i-1}\}$$

is a canonical basis of eigenvectors and generalized eigenvectors for A at λ. So in that case, $\mathrm{M}(\Delta(\lambda))$ equals $M(A;\lambda)$, the algebraic multiplicity of A at λ defined as the dimension of the generalized eigenspace.

So, the definition of the *algebraic multiplicity* in terms of Jordan chains is a proper extension to the case of nonlinear dependence on z. Note that, in general,

$$\dim \mathcal{N}(\Delta(\lambda)^q) \ne \mathrm{M}(\Delta(\lambda)), \qquad q = \max\{r_j \mid j=1,\ldots,p\}.$$

The following result will be useful to compute the Jordan chains for a holomorphic matrix-valued function $\Delta : \Omega \to \mathcal{L}(\mathbb{C}^n)$ with $\det\Delta \not\equiv 0$.

Theorem 4.9. *Let $\Delta : \Omega \to \mathcal{L}(\mathbb{C}^n)$ with* $\det\Delta \not\equiv 0$ *be a holomorphic matrix-valued function. For $\lambda \in \Omega$ there exist a neighbourhood $\mathcal{U}$ of λ and holomorphic matrix-valued functions E and F on $\mathcal{U}$ whose values are bijective operators such that*

$$\Delta(z) = F(z)D(z)E(z), \quad z \in \mathcal{U}, \tag{4.12}$$

where

$$D(z) = \mathrm{diag}\big((z-\lambda)^{\nu_1},\ldots,(z-\lambda)^{\nu_n}\big), \quad z \in \mathcal{U}. \tag{4.13}$$

The integers $\{\nu_1, \ldots, \nu_n\}$ are uniquely determined by Δ and the diagonal matrix D is called the *local Smith form* for Δ at λ.

To outline the proof of the theorem, we describe the method to construct D, E and F. We restrict ourselves to the case that Δ is given by a 2×2 matrix. The general case follows by induction using elementary matrix operations.

For every $\lambda \in \mathbb{C}$ we define A_λ to be the set of equivalence classes of functions f which are analytic in some neighbourhood of λ, with the equivalence relation $f \sim g$ if $f = g$ in a neighbourhood of λ. For every $f \in A_\lambda$, there exists a k such that

$$f(z) = (z - \lambda)^k u(z) \qquad \text{with} \quad u(\lambda) \neq 0,$$

and for every $g \in A_\lambda$, there exist $q, r \in A_\lambda$ such that $g = qf + r$, where r is a polynomial of degree less than k. Therefore we can define the *greatest common divisor* $\gcd(f, g)$ of f and g in A_λ, that is, the common divisor of f and g that is divisible by any other common divisor of f and g. The $\gcd(f, g)$ has the property that there exist $a_1, a_2 \in A_\lambda$ and a neighbourhood $\mathcal{U}$ of λ such that for $z \in \mathcal{U}$

$$a_1(z)f(z) + a_2(z)g(z) = \gcd(f, g)(z).$$

We can also define the *least common multiple* $\operatorname{lcm}(f, g)$ of two elements f and g in A_λ, that is, the common multiple of f and g that is a divisor of any other common multiple of f and g. The $\operatorname{lcm}(f, g)$ has the property that there exist $b_1, b_2 \in A_\lambda$ and a neighbourhood $\mathcal{U}$ of λ such that for $z \in \mathcal{U}$

$$\begin{aligned} b_1(z)f(z) + b_2(z)g(z) &= 2\operatorname{lcm}(f, g)(z), \\ b_1(z)f(z) - b_2(z)g(z) &= 0. \end{aligned}$$

Therefore, for $f, g \in A_\lambda$, we can define $p, q, r, s \in A_\lambda$ so that

$$\begin{aligned} pf + qg &= \gcd(f, g), \\ rf - sg &= \operatorname{lcm}(f, g), \\ rf + sg &= 0, \end{aligned} \tag{4.14}$$

where we have suppressed the dependence of z in the notation and adjusted the signs and the factor 2 to facilitate certain computations below.

Exercise 4.10. Prove that the matrix

$$\begin{pmatrix} p & r \\ q & s \end{pmatrix} \tag{4.15}$$

is invertible.
Hint: Define f_1, g_1, c_1, c_2, u in A_λ such that

$$f = \gcd(f, g) f_1, \quad g = \gcd(f, g) g_1, \quad 1 = c_1 r + c_2 s$$

and verify that

$$\begin{pmatrix} f_1 & g_1 \\ c_1 - \alpha f_1 & c_2 - \alpha g_1 \end{pmatrix} \begin{pmatrix} p & r \\ q & s \end{pmatrix} = \begin{pmatrix} 1 & 0 \\ 0 & 1 \end{pmatrix} \qquad \text{in } A_\lambda,$$

where $\alpha = c_1 p + c_2 q$.

Exercise 4.11. Let $\mathcal{U}$ be a neighbourhood of λ such that

$$f(z) = (z-\lambda)^6 a(z), \quad a(\lambda) \neq 0, \qquad g(z) = (z-\lambda)^2 b(z), \quad b(\lambda) \neq 0.$$

Find p, q, r and s in A_λ such that (4.14) holds and verify the invertibility of (4.15).

The next exercise provides the construction of E and F.

Exercise 4.12. Let

$$\Delta = \begin{pmatrix} \Delta_{11} & \Delta_{12} \\ \Delta_{21} & \Delta_{22} \end{pmatrix}.$$

(i) The p, q, r and s in A_λ such that (4.14) holds for $f = \Delta_{11}$ and $g = \Delta_{12}$ are indicated by an index 0. Set

$$G_0 = \begin{pmatrix} p_0 & r_0 \\ q_0 & s_0 \end{pmatrix}.$$

Verify that

$$\Delta G_0 = \begin{pmatrix} h_0 & 0 \\ p_0 \Delta_{21} + q_0 \Delta_{22} & r_0 \Delta_{21} + s_0 \Delta_{22} \end{pmatrix},$$

where

$$h_0 = \gcd(\Delta_{11}, \Delta_{12}).$$

(ii) The p, q, r and s in A_λ such that (4.14) holds for $f = h_0$ and $g = p_0 \Delta_{21} + q_0 \Delta_{22}$ are indicated by an index 1. Set

$$G_1 = \begin{pmatrix} p_1 & q_1 \\ r_1 & s_1 \end{pmatrix}.$$

Verify that

$$G_1 \Delta G_0 = \begin{pmatrix} h_1 & q_1(r_0 \Delta_{21} + s_0 \Delta_{22}) \\ 0 & s_1(r_0 \Delta_{21} + s_0 \Delta_{22}) \end{pmatrix},$$

where

$$h_1 = \gcd(h_0, p_0 \Delta_{21} + q_0 \Delta_{22}).$$

(iii) The p, q, r and s in A_λ such that (4.14) holds for $f = h_1$ and $g = q_1(r_0 \Delta_{21} + s_0 \Delta_{22})$ are indicated by an index 2. Set

$$G_2 = \begin{pmatrix} p_2 & r_2 \\ q_2 & s_2 \end{pmatrix}.$$

Verify that

$$G_1 \Delta G_0 G_2 = \begin{pmatrix} h_2 & 0 \\ q_2 s_1(r_0\Delta_{21} + s_0\Delta_{22}) & s_2 s_1(r_0\Delta_{21} + s_0\Delta_{22}) \end{pmatrix},$$

where

$$h_2 = \gcd(h_1, q_1(r_0\Delta_{21} + s_0\Delta_{22})).$$

(iv) The p, q, r and s in A_λ such that (4.14) holds for $f = h_2$ and $g = q_2 s_1(r_0\Delta_{21} + s_0\Delta_{22})$ are indicated by an index 3. Prove that $q_3 = 0$.

(v) Set

$$G_3 = \begin{pmatrix} p_3 & 0 \\ r_3 & s_3 \end{pmatrix}$$

and verify that

$$G_3 G_1 \Delta G_0 G_2 = \begin{pmatrix} h_2 & 0 \\ 0 & s_3 s_2 s_1(r_0\Delta_{21} + s_0\Delta_{22}) \end{pmatrix}.$$

Exercise 4.13. Complete the proof of Theorem 4.9 for $n = 2$.

The next exercise is intended to provide a simple illustration.

Exercise 4.14. Find the local Smith form for

$$A(z) = \begin{pmatrix} z & 1 \\ 0 & z \end{pmatrix}.$$

Hint: Due to the algorithmic nature of the proof of the local Smith form one can use computer algebra to compute the local Smith form when the entries of the matrix are polynomials. In Maple, just type "smith(A(z),z)". See also the next exercise.

Define a *minor* of order j of Δ to be a $j \times j$ subdeterminant of Δ, i.e., remove $n - j$ rows and columns of Δ and take the determinant of the resulting matrix. The following exercise (see [160]) enables us to compute the local Smith form directly.

Exercise 4.15. Let $\Delta : \Omega \to \mathcal{L}(\mathbb{C}^n)$ be a holomorphic matrix-valued function with $\det \Delta \not\equiv 0$. Define d_j to be the greatest common divisor of all minors of Δ of order j, $j = 1, 2, \ldots, n$, and set $d_0 = 1$. Show the following:

(i) The d_j is divisible by d_{j-1}, $j = 1, 2, \ldots, n$.

(ii) Let a_j be divisible by a_{j-1} for $j = 1, 2, \ldots, n$. If $\Delta = \operatorname{diag}(a_1, \ldots, a_n)$, then

$$d_1 = a_1, \quad d_2 = a_1 a_2, \quad \ldots, d_n = \prod_{j=1}^{n} a_j.$$

(iii) If G is an invertible matrix and $\widetilde{\Delta}(z) = G\Delta(z)G^{-1}$ for $z \in \mathbb{C}$, then

$$\tilde{d}_j = d_j, \qquad j = 1, 2, \ldots, n.$$

(iv) The local Smith form of Δ is given by

$$\text{(4.16)} \qquad \operatorname{diag}(i_1, \ldots, i_n), \qquad i_j = \frac{d_j}{d_{j-1}}, \quad j = 1, \ldots, n.$$

For the local Smith form D as given in (4.13) the Jordan chains are easily determined and it is clear that the set of zero multiplicities is $\{\nu_1, \ldots, \nu_n\}$. An immediate corollary of the local Smith form is the following:

Corollary 4.16. *The algebraic multiplicity of Δ at λ equals the multiplicity of λ as zero of* $\det \Delta$.

Proof. Let

$$D(z) = \operatorname{diag}\big((z-\lambda)^{\nu_1}, \ldots, (z-\lambda)^{\nu_n}\big)$$

denote the local Smith form at λ. From (4.12) and Lemma 4.7 it follows that the zero multiplicities of Δ at λ are given by $\{\nu_1, \ldots, \nu_n\}$. So, by definition, the algebraic multiplicity of λ is given by

$$\text{(4.17)} \qquad \mathrm{M}(\Delta(\lambda)) = \sum_{l=1}^{n} \nu_l.$$

On the other hand, equivalence (4.12) yields

$$\begin{aligned} \det \Delta(z) &= \det F(z) \det D(z) \det E(z) \\ &= \det F(z)(z-\lambda)^{\sum_{l=1}^{n} \nu_l} \det E(z). \end{aligned}$$

Since $\det E(\lambda) \neq 0$ and $\det F(\lambda) \neq 0$, the corollary follows. □

Our next step is to make precise what we suggested before, i.e., the equivalence with holomorphic functions which are essentially matrix valued.

Definition 4.17. Let A be an unbounded operator on a Banach space X and Ω an open set in the complex plane. We call a holomorphic matrix function $\Delta : \Omega \to \mathcal{L}(\mathbb{C}^n)$ a *characteristic matrix* for A on Ω if there exist a Banach space Y and holomorphic operator functions $E : \Omega \to \mathcal{L}(\mathbb{C}^n \times Y, X_A)$ and $F : \Omega \to \mathcal{L}(X, \mathbb{C}^n \times Y)$, whose values are bijective operators, such that

$$\text{(4.18)} \qquad \begin{pmatrix} \Delta(z) & 0 \\ 0 & I_Y \end{pmatrix} = F(z)(z\tilde{I} - \tilde{A})E(z), \quad z \in \Omega.$$

Here $\tilde{A} : X_A \to X$ and $\tilde{I} : X_A \to X$ are the bounded operators induced by A and I, where X_A denotes the Banach space $\mathcal{D}(A)$ provided with the graph norm $\|\cdot\|_A = \|\cdot\| + \|A\cdot\|$. The operator function appearing on the left hand side of (4.18) is called the *Y-extension* of Δ.

The next theorem justifies the terminology introduced above.

Theorem 4.18. *Let A be a closed unbounded operator on a Banach space X and let Δ be a characteristic matrix for A on Ω such that* $\det \Delta \not\equiv 0$. *Then*

(i) *The set $\sigma(A) \cap \Omega$ consists of eigenvalues of finite type and*

$$\sigma(A) \cap \Omega = \{z \in \Omega \mid \det \Delta(z) = 0\}.$$

(ii) *For $\lambda \in \sigma(A) \cap \Omega$, the partial multiplicities of λ as an eigenvalue of A are equal to the zero multiplicities of Δ at λ.*

(iii) *For $\lambda \in \sigma(A) \cap \Omega$, the algebraic multiplicity of λ as an eigenvalue of A equals $m = m(\lambda, \Delta)$, the order of λ as a zero of* $\det \Delta$.

(iv) *For $\lambda \in \sigma(A) \cap \Omega$, the ascent of λ as an eigenvalue of A equals $q = q(\lambda, \Delta)$, the order of λ as a pole of Δ^{-1}.*

Proof. Define $\Sigma = \{z \in \Omega \mid \det \Delta(z) \neq 0\}$; then

$$(4.19) \qquad (z\tilde{I} - \tilde{A})^{-1} = E(z) \begin{pmatrix} \Delta(z)^{-1} & 0 \\ 0 & I_X \end{pmatrix} F(z), \quad z \in \Sigma.$$

Since the values of E and F are bijective operators, the function $z \mapsto (z\tilde{I} - \tilde{A})^{-1}\tilde{I}$ is finite meromorphic on Ω, that is, for every $\lambda \in \sigma(A) \cap \Omega$ the Laurent expansion in a neighbourhood of λ has the form

$$(4.20) \qquad (zI - A)^{-1} = (z\tilde{I} - \tilde{A})^{-1}\tilde{I} = \sum_{l=-n}^{\infty} (z-\lambda)^l R_l$$

with $R_{-1}, \ldots, R_{-n}$ operators of finite rank. In particular, R_{-1} has finite rank and (i) follows.

To prove (ii) let

$$x_{1,0}, \ldots, x_{1,\nu_1-1}, \ldots, x_{p,0}, \ldots, x_{p,\nu_p-1}$$

with $\nu_1 \leq \cdots \leq \nu_p$ be a canonical system of Jordan chains for Δ at λ. For $i = 1, \ldots, p$ consider the root function

$$\varphi_i(z) = x_{i,0} + (z-\lambda)x_{i,1} + \cdots + (z-\lambda)^{\nu_i-1}x_{i,\nu_i-1}.$$

We know that

$$(4.21) \qquad \Delta(z)\varphi_i(z) = O\big((z-\lambda)^{\nu_i}\big), \qquad |z-\lambda| \to 0.$$

Put

$$\psi_i(z) = E(z)\begin{pmatrix}\varphi_i(z)\\ 0\end{pmatrix}$$
$$= y_{i,0} + (z-\lambda)y_{i,1} + \cdots + (z-\lambda)^{\nu_i-1}y_{i,\nu_i-1} + O\big((z-\lambda)^{\nu_i}\big)$$

for $|z-\lambda| \to 0$. From (4.21) and the equivalence (4.18), it follows that

$$(zI-A)\psi_i(z) = O\big((z-\lambda)^{\nu_i}\big)$$

for $|z-\lambda| \to 0$ and thus

$$(A-\lambda I)y_{i,0} = 0, \ldots, (A-\lambda I)y_{i,\nu_i-1} = y_{i,\nu_i-2}. \tag{4.22}$$

We shall prove that

$$y_{1,0}, \ldots, y_{1,\nu_1-1}, \ldots, y_{p,0}, \ldots, y_{p,\nu_p-1} \tag{4.23}$$

is a canonical basis of eigenvectors and generalized eigenvectors of A at λ. Note that $N : \mathbb{C}^n \to X$ defined by

$$c \mapsto E(\lambda)\begin{pmatrix}c\\ 0\end{pmatrix}$$

maps $\mathrm{Ker}\,\Delta(\lambda)$ in a one-to-one way onto $\mathcal{N}(\lambda I - A)$. It follows that the vectors $y_{1,0}, \ldots, y_{p,0}$ are linearly independent. But then we can use (4.22) to show that the set of vectors (4.23) is linearly independent. It remains to verify that the vectors (4.23) span the generalized eigenspace of A at λ. This can be done by showing that $M(A;\lambda)$, the algebraic multiplicity of A at λ, is equal to $M(\Delta(\lambda))$, which we shall do next.

First, note that the range of the spectral projection P_λ is a finite dimensional subspace contained in $\mathcal{D}(A)$ with dimension $M(A;\lambda)$. The trace of the spectral projection P_λ is, by definition, equal to the sum of the eigenvalues of P_λ which is equal to the dimension of the range of P_λ. So, $M(A;\lambda)$ equals $\mathrm{tr}\,P_\lambda$, the trace of the spectral projection P_λ. Let

$$D(z) = \mathrm{diag}\big((z-\lambda)^{\nu_1}, \ldots, (z-\lambda)^{\nu_n}\big) \tag{4.24}$$

denote the local Smith form for Δ in a neighbourhood $\mathcal{U} \subset \Omega$ of λ; then there exist holomorphic matrix-valued functions E_1 and F_1 on $\mathcal{U}$ whose values are bijective operators, such that

$$\Delta(z) = F_1(z)D(z)E_1(z), \quad x \in \mathcal{U}. \tag{4.25}$$

Substituting (4.25) into (4.18) we find operator functions E_2 and F_2, holomorphic on $\mathcal{U}$, whose values are bijective operators, such that

$$\begin{pmatrix}D(z) & 0\\ 0 & I_Y\end{pmatrix} = F_2(z)(z\tilde{I} - \tilde{A})E_2(z), \quad z \in \mathcal{U}. \tag{4.26}$$

Let

$$\mathrm{S}\big[L\big](\lambda) = \sum_{l=-n}^{-1} (z-\lambda)^l L_l$$

denote the singular part in the Laurent expansion of a meromorphic operator function L at λ. From (4.26) we derive

$$\begin{aligned}
\operatorname{tr}\mathrm{S}\big[D(z)^{-1}\frac{dD}{dz}(z)\big](\lambda)& \\
&= \operatorname{tr}\mathrm{S}\big[E_2(z)^{-1}L(z)^{-1}F_2(z)^{-1}\frac{dF_2}{dz}(z)L(z)E_2(z)\big](\lambda) \\
&\quad + \operatorname{tr}\mathrm{S}\big[E_2(z)^{-1}L(z)^{-1}\frac{dL}{dz}(z)E_2(z)\big](\lambda) \\
&\quad + \operatorname{tr}\mathrm{S}\big[E_2(z)^{-1}\frac{dE_2}{dz}(z)\big](\lambda),
\end{aligned} \tag{4.27}$$

where $L(z) = (z\tilde{I} - \tilde{A})$. The operator function

$$z \mapsto E_2(z)^{-1}\frac{dE_2}{dz}(z)$$

is holomorphic in a neighbourhood of λ. So the third term on the right hand side vanishes. To analyse the remaining terms we shall use that, with respect to the trace, products commute, i.e.,

$$\operatorname{tr}\mathrm{S}\big[LM\big](\lambda) = \operatorname{tr}\mathrm{S}\big[ML\big](\lambda). \tag{4.28}$$

(See Gohberg and Sigal [90].)

Because of (4.28), the first term on the right hand side of (4.27) equals

$$\operatorname{tr}\mathrm{S}\big[F_2(z)^{-1}\frac{dF_2}{dz}(z)\big](\lambda)$$

and hence vanishes. We conclude that

$$\operatorname{tr}\mathrm{S}\big[D(z)^{-1}\frac{dD}{dz}(z)\big](\lambda) = \operatorname{tr}\mathrm{S}\big[(z\tilde{I} - \tilde{A})^{-1}\tilde{I}\big](\lambda). \tag{4.29}$$

Now we use that D is a diagonal matrix given by (4.24) to calculate that the left hand side of (4.29) equals

$$\sum_{l=1}^{n} \nu_l (z-\lambda)^{-1}.$$

Recalling (4.20) we can now rewrite (4.29) in the form

$$\sum_{l=1}^{n} \nu_l (z-\lambda)^{-1} = \operatorname{tr}\Big(\sum_{l=-q}^{-1} (z-\lambda)^l R_l\Big), \qquad z \in \mathcal{U},$$

and we conclude that necessarily

$$\operatorname{tr} R_{-1} = \sum_{l=1}^{n} \nu_l = M(\Delta(\lambda)). \tag{4.30}$$

Since the spectral projection P_λ equals the residue R_{-1}, this completes the proof of (ii).

To conclude that (iii) holds, it suffices to remark that, for a holomorphic matrix-valued function Δ with $\det \Delta \not\equiv 0$, the algebraic multiplicity $M(\Delta(\lambda))$ equals the multiplicity of λ as a zero of $\det \Delta$ (see Corollary 4.14).

For the proof of (iv) note that the ascent of λ equals the order of λ as a pole of the resolvent $z \mapsto (zI - A)^{-1}$. But the order of a pole is invariant under equivalence and (iv) follows from (4.19). □

Remark **4.19**. Note that in the proof of Theorem 4.18 we computed a canonical bases of eigenvectors and generalized eigenvectors of A at λ.

IV.5 The semigroup action on spectral subspaces for delay equations

In this section we shall repeat the construction of an explicit representation of the resolvent in the case of RFDE (cf. Theorem 3.1), but now in such a way that it becomes clear that Δ is a characteristic matrix for $A^{\odot *}$.

Recall from Chapter III that the retarded functional differential equation

$$\dot{x}(t) = \int_0^h d\zeta(\theta) x(t-\theta), \qquad t \geq 0, \quad x_0 = \varphi, \tag{5.1}$$

with $\varphi \in X = C([-h,0],\mathbb{C}^n)$, can be realised as an abstract Cauchy problem on the space $X^{\odot *} = \mathbb{C}^n \times L^\infty([-h,0],\mathbb{C}^n)$,

$$\frac{du}{dt}(t) = A^{\odot *}u(t) = A_0^{\odot *}u(t) + Bu(t) \tag{5.2}$$

with $u(0) = u_0 \in X^{\odot *}$ and where

$$\mathcal{D}(A_0^{\odot *}) = \{(\alpha,\varphi) \mid \varphi \in \operatorname{Lip}(\alpha)\}, \quad A_0^{\odot *}(\alpha,\varphi) = (0,\dot{\varphi}),$$

and

$$\mathcal{D}(B) = \{(\alpha,\varphi) \mid \varphi \in C(\alpha)\}, \quad B(\alpha,\varphi) = (\langle \zeta,\varphi\rangle_n, 0).$$

Note that the duality construction has provided us in a natural manner with a product of $\mathbb{C}^n$ and an infinite dimensional space. We shall exploit this fact in our construction of E and F (cf. Definition 4.15).

Set $L^\infty = L^\infty\big([-h,0],\mathbb{C}^n\big)$ and define $D : L^\infty \to L^\infty$ to be the unbounded operator with domain $\mathcal{D}(D) = \text{Lip}$, the set of equivalence classes containing a Lipschitz continuous function, and action

$$D\varphi = \dot{\varphi}.$$

For every $z \in \mathbb{C}$ we define a pseudo-inverse

$$\begin{aligned}\big(Ps(zI - D)^{-1}\varphi\big)(\tau) &= -\int_0^\tau e^{(\tau-\sigma)z}\varphi(\sigma)d\sigma \\ &= -\exp_z * \varphi(\tau),\end{aligned}$$

where $\exp_z(\tau) = e^{z\tau}$. Note that indeed $(zI - D)Ps(zI - D)^{-1} = I$, but that

$$\big(Ps(zI - D)^{-1}(zI - D)\varphi\big)(\tau) = \varphi(\tau) - \exp_z(\tau)\varphi(0) \quad \text{for } \varphi \in \mathcal{D}(D).$$

A matrix representation of the unperturbed operator $A_0^{\odot *}$ is given by

$$\mathcal{D}(A_0^{\odot *}) = \Big\{\begin{pmatrix}\alpha\\ \varphi\end{pmatrix} \mid \varphi \in \mathcal{D}(D), \alpha = \varphi(0)\Big\}, \qquad A_0^{\odot *} = \begin{pmatrix} 0 & 0 \\ 0 & D\end{pmatrix}.$$

A straightforward calculation yields

$$(zI - A_0^{\odot *})^{-1} = \begin{pmatrix} z^{-1} & 0 \\ z^{-1}\exp_z & -(\exp_z * \cdot)\end{pmatrix}. \tag{5.3}$$

In the same spirit, the perturbation is given by

$$B = \begin{pmatrix} 0 & \zeta \\ 0 & 0\end{pmatrix}, \tag{5.4}$$

where ζ denotes the functional $\varphi \mapsto \langle \zeta, \varphi\rangle$. Using the above representations for the operators we find the following theorem.

Theorem 5.1. *Let, as before, the matrix-valued function $\Delta : \mathbb{C} \to \mathcal{L}(\mathbb{C}^n)$ be defined by*

$$\Delta(z) = zI - \int_0^h e^{-z\theta} d\zeta(\theta); \tag{5.5}$$

then there exist holomorphic operator-valued functions E and F whose values are bijective operators such that

$$\begin{pmatrix} \Delta(z) & 0 \\ 0 & I\end{pmatrix} = F(z)(zI - A_0^{\odot *} - B)E(z), \tag{5.6}$$

where $E : \mathbb{C} \to \mathcal{L}(X^{\odot *}, X^{\odot *}_{A_0^{\odot *}+B})$ is defined by

$$E(z) = \begin{pmatrix} I & 0 \\ \exp_z & -(\exp_z * \cdot) \end{pmatrix},$$

and $F : \mathbb{C} \to \mathcal{L}(X^{\odot *}, X^{\odot *})$ *by*

$$F(z) = \begin{pmatrix} I & -\langle \zeta, (\exp_z * \cdot) \rangle \\ 0 & I \end{pmatrix}.$$

Proof. For the perturbed operator we find the identity

$$\left(zI - A^{\odot *}\right)^{-1} = \left(zI - A_0^{\odot *}\right)^{-1} \left(I - B\left(zI - A_0^{\odot *}\right)^{-1}\right)^{-1}.$$

Combining (5.3) and (5.4) we can write

$$I - B\left(zI - A_0^{\odot *}\right)^{-1} = F(z)^{-1} \begin{pmatrix} z^{-1}\Delta(z) & 0 \\ 0 & I \end{pmatrix}$$

and

$$\left(zI - A_0^{\odot *}\right)^{-1} = E(z) \begin{pmatrix} z^{-1} & 0 \\ 0 & I \end{pmatrix},$$

where

$$F(z) = \begin{pmatrix} I & -\langle \zeta, (\exp_z * \cdot) \rangle \\ 0 & I \end{pmatrix}$$

and

$$E(z) = \begin{pmatrix} I & 0 \\ \exp_z & -(\exp_z * \cdot) \end{pmatrix}.$$

This proves the equivalence (5.6). □

Corollary 5.2. *Δ is a characteristic matrix for A on $\mathbb{C}$.*

Exercise 5.3. Prove the above corollary.
Hint: Take $Y = X = C([-h, 0], \mathbb{C}^n)$ and rewrite (5.6) in the form

$$\begin{pmatrix} \Delta(z) & 0 \\ 0 & I \end{pmatrix} = F(z) j (z\widetilde{I} - \widetilde{A}) j^{-1} E(z), \tag{5.7}$$

where now the domain of $E(z)$ and the range of $F(z)$ are taken to be the subspace $\mathbb{C}^n \times Y$ of $X^{\odot *}$ and where j denotes the natural embedding of X into $X^{\odot *}$ (see Corollary 3.16 of Appendix II).

Corollary 5.4. *The resolvent $(zI - A^{\odot *})^{-1}$ has the following representation:*

$$(zI - A^{\odot *})^{-1} \begin{pmatrix} c \\ \varphi \end{pmatrix} = \begin{pmatrix} \psi(0; c, \varphi, z) \\ \psi(\,\cdot\,; c, \varphi, z) \end{pmatrix}, \tag{5.8}$$

where

$$\begin{aligned}\psi(\theta; c, \varphi, z) = e^{z\theta}\{\Delta(z)^{-1}[c + \int_0^h d\zeta(\tau) \int_0^\tau e^{-z\sigma}\varphi(\sigma - \tau)\, d\sigma] \\ + \int_\theta^0 e^{-z\sigma}\varphi(\sigma)\, d\sigma\}, \qquad -h \le \theta \le 0.\end{aligned} \tag{5.9}$$

Theorem 5.5. *The spectrum of the generator* $A_0^{\odot *} + B$ *consists of eigenvalues of finite type only,*

$$\sigma(A_0^{\odot *} + B) = \{\lambda \mid \det \Delta(\lambda) = 0\}.$$

For $\lambda \in \sigma(A_0^{\odot *} + B)$, *the algebraic multiplicity of the eigenvalue* λ *equals the order of* λ *as a zero of* $\det \Delta$, *the partial multiplicities of the eigenvalue* λ *are equal to the zero multiplicities of* λ *as a characteristic value of* Δ, *and the largest partial multiplicity (i.e., the ascent) of* λ *equals the order of* λ *as a pole of* Δ^{-1}. *Furthermore, a canonical basis of eigenvectors and generalized eigenvectors for* $A_0^{\odot *} + B$ *at* λ *may be obtained in the following way: If*

$$\{(\gamma_{i,0}, \dots, \gamma_{i,k_i-1}) \mid i = 1, \dots, p\}$$

is a canonical system of Jordan chains for Δ *at* $\lambda \in \Omega$, *then*

$$\{(\gamma_{i,0}, \chi_{i,0}), \dots, (\gamma_{i,k_i-1}, \chi_{i,k_i-1}) \mid i = 1, \dots, p\},$$

where

$$\chi_{i,\nu}(\theta) = e^{\lambda\theta} \sum_{l=0}^{\nu} \gamma_{i,\nu-l} \frac{\theta^l}{l!} \tag{5.10}$$

is a canonical basis for $A_0^{\odot *} + B$ *at* λ.

Proof. We first show that $\det \Delta \not\equiv 0$. From the representation (5.5) for Δ it follows that it suffices to prove that

$$|z^{-1} \int_0^h e^{-z\theta} d\zeta(\theta)| \to 0$$

as $\operatorname{Re} z \to \infty$. But this is obvious since ζ is of bounded variation.

Next we prove the representation for the canonical basis for $A_0^{\odot *} + B$ at λ. Recalling the proof of Theorem 4.18(ii), we set out to compute Jordan chains for $zI - A_0^{\odot *} - B$ from Jordan chains for Δ. Let $(\gamma_0, \dots, \gamma_{k-1})$ be a Jordan chain for Δ at λ of length k. Since

$$\begin{aligned} E(z) &\begin{pmatrix} \gamma_0 + \gamma_1(z-\lambda) + \cdots + \gamma_{k-1}(z-\lambda)^{k-1} \\ 0 \end{pmatrix} \\ &= \begin{pmatrix} \gamma_0 + \gamma_1(z-\lambda) + \cdots + \gamma_{k-1}(z-\lambda)^{k-1} \\ (\gamma_0 + \gamma_1(z-\lambda) + \cdots + \gamma_{k-1}(z-\lambda)^{k-1}) \exp_z \end{pmatrix} \end{aligned}$$

and

$$\exp_z(\theta) = e^{\lambda\theta}\big[1 + \theta(z-\lambda) + \cdots + \frac{\theta^{k-1}}{(k-1)!}(z-\lambda)^{k-1} + O\big((z-\lambda)^k\big)\big],$$

we define

$$\chi_i(\theta) = e^{\lambda\theta}\sum_{l=0}^{i}\gamma_{i-l}\frac{\theta^l}{l!}$$

and elements (γ_i, χ_i) of $X^{\odot *}$ belonging to $\mathcal{D}(A_0^{\odot *})$, and write

$$E(z)\begin{pmatrix}\sum_{i=0}^{k-1}\gamma_i(z-\lambda)^i \\ 0\end{pmatrix} = \sum_{i=0}^{k-1}(\gamma_i, \chi_i)(z-\lambda)^i + O\big((z-\lambda)^k\big).$$

The equivalence relation (5.6) now requires that

$$(zI - A_0^{\odot *} - B)\sum_{i=0}^{k-1}(\gamma_i, \chi_i)(z-\lambda)^i = O\big((z-\lambda)^k\big)$$

for $|z-\lambda| \to 0$. So the Jordan chain for $(zI - A_0^{\odot *} - B)$ at λ becomes $\big\{(\gamma_0, \chi_0), \ldots, (\gamma_{k-1}, \chi_{k-1})\big\}$. All the other arguments in the proof of part (ii) of Theorem 4.18 carry over and show that this procedure transforms a canonical system of Jordan chains for Δ into a canonical basis for $A_0^{\odot *} + B$ at λ. □

Since the spectral data of A and $A_0^{\odot *} + B$ are the same, the eigenvectors and generalized eigenvectors of A at λ can be computed from the eigenvectors and generalized eigenvectors of $A_0^{\odot *} + B$ at λ.

Exercise 5.6. Rederive Theorem 3.1 by using the embedding of X into $X^{\odot *}$.

Exercise 5.7. Give a basis of eigenvectors and generalized eigenvectors of A at λ.

Exercise 5.8. Show that

$$(zI - A^{\odot *})^{-1} r_i^{\odot *} = \exp_z \Delta(z)^{-1} e_i,$$

where e_i is the ith unit vector in $\mathbb{C}^n$.

Theorem 5.9. *The spectrum of the generator A^* consists of eigenvalues of finite type only,*

$$\sigma(A^*) = \{\lambda \mid \det \Delta(\lambda) = 0\}.$$

For $\lambda \in \sigma(A^*)$, the algebraic multiplicity of the eigenvalue λ equals the order of λ as a zero of $\det \Delta$, the partial multiplicities of the eigenvalue λ are equal to the zero multiplicities of λ as a characteristic value of Δ, *and the largest partial multiplicity (ascent) of λ equals the order of λ as a pole of Δ^{-1}. Furthermore, a canonical basis of eigenvectors and generalized eigenvectors for A^* at λ is obtained in the following way: If*

$$\{(\beta_{i,0},\dots,\beta_{i,k_i-1}) \mid i=1,\dots,p\}$$

is a canonical system of Jordan chains for Δ^T at $\lambda \in \Omega$, then

$$\{\xi_{i,0}\dots,\xi_{i,k_i-1} \mid i=1,\dots,p\},$$

where for $\theta > 0$

$$\xi_{i,\nu}(\theta) = \beta_{i,\nu} + \sum_{l=0}^{\nu} \int_\theta^0 \Big(\int_\sigma^h e^{\lambda(\sigma-\tau)} \frac{(\sigma-\tau)^l}{l!}\, d\zeta^T(\tau)\Big)\, d\sigma\, \beta_{i,\nu-l} \tag{5.11}$$

is a canonical basis for A^* at λ.

Proof. Let $(\beta_0,\dots,\beta_{k-1})$ be a Jordan chain for Δ^T at λ of length k and let $\alpha(z) = \beta_0 + \beta_1(z-\lambda) + \dots + \beta_{k-1}(z-\lambda)^{k-1}$ denote the corresponding root function. From the properties of the adjoint operation and equivalence (5.6) we have that

$$(zI - A^*) \int_\theta^0 \Big(\int_\sigma^h e^{z(\sigma-\tau)}\, d\zeta^T(\tau)\Big)\, d\sigma\, \alpha(z) = O\big((z-\lambda)^k\big) \tag{5.12}$$

for $|z-\lambda| \to 0$. Therefore Lemma 4.7 implies that we have to expand

$$z \mapsto e^{(\sigma-\tau)z}\alpha(z)$$

up to order k in a neighbourhood of λ. The formula for the ξ's now follows from a straightforward computation. □

Exercise 5.10. Compute $F(z)^*$ and verify (5.12).

Exercise 5.11. This is a sophisticated generalization of Exercise 3.12. Set

$$P_j = P_j(z) = \frac{\Delta^{(j-1)}(z)}{(j-1)!}$$

and define $A_k = A_k(z)$ to be the $(nk)\times(nk)$-matrix

$$A_k = \begin{pmatrix} P_1 & 0 & \cdots & 0 \\ P_2 & P_1 & \cdots & 0 \\ \vdots & & \ddots & \vdots \\ P_k & P_{k-1} & \cdots & P_1 \end{pmatrix}, \qquad k = 1,2,\dots. \tag{5.13}$$

Prove that $(\chi_0,\dots,\chi_{k-1})$ is a Jordan chain of length k for Δ at λ if and only if

$$A_k(\lambda)\begin{pmatrix} \chi_0 \\ \vdots \\ \chi_{k-1} \end{pmatrix} = 0.$$

Before formulating the next corollary, which rederives the result of Exercise 3.12 and Theorem 2.5(vi), we repeat that we restrict attention to real-valued kernels ζ.

Corollary 5.12. *Let λ be a simple zero of* $\det \Delta$. *The Jordan chain for Δ at λ has rank one and is given by p_λ with $\Delta(\lambda)p_\lambda = 0$. The corresponding eigenvector of A is given by*

$$\phi_\lambda(\theta) = p_\lambda e^{\lambda\theta}, \qquad -h \le \theta \le 0. \tag{5.14}$$

The Jordan chain for Δ^T at λ has rank one and is given by q_λ with $\Delta(\lambda)^T q_\lambda = 0$. The corresponding eigenfunction of A^* is given by

$$\psi_\lambda(\tau) = q_\lambda + \int_0^\tau \Big(\int_\sigma^h e^{\lambda_1(\sigma-s)} d\zeta^T(s)\Big)\, d\sigma\, q_\lambda, \qquad 0 < \tau \le h. \tag{5.15}$$

Furthermore,

$$\langle \psi_\lambda, \phi_\lambda \rangle = q_\lambda \Delta'(\lambda) p_\lambda \ne 0, \tag{5.16}$$

where Δ' denotes the derivative of Δ.

Proof. The first statements and representation (5.14) and (5.15) immediately follow from Theorem 5.5. To prove (5.16) we observe that

$$\begin{aligned}
\langle \psi_\lambda, \phi_\lambda \rangle &= \int_0^h d\psi_\lambda(\tau)\phi_\lambda(-\tau) \\
&= q_\lambda p_\lambda + q_\lambda \int_0^h \Big(\int_\tau^h e^{-\lambda s} d\zeta(s)\Big)\, d\tau\, p_\lambda \\
&= q_\lambda \big(I + \int_0^h s e^{-\lambda s} d\zeta(s)\big) p_\lambda,
\end{aligned}$$

where we have used Fubini's theorem to reverse the order of integration. This proves the first equality in (5.16) and it only remains to prove that this expression cannot be zero. Since the Jordan chain for Δ at λ has rank one, it follows from Exercise 5.12 that

$$\Delta(\lambda)\gamma + \Delta'(\lambda)p_\lambda \ne 0, \qquad \text{for all } \gamma \in \mathbb{C}^n.$$

So $\Delta'(\lambda)p_\lambda \ne 0$ and

$$\Delta'(\lambda)p_\lambda \notin \mathcal{R}\big(\Delta(\lambda)\big) = \mathcal{N}\big(\Delta(\lambda)^T\big)^\perp.$$

But $\mathcal{N}\big(\Delta(\lambda)^T\big)$ is spanned by q_λ and so necessarily

$$q_\lambda \Delta'(\lambda) p_\lambda \ne 0.$$

□

Next we illustrate how to compute the Jordan chains in a simple example. See also Exercise 3.15.

Example 5.13. Consider the differential-difference equation

$$\dot{x}(t) = x(t) - x(t-1), \quad t \geq 0, \tag{5.17}$$

on $\mathcal{C} = \mathrm{C}[-1,0]$. The associated infinitesimal generator A is given by

$$\mathcal{D}(A) = \{\varphi \in \mathcal{C} \mid \varphi \in C^1[-1,0],\ \dot{\varphi}(0) = \varphi(0) - \varphi(-1)\}, \quad A\varphi = \dot{\varphi}.$$

According to Theorem 5.1 the scalar function

$$\Delta(z) = z - 1 + e^{-z}$$

is a characteristic matrix for A. The set of zeros of Δ is infinite and $\operatorname{Re}\lambda_j \to -\infty$ if λ_j denotes the sequence of zeros of Δ. Let us look for the finitely many zeros in the right half-plane $\operatorname{Re} z \geq 0$. This yields $\lambda = 0$. Next we can find a canonical basis of eigenvectors and generalized eigenvectors of A at $\lambda = 0$. To do this we first calculate the Jordan chains for Δ at $\lambda = 0$.

The set $(x_0, \ldots, x_{k-1})$ is a Jordan chain of Δ at λ if and only if $x_0 \neq 0$ and

$$\Delta(z)(x_0 + x_1 z + \cdots + x_{k-1} z^{k-1}) = O(z^k).$$

Since $\Delta(0) = 0$ and $\Delta'(0) = 0$, the expression

$$(z^2/2! - z^3/3! + \cdots)x_0 + \cdots + (z^{k+1}/2! - z^{k+2}/3! + \cdots)x_{k-1}$$

must be of order $O(z^k)$. So $x_0 = x_1 = \cdots = x_{k-3} = 0$, and $x_{k-2}, x_{k-1} \in \mathbb{C}$ are arbitrary. But x_0 must be different from zero. So the canonical system of Jordan chains of Δ at λ consists of one chain of length 2. For example, we may take the chain $(1,0)$. It follows (apply Theorem 5.3) that

$$\mathcal{N}(A^2) = \operatorname{span}\{\varphi_1, \varphi_2\},$$

where

$$\varphi_1(\theta) = 1, \quad \varphi_2(\theta) = \theta, \quad -1 \leq \theta \leq 0.$$

Next we determine the action of the semigroup on the generalized eigenspace using Theorem 2.15. Let $\Phi = (\varphi_1, \varphi_2)$. Then

$$A\Phi = \Phi \begin{pmatrix} 0 & 1 \\ 0 & 0 \end{pmatrix}.$$

Therefore

$$M = \begin{pmatrix} 0 & 1 \\ 0 & 0 \end{pmatrix}, \qquad e^{tM} = \sum_{k=0}^{\infty} \frac{t^k}{k!} M^k = \begin{pmatrix} 1 & t \\ 0 & 1 \end{pmatrix}.$$

Note that indeed

$$\Phi(\theta) = (1,\theta) = (1,0)\begin{pmatrix} 1 & \theta \\ 0 & 1 \end{pmatrix} = \Phi(0)e^{\theta M}$$

and that $x(t;\varphi_1) = 1$ and $x(t;\varphi_2) = t$ from which it follows that

$$\begin{aligned} T(t)\Phi &= (\varphi_1, \varphi_2 + t\varphi_1) \\ &= (\varphi_1,\varphi_2)\begin{pmatrix} 1 & t \\ 0 & 1 \end{pmatrix} \\ &= \Phi e^{tM}. \end{aligned}$$

We now present a series of exercises which are a continuation of earlier exercises on age-dependent population models, in particular III.4.11 and III.4.12, to which we refer for notation, etc. The aim is to show that, for these problems as well, all spectral information is contained in a characteristic matrix.

Exercise 5.14. Let now

$$\Delta(z) = I - \int_0^h \beta(s)e^{-zs}\,ds. \tag{5.18}$$

(i) Derive the following explicit representation for the resolvent:

$$\begin{aligned} ((zI-A)^{-1}\varphi)(a) &= e^{-za}\Delta(z)^{-1}\int_0^h \beta(s)\int_0^s e^{-z(s-\tau)}\varphi(\tau)\,d\tau ds \\ &\quad + \int_0^a e^{-z(a-s)}\varphi(s)\,ds, \qquad 0 \le a \le h. \end{aligned} \tag{5.19}$$

(ii) Show that

$$\sigma(A) = \sigma_p(A) = \{\lambda \in \mathbb{C} \mid \det \Delta(z) = 0\}.$$

Hint: See Theorem 3.1 and Corollary 3.3 for inspiration.

Exercise 5.15. Derive the representation

$$((zI - A_0^{\odot *})^{-1}\varphi)(a) = \int_0^a e^{-z(a-s)}\varphi(s)\,ds, \qquad 0 \le a \le h, \tag{5.20}$$

and conclude that $\rho(A_0^{\odot *}) = \mathbb{C}$ and that $z \mapsto (zI - A_0^{\odot *})^{-1}$ is an entire function.

Exercise 5.16. Let $j : X \to X^{\odot *}$ denote the embedding

$$(j\varphi)(a) = \int_0^a \varphi(\alpha)\,d\alpha. \tag{5.21}$$

Show that

$$(j^{-1}(zI - A_0^{\odot *})^{-1}\varphi)(a) = \varphi(a) - z\int_0^a e^{-z(a-s)}\varphi(s)\,ds$$

and observe that if φ is constant, $\varphi(a) = c$, then the right hand side reduces to $e^{-za}c$.

As in the proof of Theorem 5.1 we shall use the identity

$$(zI - A^{\odot *})^{-1} = (zI - A_0^{\odot *})^{-1}\big(I - Bj^{-1}(zI - A_0^{\odot *})^{-1}\big)^{-1}. \tag{5.22}$$

So we have to compute the inverse of $I - Bj^{-1}(zI - A_0^{\odot *})^{-1}$.

Exercise 5.17. Let

$$\varphi = (I - Bj^{-1}(zI - A_0^{\odot *})^{-1})^{-1}\psi.$$

Show that $\varphi = \psi + c$ where c is the vector given by

$$\begin{aligned} c &= \Delta(z)^{-1}\{\int_0^h \beta(\alpha)\psi(\alpha)\,d\alpha - z\int_0^h \beta(\alpha)\int_0^\alpha e^{-z(\alpha-s)}\psi(s)\,dsd\alpha\} \\ &= \Delta(z)^{-1}\{\int_0^h \beta(\alpha)(\psi(0+)e^{-z\alpha} + \int_0^\alpha e^{-z(\alpha-s)}\,d\psi(s))\,d\alpha\}. \end{aligned}$$

In order to formulate the equivalence, we single out the jump at zero by noting that

$$\begin{aligned} \mathrm{NBV}\big([0,h),\mathbb{C}^n\big) &= \operatorname{span}\{\delta_j \mid j = 1,\dots,n\} \oplus \mathrm{NBV}\big((0,h),\mathbb{C}^n\big) \\ &\cong \mathbb{C}^n \times \mathrm{NBV}\big((0,h),\mathbb{C}^n\big). \end{aligned}$$

Here $\mathrm{NBV}\big((0,h),\mathbb{C}^n\big)$ denotes the space of normalized functions of bounded variation that are continuous both at 0 and h. The isometric isomorphism is given by

$$\psi \mapsto \big(\psi(0+), \psi - \psi(0+)\big)$$

and we shall use coordinates

$$\begin{pmatrix} \psi_0 \\ \psi \end{pmatrix} \in \mathbb{C}^n \times \mathrm{NBV}\big((0,h),\mathbb{C}^n\big)$$

to represent the element

$$\sum_{j=1}^n \psi_{0j}\delta_j + \psi_1$$

of $\mathrm{NBV}\big([0,h),\mathbb{C}^n\big)$.

Exercise 5.18. Reformulate the statement of Exercise 5.17 in terms of the coordinates. More precisely, show that if

$$\begin{pmatrix} \varphi_0 \\ \varphi \end{pmatrix} = (I - Bj^{-1}(zI - A_0^{\odot *})^{-1})^{-1}\begin{pmatrix} \psi_0 \\ \psi \end{pmatrix},$$

then

$$\varphi_0 = \Delta(z)^{-1}\{\psi_0 + \int_0^h \beta(\alpha) \int_0^\alpha e^{-z(\alpha-s)}\, d\psi_1(s) d\alpha\}, \qquad \varphi = \psi.$$

Exercise 5.19. Show that

$$(I - Bj^{-1}(zI - A_0^{\odot *})^{-1})^{-1} = \begin{pmatrix} \Delta(z)^{-1} & 0 \\ 0 & I \end{pmatrix} F(z),$$

where

$$F(z) = \begin{pmatrix} I & \int_0^h \beta(\alpha) \int_0^\alpha e^{-z(\alpha-s)}\, d[\cdot]ds \\ 0 & I \end{pmatrix}. \tag{5.23}$$

Show that $F : \mathbb{C} \to \mathcal{L}(X^{\odot *})$ is a holomorphic operator valued function whose values are bijective operators. Give a representation for $F(z)^{-1}$.

Exercise 5.20. Define $E : \mathbb{C} \to \mathcal{L}(X^{\odot *}, X^{\odot *}_{A^{\odot *}})$ by

$$E(z) = (zI - A_0^{\odot *})^{-1}. \tag{5.24}$$

Verify that E is represented by the matrix

$$\begin{pmatrix} 0 & 0 \\ z^{-1}(1 - \exp_{-z}) & \exp_{-z} * \cdot \end{pmatrix},$$

where $\exp_{-z}(a) = e^{-za}$ for $0 \le a \le h$. Show that E is a holomorphic operator valued function whose values are bijective operators. Give a representation for $E(z)^{-1}$ (note that many are possible).

Exercise 5.21. Prove that

$$\begin{pmatrix} \Delta(z) & 0 \\ 0 & I \end{pmatrix} = F(z)(zI - A^{\odot *})E(z).$$

Exercise 5.22. Prove that Δ is a characteristic matrix for A on $\mathbb{C}$.
Hint: See Exercise 5.3.

Exercise 5.23. Prove a version of Theorem 5.5 for age-dependent population models.
Hint: See Theorem 4.18.

Exercise 5.24. Prove the analogue of Corollary 5.12 for age-dependent population models.

Exercise 5.25. Formulate results about the asymptotic behaviour of the semigroup $T(t)$ corresponding to age-dependent population growth. Deduce the proofs from the abstract theory in Section 2.
Hint: $T_0(t)$ is eventually compact.

IV.6 Comments

The lecture notes [202] edited by Nagel provide a rich source of results on linear semigroups in general and spectral theory in particular.

Rather than by duality (see Hale [102, 104]) we have chosen to compute the spectral projections directly from the Dunford integral using residue calculus. Together with the notions of Jordan chain and equivalence (see Bart, Gohberg and Kaashoek [15]) this yields a powerful method.

Jordan chains have also been used in bifurcation theory (see Rabier [244]).

The definition of characteristic matrix comes from Kaashoek and Verduyn Lunel [142]. There exists a large class of unbounded operators for which one can construct a characteristic matrix. In [142] an abstract scheme is presented and several examples are worked out explicitly. For further references to the literature we also refer to [142].

Chapter V

Completeness or small solutions?

V.1 Introduction

In Section I.5 the large time behaviour of solutions of a linear retarded functional differential equation

$$(1.1)\qquad \dot{x}(t) = \int_0^h d\zeta(\theta)x(t-\theta), \qquad x_0 = \varphi$$

was studied. Using the renewal equation and the inverse Laplace transform, we found the following representation [see (I.5.3)] for the solution of (1.1):

$$(1.2)\quad x(t) = \sum_{j=1}^{l} p_j(t)e^{\lambda_j t} + \frac{1}{2\pi i}\int_{L(\gamma)} e^{zt}\Delta(z)^{-1}\big(f(0) + \int_0^h e^{-z\theta}\,df(\theta)\big)\,dz,$$

where

$$f(t) = \varphi(0) + \int_0^t [\zeta(t+\sigma) - \zeta(\sigma)]\varphi(-\sigma)\,d\sigma.$$

In this chapter we are interested in the question of whether we can obtain a convergent series by letting $\gamma \to -\infty$ in (1.2). In other words, what is, for a given function f, the behaviour of

$$(1.3)\qquad \lim_{\gamma\to-\infty} \frac{1}{2\pi i}\int_{L(\gamma)} e^{zt}\Delta(z)^{-1}\big(f(0) + \int_0^h e^{-z\theta}\,df(\theta)\big)\,dz$$

for $t > 0$. To analyse this limit we shall need more refined techniques than developed in Chapter I. In this chapter we shall present a theory that can be used not only to analyse the limit in (1.3) but also more general problems. Note that this approach, which starts from representation (1.2), only yields a series expansion for the solution for $t > 0$. In order to extend the series expansion problem such that it makes sense for initial data as well, we introduce a special class of solutions. Solutions of RFDE (1.1) of the form

$$x(t) = p(t)e^{\lambda t}$$

where p is a polynomial and $\lambda \in \mathbb{C}$, exist for all time and are called *elementary solutions.*

In Chapter IV we have seen that the eigenvectors and generalized eigenvectors of the generator associated with RFDE (1.1) are precisely the initial data corresponding to the elementary solutions. In particular, the question of whether the solution or an initial condition of RFDE (1.1) can be expanded into elementary solutions can be rephrased as a question concerning the convergence of the spectral projections of the generator.

In this chapter, we shall analyse when

$$T(t)\varphi = \sum_{\lambda \in \sigma(A)} T(t)P_\lambda \varphi, \tag{1.4}$$

where the convergence is in the state space $\mathcal{C}$. The initial question, whether every solution of RFDE (1.1) has a series expansion into elementary solutions, is contained in this problem. Indeed, every solution of RFDE (1.1) has a convergent series expansion if and only if (1.4) holds for every $t > h$. In order to analyse series of spectral projections of the generator A corresponding to the RFDE (1.1), we first recall some basic facts.

The spectrum of A consists of eigenvalues of finite type which are the poles of the resolvent $(zI - A)^{-1}$. Furthermore, for $\lambda \in \sigma(A)$ the space $\mathcal{C}$ decomposes into

$$\mathcal{C} = \mathcal{R}\big((\lambda I - A)^{m_\lambda}\big) \oplus \mathcal{N}\big((\lambda I - A)^{m_\lambda}\big)$$

with corresponding Riesz projection P_λ onto $\mathcal{N}\big((\lambda I - A)^{m_\lambda}\big)$ given by

$$P_\lambda = \frac{1}{2\pi i}\int_{\Gamma_\lambda} (zI - A)^{-1}\, dz, \tag{1.5}$$

where Γ_λ is a simple closed rectifiable curve enclosing λ, but no other points in the spectrum of A. The range of P_λ is denoted by $\mathcal{M}_\lambda$ and is called the generalized eigenspace at λ. The linear space spanned by all generalized eigenspaces $\mathcal{M}_\lambda$, $\lambda \in \sigma(A)$, will be denoted by $\mathcal{M}$ and is called the *generalized eigenspace* of A. The system of eigenvectors and generalized eigenvectors is called *complete* if $\overline{\mathcal{M}} = \mathcal{C}$.

To analyse the behaviour of sums of spectral projections, we shall use the Riesz projection and the Cauchy theorem on residues. In order to do so, we need good estimates for the resolvent of A near infinity. From the equivalence relation IV.5.6 or Theorem IV.3.1 we find the following representation for the resolvent of A:

$$(zI - A)^{-1}\varphi = \frac{1}{\det \Delta(z)} P(z)\varphi, \tag{1.6}$$

where

$$(P(z)\varphi)(\theta) = e^{z\theta}\{\operatorname{adj}\Delta(z)[\varphi(0) + \int_0^h d\zeta(\tau)\int_0^\tau e^{-z\sigma}\varphi(\sigma-\tau)\,d\sigma] \\ + \det\Delta(z)\int_\theta^0 e^{-z\sigma}\varphi(\sigma)\,d\sigma\}. \tag{1.7}$$

In the next section we present material from complex analysis which we need in order to estimate $\|(zI - A)^{-1}\varphi\|$ near infinity.

V.2 Exponential type calculus

Let X be a complex Banach space. An entire function $F : \mathbb{C} \to X$ is of *order* ρ if and only if

$$\limsup_{r\to\infty} \frac{\log\log M(r)}{\log r} = \rho,$$

where

$$M(r) = \max_{0\le\theta\le 2\pi} \{\|F(re^{i\theta})\|\}.$$

An entire function of order at most 1 is of *exponential type* if and only if

$$\limsup_{r\to\infty} \frac{\log M(r)}{r} = \mathrm{E}(F),$$

where $0 \le \mathrm{E}(F) < \infty$. In that case, $\mathrm{E}(F)$ is called the exponential type of F.

We start with a special case of the Paley-Wiener theorem [17, 6.9.1].

Theorem 2.1. *Let $F : \mathbb{C} \to \mathbb{C}$ be an entire function which is uniformly bounded in the closed right half-plane* $\operatorname{Re} z \ge 0$. *Then F is of exponential type α and L^2-integrable along the imaginary axis if and only if*

$$F(z) = \int_0^\alpha e^{-z\tau}\varphi(\tau)\,d\tau, \tag{2.1}$$

where $\varphi \in L^2([0,\alpha],\mathbb{C})$ and φ does not vanish a.e. in any neighbourhood of α.

For a large class of entire functions of exponential type, one can describe quite precisely the behaviour at infinity. The first result is known as the Ahlfors-Heins theorem. (See [17, Thm. 7.26].)

Theorem 2.2. *Let $F : \mathbb{C} \to \mathbb{C}$ be an entire function of exponential type. If*

$$\int_1^\infty \frac{\log|F(\pm iy)|}{y^2}dy \quad \text{exist,} \tag{2.2}$$

then for almost all $\theta \in (\frac{\pi}{2}, \frac{3\pi}{2})$

$$\lim_{r\to\infty} \frac{\log|F(re^{i\theta})|}{r} = -\mathrm{E}(F)\cos\theta. \tag{2.3}$$

For every fixed $\theta_0 \in (0, \frac{\pi}{2})$, there exists a sequence r_j, such that $r_j \to \infty$ and

$$\lim_{j\to\infty} \frac{\log|F(r_j e^{i\theta})|}{r_j} = -\mathrm{E}(F)\cos\theta,$$

uniformly in $\frac{\pi}{2} + \theta_0 \le \theta \le \frac{3\pi}{2} - \theta_0$.

As a corollary, we have a class of entire functions of exponential type for which there is a simple calculus of exponential types.

Corollary 2.3. *Let $F_1 : \mathbb{C} \to \mathbb{C}$ and $F_2 : \mathbb{C} \to \mathbb{C}$ be entire functions of exponential type. If F_1 and F_2 are polynomially bounded in the right half-plane $\operatorname{Re} z \ge 0$, then*

$$\mathrm{E}(F_1F_2) = \mathrm{E}(F_1) + \mathrm{E}(F_2). \tag{2.4}$$

If, in addition, F_1/F_2 is entire, then

$$\mathrm{E}(F_1/F_2) = \mathrm{E}(F_1) - \mathrm{E}(F_2). \tag{2.5}$$

From Theorem IV.5.1, it follows that the characteristic matrix corresponding to the RFDE (1.1) is given by

$$\Delta(z) = zI - \int_0^h e^{-z\tau}d\zeta(\tau). \tag{2.6}$$

From the convolution property of the Laplace transform, it follows that $\det\Delta$ and the entries of $\operatorname{adj}\Delta$ are entire functions of the form

$$a_0(z)z^l + a_1(z)z^{l-1} + \cdots + a_l(z), \tag{2.7}$$

where, for $j = 0, \ldots, l$, the coefficient a_j is a finite Laplace transform of a function of bounded variation, i.e.,

$$a_j(z) = \int_0^{\alpha_j} e^{-z\tau}d\eta_j(\tau). \tag{2.8}$$

In particular, $\det\Delta$ has representation (2.7)–(2.8) with $a_0 = 1$, $l = n$ and $\alpha_j \le jh$.

Exercise 2.4. Compute the exponential type of the determinant of the characteristic matrix for the following systems:

(i)
$$\dot{x}_1(t) = x_2(t),$$
$$\dot{x}_2(t) = -x_1(t) - x_2(t) - 3x_2(t-1).$$

(ii)
$$\dot{x}_1(t) = x_2(t-1),$$
$$\dot{x}_2(t) = -x_1(t-1) - x_2(t) - 3x_2(t-1).$$

(iii)
$$\dot{x}_1(t) = x_1(t-1) + x_2(t-1),$$
$$\dot{x}_2(t) = x_1(t-1) - x_2(t-1).$$

The entries of adj $\Delta(z)$ are $(n-1)\times(n-1)$ subdeterminants of Δ and Corollary 2.3 therefore implies that the entries of adj $\Delta(z)$ are entire functions of exponential type at most $(n-1)h$. More precisely, we can write

$$(2.9)\qquad \big(\text{adj } \Delta(z)\big)_{ij} = \delta_{ij} z^{n-1} + \sum_{k=1}^{n-1} \int_0^{\alpha_{ijk}} e^{-z\tau}\, d\eta_{ijk}(\tau) z^{n-1-k},$$

where

$$\delta_{ij} = \begin{cases} 1 & \text{for } i = j, \\ 0 & \text{for } i \neq j \end{cases}$$

and $\alpha_{ijk} \leq kh$ and $\eta_{ijk} \in \text{NBV}\big(\mathbb{R}_+, \mathbb{R}\big)$, $i, j = 1, 2, \ldots, n$. So $\det \Delta$ and the entries of adj Δ are entire functions of exponential type which are polynomially bounded in the right half-plane. Therefore, Corollary 2.3 applies and we have the following result.

Corollary 2.5. *Let Δ be given by equation (2.6) and define*

$$\text{E}(\text{adj}\, \Delta) = \max_{1 \leq i,j \leq n} \text{E}\big((\text{adj}\, \Delta)_{ij}\big).$$

Then $\det \Delta$ *and the entries of* adj Δ *are entire functions of exponential type which are polynomially bounded in the right half-plane, and such that*

$$\text{E}(\det \Delta) \leq nh, \qquad \text{E}(\text{adj}\, \Delta) \leq (n-1)h.$$

The following lemma will be useful in the sequel.

Lemma 2.6. *The inequality*

$$\text{E}(\text{adj}\, \Delta) + h > \text{E}(\det \Delta)$$

holds if and only if $\text{E}(\det \Delta) < nh$.

Proof. From the definition $\det \text{adj}\, \Delta = (\det \Delta)^{n-1}$ and hence an application of Corollary 2.3 yields

$$\text{(2.10)} \qquad \mathrm{E}(\det \operatorname{adj} \Delta) = (n-1)\mathrm{E}(\det \Delta).$$

Suppose that $\mathrm{E}(\operatorname{adj} \Delta) + h \leq \mathrm{E}(\det \Delta)$. Then

$$\mathrm{E}(\det \operatorname{adj} \Delta) \leq n\mathrm{E}(\operatorname{adj} \Delta(z)) \leq n\mathrm{E}(\det \Delta) - nh$$

and, together with (2.10), this implies $\mathrm{E}(\det \Delta) \geq nh$ which is a contradiction if $\mathrm{E}(\det \Delta) < nh$.

On the other hand, it is easy to see that if $\mathrm{E}(\det \Delta) = nh$, then $\mathrm{E}(\operatorname{adj} \Delta) = (n-1)h$. So $\mathrm{E}(\operatorname{adj} \Delta) + h = \mathrm{E}(\det \Delta)$. □

We end this section with a simple application of the Paley-Wiener theorem needed in the sequel.

Lemma 2.7. *Let $F : \mathbb{C} \to \mathbb{C}$ be an entire function of exponential type zero. If there exists a γ such that $|F(z)|$ is uniformly bounded in $\{z \mid \mathrm{Re} \geq \gamma\}$ and $|F(z)| \to 0$ as $\mathrm{Re}\, z \to \infty$, then $F \equiv 0$.*

Proof. Set

$$G(z) = \frac{F(z+\gamma) - F(\gamma)}{z}.$$

Then G is an entire function of exponential type which is L^2-integrable along the imaginary axis and bounded in the half-plane $\mathrm{Re}\, z \geq 0$. Since G has exponential type zero, Theorem 2.1 implies that $G \equiv 0$. Therefore, $F(z) \equiv F(\gamma)$. But $|F(z)| \to 0$ as $\mathrm{Re}\, z \to \infty$ and this implies $F \equiv 0$. □

V.3 Completeness

Combining the results from Section 2 with the explicit representation for the resolvent of A given in Section 1, we can estimate the resolvent of A near infinity.

First, we assume that the exponential type of the determinant of the characteristic matrix is maximal, that is, $\mathrm{E}(\det \Delta) = nh$.

Let us compute the exponential type of $P(z)\varphi$, defined in (1.7), by the explicit formula

$$\text{(3.1)} \qquad \begin{aligned} (P(z)\varphi)(\theta) = e^{z\theta}\{&\operatorname{adj} \Delta(z)[\varphi(0) + \int_0^h d\zeta(\tau) \int_0^\tau e^{-z\sigma}\varphi(\sigma-\tau)\, d\sigma] \\ &+ \det \Delta(z) \int_\theta^0 e^{-z\sigma}\varphi(\sigma)\, d\sigma\}. \end{aligned}$$

Lemma 3.1. *For every $\varphi \in C$, we have*

$$\mathrm{E}(z \mapsto P(z)\varphi) \leq nh.$$

Proof. To prove the lemma, we have to rewrite the expression for $P(z)\varphi$ given in (3.1). Set

$$P(z)\varphi = \operatorname{adj} \Delta(z) C(z)\varphi, \tag{3.2}$$

where

$$\begin{aligned}(C(z)\varphi)(\theta) &= \varphi(0)e^{z\theta} + \int_0^h d\zeta(\tau) \int_0^\tau e^{-z(\sigma-\theta)}\varphi(\sigma-\tau)\,d\sigma \\ &\quad + \Delta(z)\int_0^{-\theta} e^{-z\sigma}\varphi(\sigma+\theta)\,d\sigma \\ &= \varphi(0)e^{z\theta} + \int_0^h d\zeta(\tau)\int_0^\tau e^{-z(\sigma-\theta)}\varphi(\sigma-\tau)\,d\sigma \\ &\quad + z\int_0^{-\theta} e^{-z\sigma}\varphi(\sigma+\theta)\,d\sigma \\ &\quad - \int_0^h d\zeta(\tau)e^{-z\tau}\int_0^{-\theta} e^{-z\sigma}\varphi(\sigma+\theta)\,d\sigma.\end{aligned}$$

Changing the order of integration and substituting $s = \sigma - \theta$, we find for the first integral above

$$\begin{aligned}\int_0^h d\zeta(\tau)\int_0^\tau e^{-z(\sigma-\theta)}\varphi(\sigma-\tau)\,d\sigma &= \int_0^h e^{-z(\sigma-\theta)}\int_\sigma^h d\zeta(\tau)\varphi(\sigma-\tau)\,d\sigma \\ &= \int_{-\theta}^{h-\theta} e^{-zs}\int_{s+\theta}^h d\zeta(\tau)\varphi(s+\theta-\tau)\,ds.\end{aligned}$$

Similarly, in the last integral, we take $s = \tau + \sigma$ and τ as new integration variables and find

$$\begin{aligned}&\int_0^h d\zeta(\tau)e^{-z\tau}\int_0^{-\theta} e^{-z\sigma}\varphi(\sigma+\theta)\,d\sigma \\ &\quad = \int_0^h\int_0^{-\theta} e^{-z(\tau+\sigma)}d\zeta(\tau)\varphi(\sigma+\theta)\,d\sigma \\ &\quad = \int_0^{-\theta} e^{-zs}\Big(\int_0^s d\zeta(\tau)\varphi(s+\theta-\tau)\Big)\,ds \\ &\qquad + \int_{-\theta}^{h-\theta} e^{-zs}\Big(\int_{s+\theta}^s d\zeta(\tau)\varphi(s+\theta-\tau)\Big)\,ds.\end{aligned}$$

So, we find [recall that $\zeta(\tau)$ is constant for $\tau \geq h$]

$$\begin{aligned}(C(z)\varphi)(\theta) &= \varphi(0)e^{z\theta} + z\int_0^{-\theta} e^{-z\sigma}\varphi(\sigma+\theta)\,d\sigma \\ &\quad - \int_0^{-\theta} e^{-zs}\Big(\int_0^s d\zeta(\tau)\varphi(s+\theta-\tau)\Big)\,ds \\ &\quad + \int_{-\theta}^h e^{-zs}\Big(\int_s^h d\zeta(\tau)\varphi(s+\theta-\tau)\Big)\,ds.\end{aligned} \tag{3.3}$$

Using Theorem 2.1, it follows from expression (3.3) for $C(z)\varphi$ that the exponential type of $z \mapsto C(z)\varphi$ equals $-\theta$ and is therefore less than or equal to h. Since $\mathrm{E}(\operatorname{adj} \Delta) \leq (n-1)h$, the lemma follows from (3.2). □

Definition 3.2. Let X be a complex Banach space. If W is a subspace of X, then the *annihilator* of W, denoted by $W^{\perp}$, is the subspace of X^* given by

$$W^{\perp} = \{x^* \in X^* \mid \langle x^*, w \rangle = 0 \quad \text{for all } w \in W\}.$$

Similarly, if W^* is the subspace of X^*, the annihilator of W^*, denoted by ${}^{\perp}W^*$, is a subspace of X given by

$${}^{\perp}W^* = \{x \in X \mid \langle w^*, x \rangle = 0 \quad \text{for all } w^* \in W^*\}.$$

Exercise 3.3. Let Y be a subspace of a Banach space X.

(i) Verify the following identities:

$${}^{\perp}(Y^{\perp}) = \overline{Y}, \quad Y^{\perp} = \overline{Y}^{\perp} \quad \text{and} \quad ({}^{\perp}Y^*)^{\perp} = \text{weak}^*\text{-closure}(Y).$$

Let Y_m, $m = 1, 2, \ldots$, be a family of closed subspaces of a complex Banach space X.

(ii) Show that

$$\big(\bigcap_{m=1}^{\infty} Y_m\big)^{\perp} = \text{weak}^*\text{-closure}\big(\bigoplus_{m=1}^{\infty} Y_m^{\perp}\big)$$

and

$$\big(\bigoplus_{m=1}^{\infty} Y_m\big)^{\perp} = \bigcap_{m=1}^{\infty} Y_m^{\perp}.$$

[If X is reflexive the weak*-closure in (ii) equals the norm closure.]
Hint: Use the Hahn-Banach theorem.

Exercise 3.4. Let L be a bounded linear operator on a complex Banach space X. Show that

(i) $\mathcal{R}(L)^{\perp} = \mathcal{N}(L^*)$;

(ii) $\mathcal{N}(L)^{\perp} = \text{weak}^*\text{-closure}(\mathcal{R}(L^*))$.

Lemma 3.5. *The annihilator of the generalized eigenspace of A is given by*

$$\mathcal{M}^{\perp} = \{y^* \mid z \mapsto (zI - A^*)^{-1}y^* \quad \textit{is entire}\}.$$

Proof. Let $y^* \in \overline{\mathcal{M}}^{\perp}$. It suffices to prove that for every $\varphi \in \mathcal{C}$, the function

$$z \mapsto \langle (zI - A^*)^{-1}y^*, \varphi \rangle \tag{3.4}$$

is an entire function (see Definition 1.1 of Appendix III). Since

$$\overline{\mathcal{M}}^{\perp} = \bigcap_{\lambda \in \sigma(A)} \mathcal{N}(P_\lambda^*),$$

it follows from the Laurent series of the resolvent of A^* [see formula (3.3) and Exercise 3.3 of Appendix III] that the singular part of the function in (3.4) vanishes. Therefore, for every $\varphi \in \mathcal{C}$, the function in (3.4) is indeed entire. □

We are now ready to prove the first theorem.

Theorem 3.6. *If* $\mathrm{E}(\det \Delta) = nh$, *then the system of eigenvectors and generalized eigenvectors of the generator* A *associated to the RFDE* (1.1) *is complete.*

Proof. Suppose that $\overline{\mathcal{M}} \neq \mathcal{C}$. Then there exists a nonzero $y^* \in \overline{\mathcal{M}}^{\perp}$. Fix $\varphi \in \mathcal{C}$ and let F denote the entire function associated with y^*, as defined in (3.4). Since the exponential type of $\det \Delta$ equals nh, it follows from Lemma 3.1 and Corollary 2.3 that F has zero exponential type. Furthermore, from the representation for $(zI - A)^{-1}\varphi$, it follows that there exists a γ such that F is uniformly bounded on the half-plane $\{z \mid \operatorname{Re} z \geq \gamma\}$, and $|F(z)| \to 0$ as $\operatorname{Re} z \to \infty$. Therefore, Lemma 2.7 implies that F is identically zero. So for $z \in \rho(A)$, we find

$$\langle (zI - A^*)^{-1} y^*, \varphi \rangle = 0 \quad \text{for all } \varphi \in \mathcal{C}.$$

This implies that $(zI - A^*)^{-1} y^* = 0$ and hence that $y^* = 0$, which contradicts the assumption that $y^* \neq 0$. □

Next we shall consider the case that $\mathrm{E}(\det \Delta) < nh$.

Definition 3.7. Let the set $\mathcal{E}$ be defined by

$$\mathcal{E} = \{ \varphi \in \mathcal{C} \mid \mathrm{E}(z \mapsto P(z)\varphi) \leq E(\det \Delta) \}. \tag{3.5}$$

Here $P(z) : \mathcal{C} \to \mathcal{C}$ and $\det \Delta$ are defined by (1.6).

Since $z \mapsto P(z)\varphi$ is linear with respect to φ, we derive that, for $\varphi, \psi \in \mathcal{C}$ and $\mu \in \mathbb{C}$,

$$\begin{aligned} \mathrm{E}\big(z \mapsto P(z)(\varphi + \mu\psi)\big) &= \mathrm{E}\big(z \mapsto P(z)\varphi + \mu P(z)\psi\big) \\ &\leq \max\big\{\mathrm{E}\big(z \mapsto P(z)\varphi\big), \mathrm{E}\big(z \mapsto \mu P(z)\psi\big)\big\} \\ &\leq \mathrm{E}(\det \Delta). \end{aligned}$$

So the set $\mathcal{E}$ is a linear subspace of $\mathcal{C}$. We first prove some properties of $\mathcal{E}$.

Lemma 3.8. *The linear subspace* $\mathcal{E}$ *given by* (3.5) *is an invariant subspace of the resolvent of* A. *In addition, for any* $z \in \rho(A)$, $(zI - A)^{-1} f \in \mathcal{E}$ *if and only if* $f \in \mathcal{E}$.

Proof. Fix $z_0 \in \rho(A)$ and $\varphi \in \mathcal{E}$. We have to prove that

$$\psi = (z_0 I - A)^{-1}\varphi \in \mathcal{E}.$$

From the resolvent equation we find

$$(zI - A)^{-1}(z_0 I - A)^{-1}\varphi = \frac{(zI - A)^{-1}\varphi - (z_0 I - A)^{-1}\varphi}{z_0 - z}.$$

Using

$$(zI - A)^{-1}\varphi = \frac{1}{\det \Delta(z)} P(z)\varphi,$$

we obtain

$$P(z)\psi = \frac{P(z)\varphi - \det\, \Delta(z)\psi}{z_0 - z}$$

and since $\mathrm{E}\big(z \mapsto P(z)\varphi\big) \le \mathrm{E}(\det \Delta)$, we obtain

$$\begin{aligned} \mathrm{E}\big(z \mapsto P(z)\psi\big) &= \mathrm{E}\big(z \mapsto P(z)\varphi - \det\, \Delta(z)\psi\big) \\ &\le \max\big\{\mathrm{E}\big(z \mapsto P(z)\varphi\big), \mathrm{E}(\det \Delta)\big\} \\ &\le \mathrm{E}(\det \Delta). \end{aligned}$$

Thus we have proved that $\psi \in \mathcal{E}$. Furthermore, we can rewrite the equation for $P(z)\psi$ to derive from

$$P(z)\varphi = (z_0 - z)P(z)\psi + \det\, \Delta(z)\psi$$

that $\varphi \in \mathcal{E}$ if $\psi \in \mathcal{E}$. □

Lemma 3.9. *$\mathcal{E}$ is a closed subspace of $\mathcal{C}$.*

Proof. Let φ_k be a Cauchy sequence in $\mathcal{E}$ such that $\varphi_k \to \varphi$ in $\mathcal{C}$. We have to prove that φ belongs to $\mathcal{E}$ as well.

For $g \in \mathcal{C}^* = \mathrm{NBV}$ and $\varphi \in \mathcal{C}$ define $\tau(g, \varphi)$ to be the exponential type of $z \mapsto \langle g, P(z)\varphi\rangle$, i.e.,

$$\tau(g, \varphi) = \mathrm{E}\big(z \mapsto \langle g, P(z)\varphi\rangle\big).$$

From the representation (3.2) for $P(z)\varphi$, it follows that for every $g \in \mathcal{C}^*$

$$\begin{aligned} \langle g, P(z)\varphi\rangle &= \sum_{j=1}^{n} \int_{-h}^{0} dg_j(\theta)\big(P(z)\varphi\big)(\theta)_j \\ &= \sum_{j=1}^{n} \int_{-h}^{0} dg_j(\theta) \sum_{i=1}^{n} \mathrm{adj}\, \Delta(z)_{ji}\big(C(z)\varphi(\theta)\big)_i \\ &= \sum_{j=1}^{n} \sum_{i=1}^{n} \mathrm{adj}\, \Delta(z)_{ji} \int_{-h}^{0} dg_j(\theta)\big(C(z)\varphi(\theta)\big)_i. \end{aligned}$$

Next we rewrite, using (3.3),

$$\int_{-h}^{0} dg_j(\theta)\big(C(z)\varphi(\theta)\big)_i = \int_0^h e^{-zs} dg_j(-s)\varphi_i(0) + z\int_0^h e^{-zs}\beta_{ij}(s,\varphi,g)\,ds$$
$$+\int_0^h e^{-zs}\alpha_{ij}(s,\varphi,g)\,ds,$$

where

$$\alpha_{ij}(s,\varphi,g) = \int_{-h}^{-s} dg_j(\theta)\Big(\sum_{k=1}^{n}\int_0^s d\zeta_{ik}(\tau)\varphi_k(s+\theta-\tau)\Big)$$
$$+\int_{-s}^{0}\Big(\sum_{k=1}^{n}\int_s^h d\zeta_{ik}(\tau)\varphi_k(s+\theta-\tau)\Big)$$

and

$$\beta_{ij}(s,\varphi,g) = \int_{-h}^{-s} dg_j(\theta)\varphi_i(s+\theta).$$

Note that there exist positive constants K_{ij} such that

$$|\alpha_{ij}(s,\varphi,g)| \le K_{ij}\|\varphi\|\,\|g\|, \qquad |\beta_{ij}(s,\varphi,g)| \le K_{ij}\|\varphi\|\,\|g\|.$$

Using the convolution property of the Laplace transform and representation (2.9) for adj Δ, it follows that

$$\operatorname{adj}\Delta(z)_{ij}\int_{-h}^{0} dg_j(\theta)\big(C(z)\varphi(\theta)\big)_i = \sum_{k=0}^{n}\int_0^{\sigma_{ijk}} e^{-zs}\,d\xi_{ijk}(s,\varphi,g)\,z^{n-k},$$

where the functions ξ_{ijk} are of bounded variation and there are positive constants M_{ijk} such that Var $\xi_{ijk}(s,\varphi,g) \le M_{ijk}\|\varphi\|\,\|g\|$. Together, this implies the following representation for $\langle g, P(z)\varphi\rangle$:

$$\langle g, P(z)\varphi\rangle = z^n\int_0^{\tau(g,\varphi)} e^{-zs}\,dA_n(s,\varphi,g) + \cdots$$
$$+\int_0^{\tau(g,\varphi)} e^{-zs}\,dA_0(s,\varphi,g), \tag{3.6}$$

where the functions A_j are of bounded variation and there are positive constants M_j such that Var $A_j(s,\varphi,g) \le M_j\|\varphi\|\,\|g\|$ for $j=1,\ldots,n$.

Using the representation (3.6) for $\langle g, P(z)\varphi\rangle$ we can make the following estimate. For every $g \in \mathcal{C}^* = \mathrm{NBV}$ and $\varphi \in \mathcal{C}$, there exists a constant M such that

$$\big|\langle g, P(z)\varphi\rangle\big| \le M\|g\|\,\|\varphi\|\,|z|^n \max\{1, |e^{-z\tau(g,\varphi)}|\}. \tag{3.7}$$

In particular, if $\mathrm{E}(z \mapsto P(z)\varphi) = \sigma$, then for every $g \in \mathcal{C}^*$ we have $\tau(g,\varphi) \le \sigma$ and

$$\|P(z)\varphi\| \leq M\|\varphi\|\,|z|^n \max\{1, |e^{-z\sigma}|\}.$$

Let τ denote the exponential type of $\mathrm{E}(\det \Delta)$. For $\psi \in \mathcal{E}$, we have $\mathrm{E}(z \mapsto P(z)\psi) \leq \tau$. So, for the sequence φ_k there exists a constant C, independent of φ_k, such that

$$\|P(z)\varphi_k\| \leq C|z|^n \max\{1, |e^{-z\tau}|\}.$$

Since the convergence of $\varphi_k \to \varphi$ in $\mathcal{C}$ implies that

$$P(z)\varphi_k \to P(z)\varphi \qquad \text{as } k \to \infty,$$

uniformly on compact sets, we obtain

$$\|P(z)\varphi\| \leq C|z|^n \max\{1, |e^{-z\tau}|\}.$$

This implies that $\mathrm{E}(z \mapsto P(z)\varphi) \leq \tau$ and therefore that $\varphi \in \mathcal{E}$. □

Note that if $\mathrm{E}(\det \Delta) = nh$, then Lemma 3.1 implies that $\mathcal{E} = \mathcal{C}$. The following theorem "explains" the case $\mathrm{E}(\det \Delta) < nh$.

Theorem 3.10. *The closure of the generalized eigenspace of the generator A corresponding to RFDE* (1.1) *is given by*

$$\overline{\mathcal{M}} = \mathcal{E}.$$

Proof. First we prove the inclusion $\overline{\mathcal{M}} \subset \mathcal{E}$. For $\varphi \in \mathcal{M}$ the function $z \mapsto (zI - A)^{-1}\varphi$ is rational. Therefore

$$\mathrm{E}(z \mapsto P(z)\varphi) = \mathrm{E}(\det \Delta).$$

So $\mathcal{M} \subset \mathcal{E}$ and Lemma 3.9 implies that $\overline{\mathcal{M}} \subset \mathcal{E}$.

Next we consider the part of A in $\mathcal{E}$; that is, we define $\widehat{A} : \mathcal{D}(\widehat{A}) \to \mathcal{E}$

$$\mathcal{D}(\widehat{A}) = \{\varphi \in \mathcal{E} \mid A\varphi \in \mathcal{E}\}, \qquad \widehat{A}\varphi = A\varphi.$$

Because $\mathcal{E}$ is an invariant subspace of the resolvent of A, we have

$$(zI - \widehat{A})^{-1} = (zI - A)^{-1}|_{\mathcal{E}}. \tag{3.8}$$

Since $\overline{\mathcal{M}} \subset \mathcal{E}$, it follows that $\sigma_p(A) = \sigma(\widehat{A})$, and if P_λ and $\widehat{P}_\lambda$ denote the spectral projection for A and $\widehat{A}$, respectively, then

$$\mathcal{R}(\widehat{P}_\lambda) = \mathcal{R}(P_\lambda), \qquad \widehat{P}_\lambda = P_\lambda|_{\mathcal{E}}.$$

Therefore if $\mathcal{M}(\widehat{A})$ denote the generalized eigenspace corresponding to $\widehat{A}$, then $\mathcal{M}(\widehat{A}) = \mathcal{M}$. Suppose that $\overline{\mathcal{M}(\widehat{A})} \neq \mathcal{E}$. Then there exists a $y^* \in \overline{\mathcal{M}(\widehat{A})}^{\perp}$, $y^* \neq 0$, such that

$$z \mapsto \langle y^*, (zI - \widehat{A})^{-1}\varphi\rangle$$

is an entire function for every $\varphi \in \mathcal{E}$ (see Lemma 3.5). Using (3.8), we can give precisely the same estimates as in the proof of Theorem 3.6. So, we conclude that $y^* = 0$ and

$$\overline{\mathcal{M}(\widehat{A})}^{\perp} = \{0\}.$$

This implies $\overline{\mathcal{M}(\widehat{A})} = \mathcal{E}$ and hence $\overline{\mathcal{M}} = \mathcal{E}$. □

Theorem 3.10 is interesting in the sense that it gives a complete characterisation of the closure of the generalized eigenspace of A in terms of conditions on the resolvent.

Exercise 3.11. Characterise the closure of the generalized eigenspace of the generator corresponding to the following system:

$$\dot{x}_1(t) = x_2(t-1),$$
$$\dot{x}_2(t) = x_1(t).$$

There is more of interest in Theorem 3.10, since it will allow us to prove a noncompleteness result.

Lemma 3.12. *If* $\mathrm{E}(\det \Delta) < nh$, *then* $\mathcal{E} \neq \mathcal{C}$.

Proof. Suppose that $\mathrm{E}(\det \Delta) < nh$. We have to prove that there exists a $\varphi \in \mathcal{C}$ such that

$$\mathrm{E}(z \mapsto P(z)\varphi) > \mathrm{E}(\det \Delta).$$

Recall from the proof of Lemma 3.1 that we can write

$$P(z)\varphi = \operatorname{adj} \Delta(z) C(z)\varphi,$$

where $C(z)\varphi$ is given by equation (3.3). Let $\theta = -h$ and let φ be differentiable on $[-h, 0]$. We can rewrite (3.3) as follows:

$$(3.9)\quad (C(z)\varphi)(-h) = \varphi(-h) + \int_0^h e^{-zs}[\dot{\varphi}_{-h}(s) - \int_0^s d\zeta(\tau)\varphi_{-h}(s-\tau)]\, ds.$$

From Theorem 2.1 and Corollary 2.3, it follows that there exists a $c \in \mathbb{C}^n$ and $\psi \in L^2([0,h], \mathbb{C}^n)$ such that

$$\mathrm{E}\big(z \mapsto \operatorname{adj} \Delta(z)\big[c + \int_0^h e^{-zs}\psi(s)\, ds\big]\big) > \mathrm{E}(\det \Delta).$$

To prove the lemma, it suffices to prove that we can solve the following equations for φ:

$$c = \varphi(-h),$$
$$\psi(s) = \dot{\varphi}_{-h}(s) - \int_0^s d\zeta(\tau)\varphi_{-h}(s-\tau).$$

This is a renewal equation which can be solved for every ψ given on the interval $[0,h]$. Indeed, integration by parts yields

$$\dot{\varphi}_{-h} - \zeta * \dot{\varphi}_{-h} = \psi + \zeta c, \tag{3.10}$$

where

$$\zeta * \dot{\varphi}_{-h}(t) = \int_0^t \zeta(\tau)\dot{\varphi}(s-\tau-h)\,d\tau.$$

The solution to equation (3.10) is explicitly given by

$$\dot{\varphi}_{-h} = (\psi + \zeta c) - R * (\psi + \zeta c),$$

where R is the unique matrix solution to $R = \zeta * R - \zeta$. □

A combination of Theorem 3.10 and Lemma 3.12 yields a necessary and sufficient condition for completeness.

Theorem 3.13. *The system of eigenvectors and generalized eigenvectors of A is complete if and only if* $\mathrm{E}(\det \Delta) = nh$.

The conditions are easy to verify in examples. For example, the system of eigenvectors and generalized eigenvectors of the generator corresponding to a scalar equation

$$\dot{x}(t) = \int_0^h d\zeta(\theta)x(t-\theta)$$

where ζ has a jump at h is complete. If ζ has no jump at h, the completeness depends on whether the exponential type of $\det \Delta$ equals h. For example, if $a : [0,h] \to \mathbb{R}$ is an integrable function, then the generator associated with

$$\dot{x}(t) = \int_0^h a(\theta)x(t-\theta)\,d\theta$$

has a complete system of eigenvectors and generalized eigenvectors if and only if a does not vanish a.e. in any neighbourhood of h.

In the case of systems, it is easy to construct "counter" examples. The generator corresponding to the system

$$\dot{x}(t) = Ax(t) + Bx(t-1) + Cx(t-2) \tag{3.11}$$

has a complete set of eigenfunctions and generalized eigenfunctions if and only if $\det C \neq 0$.

Since we now know that completeness can actually fail, it is natural to ask whether we can characterise the complement of $\overline{\mathcal{M}}$ in $\mathcal{C}$. The following result provides the answer.

Theorem 3.14. *Let*

$$\mathcal{S} = \{\varphi \in C \mid z \mapsto (zI - A)^{-1}\varphi \quad \textit{is entire}\}. \tag{3.12}$$

Then

$$\mathcal{C} = \overline{\overline{\mathcal{M}} \oplus \mathcal{S}}.$$

Proof. First, note that from the characterisation for $\overline{\mathcal{M}}$, we have

$$\overline{\mathcal{M}} \cap \mathcal{S} = \{0\}. \tag{3.13}$$

Indeed, if $x \in \overline{\mathcal{M}} \cap \mathcal{S}$, $x \neq 0$, then for every $x^* \in X^*$

$$z \mapsto \langle x^*, (zI - A)^{-1}x\rangle$$

is an entire function of exponential type zero, uniformly bounded in the right half-plane which tends to zero along the positive real axis. So Lemma 2.7 yields that $x = 0$ and this proves (3.13).

To prove that $\overline{\mathcal{M}} \oplus \mathcal{S}$ is dense, we shall use duality. From Exercises 3.2 and 3.3, it follows that

$$\big(\overline{\mathcal{M}} \oplus \mathcal{S}\big)^{\perp} = \text{weak}^*\text{-closure}\big(\mathcal{M}^*\big) \cap \mathcal{S}^*,$$

where $\mathcal{M}^*$ denotes the generalized eigenspace of A^* and

$$\mathcal{S}^* = \{f \in \text{NBV} \mid z \mapsto (zI - A^*)^{-1}f \quad \text{is entire}\}. \tag{3.14}$$

So in order to prove the theorem it suffices to prove that

$$\text{weak}^*\text{-closure}\big(\mathcal{M}^*\big) \cap \mathcal{S}^* = \{0\}. \tag{3.15}$$

Since $\|(zI - A)^{-1}\| = \|(zI - A^*)^{-1}\|$ and $\|P(z)\| = \|P(z)^*\|$, it follows that

$$(zI - A^*)^{-1}g = \frac{P(z)^*g}{\det \Delta(z)},$$

and $\|(zI - A^*)^{-1}\| \leq M$ for $\operatorname{Re} z > 0$. Furthermore, if μ is a real, then $\|(\mu I - A^*)^{-1}\| \to 0$ as $\mu \to \infty$. Let

$$\mathcal{E}^* = \big\{g \in \text{NBV} \mid \text{E}(z \mapsto P(z)^*g) \leq \text{E}(\det \Delta)\big\}.$$

The same argument as used in the first part of the proof to show that $\mathcal{E} \cap \mathcal{S} = \{0\}$ yields that $\mathcal{E}^* \cap \mathcal{S}^* = \{0\}$. Therefore, in order to prove (3.13), it suffices to prove that weak*-closure$\big(\mathcal{M}^*\big) \subset \mathcal{E}^*$.

First, we shall prove that $\mathcal{E}^*$ is weak*-closed. Let g_j, $j = 1, 2, \ldots$, be a sequence in $\mathcal{E}^*$ such that for every $\varphi \in C$

$$\langle g_j, \varphi\rangle \to \langle g, \varphi\rangle \qquad \text{as } j \to \infty. \tag{3.16}$$

We must prove that g belongs to $\mathcal{E}^*$ as well. Without loss of generality we can assume that $\max_j \|g_j\|$ is bounded. Let $\tau = \mathrm{E}(\det \Delta)$. Then $\mathrm{E}(z \mapsto P(z)^* g_j) \leq \tau$ and for every $\varphi \in \mathcal{C}$ we have $\mathrm{E}(z \mapsto \langle P(z)^* g_j, \varphi\rangle) \leq \tau$. Therefore inequality (3.7) implies that there exists a positive constant K_0 such that for every $\varphi \in C$

$$\begin{aligned} |\langle P(z)^* g_j, \varphi\rangle| &\leq K\|g_j\| \, \|\varphi\| \, |z|^n \max\{1, |e^{-\tau z}|\} \\ &\leq K \max_j \|g_j\| \, \|\varphi\| \, |z|^n \max\{1, |e^{-\tau z}|\}. \end{aligned}$$

Inequality (3.7) also implies that $\|P(z)\varphi\|$ is uniformly bounded on compact subsets of $\mathbb{C}$. So (3.16) implies that $\langle P(z)^* g_j, \varphi\rangle \to \langle P(z)^* g, \varphi\rangle$ uniformly on compact sets of $\mathbb{C}$. Hence for every $\varphi \in \mathcal{C}$

$$|\langle P(z)^* g, \varphi\rangle| \leq M_1 \|\varphi\| \, |z|^n \max\{1, |e^{-\tau z}|\}.$$

This implies that $\mathrm{E}(z \mapsto P(z)^* g) \leq \tau$ and hence $g \in \mathcal{E}^*$. Since $\mathcal{M}^* \subset \mathcal{E}^*$, the weak*-closure of $\mathcal{M}^*$ is contained in $\mathcal{E}^*$. This completes the proof of the theorem. □

Exercise 3.15. Show that

$$\text{weak}^*\text{-closure}(\mathcal{M}^*) = \mathcal{E}^*$$

and conclude

$$\mathrm{NBV} = \overline{\overline{\mathcal{M}^*} \oplus \mathcal{S}^*},$$

where $\mathcal{S}^*$ is given by (3.14).
Hint: Adapt the argument used in the proof of Theorem 3.10.

V.4 Small solutions

Definition 4.1. A solution x of RFDE (1.1) is called a *small solution* if

$$\lim_{t\to\infty} x(t)e^{kt} = 0 \qquad \text{for every } k \in \mathbb{R}. \tag{4.1}$$

Small solutions that are not identically zero are called *nontrivial.*

In the results so far, we have not used the fact that A is the generator of a $\mathcal{C}_0$-semigroup. If ω_0 denotes the growth bound of $T(t)$, then in the half-plane $\operatorname{Re} z > \omega_0$, the resolvent of the generator is given by the Laplace transform of the semigroup

$$(zI - A)^{-1}\varphi = \int_0^\infty e^{-zt} T(t)\varphi \, dt, \qquad \varphi \in \mathcal{C}. \tag{4.2}$$

To characterise the small solutions of RFDE (1.1) we are going to exploit the interplay between the representations (1.6) and (4.2) for the resolvent of the generator.

Let $\varphi \in C$ be such that $x(\,\cdot\,;\varphi)$ is a small solution. Then

$$\lim_{t\to\infty} e^{kt} \|T(t)\varphi\| = 0 \qquad \text{for every } k \in \mathbb{R}$$

and hence

$$z \mapsto \int_0^\infty e^{-zt} T(t)\varphi \, dt$$

defines an entire function.

Lemma 4.2. *Let*

$$\eta = \sup_{\varphi \in \mathcal{C}} \ \mathrm{E}(z \mapsto P(z)\varphi) - \mathrm{E}(\det \Delta) \tag{4.3}$$

and

$$\mathcal{S} = \{\varphi \in \mathcal{C} \,|\, z \mapsto (zI - A)^{-1}\varphi \quad \textit{is entire}\}.$$

Then

$$\mathcal{N}\big(T(t)\big) = \mathcal{S} \qquad \textit{for } t \geq \eta.$$

Proof. It is obvious that $\mathcal{N}\big(T(\eta)\big) \subset \mathcal{S}$. On the other hand, if $\varphi \in \mathcal{S}$, then for every $y^* \in \mathcal{C}^*$

$$z \mapsto \int_0^\infty e^{-zt} \langle y^*, T(t)\varphi\rangle \, dt \tag{4.4}$$

is an entire function of exponential type which is square-integrable along the imaginary axis. From (4.2) and (1.6) it follows that

$$\int_0^\infty e^{-zt} \langle y^*, T(t)\varphi\rangle \, dt = \frac{1}{\det \Delta(z)} \langle y^*, P(z)\varphi\rangle,$$

and hence for every $y^* \in \mathcal{C}^*$, the entire function in (4.4) has exponential type less than or equal to η. Therefore Theorem 2.1 implies that for every $y^* \in \mathcal{C}^*$

$$\langle y^*, T(t)\varphi\rangle = 0 \qquad \text{for } t \geq \eta.$$

This proves that if $\varphi \in \mathcal{S}$, then $\varphi \in \mathcal{N}\big(T(\eta)\big)$. $\square$

Note that Lemma 4.2 implies that the small solutions of RFDE (1.1) are identically zero after finite time. The following theorem is a consequence of Theorem 3.14 and the characterisation of $\overline{\mathcal{M}}$ given in Theorem 3.10.

Theorem 4.3. *The n-dimensional RFDE* (1.1) *has no nontrivial small solutions if and only if the exponential type of the characteristic function* $\det \Delta$ *is maximal, i.e.,* $\mathrm{E}(\det \Delta) = nh$. *Furthermore, the system of eigenvectors and generalized eigenvectors of A is complete if and only if there are no nontrivial small solutions.*

There are more simple consequences of the characterisation for $\overline{\mathcal{M}}$. We first give a definition.

Definition 4.4. Define the *ascent* α of the semigroup $T(t)$ by

$$\alpha = \inf \{ t \mid \text{for all } \epsilon > 0 : \mathcal{N}(T(t)) = \mathcal{N}(T(t+\epsilon)) \}. \tag{4.5}$$

Define nonnegative numbers ϵ and σ by the relations

$$nh - \epsilon = \mathrm{E}(\det \Delta) \tag{4.6}$$

and

$$(n-1)h - \sigma = \mathrm{E}(\operatorname{adj} \Delta). \tag{4.7}$$

The definitions imply that $\mathrm{E}(\operatorname{adj} \Delta) + h = \mathrm{E}(\det \Delta) + \epsilon - \sigma$ (note that we can now rephrase Lemma 2.6 as $\epsilon - \sigma > 0$ iff $\epsilon > 0$). In the proof of Lemma 3.12 it was shown that $\sup_{\varphi} \mathrm{E}(z \mapsto P(z)\varphi) = \mathrm{E}(\operatorname{adj} \Delta) + h$. So, with η defined by (4.3) we have $\eta = \epsilon - \sigma$. In particular, the ascent of $T(t)$ is less than or equal to $\epsilon - \sigma$.

Theorem 4.5. *For* $t \geq \epsilon - \sigma$

$$\overline{\mathcal{R}(T(t))} = \overline{\mathcal{M}}$$

and this relation holds for no t smaller than $\epsilon - \sigma$.

Proof. Since $\mathcal{M}$ is invariant under $T(t)$, the inclusions $\mathcal{M} \subseteq \mathcal{R}(T(t))$ and $\overline{\mathcal{M}} \subseteq \overline{\mathcal{R}(T(t))}$ follow immediately. To prove the remaining inclusion, we first note that the representations (1.6) and (4.2) for the resolvent of A imply

$$P(z)T(t)\varphi = e^{zt}P(z)\varphi - \det \Delta(z) \int_0^t e^{z(t-s)}T(s)\varphi\, ds. \tag{4.8}$$

For $t \geq 0$, the functions $z \mapsto P(z)T(t)\varphi$ and $z \mapsto e^{-zt}$ are polynomially bounded in the right half-plane. So Corollary 2.3 and, in particular, (2.5) implies that

$$\mathrm{E}(z \mapsto e^{zt} P(z)\varphi) = \mathrm{E}(z \mapsto P(z)\varphi) - t.$$

Therefore formula (4.8) for $t = \eta$ yields

$$\begin{aligned}\mathrm{E}(z \mapsto P(z)T(\eta)\varphi) &= \max\{\mathrm{E}(z \mapsto P(z)\varphi) - \eta, \mathrm{E}(\det \Delta)\} \\ &\leq \mathrm{E}(\det \Delta).\end{aligned}$$

Therefore (see Definition 3.7 and Theorem 3.8) we have $T(\eta)\varphi \in \mathcal{E} = \overline{\mathcal{M}}$. Since φ is arbitrary, we find

$$\mathcal{R}(T(t)) \subset \mathcal{R}(T(\eta)) \subset \overline{\mathcal{M}} \qquad \text{for } t \geq \eta.$$

On the other hand, the definition of η implies that for $t < \eta$ we can choose φ such that $\mathrm{E}(z \mapsto P(z)T(t)\varphi) > \mathrm{E}(\det \Delta)$ and hence $T(t)\varphi \notin \mathcal{E} = \overline{\mathcal{M}}$. □

Corollary 4.6. *The ascent of the adjoint semigroup* $\{T^*(t)\}_{t\geq 0}$ *equals* $\eta = \epsilon - \sigma$.

Proof. From Exercise 3.4 we know that $\mathcal{R}(T(t))^{\perp} = \mathcal{N}(T^*(t))$. So the result follows directly from Theorem 4.5. □

In the following theorem we compute the ascent of $\{T(t)\}$ itself by characterising the maximal support for the small solutions of RFDE (1.1).

Theorem 4.7. *The ascent* α *of the* C_0*-semigroup* $T(t)$ *associated with the RFDE* (1.1) *is finite and given by*

$$\alpha = \epsilon - \sigma = \mathrm{E}(\operatorname{adj} \Delta) + h - \mathrm{E}(\det \Delta).$$

Proof. From Lemma 4.2 it follows that $\alpha \leq \epsilon - \sigma$. The proof that $\alpha \geq \epsilon - \sigma$ is a refinement of the proof of Lemma 3.12. It suffices to prove the following claim: There exists a $\varphi \in C$ such that

$$\big((zI - A)^{-1}\varphi\big)(-h) \text{ is an entire function of exponential type } \epsilon - \sigma.$$

Indeed, from (4.2) and the Paley-Wiener Theorem 2.1,

$$\big((zI - A)^{-1}\varphi\big)(-h) = \int_0^{\epsilon-\sigma} e^{-zt}\big(T(t)\varphi\big)(-h)\, dt$$

and $\big(T(t)\varphi\big)(-h)$ is not identically zero in any neighbourhood of $\epsilon - \sigma$. This shows $\alpha \geq \epsilon - \sigma$. So it suffices to prove the claim.

From (3.9) and the proof of Lemma 3.12, the claim will follow once we have constructed a constant $c \in \mathbb{C}^n$ and $\psi \in L^2\big([0,h], \mathbb{C}^n\big)$ such that

$$F(z) = \frac{\operatorname{adj} \Delta(z)}{\det \Delta(z)}\Big[c + \int_0^h e^{-zs}\psi(s)\, ds\Big] \tag{4.9}$$

is an entire function of exponential type $\epsilon - \sigma$. In the following we shall call vector-valued functions of the form

$$\int_0^\rho e^{-zs}\chi(s)\,ds,$$

where $\rho \in \mathbb{R}_+$ and $\chi \in L^2([0,\rho],\mathbb{C}^n)$, Paley-Wiener functions. Let $q(z) = \int_0^h e^{-zs}\psi(s)\,ds$. It is clear that constructing a $c \in \mathbb{C}^n$ and a $\psi \in L^2([0,h],\mathbb{C}^n)$ such that (4.9) holds is equivalent to constructing a Paley-Wiener function F of exponential type $\epsilon - \sigma$ such that

$$\Delta(z)F(z) = c + q(z), \tag{4.10}$$

where $c \in \mathbb{C}^n$ and q is a Paley-Wiener function of exponential type less than or equal to h.

Choose a column of the matrix function $\operatorname{adj}\Delta(z)$ such that one of the elements of this column is the cofactor of maximal exponential type given by $(n-1)h - \sigma$. Since the arguments given below can be repeated for all other columns we may assume that we can choose the first column

$$\begin{pmatrix} C_{11}(z) \\ \vdots \\ C_{n1}(z) \end{pmatrix} \tag{4.11}$$

of $\operatorname{adj}\Delta(z)$. Then

$$\Delta(z)\begin{pmatrix} C_{11}(z) \\ \vdots \\ C_{n1}(z) \end{pmatrix} = \begin{pmatrix} \det\Delta(z) \\ 0 \\ \vdots \\ 0 \end{pmatrix}. \tag{4.12}$$

We have to consider two cases:

I $\epsilon \le (n-1)h$;

II $(n-1)h < \epsilon \le nh$.

Case I. Suppose $\epsilon \le (n-1)h$. For $1 \le j \le n$ define the function c_j to be the Taylor expansion of C_{j1} of order $n-1$ in 0; then the function F_j defined by

$$F_j(z) = \frac{C_{j1}(z) - c_j(z)}{z^n} \tag{4.13}$$

is entire. Define d_j by

$$\Delta(z)\begin{pmatrix} c_1 \\ \vdots \\ c_n \end{pmatrix} = \begin{pmatrix} d_1 \\ \vdots \\ d_n \end{pmatrix}.$$

Integration by parts yields that the functions d_j, $1 \leq j \leq n$, are polynomials of degree n with coefficients being constants plus Paley-Wiener functions of exponential type less than or equal to h. Furthermore,

$$\Delta(z)\begin{pmatrix} F_1 \\ \vdots \\ F_n \end{pmatrix} = \frac{1}{z^n}\begin{pmatrix} \det \Delta(z) - d_1 \\ -d_2 \\ \vdots \\ -d_n \end{pmatrix}. \tag{4.14}$$

Since $\det \Delta$ is a polynomial of degree n with coefficients being constants plus Paley-Wiener functions, we have by the Paley-Wiener Theorem 2.1 that the right hand side of (4.14) can be written as

$$c + \int_0^{nh-\epsilon} e^{-zt} h(t)\,dt,$$

where $c \in \mathbb{C}^n$ and $h \in L^2\big([0, nh-\epsilon], \mathbb{C}^n\big)$. Furthermore, the cofactors of Δ are polynomials of degree $n-1$ with coefficients being constants plus Paley-Wiener functions. Hence the entire function F defined by (4.13) is a Paley-Wiener function, and by the Paley-Wiener Theorem 2.1 we have

$$F(z) = \int_0^{(n-1)h-\sigma} e^{-zt}\chi(t)\,dt,$$

where $\chi \in L^2\big([0, (n-1)h-\sigma], \mathbb{C}^n\big)$. Therefore equation (4.14) can be rewritten as

$$\Delta(z)\int_0^{(n-1)h-\sigma} e^{-zt}\chi(t)\,dt = c + \int_0^{nh-\epsilon} e^{-zt}h(t)\,dt. \tag{4.15}$$

So, the function χ satisfies the equation

$$x - \zeta * x = g,$$

where $\dot{g} = h$ and g is constant on $[nh-\epsilon, \infty)$ and hence

$$\dot{\chi} \in L^2[0, (n-1)h - \sigma].$$

We rewrite equation (4.15) as follows:

$$\begin{aligned} &\Delta(z)\int_0^{(n-1)h-\sigma} e^{-zt}\chi(t)\,dt \\ &= e^{-((n-1)h-\epsilon)z}\Delta(z)\int_0^{\epsilon-\sigma} e^{-zt}\chi((n-1)h-\epsilon+t)\,dt \\ &= c + \int_0^{nh-\epsilon} e^{-zt}h(t)\,dt - \Delta(z)\int_0^{(n-1)h-\epsilon} e^{-zt}\chi(t)\,dt. \end{aligned} \tag{4.16}$$

Since the right hand side of (4.16) has exponential type less than or equal to $nh - \epsilon$, we have by Corollary 2.3 that

$$\Delta(z) \int_0^{\epsilon-\sigma} e^{-zt}\chi((n-1)h - \epsilon + t)dt \tag{4.17}$$

has exponential type less than or equal to h. Furthermore, since

$$\dot{\chi} \in L^2[0, (n-1)h - \sigma],$$

partial integration shows that (4.17) can be rewritten as $c + q(z)$, where $c \in \mathbb{C}^n$ and q is a Paley-Wiener function of exponential type less than or equal to h. Hence, $\chi((n-1)h - \epsilon + t)$ is a small solution so that

$$\chi((n-1)h - \epsilon + t) \not\equiv 0 \text{ in any neighbourhood of } \epsilon - \sigma.$$

Case II. Suppose that $(n-1)h < \epsilon \leq nh$. In this case we multiply both sides of the equation (4.12) by

$$\int_0^{\epsilon-(n-1)h} e^{-zt}dt \tag{4.18}$$

to obtain

$$\Delta(z) \begin{pmatrix} \widetilde{C}_1 \\ \vdots \\ \widetilde{C}_n \end{pmatrix} = \begin{pmatrix} G \\ 0 \\ \vdots \\ 0 \end{pmatrix},$$

where

$$G(z) = \det \Delta(z) \int_0^{\epsilon-(n-1)h} e^{-zt}\, dt$$

and

$$\widetilde{C}_j(z) = \int_0^{\epsilon-(n-1)h} e^{-zt}dt\, C_{j1}(z) \quad \text{for } 1 \leq j \leq n.$$

Hence $\mathrm{E}(G) = h$ and $\mathrm{E}(\widetilde{C}) = \epsilon - \sigma$. The same arguments as used in Case I but now applied to the modified function $\widetilde{C}$ yield

$$\Delta(z) \int_0^{\epsilon-\sigma} e^{-zt}\tilde{\chi}(t)dt = \tilde{c} + \tilde{q}(z), \tag{4.19}$$

where $\tilde{c} \in \mathbb{C}^n$ and $\tilde{q}$ is a Paley-Wiener function with $\mathrm{E}(\tilde{q}) = h$. Therefore $\tilde{\chi}$ is a small solution so that

$$\tilde{\chi} \not\equiv 0 \text{ in any neighbourhood of } \epsilon - \sigma.$$

This proves the claim and hence Theorem 4.7. □

Example 4.8. Consider the following system of differential-difference equations:

$$\begin{aligned} \dot{x}_1(t) &= -x_2(t) + x_3(t-1), \\ \dot{x}_2(t) &= x_1(t-1) + x_3(t-\frac{1}{2}), \\ \dot{x}_3(t) &= x_3(t). \end{aligned} \tag{4.20}$$

Then the characteristic matrix becomes

$$\Delta(z) = \begin{pmatrix} z & 1 & -e^{-z} \\ -e^{-z} & z & -e^{-\frac{1}{2}z} \\ 0 & 0 & z-1 \end{pmatrix} \tag{4.21}$$

with determinant

$$\det \Delta(z) = (z-1)(z^2 + e^{-z}). \tag{4.22}$$

Therefore

$$\epsilon = 2.$$

Since the cofactor

$$\begin{aligned} C_{23}(z) &= -\begin{vmatrix} z & -e^{-z} \\ -e^{-z} & -e^{-\frac{1}{2}z} \end{vmatrix} \\ &= ze^{-\frac{1}{2}z} + e^{-2z} \end{aligned} \tag{4.23}$$

has exponential type 2, we derive that $\sigma = 0$. From Theorem 4.7 we conclude that the ascent of the system (4.20) equals two. So there exists a (nontrivial) small solution $x = x(\,\cdot\,;\varphi)$ such that $\mathrm{supp}(x) = [-1, 1]$.

Exercise 4.9. Show that for $t \geq \epsilon - \sigma$

$$\text{weak}^*\text{-closure}(\mathcal{R}(T^*(t))) = \text{weak}^*\text{-closure}(\mathcal{M}^*).$$

Use this result to give another proof of Theorem 4.7.
Hint: With Exercise 3.15 the argument is similar to the one given for Theorem 4.5.

Concluding this section, we find that Theorem 3.14 implies that the following "almost" decomposition holds:

Corollary 4.10.

$$\mathcal{C} = \overline{\overline{\mathcal{R}(T(\alpha))} \oplus \mathcal{N}(T(\alpha))},$$

where $\alpha = \epsilon - \sigma$.

In general, a state space decomposition

$$\mathcal{C} = \overline{\mathcal{M}} \oplus \mathcal{N}(T(\alpha))$$

does not hold as is shown by the following exercise.

Exercise 4.11. Consider the following RFDE:

$$\dot{x}_1(t) = x_2(t-1),$$
$$\dot{x}_2(t) = x_1(t).$$

(i) Compute the ascent α.
(ii) Show that

$$\mathcal{N}(T(\alpha)) = \{(\varphi_1, \varphi_2) \mid \varphi_1(0) = 0,\ \varphi_2(\theta) = 0 \quad \text{for } -1 \le \theta \le 0\,\}.$$

(iii) Show that

$$\overline{\mathcal{M}} = \{(\varphi_1, \varphi_2) \mid \varphi_2(\theta) = \varphi_2(0) + \int_0^\theta \varphi_1(s)\, ds \quad \text{for } -1 \le \theta \le 0\,\}.$$

(iv) Conclude that $\mathcal{C} \neq \overline{\mathcal{M}} \oplus \mathcal{N}(T(\alpha))$, but $\mathcal{C} = \overline{\overline{\mathcal{M}} \oplus \mathcal{N}(T(\alpha))}$.

V.5 Precise estimates for $|\Delta(z)^{-1}|$

From the results in the previous sections we know that the system of eigenvectors and generalized eigenvectors of the generator A associated with RFDE (1.1) is complete if and only if $\mathrm{E}(\det \Delta) = nh$. In this section we shall estimate $|\Delta(z)^{-1}|$ in the complex plane outside small circles centered around the zeros of $\det \Delta$, assuming that $\mathrm{E}(\det \Delta) = nh$.

These estimates are needed in the next section where we consider series of spectral projections in case the system of eigenvectors and generalized eigenvectors of A is complete.

We start with lower bounds for $|\det \Delta|$. In Theorem I.4.1 we have seen that for $\zeta \in \mathrm{NBV}$ there exist positive constants C_0 and C such that

$$|\det \Delta(z)| \ge \frac{1}{2}|z|^n$$

for those $z \in \mathbb{C}$ for which $|z| \ge C_0 e^{-h \operatorname{Re} z}$ and $|z| \ge C$. To analyse $|\det \Delta(z)|$ in the complement of this set, we have to make further restrictions on the kernel ζ.

A function $\eta \in \mathrm{NBV}(\mathbb{R}_+, \mathbb{R})$ has an *atom* at t_0 if $\eta(t_0-) \neq \eta(t_0)$.

Lemma 5.1. *If $\eta \in NBV(\mathbb{R}_+, \mathbb{R})$ has an atom at ω and*

$$F(z) = \int_0^\omega e^{-z\tau}\, d\eta(\tau), \tag{5.1}$$

then there exist positive constants c, m and M such that in the left half-plane $\{z \mid \operatorname{Re} z < c\}$

$$m|e^{-z\omega}| \le |F(z)| \le M \max\{1, |e^{-z\omega}|\}. \tag{5.2}$$

Proof. The upper bound follows from the fact that η is of bounded variation (see Lemma I.4.2). To prove the lower bound choose $\delta > 0$ such that the variation of η over the interval $[\alpha - \delta, \alpha)$ is smaller than ϵ. Rewrite

$$F(z) = \big(\eta(\alpha) - \eta(\alpha-)\big)e^{-\alpha z} + \int_{\alpha-\delta}^{\alpha-} e^{-z\tau}\, d\eta(\tau) + \int_0^{\alpha-\delta} e^{-z\tau}\, d\eta(\tau); \tag{5.3}$$

then

$$\big|F(z)\big| \ge |e^{-\alpha z}|\big(|\eta(\alpha) - \eta(\alpha-)| - V(\eta)(\alpha - \delta)|e^{z\delta}| - \epsilon\big).$$

So for c sufficiently small, we find

$$\big|F(z)\big| \ge m|e^{-\alpha z}|, \qquad \operatorname{Re} z < c.$$

□

Definition 5.2. For an $n \times n$-matrix function of bounded variation ζ we let $\det_* \zeta$ denote the determinant of ζ with respect to the convolution product; that is, in the case $n = 2$ we have $\det_* = \zeta_{11} * \zeta_{22} - \zeta_{12} * \zeta_{21}$ and from this we build $\det_*$ inductively for higher dimensions.

Exercise 5.3. Show that if ζ has a nonsingular atom at h, that is, $\det\big(\zeta(h) - \zeta(h-)\big) \neq 0$, then $\det_* \zeta$ has an atom at nh.

From representation (2.6)–(2.7) for $\det \Delta$, we have

$$\det \Delta(z) = z^n + \sum_{j=1}^{n} \int_0^{\alpha_j} e^{-z\tau}\, d\eta_j(\tau) z^{n-j}, \qquad \alpha_j \le jh,$$

where $\eta_n = \det_* \zeta$. In particular, we can deduce lower bounds for $|\det \Delta|$ from lower bounds for

$$\Big|\int_0^{\alpha_n} e^{-zt}\, d\eta_n(t)\Big|$$

deduced from Lemma 5.1.

The following theorem, together with Theorem I.4.1, will provide the basic estimates.

Theorem 5.4. *Let $\zeta \in NBV$ be given such that $\det_* \zeta$ has an atom at nh. Then there exist positive constants B_1, C and m such that*

$$\big|\det \Delta(z)\big| \ge \frac{1}{2} m |e^{-nhz}| \tag{5.4}$$

for $z \in W_1(B_1)$, *where*

$$W_1(B_1) = \{z \in \mathbb{C} \mid |z| \leq B_1 e^{-h\mathrm{Re}\, z},\ |z| \geq C\}. \tag{5.5}$$

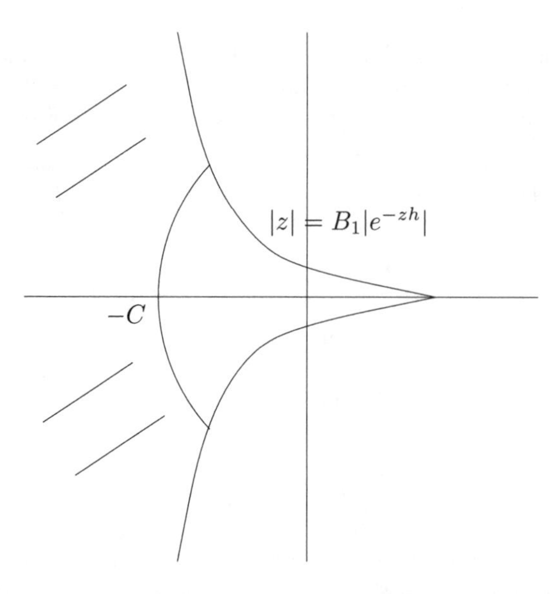

Fig. V.1. The set $W_1(B_1)$.

Proof. Write

$$\begin{aligned}|\det \Delta(z)| = |e^{-nhz}|\,\Big|z^n e^{nhz} + \sum_{j=1}^{n-1} e^{nhz} \int_0^{\alpha_j} e^{-z\tau}\, d\eta_j(\tau) z^{n-j} \\ + e^{nhz} \int_0^{nh} e^{-z\tau}\, d\eta_n(\tau)\Big|.\end{aligned}$$

Using Lemma 5.1, we find that there exists a constant m such that

$$\Big|e^{nhz} \int_0^{nh} e^{-z\tau}\, d\eta_n(\tau)\Big| \geq m \tag{5.6}$$

for $\mathrm{Re}\, z < c$ and $-c$ sufficiently large.

Since $\alpha_j \leq jh$, it follows for z satisfying (5.5) with B_1 sufficiently small that

$$\begin{aligned}&\Big|z^n e^{nhz} + \sum_{j=1}^{n-1} e^{nhz} \int_0^{\alpha_j} e^{-z\tau}\, d\eta_j(\tau) z^{n-j}\Big| \\ &\qquad \leq \Big|B_1^n + \sum_{j=1}^{n-1} \mathrm{TV}(\eta_j) B_1^{n-j}\Big| \leq \frac{1}{2} m.\end{aligned} \tag{5.7}$$

So if $C = B_1 e^{-(c+1)h}$ and B_1 is chosen such that (5.7) holds, then

$$\left|\det \Delta(z)\right| \geq \frac{1}{2} m |e^{-nhz}|.$$

□

Corollary 5.5. *If $\zeta \in NBV$ is such that* $\det_* \zeta$ *has an atom at nh, then, with at most finitely many exceptions, the zeros of* $\det \Delta$ *are located in the curvilinear strip*

$$B_1 e^{-h\operatorname{Re} z} \leq |z| \leq C_0 \max\{1, e^{-h\operatorname{Re} z}\},$$

where C_0 and B_1 are the constants introduced in Theorem I.4.1 *and Theorem* 5.4*, respectively.*

Note that the assumption that $\det_* \zeta$ has an atom at nh implies that $\mathrm{E}(\det \Delta) = nh$. So the eigenvectors and generalized eigenvectors of the generator A are complete. In order to analyse whether series of eigenvectors and generalized eigenvectors converge, we need good estimates for $|\Delta(z)^{-1}|$ for z away from the zeros of $\det \Delta$. To provide these estimates, we start with a lower bound for $\det \Delta$ inside the area where the large zeros of $\det \Delta$ are located but outside small circles containing the zeros of $\det \Delta$.

Theorem 5.6. *Let $\zeta \in NBV$ be such that* $\det_* \zeta$ *has an atom at nh. For every $\epsilon > 0$, there exists a constant m such that*

$$\left|\det \Delta(z)\right| \geq \frac{1}{2} m |e^{-nhz}| \tag{5.8}$$

for values of z outside circles of radius ϵ containing the roots of $\det \Delta$ *and inside*

$$V(B_1, C_0) = \left\{ z \in \mathbb{C} \mid B_1 e^{-h\operatorname{Re} z} \leq |z| \leq C_0 e^{-h\operatorname{Re} z},\ |z| \geq C \right\}, \tag{5.9}$$

where C_0, C and B_1 are the constants introduced in Theorem I.4.1 *and Theorem* 5.4*, respectively.*

Proof. Let $\lambda_1, \lambda_2, \ldots$ denote the zeros of $\det \Delta$. Suppose that such a constant m does not exist. Then there exists an $\epsilon > 0$ and a sequence of points inside the curvilinear strip V defined by (5.9) but outside the disks $|z - \lambda_j| \leq \epsilon$, $j = 1, 2, \ldots$, such that

$$\lim_{j \to \infty} e^{nhz_j} \det \Delta(z_j) = 0.$$

Since the z_j are inside V, we have

$$\sup_{j \geq 1} \left|\operatorname{Re}\left(z_j + \frac{1}{h} \log z_j\right)\right| < \infty.$$

Therefore the usual "subsequence argument" allows us to assume that there exists a complex number w that

$$\lim_{j\to\infty} \operatorname{Re}\left(z_j + \frac{1}{h}\log z_j\right) = w.$$

Next, we define a sequence of analytic functions by

$$F_j(z) = e^{nhz}\det \Delta(z + i\operatorname{Im} z_j - \frac{1}{h}\operatorname{Re}\log z_j). \tag{5.10}$$

Since $\det \Delta$ is bounded by $K|z|^n$ in a neighbourhood $\mathcal{U}$ of V and $|e^{zh}| \le C_0|z|^{-1}$ for $z \in \mathcal{U}$ with $|z|$ sufficiently large, it follows that the sequence $\{F_j\}$ is uniformly bounded on $\mathcal{U}$. So from Montel's theorem [50, VII.2.9], we learn that the sequence $\{F_j\}$ forms a normal family. Accordingly, there exists a subsequence $\{F_{j_k}\}$ that converges uniformly on compact subsets of V to a limit function G. Since

$$F_j(\operatorname{Re}(z_j + \frac{1}{h}\operatorname{Re}\log z_j)) \to 0 \quad \text{as } j \to \infty,$$

it follows that

$$G(w) = 0.$$

Since G does not vanish identically, Hurwitz's theorem [50, VII.2.5] implies that all but a finite number of the F_j must have a zero inside the disk $\{z \mid |z-w| < \frac{\epsilon}{2}\}$. But this contradicts the fact that $|z_j - \lambda_k| > \epsilon$ for all j and k, and the theorem is proved. □

Using the lower bounds for $\det \Delta$, we have precise information about the density of the zeros of $\det \Delta$ and the growth of $|\Delta(z)^{-1}|$.

Theorem 5.7. *Let $\zeta \in NBV$ be such that $\det_* \zeta$ has an atom at nh. Then there exist constants ϵ, C and K such that*

$$|\Delta(z)^{-1}| \le K \min\{\frac{1}{|z|}, |e^{zh}|\} \tag{5.11}$$

for $|z| \ge C$ outside circles of radius ϵ centered around the zeros of $\det \Delta$.

Proof. From representation (2.9), it follows that for $|z| \ge C\max\{1, e^{-h\operatorname{Re} z}\}$, there is a constant M_0 such that

$$|\operatorname{adj} \Delta(z)| \le M_0|z|^{n-1}.$$

So, Theorem I.4.1 yields that for $|z| \ge C\max\{1, e^{-h\operatorname{Re} z}\}$

$$|\Delta(z)^{-1}| \le 2M_0\frac{1}{|z|}. \tag{5.12}$$

For $|z| \le C\max\{1, e^{-h\operatorname{Re} z}\}$, there is a constant M_1

$$\big|\mathrm{adj}\,\Delta(z)\big| \le M_1 \big|e^{-(n-1)h\mathrm{Re}\,z}\big|,$$

and Theorems 5.4 and 5.6 yield

$$\big|\Delta(z)^{-1}\big| \le 2\frac{M_1}{m}\big|e^{h\mathrm{Re}\,z}\big|.$$

Together with (5.12) this proves the theorem. □

Theorem 5.8. *Let $\zeta \in NBV$ be such that* $\det_* \zeta$ *has an atom at nh and let $B(a,r) \subset V(B_1, C_0)$ be a disk with center a and radius r with $|a| \ge 2$ and $r \le |a|/2$. If $\partial B(a,r)$ contains no zeros of* $\det \Delta$, *then the number of zeros inside $B(a,r)$ can be bounded independently of a and r.*

Proof. The proof is based on the argument principle [50, V.3.4] and the estimates from Theorem 5.6. Since $\partial B(a,r)$ contains no zeros of $\det \Delta$, there is an $\epsilon > 0$ such that the distance of $\partial B(a,r)$ to the zero set of $\det \Delta$ is at least ϵ. Therefore we can use the estimates from Theorem 5.6.

Let $n(a,r)$ denote the number of zeros of $\det \Delta$ inside $B(a,r)$. Then the argument principle yields

$$n(a,r) = \frac{1}{2\pi i}\int_{\partial B(a,r)} \frac{d}{dz}\det \Delta(z)\big(\det \Delta(z)\big)^{-1}\,dz. \tag{5.13}$$

To estimate $n(a,r)$, we first give estimates for the integrand. Since

$$\begin{aligned}\frac{d}{dz}\det \Delta(z) = nz^{n-1} &+ \sum_{j=1}^{n}\int_0^{jh} e^{-zt}(-t)\,d\eta_j(t) z^{n-j}\\ &+ \sum_{j=1}^{n}\int_0^{jh} e^{-zt}\,d\eta_j(t)(n-j)z^{n-1-j},\end{aligned}$$

we have for $z \in V(B_1, C_0)$,

$$\begin{aligned}\Big|\frac{d}{dz}\det \Delta(z)\Big| \le \big(n + B_1^{-1}h\mathrm{TV}(\eta_1)\big)|z|^{n-1} &+ \sum_{j=2}^{n}|z|^{n-j}B_1^{-j}jh\mathrm{TV}(\eta_j)\\ &+ \sum_{j=2}^{n}|z|^{n-j}B_1^{-(j-1)}\mathrm{TV}(\eta_{j-1}).\end{aligned}$$

On the other hand, Theorem 5.6 implies that for $z \in \partial B(a,r)$, there exists a positive constant m such that

$$|\det \Delta(z)| \ge \frac{1}{2}m|e^{-nz}| \ge \frac{1}{2}mC_0^n|z|^n,$$

where we have used $B(a,r) \subset V(B_1, C_0)$.

A combination of these estimates yields that for $z \in \Gamma$,

$$\begin{aligned}
&\Big|\frac{d}{dz}\det \Delta(z)\big(\det \Delta(z)\big)^{-1}\Big| \\
&\qquad \le \frac{2C_0^n}{m|z|}\Big(\big(n + B_1^{-1}h\mathrm{TV}(\eta_1)\big) + \sum_{j=2}^{n} |z|^{1-j}B_1^{-j}jh\mathrm{TV}(\eta_j) \\
&\qquad\qquad + \sum_{j=2}^{n} |z|^{1-j}B_1^{-(j-1)}\mathrm{TV}(\eta_{j-1})\Big).
\end{aligned} \tag{5.14}$$

Since for $r \le |a|/2$,

$$\frac{1}{2\pi}\Big|\int_{\partial B(a,r)} \frac{1}{|z|}\,dz\Big| = \frac{1}{2\pi}\int_0^{2\pi} \frac{1}{|a+re^{i\theta}|} r\,d\theta \le \frac{r}{|a|-r} \le 1$$

and for $z \in \partial B(a,r)$, with $r < |a|/2$,

$$\frac{1}{|z|} \le \frac{2}{|a|},$$

it follows from (5.13)–(5.14) that for $|a| \ge 2$,

$$\begin{aligned}
n(a,r) \le \frac{2C_0^n}{m}\Big(\big(n + B_1^{-1}h\mathrm{TV}(\eta_1)\big) &+ \sum_{j=2}^{n} B_1^{-j}jh\mathrm{TV}(\eta_j) \\
&+ \sum_{j=2}^{n} B_1^{-(j-1)}\mathrm{TV}(\eta_{j-1})\Big)
\end{aligned}$$

This completes the proof. □

We end this section with the introduction of a sequence of contours that will be used to compute the complex line integral in the next section.

Using Theorem 5.8 we can construct a sequence of real numbers ρ_l, such that $\rho_l \to \infty$, and a sequence of closed contours Γ_l, $l = 0, 1, \ldots$, such that for some positive constants k, ϵ and δ:

(i) Γ_l is contained in the interior of Γ_{l+1} and there are at most k zeros between Γ_l and Γ_{l+1};
(ii) the contours have at least distance $\epsilon > 0$ from the set of zeros of $\det \Delta$;
(iii) the contour Γ_l lies along the circle $|z| = \rho_l$ outside $V(B_1, C_0)$; inside $V(B_1, C_0)$, it lies between the circle $|z| = \rho_l - \delta$ and the circle $|z| = \rho_l + \delta$;
(iv) the length of the portion of Γ_l within $V(B_1, C_0)$ is bounded for $l \to \infty$.

For any real γ we denote the part of the sequence of contours Γ_l contained in the left half-plane $\{z \mid \operatorname{Re} z \le \gamma\}$ by $\Gamma_l^-(\gamma)$.

Exercise 5.9. Show that if $\det_* \zeta$ has an atom at nh, then $\det \Delta$ has finitely many zeros of multiplicity bigger than n.
Hint: Use the Rouché theorem.

For general kernels ζ, the number of zeros of $\det \Delta$ between two contours Γ_l and Γ_{l+1} can be unbounded (see Section 9 for more information).

We end this section with an auxiliary result.

Lemma 5.10. *For any real number* γ

$$\lim_{l\to\infty} \Big| \int_{\Gamma_l^-(\gamma)} e^{zt} \min\Big(\frac{|e^{-zh}|}{|z|}, 1\Big)\, dz \Big| = 0, \qquad \textit{for } t > 0. \tag{5.15}$$

The convergence is uniform for t *in any interval* $0 < t_0 < t < t_1 < \infty$.

Proof. First we prove that inside $V(B_1, C_0)$ the contour can be modified. There exists a positive constant K such that

$$\begin{aligned} \Big| e^{zt} \min\Big(\frac{|e^{-zh}|}{|z|}, 1\Big) \Big| &\le K e^{t \operatorname{Re} z} \\ &\le K e^{-th^{-1} \log(|z| C_0^{-1})} \\ &\le K e^{-th^{-1} \log((\rho_l - \delta) C_0^{-1})}. \end{aligned}$$

Here we have used that for $z \in V(B_1, C_0)$, $\operatorname{Re} z \le -h^{-1} \log(|z| C_0^{-1})$ and $|z| \ge C$, where C is sufficiently large. So, within $V(B_1, C_0)$, the integrand tends to zero as $l \to \infty$ uniformly for $t \ge t_0 > 0$. Since the lengths of the contours Γ_l inside $V(B_1, C_0)$ are bounded, we can replace the arcs within $V(B_1, C_0)$ by circular arcs.

The integral over $\Gamma_l^-(\gamma)$ can accordingly be replaced by the integral over the full semicircle $|z| = \rho_l$, $\operatorname{Re} z \le 0$, and, if $\gamma > 0$, the part of the semicircle in $0 < \operatorname{Re} z < \gamma$. It is clear that the contribution to the integral in (5.15) corresponding to the part of the semicircle in $0 < \operatorname{Re} z < \gamma$ approaches zero, uniformly in $0 < t_0 < t < t_1 < \infty$. It remains to estimate the integral

$$\int_{\frac{\pi}{2}}^{\frac{3\pi}{2}} e^{t\rho_l \cos\theta} \min\Big(\frac{e^{-h\rho_l \cos\theta}}{\rho_l}, 1\Big) \rho_l \, d\theta. \tag{5.16}$$

Choose θ_0, $0 \le \theta_0 \le \frac{\pi}{2}$ such that

$$e^{\rho_l h \sin\theta_0} = \rho_l. \tag{5.17}$$

Then the integral becomes

$$\begin{aligned} &\int_{\frac{\pi}{2}}^{\frac{3\pi}{2}} e^{t\rho_l \cos\theta} \min\{\rho_l^{-1} e^{-h\rho_l \cos\theta}, 1\} \rho_l \, d\theta \\ &\qquad = 2 \int_0^{\frac{\pi}{2}} e^{-t\rho_l \sin\theta} \min\{e^{h\rho_l \sin\theta}, \rho_l\}\, d\theta \\ &\qquad = 2 \int_0^{\theta_0} e^{(h-t)\rho_l \sin\theta}\, d\theta + 2 \int_{\theta_0}^{\frac{\pi}{2}} e^{-t\rho_l \sin\theta} \rho_l \, d\theta. \end{aligned}$$

Note that for $t > h$, we have

$$\int_0^{\theta_0} e^{(h-t)\rho_l \sin\theta}\, d\theta \le \theta_0$$

and that for $0 < t < h$, we have, using Jordan's inequality ($\frac{2}{\pi}\theta \le \sin\theta \le \theta,\ 0 \le \theta \le \frac{\pi}{2}$),

$$\int_0^{\theta_0} e^{(h-t)\rho_l \sin\theta}\, d\theta \le \int_0^{\theta_0} e^{(h-t)\rho_l \theta}\, d\theta \le \frac{e^{(h-t)\rho_l\theta_0} - 1}{\rho_l(h-t)}.$$

Using Jordan's inequality again, we find

$$\int_{\theta_0}^{\frac{\pi}{2}} e^{-t\rho_l \sin\theta} \rho_l\, d\theta \le \int_{\theta_0}^{\frac{\pi}{2}} e^{-t\rho_l \frac{2}{\pi}\theta} \rho_l\, d\theta \le \frac{\pi}{2t}\left[e^{-t\rho_l} - e^{-t\rho_l\frac{2}{\pi}\theta_0}\right].$$

From (5.14) it follows that

$$\theta_0 = \frac{1}{h}\frac{\log \rho_l}{\rho_l}\big(1 + o(1)\big) \qquad \text{as } l \to \infty.$$

Together we have that for $t > 0$

$$\lim_{l\to\infty} \int_0^{\theta_0} e^{(h-t)\rho_l \sin\theta}\, d\theta = 0, \quad \text{and} \quad \lim_{l\to\infty} \int_{\theta_0}^{\frac{\pi}{2}} e^{-t\rho_l \sin\theta} \rho_l\, d\theta = 0.$$

This proves (5.15). □

V.6 Series expansions

We know that for $\varphi \in \mathcal{C}$ and $\operatorname{Re} z > \omega_0$, the resolvent of the generator equals the Laplace transform of the semigroup

$$(zI - A)^{-1}\varphi = \int_0^\infty e^{-zt} T(t)\varphi\, dt, \qquad \operatorname{Re} z > \omega_0.$$

Here, as usual, ω_0 denotes the growth bound of the semigroup $T(t)$. As in Chapter I, the idea is to obtain an explicit representation of $T(t)\varphi$ itself using the inverse of the Laplace transform. The following inversion formula can be found in Hille and Phillips [124, Thm. 11.6.1].

Theorem 6.1. *Let $A : \mathcal{D}(A) \to \mathcal{C}$ be the generator of a $\mathcal{C}_0$-semigroup. Then, for every $\varphi \in \mathcal{D}(A)$ and uniformly for $0 < t_0 < t < t_1 < \infty$ we have*

$$(6.1) \quad T(t)\varphi = \lim_{\omega\to\infty} \frac{1}{2\pi i} \int_{\gamma - i\omega}^{\gamma + i\omega} e^{zt}(zI - A)^{-1}\varphi\, dz, \qquad \gamma > \omega_0, \quad t > 0.$$

For $t = 0$ the limit equals $\frac{1}{2}\varphi$:

$$\frac{1}{2}\varphi = \lim_{\omega\to\infty} \frac{1}{2\pi i} \int_{\gamma - i\omega}^{\gamma + i\omega} (zI - A)^{-1}\varphi \, dz. \tag{6.2}$$

We shall use the contours Γ_l introduced in the previous section to compute the integral in (6.1). First note that $\lambda \in \sigma(A)$ is possibly a pole of $z \mapsto e^{zt}(zI - A)^{-1}\varphi$ and that

$$\operatorname*{Res}_{z=\lambda} e^{zt}(zI - A)^{-1}\varphi = T(t)P_\lambda \varphi, \tag{6.3}$$

where P_λ denotes the spectral projection onto $\mathcal{M}_\lambda$.

Exercise 6.2. Prove (6.3).
Hint: Use Theorem 6.1 and the resolvent equation.

We shall prove the following result.

Theorem 6.3. *Let $\zeta \in NBV$ be such that $\det_* \zeta$ has an atom at nh. For $\varphi \in \mathcal{D}(A)$ we have*

$$T(t)\varphi = \sum_{j=1}^{\infty} T(t)P_{\lambda_j}\varphi, \qquad t > 0. \tag{6.4}$$

Here λ_j, $j = 1, 2, \ldots$, denote the eigenvalues of A ordered according to decreasing real part, and P_λ denotes the spectral projection associated with $\lambda \in \sigma(A)$ onto the generalized eigenspace $\mathcal{M}_\lambda$. The series in (6.4) converges in norm uniformly for $0 < t_0 < t < t_1 < \infty$.

Proof. Let Γ_l denote the sequence of contours introduced after Theorem 5.8. The Cauchy theorem implies that

$$T(t)\varphi = \lim_{l\to\infty} \Big(\sum_{j=1}^{m} T(t)P_{\lambda_j}\varphi - \frac{1}{2\pi i} \int_{\Gamma_l^-(\gamma)} e^{zt}(zI - A)^{-1}\varphi \, dz \Big), \tag{6.5}$$

where $\lambda_1, \ldots, \lambda_m$ are the zeros of $\det \Delta$ inside the area enclosed by the line $\operatorname{Re} z = \gamma$ and the contour $\Gamma_l^-(\gamma)$.

To estimate the remainder integral

$$\Big\| \frac{1}{2\pi i} \int_{\Gamma_l^-(\gamma)} e^{zt}(zI - A)^{-1}\varphi \, dz \Big\|,$$

we recall the representation

$$(zI - A)^{-1}\varphi = \Delta(z)^{-1}C(z)\varphi, \tag{6.6}$$

where (cf. the proof of Lemma 3.1)

$$\begin{aligned}(C(z)\varphi)(\theta) = \varphi(0)e^{z\theta} &+ z\int_0^{-\theta} e^{-zs}\varphi(s+\theta)\,ds \\ &- \int_0^{-\theta} e^{-zs}\Big(\int_0^s d\zeta(\tau)\varphi(s+\theta-\tau)\Big)\,ds \\ &+ \int_{-\theta}^{h} e^{-zs}\Big(\int_s^h d\zeta(\tau)\varphi(s+\theta-\tau)\Big)\,ds.\end{aligned}$$

For $\varphi \in \mathcal{D}(A)$ we can write

$$\begin{aligned}(C(z)\varphi)(\theta) = \varphi(\theta) &+ \int_0^{\theta} e^{-zs}[\dot{\varphi}_\theta(s) - \int_0^s d\zeta(\tau)\varphi_{-\theta}(s-\tau)]\,ds \\ &+ \int_{-\theta}^{h} e^{-zs}\Big(\int_s^h d\zeta(\tau)\varphi(s+\theta-\tau)\Big)\,ds.\end{aligned}$$

So, it follows that for $\varphi \in \mathcal{D}(A)$ there is a constant K such that

$$\|C(z)\varphi\| \leq K \max\{1, |e^{-zh}|\}\,(\|\varphi\| + \|A\varphi\|).$$

Therefore, using representation (6.6) for the resolvent of A, we have the following estimate

$$\|(zI-A)^{-1}\varphi\| \leq K|\Delta(z)^{-1}|\,\max\{1, |e^{-zh}|\}\,(\|\varphi\| + \|A\varphi\|).$$

Because of Theorem 5.7 and the definition of the contours Γ_l there exists a constant l_0 such that we have

$$\|(zI-A)^{-1}\varphi\| \leq K \min\{\frac{|e^{-zh}|}{|z|}, 1\}\,(\|\varphi\| + \|A\varphi\|), \tag{6.7}$$

for $z \in \Gamma_l$ and $l \geq l_0$. Consequently, Lemma 5.10 implies that

$$\lim_{l\to\infty} \|\frac{1}{2\pi i}\int_{\Gamma_l^-(\gamma)} e^{zt}(zI-A)^{-1}\varphi\,dz\| = 0.$$

□

Since $T(t)P_{\lambda_j}$ can be computed explicitly, we have the following corollary.

Corollary 6.4. *Let $\zeta \in NBV$ be such that $\det_* \zeta$ has an atom at nh. Then the solution $x(\cdot\,;\varphi)$ of RFDE (1.1) has a convergent series expansion*

$$x(t) = \sum_{j=1}^{\infty} e^{\lambda_j t} p_j(t), \qquad t > 0.$$

Here p_j is a polynomial of degree $m_{\lambda_j} - 1$, where m_{λ_j} denotes the multiplicity of λ_j as a zero of $\det \Delta$. The convergence is uniform for $0 < t_0 < t < t_1 < \infty$.

So, in case the system of eigenvectors and generalized eigenvectors of A is complete, the corollary gives very precise information about the solution. Although the zeros of $\det \Delta$ can in general not be computed explicitly, one can use computer algebra and numerical methods to compute accurately the zeros of $\det \Delta$ in a given right half-plane and the corresponding coefficients of the polynomials p_j. For the latter computation, one uses the theory of Chapter IV. In particular, one uses the result that the coefficients of the polynomials of p_j can be expressed in the Jordan chains of the matrix $\Delta(z)$ at $z = \lambda_j$.

The following corollary states that there is a dense set of initial data that can be expanded into a convergent spectral projection. The result is not optimal, but in applications it usually suffices to know the convergence for a dense set. See Section 9 for a much more general result.

Theorem 6.5. *Let $\zeta \in NBV$ be such that $\det_* \zeta$ has an atom at nh. If $\varphi \in \mathcal{D}(A^3)$, then*

$$\varphi = \sum_{j=1}^{\infty} P_{\lambda_j}\varphi, \tag{6.8}$$

the convergence being in norm, uniformly for $0 < t_0 < t < t_1 < \infty$.

Proof. From the inversion formula (6.2) and the Cauchy theorem we deduce that

$$\begin{aligned}\frac{1}{2}\varphi &= \lim_{\omega\to\infty} \frac{1}{2\pi i}\int_{\gamma-i\omega}^{\gamma+i\omega} (zI-A)^{-1}\varphi\,dz \\ &= \lim_{l\to\infty} \Big(\sum_{j=1}^{m} P_{\lambda_j}\varphi - \frac{1}{2\pi i}\int_{\Gamma_l^-(\gamma)} (zI-A)^{-1}\varphi\,dz\Big),\end{aligned}$$

where $\lambda_1, \ldots, \lambda_m$ are the zeros of $\det \Delta$ inside the area enclosed by the line $\operatorname{Re} z = \gamma$ and the contour $\Gamma_l^-(\gamma)$. So it suffices to compute the integral

$$\frac{1}{2\pi i}\int_{\Gamma_l^-(\gamma)} (zI-A)^{-1}\varphi\,dz.$$

For $\varphi \in \mathcal{D}(A^2)$

$$(zI-A)^{-1}\varphi = \frac{1}{z}\varphi + \frac{1}{z^2}A\varphi + \frac{1}{z^2}(zI-A)^{-1}A^2\varphi.$$

Since

$$\frac{1}{2\pi i}\int_{\Gamma_l^-(\gamma)} \frac{1}{z}\varphi\,dz = \frac{1}{2}\varphi$$

and

$$\lim_{l\to\infty} \Big\|\frac{1}{2\pi i}\int_{\Gamma_l^-(\gamma)} \frac{1}{z^2}A\varphi\,dz\Big\| = 0,$$

it remains to prove that

$$\lim_{l\to\infty} \|\frac{1}{2\pi i}\int_{\Gamma_l^-(\gamma)} \frac{1}{z^2}(zI-A)^{-1}A^2\varphi\, dz\| = 0.$$

From estimate (6.7) and Lemma 5.10, it follows that for $\varphi \in \mathcal{D}(A)$ there is a constant $K = K(\varphi)$ such that

$$\|(zI-A)^{-1}\varphi\| \le K.$$

So, for $\varphi \in \mathcal{D}(A^3)$,

$$\lim_{l\to\infty} \|\frac{1}{2\pi i}\int_{\Gamma_l^-(\gamma)} \frac{1}{z^2}(zI-A)^{-1}A^2\varphi\, dz\| = 0$$

and this completes the proof. □

Exercise 6.6. Let $\mathcal{E}$ be defined as in Definition 3.7 and let ζ be such that $\det_* \zeta$ has an atom at nh. Use Theorem 6.5 to give a different proof for the inclusion $\mathcal{E} \subset \overline{\mathcal{M}}$.

V.7 Lower bounds and the Newton polygon

In this section we shall give lower bounds for $|\det \Delta(z)|$ when $\mathrm{E}(\det \Delta) < nh$. Although the situation becomes more involved, estimates similar to those in Sections 5 and 6 can be made. To illustrate the main idea, we consider the following example of a quasi-coupled system:

$$\begin{cases} \dot{x}_1(t) = x_1(t-2), \\ \dot{x}_2(t) = x_2(t-1). \end{cases} \tag{7.1}$$

The characteristic matrix is given by

$$\Delta(z) = \begin{pmatrix} z - e^{-2z} & 0 \\ 0 & z - e^{-z} \end{pmatrix}$$

and

$$\det \Delta(z) = z^2 - e^{-z}z - e^{-2z}z + e^{-3z}.$$

In this example, $h = 2$ and $n = 2$. So $\mathrm{E}(\det \Delta) = 3 < 4 = nh$ and the system of eigenvectors and generalized eigenvectors of the generator A associated with RFDE (7.1) is not complete. Therefore $|e^{-2z}\Delta(z)^{-1}|$ has exponential growth in the left half-plane. In particular, the lower bounds for $\det \Delta$ from Section 5 do not apply. Indeed, if $|z| > |e^{-z}|$, then $|e^{-2z}z|$ dominates $|e^{-3z}|$. This is precisely what plays an important role.

In order to derive the precise lower bounds for $\det \Delta(z)$, for $|z|$ sufficiently large, we define the following sets:

$$W_0(C_0) = \{z \in \mathbb{C} \mid |z| \geq C_0|e^{-2z}|,\ |z| \geq C\},$$
$$W_1(B_1, C_1) = \{z \in \mathbb{C} \mid C_1|e^{-z}| \leq |z| \leq B_1|e^{-2z}|,\ |z| \geq C\},$$
$$W_2(B_2) = \{z \in \mathbb{C} \mid |z| \leq B_2|e^{-z}|,\ |z| \geq C\}.$$

Here C is sufficiently large.

Exercise 7.1. Show that the sets $W_1(B_1, C_1)$ and $W_2(B_2)$ are in the left half-plane $\operatorname{Re} z < 0$ for C sufficiently large (depending on B_1, C_1 and B_2).

From Theorem I.4.1, we know that there exists a constant C_0 such that within $W_0(C_0)$

$$|\det \Delta(z)| > \frac{1}{2}|z|^2.$$

According to Exercise 7.1, there exists a constant c such that the sets $W_1(B_1, C_1)$ and $W_2(C_2)$ are contained in the left half-plane $\operatorname{Re} z \leq c$ for every C.

Within the set $W_1(B_1, C_1)$, we have that

$$|z|^2 = |z|\,|z| \leq B_1|ze^{-2z}|, \quad |e^{-3z}| \leq C_1^{-1}|ze^{-2z}|, \quad |e^z| \leq B_1^{1/2}C^{-1/2}.$$

So

$$\begin{aligned} |\det \Delta(z)| &\geq |ze^{-2z}|\left|1 - (B_1 + C_1^{-1} + B_1^{1/2}C^{-1/2})\right| \\ &\geq \frac{1}{2}|ze^{-2z}| \end{aligned}$$

for B_1 sufficiently small and C_1 and C sufficiently large. Within the set $W_2(B_2)$ we have that

$$|z|^2 \leq B_2^2|e^{-2z}|,$$
$$|z||e^{-2z} + e^{-z}| \leq B_2|e^{-3z}|\,|1 + e^z|.$$

So

$$\begin{aligned} |\det \Delta(z)| &\geq |e^{-3z}|\left|1 - B_2^2e^c - B_2(1 + e^c)\right| \\ &\geq \frac{1}{2}|e^{-3z}| \end{aligned}$$

for B_2 sufficiently small.

We conclude that in this case there are three different areas with different lower bounds for $\det \Delta$. Furthermore, the large zeros of $\det \Delta$ are located in curvilinear strips

$$V_1(B_1, C_0) = \{z \in \mathbb{C} \mid B_1|e^{-2z}| \leq |z| \leq C_0|e^{-2z}|, \quad |z| \geq C\},$$

$$V_2(B_2, C_1) = \{z \in \mathbb{C} \mid B_2|e^{-z}| \leq |z| \leq C_1|e^{-z}|, \quad |z| \geq C\}.$$

The same arguments as given in the proof of Theorem 5.6 imply that within $V_1(B_1, C_0)$ ($V_2(B_2, C_1)$) for values of z outside small circles containing the roots of $\det \Delta$, the same lower bounds hold as for $W_1(B_1, C_1)$ ($W_2(B_2)$) hold. So

$$|\det \Delta(z)| \geq \frac{1}{2}|z||e^{-2z}|$$

for $z \in W_1(B_1, C_1) \cup V_1(B_1, C_0)$ outside small circles containing the roots of $\det \Delta$, and

$$|\det \Delta(z)| \geq \frac{1}{2}|e^{-3z}|$$

for $z \in W_2(B_2) \cup V_2(B_2, C_1)$ outside small circles containing the roots of $\det \Delta$.

To see in a general situation which terms in $\det \Delta$ are dominant, we shall use the Newton polygon. Let

$$H(z) = a_n e^{-\tau_n z} z^{\nu_n} + a_{n-1} e^{-\tau_{n-1} z} z^{\nu_{n-1}} + \cdots + a_0 e^{-\tau_0 z} z^{\nu_0} \tag{7.2}$$

where a_j, $j = 0, \ldots, n$, are real numbers and $a_j \neq 0$, the exponents τ_j, $j = 0, \ldots, n$, are nonnegative real numbers, and the exponents ν_j, $j = 0, \ldots, n$, are nonnegative integers such that $\nu_n \geq \nu_{n-1} \geq \cdots \geq \nu_0 \geq 0$ and if for some j, $\nu_j = \nu_{j-1}$, then $\tau_j < \tau_{j-1}$.

For each a_j, $j = 0, \ldots, n$, assign a point A_j in the plane with coordinates (τ_j, ν_j). To define the Newton polygon of H, denoted by $N(H)$, we start at A_n. The point A_n always belongs to the Newton polygon. A line segment connecting A_n to one of the other points A_j, $j = 0, \ldots, n-1$, belongs to the Newton polygon of H if it has negative slope and there is no other line segment connecting A_n to one of the other points A_j that either has a less negative slope or a larger (Euclidean) length. If such a line segment exists we repeat the procedure with respect of the right end point of this line segment, say A_{j_1}. Now a line segment connecting A_{j_1} to one of the other points A_j, $j = 0, \ldots, n-1$, belongs to the Newton polygon of H if it has a slope less than the slope of the previous line segment (connecting A_n to A_{j_1}) and there is no other line segment connecting A_{j_1} to one of the other points A_j that either has a less negative slope or a larger (Euclidean) length. We repeat this procedure inductively until it comes to a stop. This defines the Newton polygon of H.

For example, the Newton polygon of $z^2 - ze^{-z} - ze^{-2z} + e^{-3z}$ is given in Figure V.3. The Newton polygon of $e^{-3z}z^2 + e^{-z}z + e^{-2z}$ consists only of the point $A_3 = (3, 2)$.

Let A_{j_i}, $i = 0, \ldots, m$ ($n = j_0 > j_1 > \cdots > j_m$) denote the vertices of the Newton polygon and L_k the corresponding line segments connecting $A_{j_{k-1}}$ and A_{j_k}, $k = 1, 2, \ldots, m$. The slope of the line segment L_k will be denoted by α_k.

The line segments of the Newton polygon will tell us the asymptotic (for large z) location of the roots of $H(z)$ and will give us lower bounds for $|H(z)|$. Given a real number c and a complex number z in the left

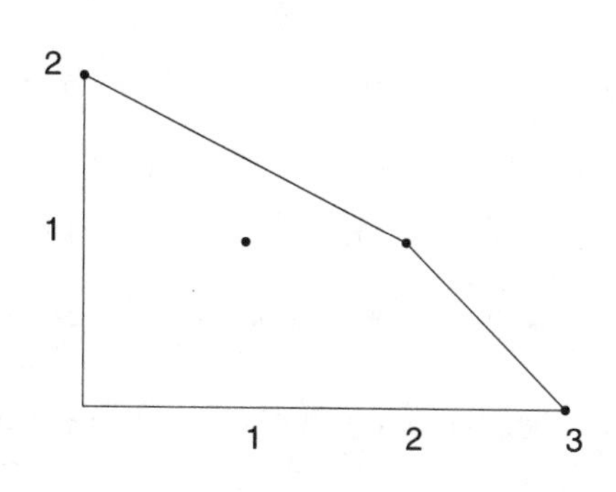

Fig. V.2. The Newton polygon for $z^2 - ze^{-z} - ze^{-2z} + e^{-3z}$.

half-plane $\{z \mid \operatorname{Re} z \leq c\}$ with $|z|$ sufficiently large, the Newton polygon will determine which term in (7.2) is dominant. The case $m = 1$ is not very interesting because the term $a_n e^{-\tau_n z} z^{\nu_n}$ is dominant in the whole left half-plane $\{z \mid \operatorname{Re} z \leq c\}$.

Exercise 7.2. Show that if the Newton polygon of H consists of a single point, there exist constants c, $d > 0$ such that

$$|H(z)| \geq d|e^{-\tau_n z}|\,|z|^{\nu_n}, \qquad \operatorname{Re} z \leq c. \tag{7.3}$$

In order to derive the precise lower bounds for $H(z)$, for $|z|$ sufficiently large, we define the following sets:

$$W_0(C_0) = \{z \in \mathbb{C} \mid |z| \geq C_0 |e^{\alpha_1^{-1} z}|,\ |z| \geq C\},$$

for $k = 1, \ldots, m-1$

$$W_k(B_k, C_k) = \{z \in \mathbb{C} \mid C_k |e^{\alpha_{k+1}^{-1} z}| \leq |z| \leq B_k |e^{\alpha_k^{-1} z}|,\ |z| \geq C\}, \tag{7.4}$$

and

$$W_m(B_m) = \{z \in \mathbb{C} \mid |z| \leq B_m |e^{\alpha_m^{-1} z}|,\ |z| \geq C\}.$$

Furthermore, define for $k = 1, \ldots, m-1$, the sets

$$V_k(B_k, C_{k-1}) = \{z \in \mathbb{C} \mid B_k |e^{\alpha_k^{-1} z}| \leq |z| \leq C_{k-1} |e^{\alpha_k^{-1} z}|,\ |z| \geq C\}. \tag{7.5}$$

In general we shall suppress in the notation the explicit dependence of W_k and V_k on constants. The following theorem provides us with the estimates in the general case.

Theorem 7.3. *Let H be given by (7.2) and suppose that the Newton polygon of H has m line segments. Then, for $k = 0, 1, \ldots, m$, there are positive constants ϵ, d_k, B_k, C_k and C such that*

$$|H(z)| \geq \frac{1}{2} d_k |e^{-\tau_{j_k} z}||z|^{\nu_{j_k}} \tag{7.6}$$

for $z \in W_k$. Here W_k is defined in (7.4). Furthermore, for $z \in V_k$, $k = 1, \ldots, m$,

$$|H(z)| \geq \frac{1}{2} d_k |e^{-\tau_{j_k} z}||z|^{\nu_{j_k}} \tag{7.7}$$

outside circles of radius ϵ centered around the zeros of H.

In a series of exercises we shall outline the proof of Theorem 7.3. The proof is based on the same ideas as given in the proofs of Theorem 5.4 and 5.6.

Exercise 7.4. For $k = 1, \ldots, m$, let L_k be the line segment of the Newton polygon of H connecting the points $A_{j_{k-1}}$ and A_{j_k}, and let α_k denote the slope of the line segment L_k. Show that for $i = j_k, j_k + 1, \ldots, j_{k-1}$

$$\frac{\nu_i - \nu_{j_k}}{\tau_{j_k} - \tau_i} \leq -\alpha_k. \tag{7.8}$$

Exercise 7.5. For $k = 1, \ldots, m$, let L_k be the line segment of the Newton polygon of H connecting the points $A_{j_{k-1}}$ and A_{j_k} and let H_k denote the part of H that corresponds to this line segment

$$H_k(z) = \sum_{i=j_k}^{j_{k-1}-1} a_i e^{\tau_i z} z^{\nu_i}. \tag{7.9}$$

Prove that there exist positive constants C, B_k, d_k and ϵ such that

$$|H_k(z)| \geq \frac{d_k}{2} |e^{-\tau_{j_k} z} z^{\nu_{j_k}}|$$

for $|z| \leq B_k |e^{\alpha_k^{-1} z}|$ and $|z| \geq C$ (outside circles of radius ϵ centered around the zeros of H_k).

Hint: First write

$$H_k(z) = e^{-\tau_{j_k} z} z^{\nu_{j_k}} \left(a_{j_k} + \sum_{i=j_k+1}^{j_{k-1}-1} a_i e^{(\tau_{j_k} - \tau_i) z} z^{\nu_i - \nu_{j_k}}\right).$$

Next use Exercise 7.4 to show that

$$\sum_{i=j_k+1}^{j_{k-1}-1} |a_i e^{(\tau_{j_k} - \tau_i) z} z^{\nu_i - \nu_{j_k}}|$$

provided $\operatorname{Re} z - \alpha_k \log |z| < -\alpha_k \log B_k < 0$ and $|z| \geq C$.

Exercise 7.6. Define

$$H_0(z) = a_n e^{-\tau_n z} z^{\nu_n} \quad \text{and} \quad H_{m+1}(z) = \sum_{i=0}^{j_m} a_i e^{\tau_i z} z^{\nu_i}.$$

Then

$$H(z) = \sum_{k=0}^{m+1} H_k(z).$$

Prove that for $k = 0, \dots, m$, there exist positive constants C, C_k ($C_m = 0$) and d_k such that

$$|H(z)| \geq d_k |H_k(z)|, \qquad \text{for} \quad z \in W_k$$

(outside circles of radius ϵ centered around the zeros of H).
Hint: Write

$$H(z) = H_k(z)(1 + \sum_{j=0,\, i\neq k}^{m+1} \frac{H_i(z)}{H_k(z)}).$$

Next use the estimate from Exercise 7.5 to show that

$$\sum_{j=0,\, i\neq k}^{m+1} |\frac{H_i(z)}{H_k(z)}| < \frac{1}{2}$$

provided $\operatorname{Re} z - \alpha_{k+1} \log|z| \geq -\alpha_{k+1} \log C_k$ and $|z| \geq C$.

Exercise 7.7. Prove Theorem 7.3 for $k = 1, 2, \dots, m$.
Hint: Use Exercises 7.5 and 7.6 to give the lower bounds in W_k. Next use an argument as given in the proof of Theorem 5.6 to give the lower bounds in V_k.

Corollary 7.8. *Let H given by (7.2). For $i = 0, 1, \dots, n$ there exist positive constants ϵ, C, and K such that*

$$|\frac{e^{-\tau_i z} z^{\nu_i}}{H(z)}| \leq K, \quad \text{for } |z| \geq C, \tag{7.10}$$

outside circles of radius ϵ centered around the zeros of H.

Proof. For every $z \in \mathbb{C}$ with $|z| \geq C$, where C is chosen as in Theorem 7.3, we have $z \in W_0$ or there exists a unique k, $k \in \{1, \dots, m\}$, such that

$$z \in V_k \cup W_k.$$

If $z \in W_0$, we set $k = 0$. It follows from Theorem 7.3 that

$$|H(z)| \geq \frac{1}{2} d_k |z|^{\nu_{j_k}} |e^{-\tau_{j_k} z}|.$$

But from the proof of Theorem 7.3 (see Exercises 7.5 and 7.6) it also follows that there exists a positive constant K_0 such that

$$\sum_{i=0, i\neq k} |a_i e^{-\tau_i z} z^{\nu_i}| \leq K_0 |z|^{\nu_{j_k}} |e^{-\tau_{j_k} z}|.$$

□

It is not difficult to prove a generalization of Theorem 7.3 that applies to a larger class of functions. In order to do this we start with a definition.

Definition 7.9. A function $\eta \in \mathrm{NBV}(\mathbb{R}_+, \mathbb{R})$ belongs to $\mathrm{SBV}(\mathbb{R}_+, \mathbb{R})$, if there exists a $t_0 > 0$ such that $\eta(t_0-) \neq \eta(t_0)$ and $\eta(t) = \eta(t_0)$ for $t \geq t_0$. (This means that η has an atom before it becomes constant.)

Let $G : \mathbb{C} \to \mathbb{C}$ be defined as

$$G(z) = \sum_{j=0}^{n} \int_0^{\tau_j} e^{-z\theta}\, d\eta_j(\theta)\, z^{\nu_j}, \tag{7.11}$$

where $\eta_j \in \mathrm{SBV}(\mathbb{R}_+, \mathbb{R})$, $\eta_n \neq 0$, τ_j are positive real numbers such that if $\tau_j > 0$, then $\eta(\tau_j-) \neq \eta(\tau_j)$, and $\eta(t) = \eta(\tau_j)$ for $t \geq \tau_j$. Furthermore, the ν_j are nonnegative integers such that $\nu_n \geq \nu_{n-1} \geq \ldots \geq \nu_0 \geq 0$.

The Newton polygon of G we define to be the Newton polygon of H, where

$$H(z) = \sum_{j=0}^{n} \big(\eta_j(\tau_j) - \eta_j(\tau_j-)\big) e^{-\tau_j z}\, z^{\nu_j}. \tag{7.12}$$

The next lemma clarifies the motivation for this definition.

Lemma 7.10. *Let $G : \mathbb{C} \to \mathbb{C}$ be given by (7.11) and let H be given by (7.12). Then there exist positive constants ϵ, m, M and c such that*

$$m \leq \left|\frac{G(z)}{H(z)}\right| \leq M, \quad \textit{for } \operatorname{Re} z < c,$$

outside circles of radius ϵ centered around the zeros of H.

Proof. From Lemma 5.1 it follows that there exist positive constants m_k, M_k and c such that

$$m_k |e^{-\tau_k z}| \leq \left|\int_0^{\tau_k} e^{-z\theta} d\eta_k(\theta)\right| \leq M_k \max\{1, |e^{-\tau_k z}|\} \tag{7.13}$$

for $\operatorname{Re} z < c$. So the lemma follows from Corollary 7.8. □

Lemma 7.10 implies that the lower bounds for H obtained in Theorem 7.3 yield lower bounds for G as well. Furthermore, the zeros of G are contained inside circles of radius ϵ centered around the zeros of H.

Theorem 7.11. *Let G be given by (7.11) and suppose that the Newton polygon of G has m line segments. Then for $k = 0, 1, \ldots, m$, there are positive constants ϵ, c, d_k, B_k, C_k and C such that*

$$|G(z)| \geq \frac{1}{2} d_k |e^{-\tau_{j_k} z}| |z|^{\nu_{j_k}} \tag{7.14}$$

for $z \in W_k$. Here W_k is defined in (7.4). Furthermore, for $z \in V_k$, $k = 1, \ldots, m$,

$$|G(z)| \geq \frac{1}{2} d_k |e^{-\tau_{j_k} z}| |z|^{\nu_{j_k}} \tag{7.15}$$

in the left half-plane $\{z \mid \operatorname{Re} z < c\}$, outside circles of radius ϵ centered around the zeros of G.

In the last part of this section, we present a result that we shall use in the next section to analyse the asymptotic behaviour of $|\Delta(z)^{-1}|$.

Let

$$F(z) = \sum_{j=0}^{n} \int_0^{\tau_j} e^{-z\theta}\, d\zeta_j(\theta)\, z^{\nu_j}, \tag{7.16}$$

where $\zeta_j \in \mathrm{NBV}(\mathbb{R}_+, \mathbb{R})$ and

$$G(z) = \sum_{j=0}^{n} \int_0^{\tau_j} e^{-z\theta}\, d\eta_j(\theta)\, z^{\nu_j}, \tag{7.17}$$

where $\eta_j \in \mathrm{SBV}(\mathbb{R}_+, \mathbb{R})$. We write

$$\mathrm{N}(G) \leq \mathrm{N}(F) \tag{7.18}$$

to denote that the subset of the plane bounded above by $\mathrm{N}(F)$ and enclosed by the vertical lines $x = 0$ and $x = \mathrm{E}(F)$, is a subset of the set bounded above by $\mathrm{N}(G)$ and enclosed by the vertical lines $x = 0$ and $x = \mathrm{E}(G)$. Using this notation we can formulate the following corollary.

Corollary 7.12. *Let F and G be given by (7.16) and (7.17), respectively. If $\mathrm{N}(F) \leq \mathrm{N}(G)$, then there exists constants c and K such that*

$$\left|\frac{F(z)}{G(z)}\right| \leq K \quad \text{for } \operatorname{Re} z \leq c, \tag{7.19}$$

outside circles of radius ϵ centered around the zeros of G.

Proof. From representation (7.16) for F it follows that we have the estimate

$$|F(z)| \leq \sum_{j=0}^{n} M_j \max\{1, |e^{-\tau_j z}|\} |z|^{\nu_j}, \tag{7.20}$$

where M_j are positive constants. So, it suffices to prove that there exists a constants c and K such that for $j = 0, 1, \ldots, n$,

$$\frac{|e^{-\tau_j z} z^{\nu_j}|}{|G(z)|} \leq \frac{K}{n M_j} \quad \text{for } \operatorname{Re} z < c. \tag{7.21}$$

The same argument as in the proof of Corollary 7.8 shows that estimate (7.21) and the condition $\mathrm{N}(F) \leq \mathrm{N}(G)$ imply that there exist constants c and K such that (7.21) holds for $j = 0, 1, \ldots, n$. □

In general, when $\mathrm{N}(F) \leq \mathrm{N}(G)$ does not hold, we have to use Theorem 7.11 directly to find lower bounds for G in different regions in the complex plane (the sets W_k for G).

V.8 Noncompleteness, series expansions and examples

In this section we present results about series expansions in case the system of eigenvectors and generalized eigenvectors of the generator is not complete. It is clear that a necessary condition for convergence of the series

$$\sum_{j=1}^{\infty} T(t) P_{\lambda_j} \varphi \tag{8.1}$$

to $T(t)\varphi$ for $t > 0$ is given by $\varphi \in \overline{\mathcal{M}}$. Because of Theorem 6.3 one might expect that the series in (8.1) converges for $t > 0$ to $T(t)\varphi$ if $\varphi \in \mathcal{D}(A) \cap \overline{\mathcal{M}}$. It turns out, however, that this is not always the case. The reason is that in the situation of Theorem 6.3, the Newton polygon of $\det \Delta$ is a straight line, whereas in the case $\mathrm{E}(\det \Delta) < nh$ the Newton polygon can be a more complicated polygon.

In the sequel we shall investigate these questions, but first we have to restrict our class of kernels ζ so that we can apply the results from Section 7. If

$$\det \Delta(z) = z^n + \sum_{j=1}^{n} \int_0^{\tau_j} e^{-z\theta}\, d\eta_j(\theta) z^{n-j},$$

we have to impose that η_j, $j = 1, \ldots, n$, belong to $\mathrm{SBV}(\mathbb{R}_+, \mathbb{R})$ (see Definition 7.9). In short, we shall say $\zeta \in \mathrm{SBV}$ if η_j, $j = 1, \ldots, n$, belong to $\mathrm{SBV}(\mathbb{R}_+, \mathbb{R})$; for example, if $n = 1$ and ζ has an atom at h, or if $n = 2$ and $\zeta_{11} + \zeta_{22} \in \mathrm{SBV}(\mathbb{R}_+, \mathbb{R})$ and $\det_* \zeta \in \mathrm{SBV}(\mathbb{R}_+, \mathbb{R})$, or if ζ consists of atoms only.

Let $\mathrm{N}(\operatorname{adj} \Delta) \leq \mathrm{N}(\det \Delta)$ denote that

$$\mathrm{N}((\operatorname{adj} \Delta)_{ij}) \leq \mathrm{N}(\det \Delta), \qquad i, j = 1, \ldots, n.$$

Observe that if $\mathrm{E}(\det \Delta) = nh$, then

$$\mathrm{N}(z \mapsto e^{-zh} \operatorname{adj} \Delta(z)) \leq \mathrm{N}(\det \Delta). \tag{8.2}$$

An application of Corollary 7.12 yields that if (8.2) holds, then there exist constants c and M such that

$$\left|e^{-zh}\Delta(z)^{-1}\right| \le M, \quad \text{for } \operatorname{Re} z < c,$$

outside circles of radius ϵ centered around the zeros of $\det \Delta$.

Together with Theorem I.4.1 (to give the estimate in the strip $c < \operatorname{Re} z < \gamma$ for $|z|$ large) this gives another proof of estimate (6.7) needed in order to prove Theorem 6.3. If $\mathrm{E}(\det \Delta) < nh$, then (8.2) does no longer hold and we need to make more precise estimates of $\|(zI - A)^{-1}\varphi\|$.

Recall the representation

$$(zI - A)^{-1}\varphi = \Delta(z)^{-1}C(z)\varphi, \tag{8.3}$$

where (cf. the proof of Lemma 3.1)

$$\begin{aligned}\big(C(z)\varphi\big)(\theta) = \varphi(0)e^{z\theta} &+ z\int_0^{-\theta} e^{-zs}\varphi(s+\theta)\,ds \\ &- \int_0^{-\theta} e^{-zs}\Big(\int_0^s d\zeta(\tau)\varphi(s+\theta-\tau)\Big)\,ds \\ &+ \int_{-\theta}^{h} e^{-zs}\Big(\int_s^h d\zeta(\tau)\varphi(s+\theta-\tau)\Big)\,ds.\end{aligned} \tag{8.4}$$

We have the following generalization of Theorem 6.3.

Theorem 8.1. *Let $\zeta \in SBV$, $\varphi \in C$ and let $C(z)\varphi$ be given by (8.4). If $\varphi \in \mathcal{D}(A)$ is such that*

$$\mathrm{N}(z \mapsto \operatorname{adj} \Delta(z)C(z)\varphi) \le \mathrm{N}(\det \Delta), \tag{8.5}$$

then

$$T(t)\varphi = \sum_{j=1}^{\infty} T(t)P_{\lambda_j}\varphi, \qquad t > 0. \tag{8.6}$$

Here λ_j, $j = 1, 2, \ldots$, denote the eigenvalues of A ordered according to decreasing real part and P_λ denotes the spectral projection associated with $\lambda \in \sigma(A)$ onto the generalized eigenspace $\mathcal{M}_\lambda$. The series in (8.6) converges in norm uniformly for $0 < t_0 < t < t_1 < \infty$.

Before we prove the theorem we shall first prove a lemma that will be useful later in this section as well. The following estimate is independent of $\mathrm{E}(\det \Delta)$.

Lemma 8.2. *There exists a positive constant K such that for every $\varphi \in \mathcal{D}(A)$*

$$\|(zI - A)^{-1}\varphi\| \le K|z|^{-1}\max\{1, |e^{-zh}|\}\big(\|\varphi\| + \|A\varphi\|\big) \tag{8.7}$$

for z in $\{z \in \mathbb{C} \mid |z| \ge C_0|e^{-z}|,\ |z| \ge C\}$. Here the constants C_0 and C are chosen as in Theorem I.4.1.

Proof. Because of Theorem I.4.1 there are positive constants C_0 and C such that

$$\left|\det \Delta(z)\right| \geq \frac{1}{2}|z|^n \tag{8.8}$$

for z in $\{z \in \mathbb{C} \mid |z| \geq C_0|e^{-z}|,\ |z| \geq C\}$. For $\varphi \in \mathcal{D}(A)$, it follows from (8.4), using integration by parts, that there exists a positive constant K_1 such that

$$\|C(z)\varphi\| \leq K_1 \max\{1, |e^{-zh}|\}\big(\|\varphi\| + \|A\varphi\|\big) \quad \text{for } z \in \mathbb{C}. \tag{8.9}$$

Using representation (2.9) for $\operatorname{adj} \Delta$, we find that there exists a constant K_2 such that

$$\|\operatorname{adj} \Delta(z) C(z)\varphi\| \leq K_2|z|^{n-1} \max\{1, |e^{-zh}|\}\big(\|\varphi\| + \|A\varphi\|\big). \tag{8.10}$$

So a combination of (8.8) and (8.10) proves (8.7). □

Proof of Theorem 8.1. Let Γ_l denote the sequence of contours introduced directly before Lemma 5.10. From equation (6.5) it follows that we have to estimate the remainder integral

$$\|\frac{1}{2\pi i}\int_{\Gamma_l^-(\gamma)} e^{zt}(zI - A)^{-1}\varphi\, dz\|. \tag{8.11}$$

To give the estimate for z in $\{z \in \mathbb{C} \mid |z| \geq C_0|e^{-z}|,\ |z| \geq C\}$ we use Lemma 8.2. To give the estimate for z in $\{z \in \mathbb{C} \mid |z| \leq C_0|e^{-z}|,\ |z| \geq C\}$, we have to use the Newton polygon condition (8.5). From Corollary 7.12, it follows that there exist constants c and K_3 such that

$$\left|\Delta(z)^{-1}C(z)\varphi\right| \leq K_3 \quad \text{for } \operatorname{Re} z < c, \quad z \in \Gamma_l^-(\gamma). \tag{8.12}$$

Therefore, if we combine (8.7) and (8.12), we have the following estimate: There exist positive constants l_0 and K_4 such that for $l \geq l_0$ and $z \in \Gamma_l^-(\gamma)$,

$$\|(zI - A)^{-1}\varphi\| \leq K_4 \min\{\frac{\max\{|e^{-zh}|, 1\}}{|z|}, 1\}\,\big(\|\varphi\| + \|A\varphi\|\big). \tag{8.13}$$

Consequently, an application of Lemma 5.10 shows that

$$\lim_{l\to\infty} \|\frac{1}{2\pi i}\int_{\Gamma_l^-(\gamma)} e^{zt}(zI - A)^{-1}\varphi\, dz\| = 0$$

and this completes the proof of the theorem. □

Since $T(t)P_{\lambda_j}$ can be computed explicitly, we have the following corollary.

Corollary 8.3. *Let $\zeta \in SBV$, $\varphi \in \mathcal{C}$ and let $C(z)\varphi$ be given by* (8.4). *If*

$$\mathrm{N}(z \mapsto \operatorname{adj} \Delta(z)C(z)\varphi) \leq \mathrm{N}(\det \Delta),$$

then the solution $x(\,\cdot\,;\varphi)$ of RFDE (1.1) *satisfies*

$$x(t;\varphi) = \sum_{j=1}^{\infty} e^{\lambda_j t} p_j(t) \quad \textit{for } t > 0,$$

uniformly for $0 < t_0 < t < t_1 < \infty$.

Next we shall exploit the observation that estimate (8.7) always holds [independent of $\mathrm{E}(\det \Delta)$].

Theorem 8.4. *Let $\zeta \in SBV$. If $\varphi \in \mathcal{D}(A)$, then*

$$T(t)\varphi = \sum_{j=1}^{\infty} T(t)P_{\lambda_j}\varphi, \qquad t > nh - \mathrm{E}(\det \Delta). \tag{8.14}$$

The series in (8.14) *converges in norm uniformly for $0 < t_0 < t < t_1 < \infty$.*

Proof. As in the proof of Theorem 8.1 we shall estimate the remainder integral in (8.11). Let $\tau = \mathrm{E}(\det \Delta)$ and define $G(z) = z^n + e^{-\tau z}$. Then $\mathrm{N}(G) \leq \mathrm{N}(\det \Delta)$ and hence Corollary 7.12 implies that there exist constants c, l_0 and K_0 such that for $l \geq l_0$,

$$\big|\det \Delta(z)\big| \geq K_0 \big|e^{-\tau z}\big| \quad \text{for } \operatorname{Re} z < c, \quad z \in \Gamma_l^-(\gamma). \tag{8.15}$$

Let C_0 be defined as in Theorem I.4.1. From (8.9) and (2.9), it follows that for $\varphi \in \mathcal{D}(A)$ and z in $\{z \in \mathbb{C} \mid |z| < C_0|e^{-z}|\}$, there exists a constant K_1 such that

$$\|\operatorname{adj} \Delta(z)C(z)\varphi\| \leq K_1 |e^{-nhz}| \big(\|\varphi\|| + \|A\varphi\|\big).$$

In combination with (8.15) this shows that there are constants K_2 and l_0 such that

$$\|e^{(\tau - nh)z} \Delta(z)^{-1} C(z)\varphi\| \leq K_2 \big(\|\varphi\|| + \|A\varphi\|\big) \tag{8.16}$$

for z in $\{z \in \mathbb{C} \mid |z| \geq C_0|e^{-z}|,\ z \in \Gamma_l^-(\gamma)\}$ and $l \geq l_0$. To estimate

$$\|e^{(\tau - nh)z} \Delta(z)^{-1} C(z)\varphi\|$$

for z in $\{z \in \mathbb{C} \mid |z| < C_0|e^{-z}|,\ \operatorname{Re} z < \gamma\}$ we use Lemma 8.2. Since for z with $\operatorname{Re} z < \gamma$

$$\big|e^{(\tau - nh)z}\big| \leq e^{(\tau - nh)\gamma},$$

it follows that

$$\begin{aligned}(8.17)\qquad &\|e^{(\tau-nh)z}(zI-A)^{-1}\varphi\|\\ &\qquad \le Ke^{(\tau-nh)\gamma}|z|^{-1}\max\{1,|e^{-zh}|\}(\|\varphi\|+\|A\varphi\|)\end{aligned}$$

for z in $\{z\in\mathbb{C}\mid |z|<C_0|e^{-z}|,\ \operatorname{Re} z<\gamma\}$. Therefore, if we combine (8.16) and (8.17), we have the following estimate. There exist positive constants l_0 and K_4 such that for $l\ge l_0$ and $z\in\Gamma_l^-(\gamma)$,

$$\begin{aligned}(8.18)\qquad &\|e^{(\tau-nh)z}(zI-A)^{-1}\varphi\|\\ &\qquad \le K_4\min\{\frac{\max\{|e^{-zh}|,1\}}{|z|},1\}\,(\|\varphi\|+\|A\varphi\|).\end{aligned}$$

So Lemma 5.10 yields that the integral in (8.11) tends to zero for $t>nh-\tau$. □

The following corollary is immediate.

Corollary 8.5. *Let $\zeta\in SBV$. For every $\varphi\in C$ the solution $x(\cdot\,;\varphi)$ of RFDE (1.1) satisfies*

$$x(t;\varphi)=\sum_{j=1}^{\infty}e^{\lambda_j t}p_j(t)\quad \textit{for } t>(n-1)h-\operatorname{E}(\det\Delta),$$

uniformly for $0<t_0<t<t_1<\infty$.

Instead of shifting in time to achieve convergence of the series expansion in (8.6), we can also restrict φ to a smaller subspace.

Theorem 8.6. *Let $\zeta\in SBV$. If $\varphi\in\overline{\mathcal{M}}\cap\mathcal{D}(A^n)$, then*

$$(8.19)\qquad T(t)\varphi=\sum_{j=1}^{\infty}T(t)P_{\lambda_j}\varphi,\qquad t>0.$$

The series in (8.19) converges in norm uniformly for $0<t_0<t<t_1<\infty$.

Proof. Let $\tau=\operatorname{E}(\det\Delta)$. Since $\varphi\in\overline{\mathcal{M}}$, we have $\operatorname{E}(z\mapsto \operatorname{adj}\Delta(z)C(z)\varphi)\le\tau$. So, representation (2.9) for adj Δ and (8.3) for $C(z)\varphi$ imply that there exists a constant K_0 such that for every $\varphi\in\mathcal{C}$,

$$(8.20)\qquad \|\operatorname{adj}\Delta(z)C(z)\varphi\|\le K_0|z|^n|e^{-\tau z}|\quad\text{for } \operatorname{Re} z<\gamma.$$

Therefore a combination with estimate (8.15) yields that for $z\in\{z\in\mathbb{C}\mid |z|<C_0|e^{-z}|,\ |z|\ge C\}\cap\Gamma_l^-(\gamma)$, we have

$$(8.21)\qquad \|(zI-A)^{-1}\varphi\|\le K_2|z|^n\|\varphi\|,\qquad \varphi\in\mathcal{C}.$$

Using Lemma 8.2 and the same argument as in the proof of Theorem 8.1, the theorem will follow once we have proved that for $\varphi \in \mathcal{D}(A^n)$ and $z \in \{z \in \mathbb{C} \mid |z| < C_0|e^{-z}|,\ |z| \geq C\} \cap \Gamma_l^-(\gamma)$

$$\|(zI - A)^{-1}\varphi\| \leq K_3, \tag{8.22}$$

for some constant K_3 depending on $A^j\varphi$, $j = 0, 1, \ldots$ [see (8.12)]. In order to achieve this estimate we shall use the resolvent equation. Fix $\lambda \in \rho(A)$. By complete induction one can prove the following identity:

$$\begin{aligned}(zI - A)^{-1}\varphi = (\lambda - z)^{-k}(zI - A)^{-1}\varphi_k \\ - \sum_{j=0}^{k-1}(\lambda - z)^{-(k-j)}(\lambda I - A)^{-1}\varphi_j,\end{aligned} \tag{8.23}$$

where $\varphi_j = (\lambda - A)^j\varphi$, $j = 0, 1, \ldots, k$. The second term on the right hand side of (8.23) is of order $|z|^{-1}$, so (8.21) and (8.23) (with $k = n$) imply that

$$\begin{aligned}\|(zI - A)^{-1}\varphi\| &\leq |\lambda - z|^{-n}\|(zI - A)^{-1}\varphi_n\| + K_4(\|A^j\varphi\|)|z|^{-1} \\ &\leq |z|^n|\lambda - z|^{-1}K_2\|\varphi_n\| + K_4(\|A^j\varphi\|)|z|^{-1} \\ &\leq K_5(\|A^j\varphi\|).\end{aligned} \tag{8.24}$$

□

Exercise 8.7. Analyse the behaviour of the norm of the inverse of the characteristic matrix for the systems of equations given in Exercise 2.4. What is your conclusion about the convergence of the series in (8.6) to the solution.

Example 8.8. The delayed friction force model is described by the equation

$$\ddot{x}(t) + a\dot{x}(t) + b\dot{x}(t - h) + cx(t) = 0. \tag{8.25}$$

To reduce this equation to a first order system introduce the variables $x_1 = x$ and $x_2 = \dot{x}$. Then (8.25) becomes

$$\begin{aligned}\dot{x}_1(t) &= x_2(t), \\ \dot{x}_2(t) &= -cx_1(t) - ax_2(t) - bx_2(t - h).\end{aligned} \tag{8.26}$$

The characteristic matrix of (8.26) is given by

$$\Delta(z) = \begin{pmatrix} z & -1 \\ c & z + a + be^{-hz} \end{pmatrix} \tag{8.27}$$

and

$$\det \Delta(z) = z^2 + (a + be^{-hz})z + c. \tag{8.28}$$

The matrix of cofactors equals

$$\operatorname{adj}\Delta(z) = \begin{pmatrix} z + a + be^{-hz} & 1 \\ -c & z \end{pmatrix}. \tag{8.29}$$

Next, we compute $C(z)\varphi$:

$$\begin{aligned}(C(z)\varphi)(\theta) = \begin{pmatrix} \varphi_1(\theta) \\ \varphi_2(\theta) \end{pmatrix} &+ \int_0^{-\theta} e^{-zs} \begin{pmatrix} \dot\varphi_1(\theta+s) \\ \dot\varphi_2(\theta+s) \end{pmatrix} ds \\ &- \int_0^{-\theta} e^{-zs} \begin{pmatrix} \varphi_2(s+\theta) \\ -c\varphi_1(s+\theta) + a\varphi_2(s+\theta) \end{pmatrix} ds \\ &+ \int_{-\theta}^{h} e^{-zs} \begin{pmatrix} 0 \\ b\varphi_2(s+\theta-1) \end{pmatrix} ds.\end{aligned}$$

Now one simply observes that

$$\overline{\mathcal{M}} = \{\varphi \in C \mid \dot\varphi_1 = \varphi_2\},$$

where A denotes the generator associated with RFDE (8.26). Furthermore, for $\varphi \in \mathcal{D}(A)$, condition (8.5) is satisfied. So Theorem 8.1 applies.

Exercise 8.9. (The delayed restoring force model.) Characterise $\overline{\mathcal{M}}$ for the generator associated with the RFDE

$$\ddot x(t) + a\dot x(t) + bx(t) + cx(t-h) = 0. \tag{8.30}$$

Prove that for $\varphi \in \overline{\mathcal{M}} \cap \mathcal{D}(A)$, the spectral projection series for $T(t)\varphi$ converges for $t > 0$.

Exercise 8.10. Consider the following differential-difference equation:

$$\begin{aligned}\dot x_1(t) &= x_1(t-1) - x_2(t-\frac{1}{4}) + x_2(t-1), \\ \dot x_1(t) &= -x_1(t-1) - x_1(t-\frac{1}{4}) - x_2(t-1).\end{aligned} \tag{8.31}$$

Let $\varphi \in \overline{\mathcal{M}} \cap \mathcal{D}(A)$, where A denotes the generator associated with RFDE (8.31). Does the spectral projection series for $T(t)\varphi$ converge for $t > 0$?

V.9 Arbitrary kernels of bounded variation

In the results about series expansions of spectral projections so far, we made the assumption that ζ was such that $\det_* \zeta$ has an atom at nh (or, in case of noncompleteness, we made the assumption that $\zeta \in \mathrm{SBV}$; see Section 8). The completeness result from Section 3, however, can be formulated for arbitrary kernels ζ. So it is natural to ask whether Theorem 6.3 holds for more general kernels ζ. In order to analyse this question as well we can no longer rely on the estimates from Section 5 (or, in general, from Section 7) to prove lower bounds for $\det \Delta$. Only the lower bound obtained in Theorem I.4.1 (and hence Lemma 8.2) remains valid.

To resolve this problem we shall prove lower bounds for $\det \Delta$ by using estimates for Blaschke products. We start with an abstract result about the density of the zeros of $\det \Delta$ which can be used to modify the construction of the contours Γ_l defined in Section 5. Let $n(r)$ denote the number of zeros of $\det \Delta$ with radius less than r.

Theorem 9.1. *The zeros of* $\det \Delta$ *have a density in the ordinary sense:*

$$\lim_{r\to\infty} \frac{n(r)}{r} = \frac{\mathrm{E}(\det \Delta)}{\pi}. \tag{9.1}$$

Proof. From the Alhfors-Heins Theorem 2.2 and the Lebesgue dominated convergence theorem we have for r tending to infinity

$$r^{-1}\int_0^{2\pi} \log|\det \Delta(re^{i\theta})|\, d\theta \to -\int_{\frac{\pi}{2}}^{\frac{3\pi}{2}} \mathrm{E}(\det \Delta)\cos\theta\, d\theta = 2\mathrm{E}(\det \Delta).$$

From Jensen's formula [50, XI.1.2]

$$\begin{aligned}\int_0^r \frac{n(s)}{s} ds &= \frac{1}{2\pi}\int_0^{2\pi} \log|\det \Delta(re^{i\theta})| d\theta \\ &\quad - \log \frac{1}{m!}\Big|\frac{d^m}{dz^m}\det \Delta(0)\Big| + m\log r,\end{aligned} \tag{9.2}$$

where $\det \Delta$ has a zero at $z = 0$ of multiplicity m, it follows that

$$r^{-1}\int_0^r \frac{n(s)}{s} ds \to \frac{\mathrm{E}(\det \Delta)}{\pi} \quad \text{as } r \to \infty. \tag{9.3}$$

For every $k > 1$ we have

$$\begin{aligned} n(r)\log k &\le \int_r^{kr} \frac{n(s)}{s}\, ds \\ &= \int_0^{kr} \frac{n(s)}{s}\, ds - \int_0^r \frac{n(s)}{s}\, ds. \end{aligned} \tag{9.4}$$

A combination of (9.3) and (9.4) implies that for every $\epsilon > 0$, we can choose r so large that

$$n(r) < \frac{k-1}{\log k}\Big(\frac{\mathrm{E}(F)}{\pi} + \epsilon\Big) r.$$

By taking the limit $k \downarrow 1$ we obtain

$$n(r) \le \left(\frac{\mathrm{E}(F)}{\pi} + \epsilon\right) r.$$

The proof that

$$n(r) \geq \left(\frac{\mathrm{E}(F)}{\pi} - \epsilon \right) r,$$

follows similar lines. □

Using Theorem 9.1 we can construct a sequence of real numbers ρ_l, such that $\rho_l \to \infty$, and a sequence of closed contours Γ_l, $l = 0, 1, \ldots$, such that for some positive constants ϵ and δ:

(i) Γ_l is contained in the interior of Γ_{l+1};
(ii) the contours have at least distance $\epsilon > 0$ from the set of zeros of $\det \Delta$;
(iii) the contour Γ_l lies along the circle $|z| = \rho_l$.

For any real γ we denote the part of the sequence of contours Γ_l contained in the left half-plane $\{z \mid \operatorname{Re} z \leq \gamma\}$ by $\Gamma_l^-(\gamma)$.

We shall prove the following theorem:

Theorem 9.2. *For every $\varphi \in \overline{\mathcal{M}} \cap \mathcal{D}(A^\infty)$, we have*

$$T(t)\varphi = \lim_{l \to \infty} \sum_{j=1}^{m_l} T(t) P_{\lambda_j} \varphi, \qquad t > 0. \tag{9.5}$$

Here λ_j, $j = 1, 2, \ldots, m_l$, are the eigenvalues of A inside the area enclosed by the line $\operatorname{Re} z = \gamma$ and the contour $\Gamma_l^-(\gamma)$, and P_{λ_j} is the spectral projection associated with $\lambda_j \in \sigma(A)$ onto the generalized eigenspace $\mathcal{M}_{\lambda_j}$. The limit in (9.5) *converges in norm uniformly for $0 < t_0 < t < t_1 < \infty$.*

Note that $\mathcal{D}(A^\infty)$ is dense in $\mathcal{C}$ (see Proposition 1.5 of Appendix II). In combination with Theorem 3.14, this result gives a complete picture in the general case. Without the condition on $\det_* \zeta$, $\det \Delta$ can have zeros of arbitrary large multiplicity. This means that, in general, the limit in (9.5) cannot be replaced by an infinite series.

General results like Theorem 3.14 and Theorem 9.2 are important in applications. For example, Theorem 4.5 and smoothness properties of the solution operator imply that if the solution x exists for all time, then $x_t \in \overline{\mathcal{M}} \cap \mathcal{D}(A^\infty)$ for every $t \in \mathbb{R}$. So, for solutions that exist for all time, Theorem 3.14 and Theorem 9.2 can be used without further assumptions.

The remaining part of this section is devoted to a proof of Theorem 9.2. We shall first prove an auxiliary result that replaces Theorem 5.6. Note that, because of Lemma 8.2, it suffices to prove estimates in the left half-plane $\{z \in \mathbb{C} \mid \operatorname{Re} z < 0\}$. In the next result, we shall give resolvent estimates for $z \in \Gamma_l^-(\gamma) \cap \{z \in \mathbb{C} \mid \operatorname{Re} z < 0\} [= \Gamma_l^-(0)]$.

Theorem 9.3. *Let $\varphi \in \mathcal{C}$. If $\mathrm{E}(z \mapsto P(z)\varphi) < \mathrm{E}(\det \Delta)$, then for l sufficiently large, there exist constants K independent of l and φ and $m = m(l)$ dependent of l but independent of φ such that, on the contour $\Gamma_l^-(0)$,*

$$\|(zI - A)^{-1}\varphi\| \leq K\|\varphi\||z|^m.$$

The proof of Theorem 9.3 is based on precise estimates for Blaschke products. It will be divided into four lemmas. For the proof of Lemma 9.4 and related results we refer to Boas [17].

Lemma 9.4. *Let* λ_j, $j = 1, 2, \ldots$, *denote the zeros of* $\det \Delta$ *in the left half-plane* $\{z \in \mathbb{C} \mid \operatorname{Re} z < 0\}$. *The infinite Blaschke product*

$$B(z) = \prod_{j=1}^{\infty} \frac{1 - \frac{z}{\lambda_j}}{1 + \frac{z}{\overline{\lambda_j}}} \tag{9.6}$$

converges uniformly on compact sets bounded away from the points $\{-\overline{\lambda}_j\}$.

Lemma 9.5. *If* B *denotes the Blaschke product corresponding to* $\det \Delta$, *then* $z \mapsto B(z)(zI - A)^{-1}$ *is an analytic function of exponential type in the left half-plane* $\{z \in \mathbb{C} \mid \operatorname{Re} z < 0\}$.

Proof. The function $B(z)(zI - A)^{-1}$ is clearly analytic for $\{z \in \mathbb{C} \mid \operatorname{Re} z < 0\}$. Furthermore, by applying the argument presented in the proof of [17, Thm. 6.4.5], we deduce that the function $z \mapsto B(z)(zI - A)^{-1}$ is of exponential type in the left half-plane. □

Lemma 9.6. *Let* $\varphi \in \mathcal{C}$. *If* $\mathrm{E}(z \mapsto P(z)\varphi) < \mathrm{E}(\det \Delta)$, *then there exists a constant* K *independent of* f *such that*

$$\|B(z)(zI - A)^{-1}\varphi\| \leq K\|\varphi\|$$

for z *in* $\{z \in \mathbb{C} \mid \operatorname{Re} z < 0\}$.

Proof. Since $|B(iy)| = 1$ there exists a constant K independent of φ such that on the imaginary axis

$$\|B(z)(zI - A)^{-1}\varphi\| \leq K\|\varphi\|.$$

Next, we estimate $\|B(z)(zI - A)^{-1}\varphi\|$ on the negative real axis. Since

$$|B(-x)| \leq 1$$

the condition $\mathrm{E}(z \mapsto P(z)\varphi) < \mathrm{E}(\det \Delta)$ and Theorem 2.2 imply that

$$\lim_{x \to \infty} \frac{\log \|B(-x)(-xI - A)^{-1}\varphi\|}{x} \leq 0.$$

An application of [17, Thm. 6.2.4] now yields

$$\|B(z)(zI - A)^{-1}\varphi\| \leq K\|\varphi\|$$

for z in $\{z \in \mathbb{C} \mid \operatorname{Re} z < 0\}$. □

Finally, we shall estimate $|B(z)^{-1}|$ on the contours $\Gamma_l^-(\gamma)$ in the left half-plane.

Lemma 9.7. *For every $\sigma > 0$ and each l sufficiently large, there exists a constant $m = m(l)$ so that on the contour $\Gamma_l^-(\gamma)$,*

$$|B(z)|^{-1} \le |z|^m e^{-\sigma \operatorname{Re} z}$$

for $z \in \Gamma_l^-(0)$.

Proof. To provide the estimate, we follow the method used by Ahlfors and Heins (see [17]). Consider the Green's function relative to $\operatorname{Re} z < 0$ with a pole at λ:

$$g(z, \lambda) = \log \left|\frac{z + \overline{\lambda}}{z - \lambda}\right|. \tag{9.7}$$

It then follows that

$$\log |B(z)| = \sum_{j=1}^{\infty} g(z, \lambda).$$

Fix l, let $z \in \Gamma_l$ and write

$$B(z)^{-1} = \prod_{j \in N_1} \frac{1 + \frac{z}{\overline{\lambda}_j}}{1 - \frac{z}{\lambda_j}} \prod_{j \in N_2} \frac{1 + \frac{z}{\overline{\lambda}_j}}{1 - \frac{z}{\lambda_j}}, \tag{9.8}$$

where

$$N_1 = N_1(l) = \{j \in \mathbb{N} : |z - \lambda|^2 \le \frac{1}{4}|\lambda|^2 \text{ for } z \in \Gamma_l\}$$

and

$$N_2 = N_2(l) = \{j \in \mathbb{N} : |z - \lambda|^2 > \frac{1}{4}|\lambda|^2 \text{ for } z \in \Gamma_l\}.$$

First, consider the set N_1. The condition $|z - \lambda|^2 \le \frac{1}{4}|\lambda|^2$ holds if and only if

$$|\frac{z}{\lambda} - 1|^2 \le \frac{1}{4}.$$

Since this condition is definitely not satisfied for

$$|z| < \frac{1}{2}|\lambda| \quad \text{or} \quad |z| > \frac{3}{2}|\lambda|,$$

we see that the set $N_1 = N_1(l)$ is finite and that for $j \in N_1$

$$\frac{1}{2}|\lambda| \le |z| \le \frac{3}{2}|\lambda|.$$

Next, rewrite (9.7):

(9.9)
$$g(z,\lambda) = \frac{1}{2}\log\bigl(1 + 4\frac{\operatorname{Re} z \operatorname{Re}\lambda}{|z-\lambda|^2}\bigr) \le 2\frac{\operatorname{Re} z \operatorname{Re}\lambda}{|z-\lambda|^2}.$$

Hence for $j \in N_2$, equation (9.9) implies

(9.10)
$$g(z,\lambda) \le 2\operatorname{Re} z \operatorname{Re}\frac{1}{\lambda}.$$

Now choose N_0 such that

$$\sum_{j=N_0}^{\infty} \operatorname{Re}\bigl(\frac{1}{\lambda}\bigr) > -\frac{1}{2}\sigma;$$

this is indeed possible since this sum converges [245, Thm. 4.9]. From (9.10)

(9.11)
$$\Bigl|\prod_{j\in N_2} \frac{1+\frac{z}{\lambda_j}}{1-\frac{z}{\lambda_j}}\Bigr| \le \prod_{j\in N_3} \frac{1+\frac{z}{\lambda_j}}{1-\frac{z}{\lambda_j}}\, e^{-\sigma\operatorname{Re}(z)},$$

where

$$N_3 = \bigl\{j \in N_2 : j \le N_0\bigr\}.$$

Hence for l sufficiently large, the finite sets N_1 and N_3 are both contained in the finite set

$$N_4 = N_4(l) = \bigl\{j \in \mathbb{N} : \frac{1}{2}|\lambda| < \min_{z\in\Gamma_l} |z|\ \bigr\}.$$

Therefore we can estimate

$$\frac{1}{B(z)} \le \prod_{j\in N_4} \frac{1+\frac{z}{\lambda_j}}{1-\frac{z}{\lambda_j}}\, e^{-\sigma\operatorname{Re}(z)},$$

where the finite product over N_4 can be estimated by

(9.12)
$$\Bigl|\prod_{j\in N_4} \frac{1+\frac{z}{\lambda_j}}{1-\frac{z}{\lambda_j}}\Bigr| \le \prod_{j\in N_4} |z|\Bigl|\frac{\frac{\lambda}{z}+\frac{\lambda}{\lambda}}{\nu}\Bigr| \le |z|^k\bigl(\frac{3}{\nu}\bigr)^k \le |z|^m,$$

for $|z| > 3\nu^{-1}$ and $m = 2k$, where k denotes the cardinality of N_4 and

$$\nu = \min_{j\ge 1}\ \bigl\{\ |z-\lambda| : z \in \Gamma_l^-(0)\bigr\}.$$

To complete the proof of the lemma, we choose l so large that for $z \in \Gamma_l^-(0)$ we have $|z| > 3\nu^{-1}$. □

This completes the proof of Theorem 9.3. Armed with these estimates for the resolvent we shall now prove Theorem 9.2.

Proof of Theorem 9.2. From equation (6.5) it follows that we have to estimate the remainder integral

$$\|\frac{1}{2\pi i}\int_{\Gamma_l^-(\gamma)} e^{zt}(zI-A)^{-1}\varphi\, dz\|. \tag{9.13}$$

To give the estimate for z in $\{z \in \mathbb{C} \mid |z| \geq C_0|e^{-z}|,\ |z| \geq C\}$ we use Lemma 8.2. To give the estimate for z in $\{z \in \mathbb{C} \mid |z| \leq C_0|e^{-z}|,\ |z| \geq C\}$, we shall use Theorem 9.3.

Since $\overline{\mathcal{M}} = \mathcal{E}$ (see Theorem 3.10), it follows that for $\varphi \in \mathcal{E} \cap \mathcal{D}(A^\infty)$, for every positive integer k and for every $\lambda_k \in \rho(A)$, there exists an element $\psi_k \in \mathcal{E}$ such that

$$\varphi = (\lambda_k - A)^{-k}\psi_k.$$

Therefore repeated application of the resolvent equation yields that for every k

$$\begin{aligned} e^{\epsilon z}(zI-A)^{-1}\varphi &= (\lambda_k I - z)^{-k} e^{\epsilon z} P(z)\psi_k(\det \Delta(z))^{-1} \\ &\quad - e^{\epsilon z}\sum_{j=0}^{k-1}(\lambda_k - z)^{-(k-j)}(\lambda_k I - A)^{k-j-1}\varphi. \end{aligned} \tag{9.14}$$

Here $\lambda_k \in \rho(A)$ will be a large positive number to be chosen later. The second term on the right hand side of (9.14) is a function of order $O(|z|^{-1})$ in the left half-plane, since

$$(\lambda_k - z)^{-(k-j)}(\lambda_k I - A)^{k-j-1} = \frac{(1 - z/\lambda_k)^{-(k-j-1)}}{(\lambda_k - z)}(I - \lambda_k^{-1}A)^{k-j-1}.$$

For the first term on the right hand side of (9.14), we would like to use Theorem 9.3. From (6.1) and the semigroup property,

$$\begin{aligned} e^{\epsilon z}(zI-A)^{-1}\psi_k &= \int_0^\infty e^{-z(t-\epsilon)}T(t)\psi_k dt \\ &= (zI-A)^{-1}T(\epsilon)\psi_k + \int_{-\epsilon}^0 e^{-zt}T(t+\epsilon)\psi_k dt. \end{aligned} \tag{9.15}$$

The integral on the right hand side of (9.15) is uniformly bounded in the left half-plane:

$$\int_{-\epsilon}^0 e^{-zt}T(t+\epsilon)\psi_k dt \leq M\|\psi_k\| \quad \text{for } \operatorname{Re} z < 0$$

where M is independent of k. Since $\psi_k \in \mathcal{E}$, the left hand side of (9.15) has no exponential growth in the left half-plane. Therefore

$$\mathrm{E}\big(z \mapsto P(z)T(\epsilon)\psi_k\big) < \mathrm{E}(\det \Delta)$$

and we can apply Theorem 9.3 to derive that, for sufficiently large l, there exist constants $m = m(l,q)$ and K independent of l and ψ_k, such that for every k,

$$\big\|e^{\epsilon z}(zI - A)^{-1}\psi_k\big\| \le K|z|^m \|T(\epsilon)\psi_k\| \quad \text{for } z \in \Gamma_l^-(0).$$

For every l, we choose $k = m+1$, and from (9.14)

$$\|e^{\epsilon z}(zI - A)^{-1}\varphi\| \le C\|\varphi\||z|^{-1}\frac{\|T(\epsilon)\psi_{m+1}\|}{(\lambda_{m+1}/|z| - 1)^{m+1}} \quad \text{for } z \in \Gamma_l^-(0).$$

Finally, we will choose $\lambda_{m+1} \in \rho(A)$ to control the norm of ψ_{m+1}. Since

$$\psi_{m+1} = (\lambda_{m+1}I - A)^{m+1}\varphi$$

we have

$$\|T(\epsilon)\psi_{m+1}\| = O(\lambda_{m+1}^{m+1}).$$

Therefore, there is a constant N independent of l such that for every m there exists a suitable $\lambda_{m+1} \in \rho(A)$ on the positive real axis so that for $z \in \Gamma_l^-(0)$

$$\frac{\|T(\epsilon)\psi_{m+1}\|}{(\lambda_{m+1}/|z| - 1)^{m+1}} \le N.$$

So we have proved the following estimate:

$$\|e^{\epsilon z}(zI - A)^{-1}\varphi\| \le KN\|\varphi\||z|^{-1} \quad \text{for } z \in \Gamma_l^-(0). \tag{9.16}$$

This estimate replaces (8.11) in the proof of Theorem 8.1, and the proof can now be completed by calling Lemma 5.10. □

V.10 Comments

The question of completeness of the system of eigenfunctions and generalized eigenfunctions of the generator associated with RFDE (1.1) has been studied by Henry [119], Delfour and Manitius [60, 61], Manitius [188] and Verduyn Lunel [279, 280, 282, 283].

Important results were obtained by Henry [119] who was the first to prove that for autonomous equations, small solutions are identically zero after finite time and, as an immediate corollary, that completeness holds if the adjoint semigroup is one-to-one.

The present presentation uses quite different techniques and allows us to obtain very precise results. The key result is Theorem 3.10 which was first proved in [280, 282]. Theorem 4.7 extends the result of Henry and was first given in [279].

Delfour and Manitius have restricted their work to kernels that consist of finitely many atoms and an absolutely continuous part. In their papers they also introduce the concept of F-completeness. The theory developed in this chapter can be used to given necessary and sufficient conditions for F-completeness.

The generalized eigenspace $\mathcal{M}$ is called *F-complete* if

$$\overline{F\mathcal{M}} = \overline{\mathcal{R}(F)},$$

where F is defined by (III.6.2). Delfour and Manitius proved that F-completeness holds if and only if

$$\mathcal{N}(T^*(\eta)) = \mathcal{N}(F^*),$$

where η denotes the ascent of $T^*(t)$ (see Corollary 4.6). From Exercise III.6.4 it follows that $\mathcal{N}(F^*) = \mathcal{N}(T^*(h))$ and F-completeness holds if and only if the ascent η of $T^*(t)$ is less than or equal to h. Therefore, using Corollary 4.6, it follows that F-completeness holds if and only if $\epsilon - \sigma \leq h$. For example, the system given in Example 4.8 is not F-complete.

The results in Sections 2, 3 and 4 have been developed over the years and partial results appeared in [280, 281, 283]. In [283] the results were extended to special classes of periodic equations.

The results in Sections 5 and 7 first appeared in [280, 281] and are generalizations of the estimates given by Bellman and Cooke [16]; for more details and further references we refer to this text.

The question whether the state x_t of the solution of RFDE (1.1) has a convergent polynomial exponential series has also been studied by Banks and Manitius [14], Levinson and McCalla [169] and Verduyn Lunel [280, 281, 284]. In [14] and [169] the kernel ζ is restricted to the class of piecewise constant kernels that have a finite number of discontinuities and such that $\zeta(h) - \zeta(h+)$ is nonsingular. The results in Section 6 are much more general.

Chapter VI

Inhomogeneous linear systems

VI.1 Introduction

When a linear delay system is subject to external "forcing", it can be described by the *inhomogeneous* equation

$$\dot{x}(t) = \langle \zeta, x_t \rangle_n + f(t) \tag{1.1}$$

with $f : \mathbb{R} \to \mathbb{C}^n$ a given (continuous) function describing the influence of the outside world. As in the foregoing chapters we shall rewrite (1.1) in the abstract form

$$\frac{du}{dt} = A^{\odot *} u + F \tag{1.2}$$

with $F : \mathbb{R} \to X^{\odot *}$ the continuous mapping defined by

$$F = F(t) = \sum_{j=1}^{n} r_j^{\odot *} f_j(t) \tag{1.3}$$

where, as before,

$$\begin{aligned}
X^{\odot *} &= \mathbb{C}^n \times L^\infty\big([-h, 0], \mathbb{C}^n\big), \\
A^{\odot *} &= A_0^{\odot *} + B, \\
\mathcal{D}\big(A^{\odot *}\big) &= \mathcal{D}\big(A_0^{\odot *}\big) = \{(\alpha, \varphi) \mid \varphi \in \mathrm{Lip}, \quad \alpha = \varphi(0)\}, \\
A_0^{\odot *}(\alpha, \varphi) &= (0, \varphi'), \\
B(\alpha, \varphi) &= (\langle \zeta, \varphi \rangle_n, 0), \\
r_j^{\odot *} &= (e_j, 0).
\end{aligned} \tag{1.4}$$

Here e_j is the jth unit vector in $\mathbb{C}^n$. But (1.2) is only a formal intermediate step on the way to the variation-of-constants expression

$$u(t) = T(t-s)u(s) + \int_s^t T^{\odot *}(t-\tau) F(\tau)\, d\tau \tag{1.5}$$

where $t \geq s$ and $u(s) \in X$ is considered as given (note that we would lose generality by taking an initial condition at time zero, since the equation now depends on time explicitly). In the literature, u given by (1.5) is called a *mild solution* of (1.2). The question of in what sense u given by (1.5) satisfies (1.2) [under various assumptions on $u(s)$ and F] is not an easy one, but many results are known (see Pazy [233, Section 4.2] as well as Clément et al. [46]). In the context of delay equations the question is irrelevant since we shall, in Section 4, establish a one-to-one correspondence between (1.5) and the solution of (1.1), thus bypassing the abstract differential equation (1.2).

Exercise 1.1. Show that, for u given by (1.5),

$$\lim_{h \downarrow 0} h^{-1}(u(t+h) - T(h)u(t)) = F(t). \tag{1.6}$$

Argue that the limits which define du/dt and $A^{\odot *}u$ may not exist but that one may intuitively interpret the limit in (1.6) as a "directional" derivative. When X is a function space and $A^{\odot *}$ a first order differential operator, this can be made more precise.

We shall first consider (1.5) for a continuous function F in general, then concentrate on the special case that F takes values in a finite dimensional subspace of $X^{\odot *}$ and finally consider delay equations, as specified by (1.1) and (1.4). Our main aim is to extend the spectral decomposition of Chapter IV to formula (1.5) and to derive an inhomogeneous ODE for the projection on a finite dimensional spectral subspace. In doing so, we introduce in the simplest possible context some ideas which will play a crucial role in the construction of invariant manifolds for nonlinear equations in Chapters VIII and IX.

VI.2 Decomposition in the variation-of-constants formula

Assume that X is $\odot$-reflexive with respect to A. Let Λ be a finite collection of isolated poles of the resolvent of A, all of finite order (or, in other words, a finite collection of eigenvalues of A of finite type). Let P denote the associated projection operator on X (see Section IV.2 or [202, 272, 313]). Likewise, let $P^{\odot}$ denote the projection operator on $X^{\odot}$ associated with Λ as a subset of $\sigma(A^{\odot})$. Let P^* and $P^{\odot *}$ be the adjoint operators of, respectively, P and $P^{\odot}$. The range of $P^{\odot *}$, $\mathcal{M}_\Lambda$, lies in (the embedding of) X (see Theorem IV.2.12) and the range of P^* in $X^{\odot}$. Moreover, $P\varphi = P^{\odot *}\varphi$ for all $\varphi \in X$ and $P^{\odot}x^{\odot} = P^*x^{\odot}$ for all $x^{\odot} \in X^{\odot}$. To verify that the adjoint of a bounded operator commutes with a weak* integral is the main step in doing the following exercise.

Exercise 2.1. Prove that

$$P^{\odot *} \int_s^t T^{\odot *}(t-\tau)F(\tau)\,d\tau = \int_s^t T(t-\tau)P^{\odot *}F(\tau)\,d\tau. \tag{2.1}$$

One has $\mathcal{M}_\Lambda \subset \mathcal{D}(A)$ and, consequently,

$$\frac{d}{dt}T(t)\varphi = AT(t)\varphi$$

for each $\varphi \in \mathcal{M}_\Lambda$. For $u(t)$ given by (1.5), define

$$v(t) = P^{\odot *}u(t) \in \mathcal{M}_\Lambda. \tag{2.2}$$

Combining (1.5) and (2.1) with (2.2) we find

$$v(t) = T(t-s)v(s) + \int_s^t T(t-\tau)P^{\odot *}F(\tau)\,d\tau. \tag{2.3}$$

We conclude that $v \in C^1$ and

$$\frac{dv}{dt} = Av + P^{\odot *}F. \tag{2.4}$$

This is summarized in

Theorem 2.2. *Let $\tau \mapsto F(\tau)$ be a norm continuous mapping from an interval $I \subset \mathbb{R}$ into $X^{\odot *}$ and let, for $s, t \in I$, $s \leq t$, $u(\cdot) \in X$ satisfy the variation-of-constants formula*

$$u(t) = T(t-s)u(s) + \int_s^t T^{\odot *}(t-\tau)F(\tau)\,d\tau.$$

If we let $v(t) = P^{\odot *}u(t)$, then $v(\cdot) \in C^1$ and satisfies (in $\mathcal{M}_\Lambda$) the differential equation (2.4).

Equation (2.4) is still coordinate free. An ODE in $\mathbb{C}^d$, where d denotes the dimension of the finite dimensional subspace $\mathcal{M}_\Lambda = P^{\odot *}X^{\odot *}$ of X, is obtained as follows. Choose a row vector Φ of elements of $\mathcal{M}_\Lambda$ which together form a basis. Recall from Section IV.2 that there exists a $d \times d$ matrix M such that

$$A\Phi = \Phi M,$$

which implies that

$$T(t)\Phi c = \Phi e^{tM}c, \qquad c \in \mathbb{C}^d.$$

We can think of Φ as the map $\mathbb{C}^d \to \mathcal{M}_\Lambda$ which assigns to $c \in \mathbb{C}^d$ the element $\Phi c \in \mathcal{M}_\Lambda$. Conversely, let $\Gamma : \mathcal{M}_\Lambda \to \mathbb{C}^d$ denote the linear coefficient map which assigns to $x \in \mathcal{M}_\Lambda$ the vector $c \in \mathbb{C}^d$ such that $x = \Phi c$. Then, $\Phi \circ \Gamma = \mathrm{Id}|_{\mathcal{M}_\Lambda}$ and $\Gamma \circ \Phi = \mathrm{Id}|_{\mathbb{C}^d}$. If we let

$$y(t) = \Gamma v(t) = \Gamma P^{\odot *} u(t), \tag{2.5}$$

then

$$\begin{aligned} \Gamma T(t-s) v(s) &= \Gamma T(t-s) \Phi \Gamma v(s) \\ &= \Gamma \Phi e^{(t-s)M} y(s) \\ &= e^{(t-s)M} y(s). \end{aligned}$$

This leads to the variation-of-constants formula for y:

$$y(t) = e^{(t-s)M} y(s) + \int_s^t e^{(t-\tau)M} \Gamma P^{\odot *} F(\tau)\, d\tau \tag{2.6}$$

and thus to

$$\dot{y} = My + \Gamma P^{\odot *} F, \tag{2.7}$$

or, in words, the coefficients of the spectral projection of $u(t)$ on a finite dimensional subspace satisfy an ODE.

Exercise 2.3. Discuss, in the spirit of Section III.8, the extension of these results to the non-$\odot$-reflexive case.

VI.3 Forcing with finite dimensional range

In this section we assume that

$$F(t) = \sum_{j=1}^{n} r_j^{\odot *} f_j(t) \tag{3.1}$$

for some set $\{r_j^{\odot *} \mid j = 1, \ldots, n\}$ of linearly independent elements of $X^{\odot *}$, and $f : \mathbb{R} \to \mathbb{C}^n$. There are now several new aspects:

(i) We can take $f \in L^1$, or $f \in L^\infty$ without worrying about measurability and integrability of functions with values in abstract spaces (recall Exercises III.1.4, III.2.2 and III.2.23).

(ii) Properties of $r^{\odot *}$ may guarantee smoothness of the integral term in (1.5) under weak conditions on f. Lemma III.4.3 in the form

$$\Big(\int_0^t T_0^{\odot *}(t-\tau) r^{\odot *} f(\tau)\, d\tau\Big)(\sigma) = \int_0^{\max(0,t+\sigma)} f(\tau)\, d\tau$$

provides a concrete example.

(iii) One can elaborate the formulas a little bit further.

Combining (3.1) and (2.7) we find

$$\frac{dy}{dt} = My + \sum_{j=1}^{n} \Gamma P^{\odot *} r_j^{\odot *} f_j(t). \tag{3.2}$$

VI.4 RFDE

In this section we shall establish the connection between, on the one hand, the explicit expression (1.5), with the spaces and the operators as defined in Section III.4 and summarized in (1.4), and, on the other hand, the inhomogeneous delay equation (1.1).

Exercise 4.1. Adapt Definition I.2.1 of a *solution* to the inhomogeneous equation (1.1). How does the smoothness of f come into play? How about giving two definitions, one for continuous f and one for $f \in L^1$?

Exercise 4.2. Show that any solution of (1.1) may be written as the sum of a solution of the homogeneous equation satisfying the initial condition and a solution of the inhomogeneous equation with zero initial condition. Conclude that we do not lose any generality by taking zero as the initial condition in the discussion below.

Let x be a solution of (1.1) for $t \geq s$ with initial condition $x(s+\theta) = 0$ for $-h \leq \theta \leq 0$. Manipulating exactly as in the beginning of Section I.2 we see that necessarily

$$\dot{x}(t) = \int_0^{t-s} \zeta(\theta)\dot{x}(t-\theta)\,d\theta + f(t), \quad t \geq s, \tag{4.1}$$

which we can rewrite as the RE

$$\dot{x}_s = \zeta * \dot{x}_s + f_s. \tag{4.2}$$

Hence $\dot{x}_s = f_s + R * f_s$ or, written in a slightly different form,

$$\dot{x}(t) = f(t) + \int_s^t R(t-\tau)f(\tau)\,d\tau, \tag{4.3}$$

from which it follows that

$$x(t) = \int_s^t f(\tau)\,d\tau + \int_s^t \int_0^{t-\sigma} R(\tau)\,d\tau f(\sigma)\,d\sigma, \qquad t \geq s, \tag{4.4}$$

or, in other symbols,

$$x(t) = \int_s^t Q(t-\sigma)f(\sigma)\,d\sigma, \qquad t \geq s, \tag{4.5}$$

with Q the fundamental matrix solution as introduced in equation (I.2.22).

Exercise 4.3. Verify the step from (4.3) to (4.4).

Exercise 4.4. Formulate, on the basis of Exercises 4.1 and 4.2, the results of Section I.2 and our manipulations above, a theorem about existence and uniqueness of a solution of (1.1) with initial condition $x(s+\theta) = \varphi(\theta)$, $-h \leq \theta \leq 0$.

In Section III.4 we found (cf. III.4.9) that for RFDE

$$(T^{\odot *}(t)r^{\odot *})(\theta) = \big(Q(t), Q(t+\theta)\big), \tag{4.6}$$

where Q is extended by zero for negative values of the argument.

So, if we define the X-valued function u by

$$u(t) = \sum_{j=1}^{n} \int_s^t T^{\odot *}(t-\tau)r_j^{\odot *} f_j(\tau)\,d\tau \tag{4.7}$$

(recall the embedding of X into $X^{\odot *}$ and Exercises III.1.4 and III.2.2), then

$$u(t)(\theta) = \int_s^{\max(t+\theta,s)} Q(t+\theta-\tau)f(\tau)\,d\tau. \tag{4.8}$$

We formulate our conclusion as a theorem:

Theorem 4.5. *Given an L^1-function $f : [s,\infty) \to \mathbb{C}^n$, let $u : [s,\infty) \to C\big([-h,0],\mathbb{C}^n\big)$ be defined by (4.7). Then $x(t) := u(t)(0)$ is the unique solution of the RFDE (1.1) satisfying the initial condition $x(s+\theta) = 0$ for $-h \leq \theta \leq 0$. Moreover, $u(t)(\theta) = x(t+\theta)$ or, in words, $u(t)$ is the segment of the solution x of (1.1) of length h preceding t.*

As an application of this representation we shall prove the Fredholm alternative for periodic solutions. Recall that the kernel ζ is, by assumption, real-valued.

Theorem 4.6. *Let $f : \mathbb{R} \to \mathbb{C}^n$ be continuous and p-periodic. The equation*

$$\dot{x}(t) = \int_0^h d\zeta(\tau)x(t-\tau) + f(t) \tag{4.9}$$

admits a p-periodic solution if and only if

$$\int_0^p z(-\tau)f(\tau)\,d\tau = 0 \tag{4.10}$$

for all p-periodic solutions of the homogeneous transposed equation

$$\dot{z}(t) = \int_0^h d\zeta^T(\tau) z(t-\tau). \tag{4.11}$$

Proof. The compactness and smoothing properties [see Exercises II.5.7 and III.3.10, Corollary III.4.7, Definition V.3.2 and Exercise V.3.4; note in particular that $\mathcal{R}(I - T(p))$ is closed since $T(p)^k = T(kp)$ is compact for k such that $pk \geq h$] guarantee that

(i) $\mathcal{R}(I - T(p)) = {}^{\perp}\mathcal{N}(I - T^*(p))$,

(ii) $\mathcal{N}(I - T^*(p)) \subset X^{\odot}$.

Now,

$$u(t) = T(t)\varphi + \sum_{j=1}^{n} \int_0^t T^{\odot *}(t-\tau) r_j^{\odot *} f_j(\tau)\, d\tau$$

is p-periodic if and only if

$$(I - T(p))\varphi = \sum_{j=1}^{n} \int_0^p T^{\odot *}(p-\tau) r_j^{\odot *} f_j(\tau)\, d\tau.$$

By (i) and (ii) above, this equation is solvable if and only if

$$\langle x^{\odot}, \sum_{j=1}^{n} \int_0^p T^{\odot *}(p-\tau) r_j^{\odot *} f_j(\tau)\, d\tau \rangle = 0$$

for all $x^{\odot}$ such that $T^{\odot}(p)x^{\odot} = x^{\odot}$. The left hand side can be rewritten as

$$\sum_{j=1}^{n} \int_0^p f_j(-\tau) \langle r_j^{\odot *}, T^{\odot}(\tau) x^{\odot} \rangle\, d\tau.$$

Now note that z defined by $z_j(\tau) = \langle r_j^{\odot *}, T^{\odot}(\tau)x^{\odot}\rangle$ is a solution of the transposed renewal equation

$$z = \zeta^T * z + z_0$$

(cf. the proof of Theorem III.3.9 and recall that for RFDE $R_0 = \zeta$). Hence z is a solution of the RFDE (4.11) for $t \geq h$ (cf. Section III.6). Since z is p-periodic, equation (4.11) is actually satisfied for all t. If, conversely, z is a p-periodic solution of (4.11), then $z_0 := z - \zeta^T * z$ defines an element of $X^{\odot}$ and $z(t) = \langle r^{\odot *}, T^{\odot}(t) z_0 \rangle$ and $T^{\odot}(p) z_0 = z_0$ (again cf. Sections III.6 and III.5). □

The following two exercises provide some more information on the representation of periodic solutions. Here we let $P_{+}^{\odot *}$ and $P_{0}^{\odot *}$ denote the projection operators corresponding to the part of the spectrum in, respectively, the strict right half-plane and the imaginary axis, and we define $P_{-}^{\odot *} = I - P_{+}^{\odot *} - P_{0}^{\odot *}$. We let

$$u_{\diamond}(t) = P_{\diamond}^{\odot *} u(t),$$

where $\diamond$ is $+$, 0 or $-$, be the components of u according to this decomposition. Using Exercise 2.1 we conclude that

$$u(t) = T(t-s)u(s) + \sum_{j=1}^{n} \int_{s}^{t} T^{\odot *}(t-\tau) r_j^{\odot *} f_j(\tau)\, d\tau \tag{4.13}$$

decomposes into

$$u_{\diamond}(t) = T(t-s)u_{\diamond}(s) + \sum_{j=1}^{n} \int_{s}^{t} T^{\odot *}(t-\tau) P_{\diamond}^{\odot *} r_j^{\odot *} f_j(\tau)\, d\tau, \tag{4.14}$$

where $\diamond$ is $+$, 0 or $-$.

Exercise 4.7. Use the estimates of Theorem IV.2.12 to show that the only bounded solution of (4.14) with $\diamond = -$ is given by

$$u_{-}(t) = \sum_{j=1}^{n} \int_{0}^{\infty} T^{\odot *}(\tau) P_{-}^{\odot *} r_j^{\odot *} f_j(t-\tau) d\tau \tag{4.15}$$

and that the only bounded solution of (4.14) with $\diamond = +$ is given by

$$u_{+}(t) = \sum_{j=1}^{n} \int_{0}^{-\infty} T^{\odot *}(\tau) P_{+}^{\odot *} r_j^{\odot *} f_j(t-\tau)\, d\tau. \tag{4.16}$$

Note that both u_{+} and u_{-} are p-periodic if f is p-periodic.

Exercise 4.8. Conclude from Sections 2 and 3 that u_0 satisfies an ODE. Use the Fredholm alternative for periodic solutions of ODE (see, e.g., [101, Lemma IV.1.1]) to give an alternative proof of Theorem 4.6.

VI.5 Comments

The material of this chapter is rather standard (e.g., Hale [102, Chap. 9]). A translation invariant space B of functions defined on the real line is called *admissible* if for every $f \in B$ there exists at least one solution $u \in B$. Apart from periodic functions, one often considers almost periodic or just bounded functions. See Prüss [242] and the references given there for recent developments concerning more general abstract equations.

Chapter VII

Semiflows for nonlinear systems

VII.1 Introduction

The present chapter contains the basic results on existence, uniqueness and smoothness of solutions to the nonlinear autonomous initial-value problems studied in this book.

Let a real Banach space X and a $\mathcal{C}_0$-semigroup of linear operators $\{T_0(t)\}_{t\geq 0}$ be given such that X is $\odot$-reflexive with respect to $T_0(t)$. Assume there is a perturbation

$$G : \mathcal{O} \to X^{\odot *}$$

of the generator A_0 of $T_0(t)$; G is defined on a subset of a product $X \times P$ so that parameters are incorporated. This is done in view of bifurcation theory, as in Chapters X and XV.

Instead of the differential equation

$$\dot{u} = A_0^{\odot *} u + G(u, p)$$

with parameter $p \in P$ and initial condition $u(0) = \varphi \in X$, we consider a more general integrated version of this problem, namely the *abstract integral equation* (AIE)

$$u(t) = T_0(t)\varphi + \int_0^t T_0^{\odot *}(t-s) G(u(s), p)\, ds \tag{1.1}$$

which can be solved by means of the iteration of a contraction mapping as in the case of ODEs.

The basic properties of solutions to ODEs, given by vector fields on open subsets in $\mathbb{R}^n$, are collected in the concept of a flow. Solutions of equation (1.1), as well as solutions to many other initial-value problems in infinite dimensional state space, do not behave as nicely. In particular, one time direction will be distinguished. In general, there are no solutions for the backward initial-value problem; when there are solutions, uniqueness for backward solutions may fail.

The appropriate abstract concept is a semiflow, introduced in Section 2. In Sections 3 and 4, maximal solutions to equation (1.1) are constructed.

They form a continuous parameterized semiflow. Already the simplest example, the RFDE

$$\dot{x} = 0$$

with the initial condition $x_0 = \varphi \in \mathcal{C}$, shows that in general there is no differentiability of the semiflow in the time direction. For smooth perturbations G we discuss derivatives of the semiflow with respect to initial states and parameters. Section 5 contains the "principle of linearized stability" which relates the stability of a stationary point to the stability of the semigroup obtained from differentiation of the semiflow. Section 6 is devoted to the special case of RFDE.

VII.2 Semiflows

Let M denote a complete metric space, with metric d.

Definition 2.1. A *semiflow* on M is a map $S : D \to M$ on an open subset

$$D \subset [0, \infty) \times M$$

with the following properties:

(i) For every $x \in M$ there exists an interval I_x, either $I_x = [0, \infty)$ or $I_x = [0, t_x)$ with some $t_x > 0$, so that

$$\{(t, x) \in [0, \infty) \times M \mid t \in I_x\} = D;$$

(ii) $S(0, x) = x$ on M;

(iii) $x \in M, s \in I_x$ and $t \in I_{S(s,x)}$ imply $t + s \in I_x$ and

$$S(t, S(s, x)) = S(t + s, x);$$

(iv) all maps

$$I_x \ni t \mapsto S(t, x) \in M, \quad x \in M,$$

are continuous;

(v) all maps

$$\{y \in M : (t, y) \in D\} \ni x \mapsto S(t, x) \in M, \quad t \geq 0,$$

are continuous.

Semiflows are sometimes also called "*semidynamical systems*".

In case $I_x = [0, \infty)$, define $t_x := +\infty$. For $t \geq 0$, we set

$$D_t = \{x \in M \mid (t, x) \in D\} = \{x \in M \mid t \in I_x\}.$$

The set D_t is open, but possibly empty. The map in (v) is usually called the *time-t-map* of the semiflow.

Proposition 2.2.

(i) *The map* $M \ni x \mapsto t_x \in \mathbb{R} \cup \{+\infty\}$ *is lower semicontinuous, i.e.,* $t < t_x$ *implies* $t < t_y$ *for all* y *in some neighbourhood of* x.

(ii) $I_{S(t,x)} = (I_x - t) \cap [0, \infty)$.

(iii) *Let* $x \in M$. *Assume* $t_x < \infty$. *Let* $K \subset M$ *be compact. Then there exists* $t_K \in I_x$ *such that*

$$S(t,x) \notin K \quad \text{on } [t_K, t_x).$$

(iv) *If* $\overline{\{S(t,x) \mid t \in I_x\}}$ *is compact then* $I_x = [0, \infty)$.

Proof. Suppose $x \in M, t < t_x$. Openness of D implies that for some $\epsilon > 0$ and for some neighbourhood N of x,

$$(t - \epsilon, t + \epsilon) \times N \subset D;$$

or, for all $y \in N, t < t + \epsilon \leq t_y$. This is (i).

Proof of (ii). If $s \in I_{S(t,x)}$, then $s + t \in I_x$. Hence $I_{S(t,x)} \subset (I_x - t) \cap [0, \infty)$. To prove the converse, note first that in case $t_{S(t,x)} = +\infty$, the above implies $t_x = +\infty$, so that the assertion holds true. In case $t_{S(t,x)} < +\infty$, we obtain either $t_{S(t,x)} + t = t_x$ $(< +\infty)$ or

$$t_{S(t,x)} + t < t_x \ (\leq +\infty). \tag{2.1}$$

We assume (2.1) and derive a contradiction. Let (t_n) denote a sequence in $I_{S(t,x)}$ with $t_n \to t_{S(t,x)}$. Then $t_n + t \in I_x$, and

$$x_n := S(t_n, S(t,x)) = S(t_n + t, x) \to S(t_{S(t,x)} + t, x) =: y,$$

by (2.1) and part (iv) of Definition 2.1. In addition,

$$\begin{aligned} 0 < t_y &\leq \liminf t_{x_n} \quad \text{[see (i)]} \\ &\leq \liminf t_{S(t,x)} - t_n \end{aligned}$$

since $I_{x_n} = I_{S(t_n, S(tx))} \subset I_{S(t,x)} - t_n$ for all n. The last inequalities contradict $t_{S(t,x)} = \lim t_n$.

Proof of (iii). Let $x \in M$, $t_x < +\infty$. Consider a compact set $K \subset M$. Assume $t_n \to t_x$ for some sequence (t_n) in I_x with $S(t_n, x) \in K$ for all n. There exists a limit point y of $(S(t_n, x))$ in K. By (i), there are a neighbourhood N of y and $\delta \in (0, t_y)$ such that for all $z \in N$, $\delta < t_z$. Choose n_0 such that for $n > n_0$, $t_n + \delta > t_x$. Then $S(t_n, x) \in N$ and $t_n + \delta > t_x$ for some n. By (ii),

$$t_x = t_{S(t_n,x)} + t_n > \delta + t_n,$$

a contradiction.

Proof of (iv). If $\overline{\{S(t,x) : t \in I_x\}}$ is compact and if $t_x < +\infty$, then (iii) yields some $t \in I_x$ such that $S(t,x) \notin \overline{\{S(s,x) : s \in I_x\}}$, a contradiction. □

Definition 2.3. A *trajectory* of the semiflow S is a map $\sigma : I \to M$, defined on an interval $I \subset R$, with positive length, so that for s and t in I with $s \leq t$,

$$t - s \in I_{\sigma(s)} \quad \text{and} \quad \sigma(t) = S(t-s, \sigma(s)).$$

The image $\sigma(I)$ of a trajectory $\sigma : I \to M$ is called the *orbit* of σ.

Trajectories are continuous. Limiting behaviour of trajectories (on unbounded intervals) occurs close to their α- and ω-limit sets, which we now define.

Definition 2.4. The *ω-limit set* of a trajectory $\sigma : I \to M$ with $\sup I = \infty$ is defined as

$$\omega(\sigma) := \{x \in M \mid \text{There exists a sequence } t_n \text{ in } I \text{ such that } t_n \to \infty \text{ and } \sigma(t_n) \to x \text{ as } n \to \infty\}.$$

The *α-limit set* of a trajectory $\sigma : I \to M$ with $\inf I = -\infty$ is defined as

$$\alpha(\sigma) := \{x \in M \mid \text{There exists a sequence } t_n \text{ in } I \text{ such that } t_n \to -\infty \text{ and } \sigma(t_n) \to x \text{ as } n \to \infty\}.$$

An immediate consequence of this definition is

$$\omega(\sigma) = \bigcap_{t \geq 0} \overline{\sigma(I \cap [t, \infty))}. \tag{2.2}$$

To prove (2.2), note first that $x \in \omega(\sigma)$, $I \ni t_n \mapsto +\infty$ and $\sigma(t_n) \to x$ altogether imply $x \in \overline{\sigma(I \cap [t, +\infty))}$ for each $t \geq 0$. On the other hand, for

$$x \in \bigcap_{t \geq 0} \overline{\sigma(I \cap [t, \infty))}$$

we have $\sup I = +\infty$, and given $n \in N$, there exists $t_n \in I$ with $d(x, \sigma(t_n)) < 1/n$. This yields $x \in \omega(\sigma)$.

Exercise 2.5. Show that

$$\alpha(\sigma) = \bigcap_{t \leq 0} \overline{\sigma(I \cap (-\infty, t]}).$$

Proposition 2.6. *Suppose $\sigma : I \to M$ is a trajectory so that $\sup I = \infty$, and $\overline{\sigma(I)}$ is compact. Then*

(i) *$\omega(\sigma)$ is nonempty, compact and connected;*

(ii) *$\text{dist}\,(\sigma(t), \omega(\sigma)) \to 0$ as $t \to +\infty$;*

(iii) *(Positive invariance.) for $x \in \omega(\sigma)$ necessarily $I_x = [0, \infty)$ and*

$$S(t,x) \in \omega(\sigma) \quad \textit{for all } t \geq 0.$$

Moreover, there exists a trajectory $\xi : \mathbb{R} \to M$ *with* $\xi(0) = x$ *and* $\xi(\mathbb{R}) \subset \omega(\sigma)$. [*For* $t \geq 0$, $\xi(t) = S(t,x)$.]

Proof. Closedness of $\omega(\sigma)$ follows from (2.2), and $\omega(\sigma) \subset \overline{\sigma(I)}$ yields compactness. If (t_n) is any sequence in I so that $t_n \to +\infty$, then $(\sigma(t_n))$ contains a convergent subsequence; its limit belongs to $\omega(\sigma)$. This proves $\omega(\sigma) \neq \emptyset$. Assertion (ii) follows easily. In order to show connectedness, let disjoint open subsets $\mathcal{O}_1$ and $\mathcal{O}_2$ of M be given such that there exist points

$$x_1 \in \omega(\sigma) \cap \mathcal{O}_1, \qquad x_2 \in \omega(\sigma) \cap \mathcal{O}_2.$$

Construct a strictly increasing sequence (t_n) in I such that $t_n \to +\infty$, $\sigma(t_{2n}) \to x_2$ and $\sigma(t_{2n+1}) \to x_1$ as $n \to \infty$. There exists a n_0 such that for $n \geq n_0$, $\sigma(t_{2n}) \in \mathcal{O}_2$ and $\sigma(t_{2n+1}) \in \mathcal{O}_1$. It follows that for some $t_n^* \in (t_{2n}, t_{2n+1})$,

$$\sigma(t_n^*) \in M \backslash (\mathcal{O}_1 \cup \mathcal{O}_2)$$

since $\sigma([t_{2n}, t_{2n+1}])$ is connected. There exists a subsequence $(t^*_{\varphi(n)})$ so that $(\sigma(t^*_{\varphi(n)}))$ converges to some $y \in \omega(\sigma)$. Hence $y \in \omega(\sigma) \backslash (\mathcal{O}_1 \cup \mathcal{O}_2)$; or, $\omega(\sigma) \not\subset \mathcal{O}_1 \cup \mathcal{O}_2$. This proves that $\omega(\sigma)$ is connected.

Proof of (iii). Let $x \in \omega(\sigma)$. Consider a sequence (t_n) in I with $t_n \to +\infty$ and $\sigma(t_n) \to x$. We have $S([0,t_x) \times \{x\}) \subset \omega(\sigma)$, since for $t \in [0, t_x)$, $t_n + t \in I$ or $t \in I_{\sigma(t_n)}$, and therefore

$$\sigma(t + t_n) = S(t, \sigma(t_n)) \to S(t,x) \quad \text{as } n \to \infty,$$

which means $S(t,x) \in \omega(\sigma)$. Proposition 2.2 and compactness of $\omega(\sigma)$ give $t_x = +\infty$. Define $\xi(t) = S(t,x)$ for $t \geq 0$. Construction of $\xi(-1)$: There exists a subsequence $(t_{\varphi_1(n)})$ with

$$\inf I < t_{\varphi_1(n)} - 1 \quad \text{for all } n,$$

and $(\sigma(t_{\varphi_1(n)} - 1))$ convergent. Define $\xi(-1)$ to be the limit. Then $\xi(-1) \in \omega(\sigma)$. In particular, $[0,\infty) \subset I_{\xi(-1)}$, and thereby

$$\begin{aligned} S(1, \xi(-1)) &= \lim_{n\to\infty} S(1, \sigma(t_{\varphi_1(n)} - 1)) = \lim_{n\to\infty} \sigma(t_{\varphi_1(n)}) \\ &= x = \xi(0). \end{aligned}$$

Construction of $\xi(-2)$: There exists a subsequence $(t_{\varphi_1 \circ \varphi_2(n)})$ with

$$\inf I < t_{\varphi_1 \circ \varphi_2(n)} - 2 \quad \text{for all } n,$$

and $(\sigma(t_{\varphi_1 \circ \varphi_2(n)} - 2))$ convergent. Define $\xi(-2)$ to be the limit. As before, $\xi(-2) \in \omega(\sigma)$ and $[0,\infty) \subset I_{\xi(-2)}$ and

$$\begin{aligned} S(1, \xi(-2)) &= \lim_{n\to\infty} S(1, \sigma(t_{\varphi_1 \circ \varphi_2(n)} - 2)) = \lim_{n\to\infty} \sigma(t_{\varphi_1 \circ \varphi_2(n)} - 1) \\ &= \xi(-1). \end{aligned}$$

Proceed by induction. One finds points $\xi(-n), n \in \mathbb{N}$, in $\omega(\sigma)$ with

$$\xi(-n+1) = S(1, \xi(-n)).$$

For $t \in (-n, -n+1)$, define $\xi(t) = S(t-(-n), \xi(-n))$ and verify the asserted properties. □

Exercise 2.7. Formulate and prove a version of Proposition 2.6 for α-limit sets.

Exercise 2.8. How should $\omega(x)$ be defined for a point $x \in M$? How would $\omega(x)$ be related to $\omega(\sigma)$ in case $x \in \sigma(I)$? Why is it not obvious how to define $\alpha(x)$?

Later, we shall have to consider parameterized semiflows.

Definition 2.9. Let P be a topological space and let $\Delta \subset [0,\infty) \times M \times P$ be open. A *parameterized semiflow* $\Sigma : \Delta \to M$ is a map such that for every $p \in P$, the map $S_p : D_p \to M$, where

$$D_p = \{(t,x) \in [0,\infty) \times M : (t,x,p) \in \Delta\}$$

and

$$S_p(t,x) = \Sigma(t,x,p)$$

is a semiflow on M.

Note that the domains D_p are open (possibly empty). In case of parameterized semiflows, we shall use the notations $t_{x,p}$ and $I_{x,p}$.

Example 2.10. (The simplest differential equation for nonlinear feedback.) Consider the Banach space $X = C([-1,0], \mathbb{R})$. Let a continuous function $f : \mathbb{R} \to \mathbb{R}$ be given and let a parameter $\alpha > 0$ be given. For every $\varphi \in X$ there exists a unique continuous function $x : [-1, +\infty) \to \mathbb{R}$, differentiable for $t > 0$, which satisfies the differential delay equation

$$\dot{x}(t) = \alpha f(x(t-1)), \quad t > 0, \tag{2.3}$$

and the initial condition $x_0 = \varphi$. This is most easily seen by repeated application of the formula

$$x(t) - x(n) = \alpha \int_{n-1}^{t-1} f\big(x(\tau)\big)\, d\tau \tag{2.4}$$

for integers $n \geq 0$ and $t \in [n, n+1]$.

There exists a right derivative at $t = 0$ which extends $\dot{x}$ to a continuous function on $\mathbb{R}^+$. We shall write $x^\varphi, x^{\varphi,\alpha}$ or $x^{\varphi,\alpha f}$ instead of x when convenient.

Exercise 2.11. Prove the following statement about continuous dependence: Given $\varphi \in X$, $\alpha > 0$, $t \geq -1$ and $\epsilon > 0$, there exists $\delta > 0$ such that

$$\max_{[-1,t]} |x^{\overline{\varphi},\overline{\alpha}} - x^{\varphi,\alpha}| < \epsilon$$

for all $\overline{\varphi} \in X$ and $\overline{\alpha} > 0$ with $\|\overline{\varphi} - \varphi\| + |\overline{\alpha} - \alpha| < \delta$.
Hint: Use formula (2.4).

Set $x_t^{\varphi,\alpha f}(\theta) = x^{\varphi,\alpha f}(t+\theta)$ for $t \geq 0$ and $\theta \in [0,1]$. The relations

$$\Sigma_f(t,\varphi,\alpha) = x_t^{\varphi,\alpha f}, \qquad t \geq 0,\ \varphi \in X,\ \alpha > 0,$$

define a parameterized semiflow which is continuous.

Exercise 2.12. Prove the following compactness property: If $B \subset [1,\infty) \times X \times (0,\infty)$ is bounded, then $\Sigma_f(B)$ has compact closure.
Hints: Choose $r > 1$ such that $B \subset \{(t,\varphi,\alpha) \in [1,\infty) \times X \times (0,\infty) \mid t \leq r,\ \|\varphi\| \leq r, \alpha \leq r\}$. Use (2.4) in order to deduce that the set

$$\{x^{\varphi,\alpha f}(s) \mid -1 \leq s \leq r,\ \|\varphi\| \leq r,\ 0 < \alpha \leq r\}$$

is bounded. Conclude that the set

$$\{\dot{x}^{\varphi,\alpha f}(s) \mid -1 \leq s \leq r,\ \|\varphi\| \leq r,\ 0 < \alpha \leq r\}$$

is bounded. Apply the Theorem of Ascoli and Arzèla to the set $\Sigma_f(B) \subset X$.

We mention two basic facts which distinguish $S = \Sigma_f(\cdot,\cdot,1)\varphi$, from, say, the flow defined by a vector field:

1. There exist initial data $\varphi \in X$ which do not define backward solutions: If φ has no time derivative at $\theta = 0$, then there is no continuous function

   $$y : [t-1,\epsilon) \to \mathbb{R}, \quad t < 0 \text{ and } \epsilon > 0,$$

   which is differentiable on (t,ε) and satisfies both

   $$\begin{aligned} \dot{y}(s) &= f(y(s-1)) \quad \text{for } t < s < \epsilon, \\ y_0 &= \varphi. \end{aligned}$$

2. There exist continuous functions $f : \mathbb{R} \to \mathbb{R}$ and initial data φ and χ in X, $\varphi \neq \chi$, so that the solutions x^φ and x^χ of (2.3) (with $\alpha = 1$) coincide on $[0,\infty)$, i.e., initially different trajectories in X flow together in finite time. This occurs, e.g., in case $f(r) = r^2$, $\varphi(\theta) = -\theta$ and $\chi(\theta) = \theta$. Also, if f is constant on some interval $[a,b]$ and if $c \in [a,b]$, then all $\varphi \in X$ with $\varphi([-1,0]) \subset [a,b]$ and $\varphi(0) = c$ determine the same solution segment on $[0,1]$, namely

   $$t \mapsto c + \alpha f(a)t.$$

VII.3 Solutions to abstract integral equations

Consider the abstract integral equation (1.1)

$$u(t) = T_0(t)\varphi + \int_0^t T_0^{\odot *}(t-s)G(u(s),p)\,ds.$$

In addition to the hypothesis from Section 1, assume that the perturbation

$$G : \mathcal{O} \to X^{\odot *}$$

is norm-continuous and that $\mathcal{O} \subset X \times P$ is an open subset. Given $(\varphi, p) \in \mathcal{O}$, we define a solution of equation (1.1) to be a continuous function

$$u : I \to X,$$

defined on some nontrivial interval $I \subset [0, \infty)$, so that (1.1) holds for all $t \in I$. [This includes $(u(s), p) \in \mathcal{O}$ on I, of course.]

Sometimes it will be convenient to write (1.1, φ, p) instead of (1.1). Let us assume from now on that the continuous map G is also locally Lipschitz continuous with respect to the state, i.e.,

(**locLip**) for every $(\varphi, p) \in \mathcal{O}$ there exist a neighbourhood N and a constant $L \geq 0$ such that $||G(\psi, q) - G(\chi, q)|| \leq L||\psi - \chi||$ for all (ψ, q) and (χ, q) in N.

The local Lipschitz condition will enable us to apply the contraction mapping principle in order to find solutions of (1.1) which are uniquely determined by initial data (and parameter).

The following remarks prepare the statement and the proof of Theorem 3.1 below which is the basic result on local existence, uniqueness and continuous dependence. If I is a compact interval and if Y is a Banach space, then $C\big(I, Y\big)$ denotes the space of continuous maps $u : I \to Y$, equipped with the maximum-norm

$$\|u\| = \max_{t \in I} \|u(t)\|.$$

Let $(\varphi, p) \in \mathcal{O}$ be given. Choose neighbourhoods N'_φ of φ and N_p of p and a constant $c = c_{\varphi,p} > 0$ so that

$$\|G(\psi, q)\| \leq c \quad \text{on } N'_\varphi \times N_p \subset \mathcal{O}$$

and, furthermore, so that the condition (locLip) holds on $N = N'_\varphi \times N_p$, with a Lipschitz constant $L = L_{\varphi,p}$. Choose $r = r_{\varphi,p} > 0$ with $\{\psi \in X \mid ||\psi - \varphi|| \leq r\} \subset N'_\varphi$. Fix $\epsilon = \epsilon_{\varphi,p} \in (0, r/4)$ and set

$$N_\varphi := \{\psi \in X \mid ||\psi - \varphi|| < \epsilon\} \subset N'_\varphi.$$

Recall from Section II.2 or Appendix II that there exist $M \geq 0$, $\omega > 0$ with $||T_0(t)|| \leq Me^{\omega t}$ on $[0, \infty)$. Choose $t = t_{\varphi,p} > 0$ so small that

$$Me^{\omega t}\epsilon + \max_{[0,t]} ||T_0(s)\varphi - \varphi|| + \frac{M}{\omega}(e^{\omega t} - 1)c \leq \frac{r}{2} \tag{3.1}$$

and

$$\frac{M}{\omega}(e^{\omega t} - 1)L < 1. \tag{3.2}$$

Theorem 3.1. *Consider* $(\varphi, p) \in \mathcal{O}$, N_φ, N_p, $t = t_{\varphi,p}$ *and* $r = r_{\varphi,p}$ *as before.*

(i) *For every* $(\psi, q) \in N_\varphi \times N_p$ *there exists a unique solution* $u = u^*_{\psi,q} \in C([0,t], X)$ *of equation* $(1.1, \psi, q)$ *which satisfies*

$$\|u(s) - \varphi\| \leq r/2 \quad \textit{for } 0 \leq s \leq t.$$

(ii) *The map*

$$N_\varphi \times N_p \to C([0,t], X), \quad (\psi, q) \mapsto u^*_{\psi,q}$$

is continuous. In case G *is* C^1*, the map* $(\psi, q) \mapsto u^*_{\psi,q}$ *is* C^1 *too.*

Proof. 1. For $t > 0$, define

$$\text{dom} := \{(\psi, q, u) \in N_\varphi \times N_p \times C([0,t], X) : ||u(s) - \varphi|| < r, \quad 0 \leq s \leq t\}.$$

For every $(\psi, q, u) \in \text{dom}$, we have a map

$$[0,t] \to X^{\odot *}, \quad s \mapsto G(u(s), q)$$

which is continuous and bounded by c. Using Lemma III.2.3, we obtain a map $A(\psi, q, u) : [0,t] \to X$,

$$A(\psi, q, u)(s) = T_0(s)\psi + \int_0^s T_0^{\odot *}(s - \tau)G(u(\tau), q)\, d\tau,$$

which is continuous and satisfies

$$\begin{aligned}
||A(\psi, q, u)(s) - \varphi|| &\leq ||T_0(s)\psi - \varphi|| + \frac{M}{\omega}(e^{\omega s} - 1)c \\
&\leq ||T_0(s)(\psi - \varphi)|| + ||T_0(s)\varphi - \varphi|| + \frac{M}{\omega}(e^{\omega s} - 1)c \\
&\leq Me^{\omega t}||\psi - \varphi|| + ||T_0(s)\varphi - \varphi|| + \frac{M}{\omega}(e^{\omega s} - 1)c \\
&\leq r/2 \qquad \text{for } 0 \leq s \leq t.
\end{aligned}$$

In particular,

$$(A(\psi, q, u)(s), q) \in N'_\varphi \times N_p \subset \mathcal{O}, \quad 0 \leq s \leq t. \tag{3.3}$$

The map

$$A : \text{dom} \to C([0,t], X), \quad (\psi, q, u) \mapsto A(\psi, q, u)$$

has range in the set

$$\{u \in C([0,t],X) \mid \|u(s)-\varphi\| \le r/2 \quad \text{for } 0 \le s \le t\},$$

and it is a "uniform contraction": For (ψ, q, u) and (ψ, q, u') in dom and for $0 \le s \le t$, we infer from Lemma III.2.3

$$\|A(\psi,q,u)(s) - A(\psi,q,u')(s)\| \le \frac{M}{\omega}(e^{\omega s}-1)L|u-u'|,$$

since

$$\max_{[0,s]} \left|G(u(s'),q) - G(u'(s'),q)\right| \le L|u-u'|.$$

Therefore

$$\|A(\psi,q,u) - A(\psi,q,u')\| \le \frac{M}{\omega}(e^{\omega t}-1)L\|u-u'\|$$

(recall 3.2). This proves (i).
2. Proof of assertion (ii).
2.1 Decomposition of the map A. We have

$$A = A_1 + A_5 \circ A_4 \circ A_3 \circ A_2$$

with maps A_i which are defined as follows:

$$\begin{aligned}
&A_1(\psi,q,u)(s) = T_0(s)\psi,\\
&A_2 : \text{dom} \to C([0,t],X) \times P, \quad A_2(\psi,q,u) = (u,q),\\
&A_3 : C([0,t],X) \times P \to C([0,t],X \times P), \quad A_3(v,q)(s) = (v(s),q).
\end{aligned}$$

The domain of A_4 is the subset

$$C([0,t],\mathcal{O}) \subset C([0,t],X \times P)$$

of elements in $C([0,t],X \times P)$ with range in $\mathcal{O}$; for such a map W,

$$A_4(w) = \tilde{G} \circ (I \times w) \in C([0,t],X^{\odot *}),$$

where $\tilde{G}(s,(\psi,q)) := G(\psi,q)$ on $[0,t] \times \mathcal{O}$. Finally,

$$A_5 : C([0,t],X^{\odot *}) \to C([0,t],X), \quad A_5(v)(s) = \int_0^s T_0^{\odot *}(s-\tau)v(\tau)\,d\tau.$$

The maps A_1, A_2, A_3 and A_5 are either linear continuous maps or restrictions of such. For A_5, see also Lemma III.2.3.
2.2 Smoothness of A_4. Note

$$A_3 \circ A_2(\text{dom}) \subset C([0,t],\mathcal{O}).$$

Lemma 1.4 of Appendix IV, on smoothness of induced maps, implies that A_4 is continuous. In case G is C^1 we infer that A_4 is C^1.
2.3 It follows that A is continuous. In case G is C^1, A is C^1 too. Apply Lemma 1.4 of Appendix IV in order to complete the proof of assertion (ii). □

For the construction of maximal solutions, another variant of uniqueness is needed.

Proposition 3.2. *Let $(\varphi, p) \in \mathcal{O}$ be given. If $u : I \to X$ is a solution of $(1.1, \varphi, p)$, then there exists $\epsilon = \epsilon_{u,\varphi,p} > 0$ such that*

$$u(s) = u^*_{\varphi,p}(s) \quad \text{for } 0 \le s < \epsilon.$$

Proof. Consider $N'_\varphi, N_p, L = L_{\varphi,p}$ and $t = t_{\varphi,p}$ as above. By continuity, there exists $\epsilon \in (0, t]$ so that

$$\big(u(s), p\big) \in N'_\varphi \times N_p \quad \text{for } 0 \le s \le \epsilon.$$

For such s,

$$\|u(s) - u^*_{\varphi,p}(s)\| \le \frac{M}{\omega}(e^{\omega t} - 1)L \max_{[0,\epsilon]} \|u(s') - u^*_{\varphi,p}(s')\|$$

so that (3.2) yields $u = u^*_{\varphi,p}$ on $[0, \epsilon]$. □

In addition, we need

Proposition 3.3. *Let $u : [0, t) \to X$ be a solution of $(1.1, \varphi, p)$. Let $t_0 \in (0, t)$, $\psi = u(t_0)$.*

(i) *Then $u_0 : [0, t - t_0) \ni s \mapsto u(t_0 + s) \in X$ is a solution of $(1.1, \psi, p)$.*

(ii) *If $\tilde{u} : [0, \tilde{t}) \to X$ is solution of $(1.1, \psi, p)$, then $\hat{u} : [0, t_0 + \tilde{t}) \to X$ given by $\hat{u} = u$ on $[0, t_0)$ and $\hat{u}(s) = \tilde{u}(s - t_0)$ on $[t_0, t_0 + \tilde{t})$ is a solution of $(1.1, \varphi, p)$.*

Proof. 1. Let $s \in [0, t - t_0)$. Then

$$\begin{aligned}
u_0(s) = u(t_0 + s) &= T_0(t_0 + s)\varphi + \int_0^{t_0} T_0^{\odot *}(t_0 + s - \tau)G(u(\tau), p)\, d\tau \\
&\quad + \int_{t_0}^{t_0+s} T_0^{\odot *}(t_0 + s - \tau)G(u(\tau), p)\, d\tau \\
&= T_0(s)T_0(t_0)\varphi + T_0^{\odot *}(s)\int_0^{t_0} T_0^{\odot *}(t_0 - \tau)G(u(\tau), p)\, d\tau \\
&\quad + \int_0^s T_0^{\odot *}(s - \tau)G(u(t_0 + \tau), p)\, d\tau \\
&= T_0(s)\big[T_0(t_0)\varphi + \int_0^{t_0} T_0^{\odot *}(t_0 - \tau)G(u(\tau), p)\, d\tau\big] \\
&\quad + \int_0^s T_0^{\odot *}(s - \tau)G(u(t_0 + \tau), p)\, d\tau \\
&= T_0(s)\psi + \int_0^s T_0^{\odot *}(s - \tau)G(u(\tau), p)\, d\tau.
\end{aligned}$$

This proves (i).
2. To prove (ii), note first that $\hat{u}$ is continuous. Let $s \in [t_0, t_0 + \tilde{t})$. Then

$$\begin{aligned}
\hat{u}(s) = \tilde{u}(s - t_0) &= T_0(s - t_0)\psi + \int_0^{s-t_0} T_0^{\odot *}(s - t_0 - \tau)G(\tilde{u}(\tau), p)\, d\tau \\
&= T_0(s - t_0)\big[T_0(t_0)\psi + \int_0^{t_0} T_0^{\odot *}(t_0 - \tau)G(u(\tau), p)\, d\tau\big] \\
&\quad + \int_0^{s-t_0} T_0^{\odot *}(s - t_0 - \tau)G(\tilde{u}(\tau), p)\, d\tau \\
&= T_0(s)\varphi + \int_0^{t_0} T_0^{\odot *}(s - t_0 + t_0 - \tau)G(\hat{u}(\tau), p)\, d\tau \\
&\quad + \int_{t_0}^{s} T_0^{\odot *}(s - t_0 + t_0 - \tau)G(\hat{u}(t_0 + \tau - t_0), p)\, d\tau \\
&= T_0(s)\varphi + \int_0^{s} T_0^{\odot *}(s - \tau)G(\hat{u}(\tau), p)\, d\tau.
\end{aligned}$$

□

Theorem 3.4. (*Maximal solutions.*) *Let $(\varphi, p) \in \mathcal{O}$. There exist an interval $I_{\varphi,p} = [0, \infty)$ or $I_{\varphi,p} = [0, t_{\varphi,p})$ for some $t_{\varphi,p} > 0$ and a solution $u_{\varphi,p} : I \to X$ of $(1.1, \varphi, p)$ so that for any other solution $u : I \to X$ of $(1.1, \varphi, p)$*

$$I \subset I_{\varphi,p} \quad \textit{and} \quad u = u_{\varphi,p}\big|_I.$$

Proof. 1. Suppose $u_i : I_i \to X$, $i \in \{1, 2\}$, are solutions of $(1.1, \varphi, p)$ and $I_1 \subset I_2$. We show $u_1 = u_2|_I$: Proposition 3.2 implies $u_1 = u_2$ on $[0, \epsilon)$ for some $\epsilon > 0$. Assume there exists a $t \in I_1$ with $u_1(t) \neq u_2(t)$. Then

$$t_* := \sup\{t \geq 0 \mid u_1 = u_2 \text{ on } [0, t]\,\}$$

is defined, with $\epsilon \leq t_* \leq t$. By continuity, $t_* < t$, and $u_1 = u_2$ on $[0, t_*]$. Set $\psi = u_1(t_*) = u_2(t_*)$. Proposition 3.3(i) implies that

$$u_{i0} : [0, t - t_*) \ni s \mapsto u_i(t_* + s) \in X$$

are both solutions of $(1.1, \psi, p)$. By Proposition 3.2, $u_{10} = u_{20}$ on $[0, \delta)$ for some $\delta > 0$. Hence $u_1 = u_2$ on $[t_*, t_* + \delta)$, a contradiction to the definition of t_*.
2. Set

$$J := \{t \geq 0 \mid\ t \in I \text{ for some solution } u : I \to X \text{ of } (1.1, \varphi, p)\,\}$$

and $I_{\varphi,p} := [0, \sup J)$, where $\sup J = \infty$ in case J is unbounded. In case J is bounded, set

$$t_{\varphi,p} := \sup J.$$

In the last case it is not hard to show that $t_{\varphi,p} \notin J$. For $t \in I_{\varphi,p}$, define $u_{\varphi,p}(t)$ to be the unique element in the set

$$\{\psi \in X \mid \text{There exists a solution } u : I \to X \text{ of } (1.1, \varphi, p) \\ \text{such that } t \in I \text{ and } u(t) = \psi\}$$

and verify the assertion for the map $u_{\varphi,p} : I_{\varphi,p} \to X$. □

Exercise 3.5. Prove the result of Theorem 3.4 in case G satisfies a global Lipschitz condition with respect to the state.

We are now ready to associate with equation (1.1) or, more precisely, with the data T_0 and G a map

$$\Sigma : \Delta \to X$$

which will turn out to be a parameterized semiflow, with additional smoothness properties. Set

$$\Delta := \{(t, \varphi, p) \in [0, \infty) \times X \times P \mid t \in I_{\varphi,p}\}$$

and

$$\Sigma(t, \varphi, p) := u_{\varphi,p}(t).$$

Corollary 3.6. *For all* $(\varphi, p) \in \mathcal{O}$, $\Sigma(0, \varphi, p) = \varphi$. *Further,* $t \in I_{\varphi,p}$ *and* $s \in I_{\Sigma(t,\varphi,p),p}$ *imply* $s + t \in I_{\varphi,p}$ *and*

$$\Sigma(s + t, \varphi, p) = \Sigma(s, \Sigma(t, \varphi, p), p).$$

Each map $I_{\varphi,p} \to X$, $t \mapsto \Sigma(t, \varphi, p)$, *is continuous.*

Proof. The first and the last assertions should be obvious. To prove the remaining one, consider maximal solutions $u = u_{\varphi,p}$ and $\tilde{u}$ of $(1.1, \varphi, p)$ and of $(1.1, \Sigma(t, \varphi, p), p)$, respectively. Let $s \in I_{\Sigma(t,\varphi,p),p}$. For some $\epsilon > 0$, $s + \epsilon \in I_{\Sigma(t,\varphi,p),p} = I_{u(t),p}$. Proposition 3.3(ii) implies that

$$\hat{u} : [0, s + \epsilon + t) \to X$$

given by $\hat{u} = u$ on $[0, t]$ and $\hat{u}(\,\cdot\,) = \tilde{u}(\,\cdot - t)$ on $[t, t + s + \epsilon)$ is a solution to $(1.1, \varphi, p)$. Therefore $s + \epsilon + t \in I_{\varphi,p}$, and $\hat{u} = u|_{[0,s+\epsilon+t)}$. In particular,

$$s + t \in I_{\varphi,p}$$

and

$$\begin{aligned}\Sigma(s + t, \varphi, p) &= u(s + t) = \hat{u}(s + t) \\ &= \tilde{u}(s) = \Sigma(s, \Sigma(t, \varphi, p), p).\end{aligned}$$

□

Exercise 3.7. Work out a simplified version of Theorem 3.1 for the case where there are no parameters.

In the next section, we verify that Σ satisfies the remaining properties of a semiflow, and more.

VII.4 Smoothness

Let Σ be as in Section 3.

Proposition 4.1.
(i) *For every $(t, \varphi, p) \in \Delta$ there are open neighbourhoods N_φ of φ and N_p of p with $[0,t] \times N_\varphi \times N_p \subset \Delta$ so that the map*

$$N_\varphi \times N_p \to C\big([0,t], X\big), \quad (\psi, q) \mapsto u_{\psi,q}\big|_{[0,t]}$$

is continuous.
(ii) *If G is C^1, then the map in* (i) *is C^1.*

For the proof, we need the result of

Exercise 4.2. For reals $a < b < c$, consider the spaces

$$C_{ab} := C\big([a,b], X\big), \quad C_{bc} := C\big([b,c], X\big), \quad C_{ac} := C\big([a,c], X\big)$$

and the closed subspace

$$C_{abc} \subset C_{ab} \times C_{bc}$$

of all (u, v) satisfying $u(b) = v(b)$. Then the map

$$j : C_{ac} \to C_{abc}, \quad j(w) = (w|[ab], w|[bc])$$

is continuous, linear and bijective. Suppose $f : U \to C_{abc}$, where U is an open subset of some real Banach space Y, has components $U \to C_{ab}$ and $U \to C_{bc}$ which are C^1. Then f is C^1, too.

Proof of Proposition 4.1. 1. Let $(\varphi, p) \in \mathcal{O}$. Theorem 3.1 implies that the set A of all $t \in I_{\varphi,p}$ so that there exist neighbourhoods as in assertion (i) is nonempty. Consider

$$t^* := \sup A \leq \sup I_{\varphi,p} \leq \infty.$$

2. Proof that $t^* = \sup I_{\varphi,p}$. Assume $t^* < \sup I_{\varphi,p}$. Then $[0, t^*] \subset I_{\varphi,p}$. Set $\varphi^* := u_{\varphi,p}(t^*)$. There are open neighbourhoods N^* of φ^* and N_p^* of p and $\epsilon > 0$ so that for all $(\psi, q) \in N^* \times N_p^*$, we have

$$\epsilon \in I_{\psi,q},$$

and the map

$$B_0 : N^* \times N_p^* \to C([0,\epsilon],X), \quad (\psi,q) \mapsto u_{\psi,q}\big|_{[0,\epsilon]}$$

is continuous. Fix $t \in A$ such that $t^* - \epsilon < t < t^*$ and $u_{\varphi,p}(t) \in N^*$. Consequently, there are neighbourhoods N_φ of φ and N_p of p so that the map

$$B_1 : N_\varphi \times N_p \to C([0,t],X), \quad (\psi,q) \mapsto u_{\psi,q}\big|_{[0,t]}$$

is defined and continuous. Applying the evaluation

$$C([0,t],X) \to X, \quad \varphi \to \varphi(t),$$

we infer that there exist neighbourhoods $\widetilde{N}_\varphi \subset N_\varphi$ of φ and $\widetilde{N}_p \subset N_p \cap N_p^*$ of p, such that for all $(\psi,q) \in \widetilde{N}_\varphi \times \widetilde{N}_p$,

$$u_{\psi,q}(t) \in N^*,$$

and, in particular,

$$\begin{aligned} &t + \epsilon \in I_{\psi,q}, \\ &u_{\psi,q}(s) = u_{u_{\psi,q}(t),q}(s-t) \quad \text{for } t \le s \le t+\epsilon. \end{aligned}$$

Consider the map

$$B_2 : \tilde{N}_\varphi \times \tilde{N}_p \to C([t,t+\epsilon],X), \quad (\psi,q) \mapsto u_{\psi,q}\big|_{[t,t+\epsilon]}.$$

For every $(\psi,q) \in \tilde{N}_\varphi \times \tilde{N}_p$,

$$B_2(\psi,q) = B_5(B_0(B_4(B_3(\psi,q)))),$$

where

$$\begin{aligned} &B_3(\psi,q) = (B_1(\psi,q),q) \in C([0,t],X) \times P, \\ &B_4 : C([0,t],X) \times P \to X \times P, \quad (v,q) \mapsto (v(t),q), \\ &B_5 : C([0,\epsilon],X) \to C([t,t+\epsilon],X), \quad B_5(v)(s) = v(s-t). \end{aligned}$$

So B_2 is continuous. We apply the result of Exercise 4.2 and deduce from the continuity of B_1 and B_2 that the map

$$\widetilde{N}_\varphi \times \widetilde{N}_p \to C([0,t+\epsilon],X), \quad (\psi,q) \mapsto u_{\psi,q}\big|_{[0,t+\epsilon]}$$

is continuous. This means $t+\epsilon \in A$, which contradicts $\sup A = t^* < t+\epsilon$.
3. Let $s \in I_{\varphi,p}$. Then $0 \le s < t$ for some $t \in A$, since $\sup A = \sup I_{\varphi,p}$. It follows that there are open neighbourhoods N_φ of φ and N_p of p so that the map

$$N_\varphi \times N_p \to C([0,t],X), \quad (\psi,q) \mapsto u_{\psi,q}\big|_{[0,t]}$$

is continuous. Applying the restriction map

$$C([0,t],X) \to C([0,s],X), \quad v \mapsto v\big|_{[0,s]},$$

we obtain $s \in A$. This proves (i).
4. The proof of (ii) is analogous. □

For $t \geq 0$, set

$$\begin{aligned}\triangle(t) :&= \{(\varphi,p) \in X \times P \mid (t,\varphi,p) \in \triangle\} \\ &= \{(\varphi,p) \in X \times P \mid t \in I_{\varphi,p}\}.\end{aligned}$$

Of course, some $\triangle(t)$ may be empty.

Corollary 4.3.

(i) *$\triangle$ is open. Σ is continuous. Each $\triangle(t)$, $t \geq 0$, is open.*

(ii) *Suppose G is C^1. Set $t \geq 0$. The maps*

$$A : \triangle(t) \to C([0,t],X), \quad A(\varphi,p) = u_{\varphi,p}\big|_{[0,t]}$$

and

$$\triangle(t) \to X, \quad (\varphi,p) \mapsto \Sigma(t,\varphi,p)$$

are C^1.

(iii) *Suppose G is C^1. Let $t \geq 0$, $(\varphi,p) \in \triangle(t)$ and $(\overline{\varphi},\overline{p}) \in X \times P$. The map*

$$[0,t] \to X, \qquad s \mapsto D^{(0,1,1)}\Sigma(s,\varphi,p)(\overline{\varphi},\overline{p})$$

is continuous.

(iv) *Suppose G is C^1. Let $t \geq 0$ and $(\varphi,p) \in \triangle(t)$. For every $\epsilon > 0$ there exists $\delta > 0$ such that for every $(\overline{\varphi},\overline{p}) \in X \times P$ with $\|\varphi-\overline{\varphi}\|+\|\overline{p}-p\| < \delta$ and for all $s \in [0,t]$, we have*

$$\|D^{(0,0,1)}\Sigma(s,\overline{\varphi},\overline{p}) - D^{(0,0,1)}\Sigma(s,\varphi,p)\| < \epsilon.$$

Proof. 1. Let $(s,\varphi,p) \in \triangle$ or, equivalently, $s \in I_{\varphi,p}$. There exists $t > s$ in $I_{\varphi,p}$, and Proposition 4.1 yields a neighbourhood $[0,t] \times N_\varphi \times N_p$ of (s,φ,p) in $\triangle$; $\triangle$ is open. We infer also that the map

$$N_\varphi \times N_p \to C([0,t],X), \quad (\psi,q) \mapsto u_{\psi,q}\big|_{[0,t]}$$

is continuous. Continuity of Σ at (s,φ,p) now follows from the inequalities

$$\begin{aligned}\|\Sigma(\overline{t},\psi,q) - \Sigma(s,\varphi,p)\| &\leq \|u_{\psi,q}(\overline{t}) - u_{\varphi,q}(\overline{t})\| + \|u_{\varphi,q}(\overline{t}) - u_{\varphi,q}(s)\| \\ &\leq \|u_{\psi,q}\big|_{0,t]} - u_{\varphi,p}\big|_{0,t]}\| + \|u_{\varphi,p}(\overline{t}) - u_{\varphi,p}(s)\|\end{aligned}$$

for $(\overline{t},\psi,q) \in [0,t] \times N_\varphi \times N_p$. Fix $p \in P$. The set

$$D(p) = \{(t, \varphi) \in \mathbb{R}^+ \times X \mid (t, \varphi, p) \in \triangle\}$$

is open (possibly empty); clearly,

$$D(p) = \{(t, \varphi) \in \mathbb{R}^+ \times X \mid t \in I_{\varphi,p}\}.$$

Continuity of the maps

$$D(p)_t \to X, \quad \varphi \mapsto S_p(t, \varphi) = \Sigma(t, \varphi, p), \; t \geq 0,$$

is a consequence of the continuity of Σ. For the remaining semiflow properties, see Corollary 3.6. Openness of the sets $\triangle(t)$, $t \geq 0$, should be obvious.
2. Suppose G is C^1. Let $t \geq 0$, $(\hat{\varphi}, \hat{p}) \in \triangle(t)$. Proposition 4.1 yields an open neighbourhood of $(\hat{\varphi}, \hat{p})$ in $\triangle(t)$ on which the map A is C^1; it follows that A is C^1. The evaluation maps

$$\delta_s : C([0, t], X) \to X, \qquad \delta_s v = v(s), \quad 0 \leq s \leq t,$$

are linear and continuous. We have

$$\Sigma(s, \varphi, p) = \delta_s \circ A(\varphi, p)$$

for all $s \in [0, t]$ and $(\varphi, p) \in \triangle(t)$. We infer that all maps

$$\triangle(t) \to X, \qquad (\varphi, p) \mapsto \Sigma(s, \varphi, p), \quad s \in [0, t],$$

are C^1. This proves (ii). Furthermore,

$$\begin{aligned} D^{(0,1,1)}\Sigma(s, \varphi, p)(\overline{\varphi}, \overline{p}) &= \delta_s\big(DA(\varphi, p)(\overline{\varphi}, \overline{p})\big) \\ &= \big(DA(\varphi, p)(\overline{\varphi}, \overline{p})\big)(s) \end{aligned}$$

for each $s \in [0, t]$ and for all $(\varphi, p) \in \triangle(t)$ and $(\overline{\varphi}, \overline{p}) \in X \times P$. The function $DA(\varphi, p)(\overline{\varphi}, \overline{p})$ is continuous. This proves (iii). Let $(\varphi, p) \in \triangle(t)$. Let $\epsilon > 0$ be given. Choose $\delta > 0$ so that for all $(\overline{\varphi}, \overline{p}) \in X \times P$ with $\|\overline{\varphi} - \varphi\| + \|\overline{p} - p\| < \delta$, we have $(\overline{\varphi}, \overline{p}) \in \triangle(t)$, and

$$\begin{aligned} \epsilon &> \|DA(\overline{\varphi}, \overline{p}) - DA(\varphi, p)\| \\ &= \sup_{(\psi,q) \in X \times P \mid \|(\psi,q)\| < 1} \Big(\sup_{0 \leq s \leq t} \|DA(\overline{\varphi}, \overline{p})(s) - DA(\varphi, p)(s)\|\Big). \end{aligned}$$

For each $s \in [0, t]$ we obtain

$$\begin{aligned} \epsilon &> \sup_{(\psi,q) \in X \times P \mid \|(\psi,q)\| < 1} \big(\|DA(\overline{\varphi}, \overline{p})(s) - DA(\varphi, p)(s)\|\big) \\ &= \|D^{0,1,1)}\Sigma(s, \overline{\varphi}, \overline{p}) - D^{0,1,1)}\Sigma(s, \varphi, p)\|. \end{aligned}$$

This proves (iv). □

Remark **4.4.**

(i) Partial derivatives of Σ with respect to the first (time) variable do not exist in general, no matter how smooth G is. To see this, consider Example 2.10 with $f(\xi) = 0$ for all $\xi \in \mathbb{R}$. Then

$$\Sigma_f(t, \varphi, \alpha) = T_0(t)\varphi \quad \text{for } t \geq 0,\ \varphi \in X = C\big([-1,0], \mathbb{R}\big),\ \alpha > 1,$$

where T_0 is the strongly continuous semigroup of Section II.2. We observe in particular that Σ_f is the semiflow of an AIE (1.1) with $G = 0$. Consider $t \in (0,1)$ and an initial function $\varphi \in X$ which is not differentiable at $t-1$. Existence of $D_1\Sigma_f(t, \varphi, \alpha)$, $\alpha > 0$, would imply

$$\frac{1}{\tau}\big(x^{\varphi}_{t+\tau} - x^{\varphi}_t\big) \to D_1\Sigma_f(t, \varphi, \alpha)1 \quad \text{as } 0 \neq \tau \to 0,$$

and thereby the existence of

$$\lim_{0 \neq \tau \to 0} \frac{1}{\tau}\big(\varphi(t-1+\tau) - \varphi(t-1)\big),$$

a contradiction.

(ii) Suppose G is C^1. The map $D^{(0,1,1)}\Sigma : \Delta \to \mathcal{L}\big(X \times P, X\big)$ is in general not continuous on all of its domain. To see this, consider Example 2.10 with $f(\xi) = 0$ for $\xi \in \mathbb{R}$ once again. The equation for Σ_f in part (i) yields

$$D^{(0,1,1)}\Sigma_f\big(t, \varphi, \alpha\big)\big(\overline{\varphi}, \overline{\alpha}\big) = T_0(t)\overline{\varphi}$$

for $t \geq 0$, $(\varphi, \alpha) \in X \times (0, \infty)$ and $(\overline{\varphi}, \overline{\alpha}) \in X \times \mathbb{R}$. Continuity would imply that the strongly continuous semigroup T_0 is continuous at points $t \in [0,1]$. This is false, however, since for $s \in [0,1] \setminus \{t\}$ and $\psi \in X$ with $\|\psi\| = 1$ and

$$\big|\psi(t-1) - \psi(s-1)\big| = 1,$$

we have

$$\begin{aligned} 1 = \|T_0(t)\psi(-1) - T_0(s)\psi(-1)\| &\leq \|T_0(t)\psi - T_0(s)\psi\| \\ &\leq \|T_0(t) - T_0(s)\|, \end{aligned}$$

which implies a contradiction to continuity.

Next we compute $D^{(0,1,1)}\Sigma$ in terms of solutions to the so-called linear variational equations along solutions of the original AIE (1.1). This is analogous to the case of ODEs. The linear variational equations are nonautonomous [unless we differentiate at a point (t, φ, p) with φ on a constant trajectory].

Proposition 4.5. *Suppose G is C^1. For every $(t,\varphi,p) \in \triangle$ and all $(\overline{\varphi},\overline{p}) \in X \times P$,*

$$D^{(0,1,1)}\Sigma(t,\varphi,p)(\overline{\varphi},\overline{p}) = T_0(t)\overline{\varphi} + \int_0^t T_0^{\odot *}(t-s)H(s,\varphi,p)(\overline{\varphi},\overline{p})\, ds,$$

where

$$\begin{aligned} H(s,\varphi,p)(\overline{\varphi},\overline{p}) &= D_1G\big(\Sigma(s,\varphi,p),p\big)D_2\Sigma(s,\varphi,p)\overline{\varphi} \\ &\quad + D_1G\big(\Sigma(s,\varphi,p),p\big)D_3\Sigma(s,\varphi,p)\overline{p} \\ &\quad + D_2G\big(\Sigma(s,\varphi,p),p\big)\overline{p} \\ &= D_1G\big(\Sigma(s,\varphi,p),p\big)D^{(0,1,1)}\Sigma(s,\varphi,p)(\overline{\varphi},\overline{p}) \\ &\quad + D_2G\big(\Sigma(s,\varphi,p),p\big)\overline{p}. \end{aligned}$$

In particular,

$$\begin{aligned} &D_2\Sigma(t,\varphi,p)\overline{\varphi} \\ &\quad = T_0(t)\overline{\varphi} + \int_0^t T_0^{\odot *}(t-s)\big\{D_1G\big(\Sigma(s,\varphi,p),p\big)D_2\Sigma(s,\varphi,p)\overline{\varphi}\big\}\, ds, \end{aligned}$$

$$\begin{aligned} D_3\Sigma(t,\varphi,p)\overline{p} &= \int_0^t T_0^{\odot *}(t-s)D_1G\big(\Sigma(s,\varphi,p),p\big)D_3\Sigma(s,\varphi,p)\overline{p}\, ds \\ &\quad + \int_0^t T_0^{\odot *}(t-s)D_2G\big(\Sigma(s,\varphi,p),p\big)\overline{p}\, ds. \end{aligned}$$

Proof. Let $t \geq 0$. It is convenient to introduce a map

$$\sigma : \triangle(t) \to X, \quad (\varphi,p) \mapsto \Sigma(t,\varphi,p),$$

i.e.,

$$\sigma(\varphi,p) = T_0(t)\varphi + \int_0^t T_0^{\odot *}(t-s)G\big(\Sigma(s,\varphi,p),p\big)\, ds$$

on $\triangle(t)$, and σ is C^1 [Corollary 4.3(ii)]. In order to decompose σ, consider maps $\sigma_1,\ldots,\sigma_6$ as follows. The map

$$\sigma_1 : \triangle(t) \to X, \quad (\varphi,p) \mapsto T_0(t)\varphi$$

is the restriction of a linear continuous map. The map

$$\sigma_2 : \triangle(t) \to C\big([0,t],X\big), \quad (\varphi,p) \mapsto u_{\varphi,p}\big|_{[0,t]}$$

is C^1; see Corollary 4.3(ii). The map

$$\sigma_3 : \triangle(t) \to P, \quad (\varphi,p) \to p$$

is again the restriction of a linear continuous map. It follows that

$$\sigma_2 \times \sigma_3 : \triangle(t) \to C\big([0,t],X\big) \times P$$

is C^1. The map

$$\sigma_4 : C\big([0,t],X\big) \times P \to C\big([0,t],X \times P\big), \quad \sigma_4(u,p)(s) = (u(s),p)$$

is linear continuous. Note that for $0 \le s \le t$,

$$\big(\sigma_4 \circ (\sigma_3 \times \sigma_2)(\varphi,p)\big)(s) = (u_{\varphi,p}(s),p) \in \mathcal{O};$$

or, $\sigma_4 \circ (\sigma_3 \times \sigma_2)$ maps $\Delta(t)$ into the subset $C\big([0,t],\mathcal{O}\big)$ of maps in $C\big([0,t],X \times P\big)$ with range in $\mathcal{O}$. Consider the map

$$\sigma_5 : C\big([0,t],\mathcal{O}\big) \to C\big([0,t],X^{\odot *}\big), \quad w \mapsto G \circ w.$$

An application of Lemma 1.4 of Appendix IV to the map G shows that σ_5 is C^1. The map

$$\sigma_6 : C\big([0,t],X^{\odot *}\big) \to C\big([0,t],X\big), \quad \sigma_6(v)(s) = \int_0^s T_0^{\odot *}(s-\tau)v(\tau)\,d\tau$$

is linear and continuous, due to Lemma III.2.3. The evaluation map $\delta_t : C\big([0,t],X\big) \to X$, $\delta_t w = w(t)$ is linear and continuous. Altogether, for every $(\varphi,p) \in \Delta(t)$,

$$\sigma(\varphi,p) = \sigma_1(\varphi,p) + \delta_t\big(\sigma_6(\sigma_5(\sigma_4(\sigma_2 \times \sigma_3(\varphi,p))))\big).$$

The chain rule yields, for $(\overline{\varphi},\overline{p}) \in X \times P$,

$$D^{(0,1,1)}\Sigma(t,\varphi,p)(\overline{\varphi},\overline{p}) = D\sigma(\varphi,p)(\overline{\varphi},\overline{p}) = \sigma_1(\overline{\varphi},p) + \delta_t \circ \sigma_6(\alpha),$$

where

$$\alpha = D([\sigma_5 \circ \sigma_4](\sigma_2 \times \sigma_3(\varphi,p)) \circ D[\sigma_2 \times \sigma_3](\varphi,p)(\overline{\varphi},\overline{p}) \in C\big([0,t],X^{\odot *}\big).$$

To compute $\alpha(s)$ for $0 \le s \le t$, consider the linear continuous evaluation map

$$\overline{\delta}_s : C\big([0,t],X^{\odot *}\big) \to X^{\odot *}, \qquad \overline{\delta}_s(v) = v(s).$$

We obtain

$$\begin{aligned}
\alpha(s) &= \overline{\delta}_s \circ D[\sigma_5 \circ \sigma_4](\sigma_2 \times \sigma_3(\varphi,p)) \circ D[\sigma_2 \times \sigma_3](\varphi,p)(\overline{\varphi},\overline{p}) \\
&= D[\overline{\delta}_s \circ \sigma_5](\sigma_4(\sigma_2 \times \sigma_3(\varphi,p))) \circ D\sigma_4(\sigma_2 \times \sigma_3(\varphi,p)) \\
&\qquad \circ D[\sigma_2 \times \sigma_3](\varphi,p)(\overline{\varphi},\overline{p}) \\
&= D[G \circ \hat{\delta}_s](\sigma_4(\sigma_2 \times \sigma_3(\varphi,p))) \circ \sigma_4 \\
&\qquad \circ D[\sigma_2 \times \sigma_3](\varphi,p)(\overline{\varphi},\overline{p}),
\end{aligned}$$

where

$$\hat{\delta}_s : C\big([0,t],\mathcal{O}\big) \to X \times P, \quad \hat{\delta}_s(w) = w(s).$$

Hence

$$\begin{aligned}\alpha(s) &= DG\big(\sigma_4(\sigma_2 \times \sigma_3(\varphi, p))(s)\big) \circ \hat{\delta}_s \circ \sigma_4 \\ &\qquad \circ D[\sigma_2 \times \sigma_3](\varphi, p)(\overline{\varphi}, \overline{p}) \\ &= DG\big(u_{\varphi,p}(s), p\big) D[\hat{\delta}_s \circ \sigma_4 \circ (\sigma_2 \times \sigma_3)](\varphi, p)(\overline{\varphi}, \overline{p}).\end{aligned}$$

Observe

$$\begin{aligned}\hat{\delta}_s \circ \sigma_4 \circ (\sigma_2 \times \sigma_3)(\tilde{\varphi}, \tilde{p}) &= \big(u_{\tilde{\varphi},\tilde{p}}(s), \tilde{p}\big) \\ &= \big(\Sigma(s, \tilde{\varphi}, \tilde{p}), \tilde{p}\big)\end{aligned}$$

for all $(\tilde{\varphi}, \tilde{p}) \in \Delta(t)$. Now it is easy to complete the calculation. □

Corollary 4.6. (*Growth bounds for derivatives.*) *Suppose that* G *is a* C^1*-function. Let* $(t_0, \varphi, p) \in \Delta$ *be given. There exist constants* $m \geq 1$ *and* $\lambda \geq 0$ *such that*

$$\|D^{(0,1,1)}\Sigma(t, \varphi, p)\| \leq m e^{\lambda t} \quad \textit{for } 0 \leq t \leq t_0.$$

Proof. Continuity and compactness imply that there is a constant $c \geq 0$ such that

$$\|DG\big(\Sigma(t, \varphi, p), p\big)\| \leq c \quad \text{for } 0 \leq t \leq t_0.$$

Proposition 4.5 yields the estimate

$$\begin{aligned}&\|D^{(0,1,1)}\Sigma(t, \varphi, p)(\overline{\varphi}, \overline{p})\| \\ &\quad \leq M e^{\omega t}\|\overline{\varphi}\| + \int_0^t M e^{\omega(t-s)} c \|D^{(0,1,1)}\Sigma(s, \varphi, p)(\overline{\varphi}, \overline{p})\| \\ &\qquad + \int_0^t M e^{\omega(t-s)} c \|\overline{\varphi}\| \, ds \\ &\quad \leq M e^{\omega t}\big(\|\overline{\varphi}\| + c\|\overline{p}\| t_0\big) + cMe^{\omega t} \int_0^t e^{-\omega s}\|D^{(0,1,1)}\Sigma(s, \varphi, p)(\overline{\varphi}, \overline{p})\| \, ds\end{aligned}$$

for $0 \leq t \leq t_0$, $\overline{\varphi} \in X$ and $\overline{p} \in P$. We multiply the last inequality by $e^{-\omega t}$, apply Gronwall's inequality (see Hale [101, I.6.6]) and find

$$e^{-\omega t}\|D^{(0,1,1)}\Sigma(s, \varphi, p)(\overline{\varphi}, \overline{p})\| \leq M\big(\|\overline{\varphi}\| + ct_0\|\overline{p}\|\big) e^{cMt}$$

for $0 \leq t \leq t_0$, $\overline{\varphi} \in X$ and $\overline{p} \in P$ which implies the assertion. □

VII.5 Linearization at a stationary point

In this section we assume that the map $G : \mathcal{O} \to X^{\odot *}$ is C^1. For simplicity, we also assume

$$\mathcal{O} = \mathcal{O}_X \times \mathcal{O}_p$$

where $\mathcal{O}_X \subset X$ and $\mathcal{O}_p \subset P$ are open subsets. We consider the parameterized semiflow $\Sigma : \triangle \to X$ defined by (1.1).

Definition 5.1. A point $\varphi_0 \in \mathcal{O}_X$ is called a *stationary point* of the parameterized semiflow Σ if $\mathbb{R}^+ \times \{\varphi_0\} \times \mathcal{O}_p \subset \triangle$ and if

$$\Sigma(t, \varphi_0, p) = \varphi_0 \quad \text{for all } t \geq 0,\ p \in \mathcal{O}_p. \tag{5.1}$$

Let a stationary point $\varphi_0 \in \mathcal{O}_X$ be given. For $p \in \mathcal{O}_p$ we let T_p denote the $\mathcal{C}_0$-semigroup defined by the solutions $u : \mathbb{R}^+ \to X$ of the AIE

$$u(t) = T_0(t)u(0) + \int_0^t T_0^{\odot *}(t-s) D_1 G(\varphi_0, p) u(s)\, ds. \tag{5.2}$$

Recall that the space $X_p^{\odot *}$ defined by T_p coincides with the space $X^{\odot *}$ defined by T_0, and that X is $\odot$-reflexive with respect to T_p.

It is often convenient to reduce the discussion of solutions of (1.1) close to a stationary point to the case where $\varphi_0 = 0$. This is done by means of the following constructions.

Define a C^1-map R with domain

$$\widetilde{\mathcal{O}} = (\mathcal{O}_x - \varphi_0) \times \mathcal{O}_p$$

and range in $X^{\odot *}$ by the equation

$$G(\varphi, p) = G(\varphi_0, p) + D_1 G(\varphi_0, p)(\varphi - \varphi_0) + R(\varphi - \varphi_0, p) \quad \text{for } (\varphi, p) \in \mathcal{O}.$$

Then

$$R(0, p) = 0 \quad \text{and} \quad D_1 R(0, p) = 0 \quad \text{for all } p \in \mathcal{O}_p.$$

It follows that the C^1-map $\widetilde{G} : \widetilde{\mathcal{O}} \to X^{\odot *}$ given by

$$\widetilde{G}(\varphi, p) = G(\varphi + \varphi_0, p) - G(\varphi_0 . p)$$

satisfies

$$\widetilde{G}(0, p) = 0 \quad \text{and} \quad D_1 \widetilde{G}(0, p) = D_1 G(\varphi_0, p) \quad \text{for all } p \in \mathcal{O}_p,$$

and we have

$$\widetilde{G}(\varphi, p) = D_1 \widetilde{G}(0, p)\varphi + R(\varphi, p) \quad \text{for all } (\varphi, p) \in \widetilde{\mathcal{O}}.$$

Proposition 5.2. *Let a function $v : [0, t_0) \to X$ be given, and consider the function $u : [0, t_0) \ni t \mapsto v(t) + \varphi_0 \in X$. Then u is a solution of* (1.1) *if and only if v is a solution of the* AIE

$$v(t) = T_0(t)\overline{\varphi} + \int_0^t T_0^{\odot *}(t-s)\widetilde{G}\big(v(s), p\big)\, ds,$$

where $\overline{\varphi} = \varphi - \varphi_0$ and $\varphi = u(0)$.

Proof. Let u be a solution of (1.1) on $[0, t_0)$. For $0 \le t < t_0$, we obtain

$$\begin{aligned}
v(t) = u(t) - \varphi_0 &= T_0(t)\varphi + \int_0^t T_0^{\odot *}(t-s)G\big(u(s), p\big)\, ds - \Sigma(t, \varphi_0, p) \\
&= T_0(t)\varphi + \int_0^t T_0^{\odot *}(t-s)\big(G(\varphi_0, p) + D_1G(\varphi_0, p)(u(s) - \varphi_0) \\
&\qquad\qquad + R(u(s) - \varphi_0, p)\big)\, ds \\
&\quad - \Big(T_0(t)\varphi_0 + \int_0^t T_0^{\odot *}(t-s)G\big(\Sigma(s, \varphi_0, p)\big)\, ds\Big) \\
&= T_0(t)(\varphi - \varphi_0) + \int_0^t T_0^{\odot *}(t-s)\big(D_1G(\varphi_0, p)v(s) + R(v(s), p)\big)\, ds \\
&= T_0(t)\overline{\varphi} + \int_0^t T_0^{\odot *}(t-s)\widetilde{G}(v(s), p)\, ds,
\end{aligned}$$

where we have used that φ_0 is stationary. The remaining part is proved analogously. □

The solutions of the AIE (5.3) define a parameterized semiflow

$$\widetilde{\Sigma} : \widetilde{\Delta} \to X.$$

Proposition 5.2 yields

Corollary 5.3. *We have*

$$\widetilde{\Delta} = \big\{(t, \overline{\varphi}, p) \in \mathbb{R}^+ \times X \times P \mid (t, \overline{\varphi} + \varphi_0, p) \in \Delta\big\}$$

and

$$\widetilde{\Sigma}(t, \overline{\varphi}, p) + \varphi_0 = \Sigma(t, \overline{\varphi} + \varphi_0, p) \quad \textit{for all } (t, \overline{\varphi}, p) \in \widetilde{\Delta}.$$

In particular, $\overline{\varphi}_0 = 0$ is a stationary point of the parameterized semiflow $\widetilde{\Sigma}$.

We return to Σ. Linearization at the stationary point φ_0 means differentiation, with respect to the state, at $\varphi = \varphi_0$. Proposition 4.5 and the definition of the semigroups T_p yield

$$D_2\Sigma(t, \varphi_0, p) = T_p(t) \quad \text{for all } t \ge 0,\ p \in \mathcal{O}_p. \tag{5.4}$$

In the sequel we give a more elementary proof of (5.4) which works also under weaker assumptions on G and which yields results on uniformity of the differentiation process with respect to the other variables. An application of the uniformity property is found in Chapter XV on bifurcation of periodic orbits from a stationary point; see the proof of Corollary XV.6.3.

Finally, we shall discuss the principle of linearized stability which links stability properties of the semigroup T_p to those of the stationary point φ_0 of the semiflow $\Sigma(\,\cdot\,,\,\cdot\,,p)$.

We need the following variation-of-constants formula, which is an equation between elements of $X^{\odot *}$, like equation (1.1):

Proposition 5.4. *For every $(t,\varphi,p) \in \triangle$ we have*

$$\Sigma(t,\varphi,p) - \varphi_0 = T_p(t)(\varphi - \varphi_0) + \int_0^t T_p^{\odot *}(t-s)R\big(\Sigma(s,\varphi,p) - \varphi_0, p\big)\,ds,$$

where, by definition, $R : \big(\mathcal{O}_X - \varphi_0\big) \times \mathcal{O}_p \to X^{\odot *}$ with

$$R(w,p) = G(\varphi_0 + w, p) - G(\varphi_0, p) - D_1G(\varphi_0,p)w.$$

Proof. Write $\Sigma(t,\varphi,p) = \varphi_0 + w(t)$, $\varphi = \varphi_0 + \psi$, and use that for all $t \geq 0$,

$$\varphi_0 = T_0(t)\varphi_0 + \int_0^t T_0^{\odot *}(\tau)\,d\tau G(\varphi_0,p),$$

since φ_0 is a stationary point, to rewrite (1.1) as

$$w(t) = T_0(t)\psi + \int_0^t T_0^{\odot *}(t-\tau)\big\{D_1G(\varphi_0,p)w(\tau) + R(w(\tau),p)\big\}\,d\tau.$$

Now apply Lemma III.2.21 with $B = D_1G(\varphi_0,p)$ and

$$f(t) = R(w(t),p)$$

to conclude that

$$w(t) = T_p(t)\psi + \int_0^t T_p^{\odot *}(t-\tau)R\big(w(\tau),p\big)\,d\tau.$$

This is exactly the required identity. □

Proposition 5.5. [*Uniform differentiability of $\Sigma(t,\,\cdot\,,p)$ at φ_0.*] *Let $t_0 \geq 0$ and let a compact set $K \subset \mathcal{O}_p$ be given. For every $\epsilon > 0$ there exists $\delta > 0$ such that*

$$\|\Sigma(t,\varphi,p) - \varphi_0 - T_p(t)(\varphi - \varphi_0)\| \leq \epsilon\|\varphi - \varphi_0\|$$

for all $t \in [0,t_0)$, all $\varphi \in X$ with $\|\varphi - \varphi_0\| < \delta$ and all $p \in K$.

Proof. (In case $\varphi_0 = 0$.)

1. The continuous map $D_1G(0,\cdot)$ is bounded on the compact set. Gronwall's lemma, applied to (5.2), yields constants $m \geq 1$, $\lambda \geq 0$ such that

$$\|T_p^{\odot *}(t)\| = \|T_p(t)\| \leq me^{\lambda t} \quad \text{for all } t \geq 0,\ p \in K.$$

2. Claim. [Uniform differentiability of $R(\cdot, p)$]. For every $\overline{\epsilon} > 0$ there exists $\overline{\delta} > 0$ such that $\varphi \in \mathcal{O}_X$ and for all $\varphi \in X$ with $\|\varphi\| \leq \overline{\delta}$ and all $p \in K$,

$$\|R(\varphi, p)\| \leq \overline{\epsilon}\|\varphi\|.$$

Proof. Let $\overline{\epsilon} > 0$. Recall that $D_1R(0,p) = 0$ and D_1R is continuous. There exists a $\delta(p) > 0$ such that $(\varphi, \overline{p}) \in \mathcal{O}$ and $\|D_1R(\varphi, \overline{p})\| \leq \overline{\epsilon}$ for all $\varphi \in X$ with $\|\varphi\| \leq \delta(p)$ and all $\overline{p} \in P$ with $\|\overline{p} - p\| \leq \delta(p)$. The compactness of K implies that there exist a finite number of points in K, say, $p_1, \ldots, p_r$, so that

$$K \subseteq \bigcup_{j=1}^{r} \{p \in P \mid \|p - p_j\| \leq \delta(p_j)\}.$$

Set

$$\overline{\delta} = \min_{j=1,\ldots,r} \delta(p_j).$$

For $\varphi \in X$ with $\|\varphi\| \leq \overline{\delta}$ and $p \in K$ there exists $j \in \{1, \ldots, r\}$ so that $\|p - p_j\| \leq \delta(p_j)$; we have $\|\varphi\| \leq \delta(p_j)$. Hence

$$\begin{aligned} \|R(\varphi, p) = \|R(\varphi, p) - R(0, p)\| &= \| \int_0^1 D_1R(s, \varphi, p)\varphi \, ds\| \\ &\leq \overline{\epsilon}\|\varphi\|. \end{aligned}$$

3. As in part 2, we find $\delta^* > 0$ such that $(t_0, \varphi, p) \in \Delta$ for all $\varphi \in K$ with $\|\varphi\| \leq \delta^*$, and all $p \in K$.
4. Let $\epsilon > 0$ be given. Choose $\overline{\epsilon} > 0$ so small that

$$2m^2 t_0 e^{\lambda t_0} \overline{\epsilon} < \epsilon$$

and

$$\frac{m}{\lambda}\left(e^{\lambda t_0} - 1\right)\overline{\epsilon} < \frac{1}{2}.$$

Choose $\overline{\delta} \in (0, \delta^*)$ according to claim 2. Set

$$\delta = \frac{\overline{\delta}}{2me^{\lambda t_0}} < \overline{\delta}.$$

Let $(\varphi, p) \in X \times K$ be given with $\|\varphi\| \leq \delta$. Continuity implies that there exists $t_1 = t_1(\varphi, p) \in (0, t_0]$ so that

$$\|\Sigma(t, \varphi, p)\| < \overline{\delta} \quad \text{for } 0 \leq t \leq t_1.$$

Claim.

$$\|\Sigma(t, \varphi, p)\| < \overline{\delta} \quad \text{for all } t \in [0, t_0].$$

Proof. Suppose the assertion is false. It follows that there exists $t \in (t_1, t_0]$ such that $\|\Sigma(t,\varphi,p)\| < \overline{\delta}$ for $0 \le t \le t$ and

$$\|\Sigma(t,\varphi,p)\| = \overline{\delta}.$$

Proposition 5.4 and the estimates in part 1 and 2 of this proof yield

$$\|\Sigma(\overline{t},\varphi,p)\| \le me^{\lambda\overline{t}}\|\varphi\| + \frac{m}{\lambda}(e^{\lambda\overline{t}}-1)\overline{\epsilon}\max_{0\le s\le\overline{t}}\|\Sigma(s,\varphi,p)\|$$

for $0 \le \overline{t} \le t$. Choose $\overline{t} \in [0,t]$ with

$$\|\Sigma(\overline{t},\varphi,p)\| = \max_{0\le s\le t}\|\Sigma(s,\varphi,p)\|.$$

It follows that

$$\max_{0\le s\le t}\|\Sigma(s,\varphi,p)\| = \|\Sigma(\overline{t},\varphi,p)\| \le me^{\lambda\overline{t}}\|\varphi\| + \frac{1}{2}\max_{0\le s\le\overline{t}}\|\Sigma(s,\varphi,p)\|$$

(by the choice of $\overline{\epsilon}$)

$$\le me^{\lambda t}\|\varphi\| + \frac{1}{2}\max_{0\le s\le t}\|\Sigma(s,\varphi,p)\|.$$

Hence

$$\begin{aligned}\|\Sigma(t,\varphi,p)\| &\le \max_{0\le s\le t}\|\Sigma(s,\varphi,p)\| \le 2me^{\lambda t}\|\varphi\| \\ &\le 2me^{\lambda t_0}\delta < \overline{\delta},\end{aligned}$$

a contradiction.

5. Arguing as in the proof of the last claim, we deduce that

$$\|\Sigma(t,\varphi,p)\| \le 2me^{\lambda t}\|\varphi\| < \overline{\delta} \tag{5.5}$$

for all $t \in [0,t_0]$, all $\varphi \in X$ with $\|\varphi\| \le \delta$ and all $p \in K$. Using Proposition 5.4, we infer

$$\begin{aligned}\|\Sigma(t,\varphi,p) - T_p(t)\varphi\| &= \|\int_0^t T_p^{\odot *}(t-s)R(\Sigma(s,\varphi,p),p)\,ds\| \\ &\le \int_0^t me^{\lambda(t-s)}\overline{\epsilon}2me^{\lambda s}\|\varphi\|\,ds\end{aligned}$$

[by part 1, part 2, and (5.5)]

$$\le 2m^2 t_0 e^{\lambda t_0}\overline{\epsilon}\|\varphi\| \le \epsilon\|\varphi\|.$$

□

For Chapter XV, notably for the proof of Corollary XV.6.3, we need a version of Proposition 5.5 for the one-parameter family of RFDE

$$\dot{x}(t) = \alpha f(x(t-1))$$

of Example 2.10, which holds under comparatively weaker smoothness conditions.

We assume that the continuous function $f : \mathbb{R} \to \mathbb{R}$ satisfies $f(0) = 0$, that $Df(0)$ exists and that $Df(0)1 \neq 1$. Observe that under these conditions the theory of Sections 3 and 4, and in particular Proposition 5.5, are not applicable. The reason is not so much that we did not yet associate an AIE with equation (2.3). This will be done in Section 6. More severe is that f is not assumed to be locally Lipschitz continuous, not to speak of continuous differentiability. Nevertheless, we may investigate the continuous parameterized semiflow

$$\Sigma_f : \mathbb{R}^+ \times X \times (0, \infty) \to X, \qquad X = C\big([-1, 0], \mathbb{R}\big)$$

of Example 2.10. The point φ_0 is stationary for σ_f. Replacing f with its derivative $Df(0)$, we obtain from (2.3) its linearization at the zero solution, namely the equation

$$\dot{x}(t) = \alpha x(t-1), \tag{5.6}$$

which is also a special case of (2.3) with $f = -\mathrm{id}$. For $\alpha > 0$ we consider the C_0-semigroup

$$T_\alpha : \mathbb{R}^+ \to \mathcal{L}_{\mathbb{C}}(X, X)$$

which is associated with the linear autonomous RFDE (5.6) as in Section III.4. Then

$$T_\alpha(t)\varphi = x_t^{\varphi, \alpha \mathrm{id}} \quad \text{for all } t \geq 0,\ \varphi \in X,\ \alpha > 0,$$

in the notation of Example 2.10.

Proposition 5.6. [*Uniform differentiability of $\Sigma_f(t, \cdot, \alpha)$ at $\varphi_0 = 0$.*] *Let $t_0 \geq 0$, $\alpha_0 > 0$ and $\epsilon > 0$ be given. There exists $\delta > 0$ such that for all $t \in [0, t_0]$, all $\varphi \in X$ with $\|\varphi\| \leq \delta$ and all $\alpha \in (0, \alpha_0]$*

$$\|\Sigma_f(t, \varphi, \alpha) - T_\alpha(t)\varphi\| \leq \epsilon \|\varphi\|.$$

Sketch of the proof. The proof mimics the proof of Proposition 5.5.
1. Let $r^{\odot *} \in X^{\odot *}$ denote the element which corresponds to the pair $(1, 0) \in \mathbb{R} \times L^\infty\big([-1, 0], \mathbb{R}\big)$, as in Section III.4. For $\xi \in \mathbb{R}$ define $r(\xi)$ by

$$f(\xi) = -\xi + r(\xi).$$

Theorem VI.4.5 yields the variation-of-constants formula

$$x_t = T_\alpha(t)\varphi + \int_0^t T_\alpha^{\odot *}(t-s)\alpha r(x(s-1)) r^{\odot *}\, ds$$

for the solution $x = x^{\varphi, \alpha f}$ of (2.3), since we can write

$$\dot{x}(t) = -\alpha x(t-1) + \alpha r(x(t-1)).$$

2. For every $\bar{\epsilon} > 0$ there exists $\bar{\delta} > 0$ such that $|\alpha r(\xi)| \le \bar{\epsilon}|\xi|$ for all $\xi \in \mathbb{R}$ with $|\xi| \le \bar{\delta}$ and all $\alpha \in (0, \alpha_0]$.
3. Proceed as in the proof of Proposition 5.5. □

Concerning stationary points, the *principle of linearized stability* asserts that the stability, with respect to the given nonlinear semiflow, can be generically inferred from the stability properties of that point with respect to the linear semigroup obtained by deleting all higher order terms in the differential or integral equation. Here "generically" means that, as to be expected, one cannot draw conclusions without inspecting the higher order terms in the critical case that the growth bound of the linear semigroup equals zero.

In the remainder of this section we shall freeze the parameter p and, therefore, we also suppress it in the notation. So consider a semiflow Σ and let φ_0 be a stationary point, that is,

$$\Sigma(t, \varphi_0) = \varphi_0 \quad \text{for all } t \ge 0.$$

Definition 5.7. We say that φ_0 is (locally) *stable* whenever for every $\epsilon > 0$ we can find $\delta > 0$ such that $\|\varphi - \varphi_0\| \le \delta$ guarantees that $[0, \infty) \times \{\varphi\} \subset \Delta$ and

$$\|\Sigma(t, \varphi) - \varphi_0\| \le \epsilon \quad \text{for all } t \ge 0.$$

When φ_0 is not stable, we say it is *unstable*. When we can find $\epsilon > 0$, $K > 0$ and $\omega < 0$ such that

$$\|\Sigma(t, \varphi) - \varphi_0\| \le K e^{\omega t}, \qquad t \ge 0,$$

for all φ with $\|\varphi - \varphi_0\| \le \epsilon$, we say that φ_0 is (locally) *exponentially stable.*

Proposition 5.8. *Assume that $\gamma > 0$ and $M \ge 1$ exist such that for $t \ge 0$*

$$\|T(t) \le M e^{-\gamma t}.$$

Then, φ_0 is (locally) exponentially stable as a stationary point of Σ.

Proof. We assume, without loss of generality, that $\varphi_0 = 0$. Choose t_0 such that $M e^{-\gamma t_0} \le 1/4$. Choose $\delta > 0$ such that for $\|\varphi\| \le \delta$ and $0 \le t \le t_0$

$$\|\Sigma(t, \varphi) - T(t)\varphi\| \le \frac{1}{4}\|\varphi\|$$

(Proposition 5.5 implies that this is possible). Then, for such φ and t we have

$$\begin{aligned}
\|\Sigma(t, \varphi)\| &\le \|\Sigma(t, \varphi) - T(t)\varphi\| + \|T(t)\varphi\| \\
&\le \frac{1}{4}\|\varphi\| + M e^{-\gamma t}\|\varphi\| \\
&\le (\frac{1}{4} + M)\|\varphi\|,
\end{aligned}$$

whereas for $t = t_0$ we find the sharper estimate

$$\|\Sigma(t_0,\varphi)\| \le \frac{1}{2}\|\varphi\|.$$

Since, in particular, $\|\Sigma(t_0,\varphi)\| < \delta$, we may iterate to obtain

$$\|\Sigma(kt_0,\varphi)\| \le \left(\frac{1}{2}\right)^k \|\varphi\|.$$

For any $t > 0$, we now introduce $\tau \in [0,\tau)$ and an integer k by requiring that

$$t = kt_0 + \tau.$$

For $\|\varphi\| \le \delta\big(1/2 + M\big)^{-1}$, we know for sure that

$$\|\Sigma(t,\varphi)\| \le \big(1/4 + M\big)\|\varphi\| \le \delta$$

and so we are allowed to write

$$\begin{aligned}\|\Sigma(t,\varphi)\| &= \|\Sigma\big(kt_0,\Sigma(\tau,\varphi)\big)\| \le \left(\frac{1}{2}\right)^k \|\Sigma(\tau,\varphi)\| \\ &\le \left(\frac{1}{2}\right)^k \big(1/4 + M\big)\|\varphi\| \le \big(1/4 + M\big)e^{-\frac{\log 2}{t_0}t}\|\varphi\|,\end{aligned}$$

which is the required exponential stability estimate. □

Exercise 5.9. Show that one can prove Proposition 5.8 even when it is not known whether or not the differentiability of $\varphi \mapsto \Sigma(t,\varphi)$ is uniform on compact t-sets, provided one can verify that for any $s \in [0,\infty)$ there exist $\epsilon > 0$ and $K \ge 1$ such that for $0 \le t \le s$ and $\|\varphi\| \le \epsilon$

$$\|\Sigma(t,\varphi) \le K\|\varphi\|.$$

The above proof has demonstrated that it is an attractive strategy to somehow reduce the continuous time dynamical system to a discrete time system. This idea will reappear in the proof of the instability part of the principle of linearized stability. We therefore start with a discrete time result.

Proposition 5.10. *Let φ_0 be a fixed point of a map $F : X \to X$. Assume that F is differentiable at φ_0 with derivative L and that X admits a decomposition*

$$X = X_- \oplus X_+$$

into closed L-invariant subspaces such that

(i) *for all $\varphi \in X_-$: $\|L\varphi\| \le \theta\|\varphi\|$;*

(ii) *for all $\varphi \in X_+$: $\|L\varphi\| \ge (\theta + \delta)\|\varphi\|$,* *where $\theta \geq 1$ and $\delta > 0$. Then, z is unstable as a fixed point of the iteration scheme*

$$x(n+1) = F(x(n)).$$

Proof. Again we assume that $\varphi_0 = 0$. Let P denote the projection on X_+ along X_-. Define

$$|||\varphi||| = \|P\varphi\| + \|(I-P)\varphi\|.$$

Then $|||\cdot|||$ is an equivalent norm on X.

Choose $\epsilon > 0$ such that for $|||\varphi||| < \epsilon$ we have

$$|||F(\varphi) - L\varphi||| \leq \frac{1}{4}\delta|||\varphi|||.$$

For φ such that $|||\varphi||| < \epsilon$ and $\|(I-P)\varphi\| \leq \|P\varphi\|$ we then have

$$\begin{aligned}
\|PF(\varphi)\| &\geq \|PL\varphi\| - \|P\big(F(\varphi) - L\varphi\big)\| \\
&\geq (\theta+\delta)\|P\varphi\| - \frac{1}{4}\delta|||\varphi||| \\
&\geq (\theta+\delta)\|P\varphi\| - \frac{1}{2}\delta\|P\varphi\|
\end{aligned}$$

since $|||\varphi||| \leq 2\|P\varphi\|$. Hence

$$\|PF(\varphi)\| \geq (\theta + \frac{1}{2}\delta)\|P\varphi\|.$$

Likewise,

$$\begin{aligned}
\|(I-P)F(\varphi)\| &\leq \|(I-P)L\varphi\| + \|(I-P)\big(F(\varphi) - L\varphi\big)\| \\
&\leq \theta\|(I-P)\varphi\| + \frac{1}{2}\|P\varphi\| \\
&\leq (\theta + \frac{1}{2}\delta)\|P\varphi\|
\end{aligned}$$

and we conclude that $F(\varphi)$ satisfies the same cone condition as φ, i.e.,

$$\|(I-P)F(\varphi)\| \leq \|PF(\varphi)\|.$$

Now suppose that $\|F^{(n)}(\varphi)\| \leq \epsilon$ for all n and $\|\varphi\|$ sufficiently small. Then

$$\|PF^{(n)}(\varphi)\| \geq (\theta + \frac{1}{2}\delta)^n\|P\varphi\|,$$

which, since $\theta + \frac{1}{2}\delta > 1$, tends to infinity as $n \to \infty$ whenever $\|P\varphi\| > 0$. Clearly, this contradicts our assumption that $\|F^{(n)}(\varphi)\| \leq \epsilon$ for all n. We conclude that nonzero initial points in the cone segment $\{\varphi : \|P\varphi\| \geq \|(I-P)\varphi\| \text{ and } |||\varphi||| \leq \epsilon\}$ lead to orbits which must leave the ϵ-ball. $\square$

Proposition 5.11. *Assume that X admits a decomposition*

$$X = X_- \oplus X_+$$

into $T(t)$-invariant subspaces such that

(i) *X_+ is finite dimensional,*

(ii) $\|T(t)|_{X_-}\| \le Me^{\omega t}$,

(iii) $\operatorname{Re}\lambda > \omega$ *for all* $\lambda \in \sigma(A_{X_+})$,

where $\omega \ge 0$ and $M \ge 1$; then φ_0 is unstable as a stationary point of Σ.

Proof. Since X_+ is finite dimensional, there exists $q \in (0,1]$ and $\gamma > \omega$ such that for all $x_+ \in X_+$

$$\|T(t)x_+\| \ge qe^{\gamma t}\|x_+\|$$

[just apply the standard exponential estimates to the semigroup $S(t) = T(-t)$; so we use the finite dimensionality of X_+ to be sure that $T(t)$ extends to a group of operators on X_+]. Now choose s so large that $qe^{\gamma s} > Me^{\omega s}$ and apply Proposition 5.10 with $L = T(s)$, $F(\varphi) = \Sigma(s, \varphi)$, $\theta = Me^{\omega s}$ and $\delta = qe^{\gamma s} - Me^{\omega s}$. Note that the instability assertion carries over immediately from the discrete time to the continuous time setting (we do not have to worry about the "pieces in between"). □

Corollary 5.12. (*Principle of Linearized Stability.*) *Let φ_0 be a stationary point of the semiflow Σ and let T denote the (uniform) derivative of Σ at φ_0. Assume that X admits a decomposition*

$$X = X_- \oplus X_+$$

into $T(t)$-invariant subspaces such that

(i) *X_+ is finite dimensional,*

(ii) *the restriction of $T(t)$ to X_- converges to zero exponentially as $t \to \infty$.*

Then φ_0 is

(i) *(locally) exponentially stable when* $\operatorname{Re}\lambda < 0$ *for all* $\lambda \in \sigma(A|_{X_+})$,

(ii) *unstable when there exists a* $\lambda \in \sigma(A|_{X_+})$ *with* $\operatorname{Re}\lambda > 0$.

VII.6 Autonomous RFDE

Consider $X = C([-h, 0], \mathbb{R}^n)$, for $h > 0$ and $n \in \mathbb{N}$. Let P be a real Banach space, the parameter space. Let a map $g : \mathcal{O} \to \mathbb{R}^n$, $\mathcal{O} \subset X \times P$, be given. A solution of the autonomous retarded functional differential equations

$$(6.1, p) \qquad \dot{x}(t) = g(x_t, p)$$

is either a differentiable function $x : \mathbb{R} \to \mathbb{R}^n$ so that $\dot{x}(t)$ and the segments

$$x_t : \theta \mapsto x(t+\theta)$$

in X satisfy $(6.1, p)$ everywhere, or a continuous function $x : [t_0 - h, t_+) \to \mathbb{R}^n$, $t_0 < t_+ \leq \infty$, which is differentiable on (t_0, t_+) and satisfies $(6.1, p)$ for $t_0 < t < t_+$. It follows easily that every solution on $\mathbb{R}$ (or $[t_0 - h, t_+)$) defines a continuous map

$$t \mapsto x_t$$

from $\mathbb{R}$ [or $[t_0 - h, t_+)$] into the state space X. Observe also that continuity of $g(\,\cdot\,, p)$ implies that every solution

$$x : [t_0 - h, t_+) \to \mathbb{R}^n$$

has a right derivative $g(x_{t_0}, p)$ at t_0: This is immediate from

$$\begin{aligned} \|x(t_0+\tau) - x(t_0) - \tau g(x_{t_0}, p)\| &= \|\int_{t_0}^{t_0+\tau} [g(x_t, p) - g(x_{t_0}, p)]\, dt\| \\ &\leq \tau \max_{[t_0, t_0+\tau]} \|g(x_t, p) - g(x_{t_0}, p)\|. \end{aligned}$$

Therefore we can define

$$\dot{x}(t_0) := \lim_{\tau \downarrow 0} \frac{1}{\tau}\big(x(t_0+\tau) - x(t_0)\big),$$

which yields a continuous extension of $\dot{x}$ to the interval $[t_0, t_+)$.

We associate a semigroup $\big(T_0(t)\big)_{t\geq 0}$ and a perturbation G with f: for $T_0(t)$, $t \geq 0$, take the operators of the $\mathcal{C}_0$-semigroup on X formed by the solutions of the initial-value problems

$$\begin{aligned} \dot{x}(t) &= 0, \\ x_0 &= \varphi, \qquad \varphi \in X. \end{aligned}$$

The Banach space X is $\odot$-reflexive with respect to $\big(T_0(t)\big)_{t\geq 0}$ (see Section II.5). The canonical basis $(e_1, \ldots, e_n)$ of $\mathbb{R}^n$, more specifically, the n elements

$$(e_i, 0) \in \mathbb{R}^n \times L^\infty\big([-h, 0], \mathbb{R}^n\big),$$

determines linearly independent elements $r_i^{\odot *} \in X^{\odot *}$. Define a map $G : \mathcal{O} \to X^{\odot *}$ by

$$G(\varphi, p) := \sum_{i=1}^{n} g_i(\varphi, p) r_i^{\odot *}.$$

Continuity of g implies continuity of G. If g is locally Lipschitz continuous with respect to the state, i.e.,

((**locLip**)) for every $(\varphi, p) \in \mathcal{O}$ there exist a neighbourhood N of (φ, p) and a constant $L \geq 0$ such that $\|g(\psi, q) - g(\chi, q)\| \leq L\|\psi - \chi\|$ for all (ψ, q) and (χ, q) in N, then G satisfies the condition (locLip) from Section 3. Also, if g is C^1, then G is C^1, too.

Assume from now on that g is continuous and satisfies condition ((locLip)). Let $\Sigma : \triangle \to X$ denote the parameterized semiflow associated with $\left(T_0(t)\right)_{t\geq 0}$ and G, given by the maximal solutions $u_{\varphi,p} : I_{\varphi,p} \to X$ of equation $(1.1, \varphi, p)$

$$u(t) = T_0(t)\varphi + \int_0^t T_0^{\odot *}(t-s)G\big(u(s), p\big)\, ds, \quad \text{for } (\varphi, p) \in \mathcal{O}.$$

We want to show that these maximal solutions are in one-to-one correspondence with $\mathbb{R}^n$-valued solutions of $(6.1, p)$. The proof of this relies on the technical Lemma III.4.3.

Proposition 6.1. *Let* $(\varphi, p) \in \mathcal{O}$.

(i) *Suppose* $x : [-h, t_+) \to \mathbb{R}^n$ *is a solution to equation* $(6.1, p)$ *and* $x_0 = \varphi \in X$. *Then* $t_+ \leq \sup I_{\varphi,p}$ *and for* $0 \leq t < t_+$,

$$x_t = u_{\varphi,p}(t) = \Sigma(t, \varphi, p).$$

(ii) *Consider the function* $\tilde{x} : [-h, 0] \cup I_{\varphi,p} \to \mathbb{R}^n$, *given by* $\tilde{x}_0 = \varphi$ *and*

$$\tilde{x}(t) = \Sigma(t, \varphi, p)(0), \quad \textit{for } t \in I_{\varphi,p}.$$

Then $\tilde{x}$ *is a solution of* $(6.1, p)$ *and*

$$\tilde{x}_t = \Sigma(t, \varphi, p), \quad \textit{for all } t \in I_{\varphi,p}.$$

Proof. 1. For x as in part (i) and for $0 \leq t < t_+$, define $u(t) := x_t$. This yields a continuous map $u : [0, t_+) \to X$. We verify $(1.1, \varphi, p)$: let $0 \leq t < t_+$ and $-h \leq \theta \leq 0$. Then

$$u(t)(\theta) - T_0(t)\varphi(\theta) = \begin{cases} 0, & \text{for } t + \theta \leq 0, \\ x(t+\theta) - \varphi(0), & \text{for } 0 < t + \theta. \end{cases}$$

For $0 < t + \theta$, an integration of equation $(6.1, p)$ yields

$$\begin{aligned} x(t+\theta) - \varphi(0) &= \int_0^{t+\theta} g(u(s), p)\, ds \\ &= \Big(\int_0^t T_0^{\odot *}(t-s)G(u(s), p)\, ds\Big)(\theta), \end{aligned}$$

by Lemma III.4.2. Applying Lemma III.4.2 also in case $t + \theta \leq 0$, we arrive at equation $(1.1, \varphi, p)$. Also, $u = u_{\varphi,p}\big|_{[0,t_+)}$.

2. The function $\tilde{x}$ of part (ii) is well defined and continuous since we have $\varphi(0) = \Sigma(0, \varphi, p)(0)$. Set $u := u_{\varphi,p}$. For $0 < t \in I_{\varphi,p}$,

$$(6.2)\qquad \begin{aligned}\tilde{x}(t) = u(t)(0) &= T_0(t)\varphi(0) + \Big(\int_0^t T_0^{\odot *}(t-s)G(u(s),p)\,ds\Big)(0)\\ &= \varphi(0) + \int_0^{t+0} g(u(s),p)\,ds,\end{aligned}$$

by Lemma III.4.2. It follows that $\tilde{x}$ is differentiable at t, and we have

$$(6.3)\qquad \dot{\tilde{x}}(t) = g(u(t),p).$$

Similarly, for $t \in I_{\varphi,p}$ and $-h \le \theta \le 0$,

$$\begin{aligned}u(t)(\theta) &= T_0(t)\varphi(\theta) + \Big(\int_0^t T_0^{\odot *}(t-s)G(u(s),p)\,ds\Big)(\theta)\\ &= \begin{cases}\varphi(t+\theta)+0, & \text{if } t+\theta \le 0,\\ \varphi(0) + \int_0^{t+\theta} g(u(s),p)\,ds, & \text{if } 0 < t+\theta.\end{cases}\end{aligned}$$

In the first case,

$$u(t)(\theta) = \varphi(t+\theta) = \tilde{x}(t+\theta).$$

In the second case,

$$u(t)(\theta) = \varphi(0) + \int_0^{t+\theta} g(u(s),p)\,ds = \tilde{x}(t+\theta),$$

by (6.2). Altogether, $\Sigma(t,\varphi,p) = u(t) = \tilde{x}_t$ for $t \in I_{\varphi,p}$, and (6.3) yields

$$\dot{\tilde{x}}(t) = g(\tilde{x}_t, p), \quad \text{for } 0 < t \in I_{\varphi,p}.$$

□

It follows that for every $(\varphi,p) \in \mathcal{O}$ there exists a unique solution

$$x^{\varphi,p} : I^{\varphi,p} \to \mathbb{R}^n$$

of the initial-value problem

$$(6.4,p,\varphi)\qquad \begin{aligned}\dot{x}(t) &= g(x_t,p),\\ x_0 &= \varphi\end{aligned}$$

so that any other solution $x : [-h,t_+) \to \mathbb{R}^n$ of this is a restriction of $x^{\varphi,p}$. We have $I^{\varphi,p} = [-h,\infty)$, or $I^{\varphi,p} = [-h,t^{\varphi,p})$ for some $t^{\varphi,p} > 0$. [Upper indices indicate that here we consider solutions to RFDE as defined at the beginning of this section, not solutions to (1.1,φ,p).]

We call $x^{\varphi,p}$ the maximal solution of the initial-value problem (6.1,p,φ). Proposition 6.1 rephrased now reads

$$(6.5)\qquad \Sigma(t,\varphi,p) = x_t^{\varphi,p} \quad \text{for all } (t,\varphi,p) \in \Delta.$$

Remark 4.4(i) shows that semiflows of RFDE with smooth g are, in general, not differentiable with respect to the time variable.

Proposition 6.2.

(i) *For* $(t,\varphi,p)\in\triangle$ *with* $t>h$, *the partial derivative* $D_1\Sigma(t,\varphi,p)$ *exists, and we have*

$$D_1\Sigma(t,\varphi,p)1=\dot{x}_t^{\varphi,p}\in X.$$

(ii) *At points* $(h,\varphi,p)\in\triangle$, *there is a right derivative* $D_1^+\Sigma(h,\varphi,p)\in\mathcal{L}(\mathbb{R},X)$. *We have*

$$D_1^+\Sigma(h,\varphi,p)1=\dot{x}_h^{\varphi,p}\in X,$$

with the continuous extension of $\dot{x}^{\varphi,p}$ *to* $[0,\infty)\cap I^{\varphi,p}$.

(iii) *The map*

$$\{(t,\varphi,p)\in\triangle\mid t\geq h\}\to\mathcal{L}(\mathbb{R},X)$$

given by $D_1\Sigma(t,\varphi,p)$ *for* $t>h$ *and by* $D_1^+\Sigma(h,\varphi,p)$ *at points* $(h,\varphi,p)\in\triangle$ *is continuous.*

Proof. For $(t,\varphi,p)\in\triangle$ with $t>h$, set $x:=x^{\varphi,p}$. Then we have an element $\dot{x}_t\in X$. Note

$$\dot{x}_t(\theta)=\dot{x}(t+\theta)=g(x_{t+\theta},p)\quad\text{for }-h\leq\theta\leq 0.$$

For $\epsilon>0$ with $h<t-\epsilon<t+\epsilon<\sup I^{\varphi,p}=\sup I_{\varphi,p}$ and for $-\epsilon<\tau<\epsilon$, $-h\leq\theta\leq 0$, we have

$$\begin{aligned}\{\Sigma(t+\tau,\varphi,p)-\Sigma(t,\varphi,p)-\tau\dot{x}_t\}(\theta)&=x(t+\tau+\theta)-x(t+\theta)-\tau\dot{x}(t+\theta)\\&=\int_t^{t+\tau}\big(\dot{x}(s+\theta)-\dot{x}(t+\theta)\big)\,ds\\&=\int_t^{t+\tau}\big(g(x_{s+\theta},p)-g(x_{t+\theta},p)\big)\,ds.\end{aligned}$$

Use continuity of the map $[0,t+\epsilon]\to\mathbb{R}^n,\ s\mapsto g(x_s,p)$ to complete the proof of assertion (i). Assertion (ii) follows in the same way.
Proof of (iii). Let $(t,\varphi,p)\in\triangle$, $t\geq h$. Choose $\epsilon>0$ with

$$t+\epsilon<\sup I_{\varphi,p}.$$

For $h\leq s<\sup I_{\varphi,p}$, the element $D_1\Sigma(s,\varphi,p)1\in X$ is given by

$$\theta\mapsto g\big(\Sigma(s+\theta,\varphi,p),p\big).$$

The map

$$\triangle(t+\epsilon)\to C\big([0,t+\epsilon],X\big),\quad(\psi,q)\mapsto u_{\psi,q}\big|_{[0,t+\epsilon]}$$

is continuous (see Section 4); $\triangle(t+\epsilon)$ is an open neighbourhood of (φ,p). Arguing as in Section 4, one finds that the map

$$\alpha:\triangle(t+\epsilon)\to C\big([0,t+\epsilon],\mathbb{R}^n\big),\quad\alpha(\psi,q)(s)=g\big(\Sigma(s,\psi,q),q\big)$$

is continuous. (This involves Lemma 1.4 of Appendix IV.)

Let a sequence of points (t_k, φ_k, p_k) be given such that for all k,

$$h \leq t_k < t + \epsilon, \quad (\varphi_k, p_k) \in \Delta(t+\epsilon)$$

and

$$(t_k, \varphi_k, p_k) \to (t, \varphi, p) \quad \text{as } k \to \infty.$$

For $-h \leq \theta \leq 0$ and for all k, we have

$$\begin{aligned}
&\|(D_1\Sigma(t_k, \varphi_k, p_k)1 - D_1\Sigma(t, \varphi, p)1)(\theta)\| \\
&\quad = \|g(\Sigma(t_k + \theta, \varphi_k, p_k), p_k) - g(\Sigma(t + \theta, \varphi, p), p)\| \\
&\quad \leq \|g(\Sigma(t_k + \theta, \varphi_k, p_k), p_k) - g(\Sigma(t_k + \theta, \varphi, p), p)\| \\
&\qquad\qquad + \|g(\Sigma(t_k + \theta, \varphi, p), p) - g(\Sigma(t + \theta, \varphi, p), p)\| \\
&\quad \leq \sup_{s \in [0, t+\epsilon]} \|g(\Sigma(s, \varphi_k, p_k), p_k) - g(\Sigma(s, \varphi, p), p)\| \\
&\qquad\qquad + \|g(\Sigma(t_k + \theta, \varphi, p), p) - g(\Sigma(t + \theta, \varphi, p), p)\| \\
&\quad = \|\alpha(\varphi_k, p_k) - \alpha(\varphi, p)\| + \|g(\Sigma(t_k + \theta, \varphi, p), p) \\
&\qquad\qquad - g(\Sigma(t + \theta, \varphi, p), p)\|.
\end{aligned}$$

Now, continuity of α and uniform continuity of $u_{\varphi,p}$ on $[0, t+\epsilon]$ imply that

$$D_1\Sigma(t_k, \varphi_k, p_k)1 \to D_1\Sigma(t, \varphi, p)1 \quad \text{as } k \to \infty.$$

□

Next we consider partial derivatives with respect to state and parameters. According to Corollary 4.3(ii), the derivatives $D^{(0,0,1)}\Sigma(t, \varphi, p)$ exist on all of Δ, provided G is C^1. Remark 4.4(ii) shows that even for the simplest RFDE

$$\dot{x} = 0$$

with the state space $X = C([-h, 0], \mathbb{R})$, $h = 1$ and arbitrary parameter space P, $\mathcal{O} = X \times P$, the map $D^{(0,0,1)}\Sigma : \Delta \to \mathcal{L}_{\mathbb{C}}(X \times P, X)$ is not continuous at points (t, φ, p) with $0 \leq t \leq h$. In the sequel we shall prove that $D^{(0,0,1)}\Sigma$ is continuous at points (t, φ, p) with $t > h$.

First we compute $D^{(0,0,1)}\Sigma$ in terms of solutions to the linear variational equations along solutions of the original RFDE $(6.1, p)$.

Proposition 6.3. *Suppose g is C^1. Let $(t, \varphi, p) \in \Delta$, $(\overline{\varphi}, \overline{p}) \in X \times P$. Then*

$$D^{(0,1,1)}\Sigma(t, \varphi, p)(\overline{\varphi}, \overline{p}) = y_t$$

where the continuous function

$$y : [-h, t_{\varphi,p}) \to \mathbb{R}^n$$

is differentiable for $0 < t < t_{\varphi,p}$ *and satisfies*

$$(6.1, p, \varphi, \overline{\varphi}, \overline{p}) \qquad \begin{aligned} \dot{y}(t) &= D_1 g\big(\Sigma(t,\varphi,p),p\big) y_t + D_2 g\big(\Sigma(t,\varphi,p),p\big)\overline{p}, \\ y_0 &= \overline{\varphi}. \end{aligned}$$

Proof. Recall from Proposition 4.5 that, in general, for abstract integral equations with a C^1 function G

$$D^{(0,1,1)}\Sigma(t,\varphi,p)(\overline{\varphi},\overline{p}) = T_0(t)\overline{\varphi} + \int_0^t T_0^{\odot *}(t-s) f(s)\, ds,$$

where

$$\begin{aligned} f(s) = {} & D_1 G\big(\Sigma(s,\varphi,p),p\big) D^{(0,1,1)}\Sigma(s,\varphi,p)(\overline{\varphi},\overline{p}) \\ & + D_2 G\big(\Sigma(s,\varphi,p),p\big)\overline{p}. \end{aligned}$$

So when G is of the form

$$G(\varphi,p) = \sum_{i=j}^{n} g_j(\varphi,p) r_j^{\odot *},$$

then f is of the form

$$\begin{aligned} f(s) = \sum_{j=1}^{n} \big\{ & D_1 g_j\big(\Sigma(s,\varphi,p),p\big) D^{(0,1,1)}\Sigma(s,\varphi,p)(\overline{\varphi},\overline{p}) \\ & + D_2 g_j\big(\Sigma(s,\varphi,p),p\big)\overline{p}\big\} r_j^{\odot *}. \end{aligned}$$

For delay equations we can now infer from Lemma III.4.2 that

$$D^{(0,1,1)}\Sigma(t,\varphi,p)(\overline{\varphi},\overline{p})(\theta) = \begin{cases} \overline{\varphi}(t+\theta), & t+\theta \le 0, \\ y(t+\theta), & 0 \le t+\theta < t_{\varphi,p}, \end{cases}$$

where for $0 \le t < t_{\varphi,p}$

$$\begin{aligned} y(t) = \overline{\varphi}(0) + \sum_{j=1}^{n} e_j \int_0^t \big\{ & D_1 g_j\big(\Sigma(s,\varphi,p),p\big) D^{(0,1,1)}\Sigma(s,\varphi,p)(\overline{\varphi},\overline{p}) \\ & + D_2 g_j\big(\Sigma(s,\varphi,p),p\big)\overline{p}\big\}\, ds. \end{aligned}$$

So y is differentiable for $t > 0$ and satisfies

$$\dot{y}(t) = D_1 g\big(\Sigma(t,\varphi,p),p\big) y_t + D_2 g\big(\Sigma(t,\varphi,p),p\big)\overline{p},$$

provided we extend the definition of y to the interval $[-h, 0]$ by requiring that $y_0 = \overline{\varphi}$. □

Proposition 6.4. *Suppose g is C^1.*

(i) *For every $(\varphi, p) \in \mathcal{O}$, the map*

$$[h, t_{\varphi,p}) \ni t \mapsto D^{(0,1,1)}\Sigma(t, \varphi, p) \in \mathcal{L}_{\mathbb{C}}(X \times P, X)$$

is continuous.

(ii) *The restricted map*

$$D^{(0,1,1)}\Sigma|_{\{(t,\varphi,p)\in\triangle|t\geq h\}}$$

is continuous.

(iii) *The restriction of the parameterized semiflow Σ to the set $\{(t, \varphi, p) \in \triangle \mid t > h\}$ is C^1.*

Proof. 1. In order to prove (i) we employ the nonautonomous RFDE from Proposition 6.3. Let $(\varphi, p) \in \mathcal{O}$ and $h \leq s < t_{\varphi,p}$. Choose a point $t_0 \in (s, t_{\varphi,p})$. Corollary 4.6 yields a constant $C \geq 0$ such that

$$\|D^{(0,1,1)}\Sigma(t, \varphi, p)\| \leq C \quad \text{for } 0 \leq t \leq t_0.$$

Set

$$\overline{C} = \max_{0\leq t\leq t_0} \|Dg(\Sigma(t, \varphi, p), p)\|.$$

For every $(\overline{\varphi}, \overline{p}) \in X \times P$ with $\|(\overline{\varphi}, \overline{p})\| \leq 1$ and for $0 < t \leq t_0$, we obtain, for the solution $y : [-h, t_{\varphi,p}) \to \mathbb{R}^n$ of the initial-value problem $(6.1, p, \varphi, \overline{\varphi}, \overline{p})$ from Proposition 6.3,

$$\begin{aligned}|\dot{y}(t)| &= |Dg(\Sigma(t, \varphi, p), p)D^{(0,1,1)}\Sigma((t, \varphi, p)(\overline{\varphi}, \overline{p}), \overline{p})| \\ &\leq \overline{C}(C + 1);\end{aligned}$$

it follows that

$$|y(t) - y(\overline{t})| \leq \overline{C}(C + 1)|t - \overline{t}| \quad \text{for } 0 \leq t \leq t_0,\ 0 \leq \overline{t} \leq t_0.$$

For $h \leq t \leq t_0$, we get

$$\begin{aligned}&\|D^{(0,1,1)}\Sigma(t, \varphi, p)(\overline{\varphi}, \overline{p}) - D^{(0,1,1)}\Sigma(s, \varphi, p)(\overline{\varphi}, \overline{p})\| \\ &\quad = \|y_t - y_s\| = \max_{-h\leq\theta\leq 0} |y(t + \theta) - y(s + \theta)| \\ &\leq \overline{C}(C + 1)|t - s|.\end{aligned}$$

Therefore

$$\|D^{(0,1,1)}\Sigma(t, \varphi, p) - D^{(0,1,1)}\Sigma(s, \varphi, p)\| \leq \overline{C}(C + 1)|t - s|$$

for $h \leq t \leq t_0$. This proves assertion (i).

2. Proof of (ii). Let $(t, \varphi, p) \in \triangle$ with $t \geq h$ be given. Choose $t_0 \in (t, t_{\varphi,p})$. For each $(s, \psi, q) \in \triangle$ with $h \leq s \leq t_0$, we have

$$
\begin{aligned}
&\|D^{(0,1,1)}\Sigma(s,\psi,q) - D^{(0,1,1)}\Sigma(t,\varphi,p)\| \\
&\quad \le \|D^{(0,1,1)}\Sigma(s,\psi,q) - D^{(0,1,1)}\Sigma(s,\varphi,p)\| \\
&\qquad + \|D^{(0,1,1)}\Sigma(s,\varphi,p) - D^{(0,1,1)}\Sigma(t,\varphi,p)\|.
\end{aligned}
$$

Apply Corollary 4.3(iv) on uniform continuity at the point (φ, p) to the maps

$$
D^{(0,1,1)}\Sigma(s,\,\cdot\,,\,\cdot\,), \qquad 0 \le s \le t_0,
$$

and apply assertion (i) to the second term of the last sum.
3. Assertion (iii) follows from assertion (ii) and Proposition 6.2. □

An important property of semiflows associated with RFDE is compactness. The basic results is

Proposition 6.5. *Assume g is bounded, $t \ge h$ and $B \subset \mathcal{O}$ is a bounded set. Then the closure of the set*

$$
\{\Sigma(s,\varphi,p) \in X \mid h \le s \le t,\ (\varphi,p) \in B \cap \Delta(t)\}
$$

is compact.

Proof. Define

$$
\begin{aligned}
c &:= \sup_{\mathcal{O}} \|g(\varphi,p)\| < \infty, \\
d &:= \sup_{(\varphi,p)\in B} \|\varphi(0)\| < \infty.
\end{aligned}
$$

If $h \le s \le t$ and $(\varphi,p) \in B \cap \Delta(t)$, then

$$
\Sigma(s,\varphi,p) = x_s^{\varphi,p}.
$$

Set $x := x^{\varphi,p}$. For $0 \le \tau \le t$

$$
\|\dot{x}(\tau)\| \le c.
$$

In particular, for $-h \le \theta \le 0$

$$
\begin{aligned}
\|x_s(\theta)\| = \|x(s+\theta)\| &\le \|x(0)\| + \|\int_0^{s+\theta} \dot{x}(\tau)\,d\tau\| \\
&\le d + ct
\end{aligned}
$$

and

$$
\|\dot{x}_s(\theta)\| = \|\dot{x}(s+\theta)\| \le c.
$$

It follows that the set

$$
\{\Sigma(s,\varphi,p) \in X \mid h \le s \le t,\ (\varphi,p) \in B \cap \Delta(t)\}
$$

is bounded and equicontinuous; the theorem of Ascoli-Arzèla yields the assertion. □

Exercise 6.6. Give a more abstract proof of this result (cf. Exercise III.2.5).

Remark **6.7**. It follows that the time-t-maps

$$\varphi \mapsto \Sigma(t, \varphi, p), \quad t \geq h, \ p \in P,$$

are compact (in the sense that bounded sets are mapped into sets with compact closure). Observe that this excludes the existence of continuous inverse maps defined on open sets: such an inverse would imply that the identity $I : X \to X$ maps an open set into a compact set, which contradicts the fact that X is not finite dimensional. In particular, smooth time t-maps, $t \geq h$ and $p \in P$, of semiflows associated with RFDE are never diffeomorphisms.

In conclusion of this section we specialize Corollary 5.12, i.e., the Principle of Linearized Stability, to the case of delay equations.

Consider the autonomous retarded functional differential equation

$$\dot{x}(t) = g(x_t)$$

and assume that

$$g(\varphi_0) = 0.$$

Then φ_0 is a stationary point and with it we can associate the linearized problem

$$\dot{y}(t) = Dg(\varphi_0)y_t$$

and, subsequently, the characteristic equation

$$\det\left(zI - Dg(\varphi_0)e_z\right) = 0,$$

where

$$e_z(\theta) = e^{z\theta} \quad \text{for } -h \leq \theta \leq 0.$$

Theorem 6.8. *The stationary point* φ_0 *is*

(i) *unstable if* $\operatorname{Re}\lambda > 0$ *for some root* λ *of the characteristic equation,*

(ii) *(locally) exponentially stable if* $\operatorname{Re}\lambda < 0$ *for all roots* λ *of the characteristic equation.*

Proof. All we have to do is to combine the spectral theory of the linear problem, as presented in Section IV.3, with Corollary 5.12. □

VII.7 Comments

This chapter generalizes the basic results on existence, uniqueness and smooth dependence on data for solutions to initial-value problems for ODEs given by vector fields on open subsets of $\mathbb{R}^n$. The introduction of semiflows in Section 2 is partly parallel to Section II.10 of Amann [5] on flows. Parts of Sections 3 and 4 are inspired by the presentation by Henry [121] on perturbations of analytic semigroups. We need a more general theory, however, since $\mathcal{C}_0$-semigroups defined by linear autonomous RFDE are never analytic.

Additional results on smoothness, e.g., a C^k-version, $k \geq 2$, of Corollary 4.3(ii), would require a larger technical apparatus, including more general versions of Lemma 1.4 of Appendix IV and of Exercise 4.2. For the case of analytic semigroups, see Henry [121].

More generality also seems possible with regard to the parameter dependence: the perturbations G, or suitable subsets of such perturbations, might be considered as parameters. Compare, e.g., Abraham and Robbin [1]. An alternative idea to handle parameterized problems would be to add to (1.1) the equation

$$\dot{p} = 0$$

and to reduce the discussion of parameterized initial-value problems to one without parameters. For another presentation of the basic theory, in case of vector fields, see Arnold [12].

Chapter VIII

Behaviour near a hyperbolic equilibrium

VIII.1 Introduction

In this chapter we start to investigate the behaviour of the nonlinear semiflow near an equilibrium. Throughout this chapter, and also in Chapters IX and X, we consider X over $\mathbb{R}$. For the linearized semiflow $\{T(t)\}$, the time asymptotic behaviour near the equilibrium $u \equiv 0$ is described by the decomposition of the state space according to the spectrum of the generator of the semigroup and the accompanying exponential dichotomy. In this chapter we will assume that there is no spectrum on the imaginary axis. In the case of RFDE, the spectrum in the right half-plane consists of finitely many, say k, eigenvalues (counting multiplicity) with a positive real part. From Chapter IV we recall that in this case we can decompose X as

$$X = X_- \oplus X_+.$$

If $\varphi \in X_+$ then $T(t)\varphi$ is defined for all negative t and $\lim_{t\to-\infty} T(t)\varphi = 0$. Actually, this property characterises X_+ (cf. Lemma 2.2 below), which is therefore called the *unstable subspace*. More importantly, the characterization "survives" a nonlinear perturbation, albeit with an invariant unstable manifold replacing the linear subspace X_+.

Similarly, elements φ of the *stable subspace* X_- are characterized by $\lim_{t\to+\infty} T(t)\varphi = 0$ and this characterization can be taken as the starting point for the construction of an invariant stable manifold for the nonlinear semiflow associated with the AIE

$$(1.1)\quad u(t) = T(t-s)u(s) + \int_s^t T^{\odot *}(t-\tau)R(u(\tau))\,d\tau, \quad -\infty < s \le t < \infty,$$

where $R(0) = 0$ and $DR(0) = 0$. These invariant manifolds play a crucial role in the description of the orbit structure near the equilibrium. (Note that the restriction to a neighbourhood of the equilibrium is essential, as R may have a large impact away from the origin.)

We will prove that, under suitable hypotheses on the nonlinear terms, there exists an open neighbourhood $\mathcal{V}$ of the origin in X and two manifolds

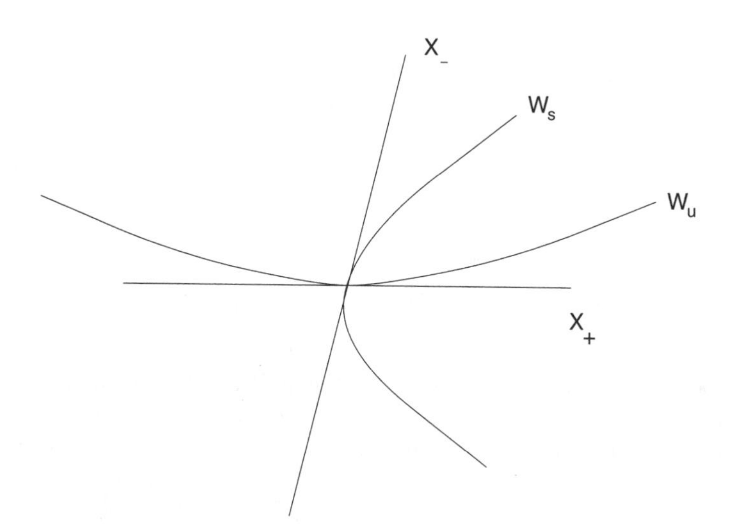

Fig. VIII.1. The stable and unstable subspace/manifold for the linear and the nonlinear equation.

$\mathcal{W}_u$ and $\mathcal{W}_s$ in X of dimension and codimension k, respectively, such that $\mathcal{W}_u \cap \mathcal{W}_s = \{0\}$, $\mathcal{W}_s$ is positively invariant under the flow of (1.1) and if $\varphi \in \mathcal{W}_u$, then there exists a unique orbit through φ which is defined for all negative time and lies in $\mathcal{W}_u$. Conversely, if a solution to equation (1.1) through $\varphi \in X$ lies in $\mathcal{V}$ for all positive (negative) time, then $\varphi \in \mathcal{W}_s$ ($\mathcal{W}_u$). Moreover, the orbit with initial value $\varphi \in \mathcal{W}_s$ converges to the origin as $t \to \infty$, and if $\varphi \in \mathcal{W}_u$, then the orbit through φ defined for all negative time converges to the origin as $t \to -\infty$. These statements can be summarized by saying that for this class of nonlinearities the saddle point property holds.

VIII.2 Spectral decomposition

We recall from Chapter IV, in particular Sections IV.2 and IV.3, the facts about spectral decomposition that we use in this chapter. We will need this decomposition in the large space $X^{\odot *}$, as we assume that the nonlinearity maps X into $X^{\odot *}$. As before, we identify X with its embedding into $X^{\odot *}$ (or, in other words, we omit the embedding operator in all that follows) and we assume $\odot$-reflexivity (mostly for ease of formulation; the same techniques yield analogous results when, for instance, we do not have $\odot$-reflexivity but still solutions of the AIE define a nonlinear semigroup on X; compare Section III.8).

We make the following assumptions:

H1 $X^{\odot *} = X_-^{\odot *} \oplus X_+$, where $X_-^{\odot *}$ and X_+ are closed subspaces of $X^{\odot *}$, with X_+ contained in X.

H2 $X_-^{\odot *}$ and X_+ are invariant under $T^{\odot *}(t)$.

H3 $T(t)$ can be extended to a one-parameter group on X_+.

H4 The decomposition is an exponential dichotomy on $\mathbb{R}$ with respect to 0, i.e., there exist real numbers a and b, $a < 0 < b$, and a positive constant K such that

$$\begin{aligned} \|T(t)\varphi\| &\leq K e^{bt} \|\varphi\|, \qquad t \leq 0,\ \varphi \in X_+, \\ \|T^{\odot *}(t)\varphi^{\odot *}\| &\leq K e^{at} \|\varphi^{\odot *}\|, \quad t \geq 0,\ \varphi^{\odot *} \in X_-^{\odot *}. \end{aligned} \tag{2.1}$$

Remark **2.1**. These assumptions certainly hold if $T(t)$ is eventually compact, in particular for RFDE, provided there is no spectrum on the imaginary axis. In that case, X_+ is finite dimensional. More precisely, we have that

$$X_+ = \bigoplus_{\lambda \in \Lambda_+} \mathcal{R}(P_\lambda^{\odot *}), \qquad X_-^{\odot *} = \bigcap_{\lambda \in \Lambda_+} \mathcal{N}(P_\lambda^{\odot *}),$$

where

$$\Lambda_+ = \{\lambda \in \sigma(A^{\odot *}) \mid \operatorname{Re}\lambda > 0 \text{ and } \operatorname{Im}\lambda \geq 0\}.$$

Here $P_\lambda^{\odot *}$ is the spectral projection operator on X (over $\mathbb{R}$) associated with $\lambda \cup \bar{\lambda}$. See Chapter IV Corollary 2.19 and Exercises 2.20-2.22. If we let

$$\begin{aligned} \gamma_+ &= \inf\{\operatorname{Re}\lambda \mid \lambda \in \Lambda_+\}, \\ \gamma_- &= \sup\{\operatorname{Re}\lambda \mid \lambda \in \sigma(A^{\odot *}) \text{ and } \operatorname{Re}\lambda < 0\}, \end{aligned}$$

then for any positive number ϵ we can take $b = \gamma_+ - \epsilon$ and $a = \gamma_- + \epsilon$.

Notation. We will denote the projection operator with range X_+ and kernel $X_-^{\odot *}$ by $P_+^{\odot *}$. If we restrict the domain to X, then we shall write P_+ instead. Whenever we use the constant K in this chapter and the next one, we mean the constant K in the exponential dichotomy above. To give an equivalent geometric description of the stable and the unstable subspace, we introduce the following function spaces:

Definition 2.2. Let E be a Banach space, then we define two Banach spaces by

$$\begin{aligned} BC(\mathbb{R}_+, E) &= \{F \in C(\mathbb{R}_+, E) \mid \sup_{t \in \mathbb{R}_+} \|F(t)\| < \infty\}, \\ BC(\mathbb{R}_-, E) &= \{F \in C(\mathbb{R}_-, E) \mid \sup_{t \in \mathbb{R}_-} \|F(t)\| < \infty\}, \end{aligned}$$

where in each case the norm is given by the supremum.

We are now ready to characterize the stable subspace and the unstable subspace in terms of bounds on the orbits.

Lemma 2.3.
(i) $X_- = \{\varphi \in X \mid t \mapsto T(t)\varphi \in BC(\mathbb{R}_+, X)\}$,
(ii) $X_+ = \{\varphi \in X \mid t \mapsto T(t)\varphi \in BC(\mathbb{R}_-, X)\}$.

Proof. We prove only the first part. If $\varphi \in X_-$, then the second estimate in (2.1) implies that $t \mapsto T(t)\varphi$ is bounded on $\mathbb{R}_+$. Conversely, assume that φ is such that $t \mapsto T(t)\varphi$ is bounded on $\mathbb{R}_+$. First note that

$$P_+T(t)\varphi = T(t)P_+\varphi.$$

Next apply $T(-t)$ to this identity and use the first estimate of (2.1) to obtain

$$||P_+\varphi|| \leq Ke^{-bt}||P_+T(t)\varphi||$$

which can be rewritten as

$$||P_+T(t)\varphi|| \geq \frac{1}{K}e^{bt}||P_+\varphi||.$$

If $||P_+\varphi|| \neq 0$, the right hand side increases beyond any bound for $t \to \infty$ which contradicts the boundedness of the left hand side. So, we conclude that necessarily $||P_+\varphi|| = 0$. □

Exercise 2.4. Prove the second part of the previous lemma.

In fact we can improve this result.

Definition 2.5. Let E be a Banach space and $\eta \in \mathbb{R}$. We define two Banach spaces by

$$BC^\eta(\mathbb{R}_+, E) = \{F \in C(\mathbb{R}_+, E) \mid \sup_{t\in\mathbb{R}_+} e^{-\eta t}||F(t)|| < \infty\},$$
$$BC^\eta(\mathbb{R}_-, E) = \{F \in C(\mathbb{R}_-, E) \mid \sup_{t\in\mathbb{R}_-} e^{-\eta t}||F(t)|| < \infty\},$$

where in each case the norm is given by the weighted supremum norm $||F||_\eta = \sup e^{-\eta t}||F(t)||$. We will write $||F||$ instead of $||F||_\eta$, whenever this does not cause confusion.

Exercise 2.6. Show that for any $\eta \in (a, b)$ we may replace BC by BC^η in Lemma 2.3.

An alternative formulation of Lemma 2.3 would be in terms of bounds for the solutions of the homogeneous linear equation

$$u(t) = T(t-s)u(s), \quad -\infty < s \leq t < \infty. \tag{2.2}$$

In the next section we shall analyse the inhomogeneous variant of (2.2) in this spirit.

VIII.3 Bounded solutions of the inhomogeneous linear equation

If u is a bounded solution of (1.1) then u is a bounded solution of the linear inhomogeneous equation

$$u(t) = T(t-s)u(s) + \int_s^t T^{\odot *}(t-\tau)F(\tau)\,d\tau, \tag{3.1}$$

with $F(t) = R(u(t))$. As a next step towards the nonlinear problem we characterise the bounded solutions of the linear inhomogeneous equation.

Definition 3.1. We define, formally at first, for $t \in \mathbb{R}_+$ ($t \in \mathbb{R}_-$) and for F a mapping from $\mathbb{R}_+$ ($\mathbb{R}_-$) into $X^{\odot *}$,

$$(\mathcal{K}_s F)(t) = \int_0^t T^{\odot *}(t-\tau)(I - P_+^{\odot *})F(\tau)\,d\tau + \int_\infty^t T^{\odot *}(t-\tau)P_+^{\odot *}F(\tau)\,d\tau \tag{3.2}$$

and

$$(\mathcal{K}_u F)(t) = \int_0^t T^{\odot *}(t-\tau)P_+^{\odot *}F(\tau)\,d\tau + \int_{-\infty}^t T^{\odot *}(t-\tau)(I - P_+^{\odot *})F(\tau)\,d\tau. \tag{3.3}$$

Below we shall use the symbols $\mathcal{K}_s$ and $\mathcal{K}_u$ only in a context which guarantees that the integrals do indeed exist. Note that the s and u refer to stable and unstable respectively.

Lemma 3.2.

(i) *For each $\eta \in (a,b)$, formula (3.2) defines a bounded linear mapping $F \mapsto \mathcal{K}_s F$ from $BC^\eta(\mathbb{R}_+, X^{\odot *})$ into $BC^\eta(\mathbb{R}_+, X)$. $\mathcal{K}_s F$ is the unique solution of* (3.1) *in* $BC^\eta(\mathbb{R}_+, X)$ *with vanishing X_- component at* $t = 0$.

(ii) *For each $\eta \in (a,b)$, formula (3.3) defines a bounded linear mapping $F \mapsto \mathcal{K}_u F$ from $BC^\eta(\mathbb{R}_-, X^{\odot *})$ into $BC^\eta(\mathbb{R}_-, X)$. $\mathcal{K}_u F$ is the unique solution of* (3.1) *in* $BC^\eta(\mathbb{R}_-, X)$ *with vanishing X_+ component at* $t = 0$.

Proof. We prove only (ii), the remaining part being left as an exercise. Fix $\eta \in (a,b)$. It follows from (2.1) that for $F \in BC^\eta(\mathbb{R}_-, X^{\odot *})$ and $t \in \mathbb{R}_-$, we can estimate

$$
\begin{aligned}
&||e^{-\eta t}(\mathcal{K}_u F)(t)|| \\
&\quad \le e^{-\eta t}(||\int_0^t T^{\odot *}(t-\tau)P_+^{\odot *}F(\tau)\,d\tau|| \\
&\qquad\qquad + ||\int_{-\infty}^t T^{\odot *}(t-\tau)P_-^{\odot *}F(\tau)\,d\tau||) \\
&\quad \le K||F||_\eta(-\int_0^t e^{(a-\eta)(t-\tau)}\,d\tau + \int_{-\infty}^t e^{(b-\eta)(t-\tau)}\,d\tau) \\
&\quad \le K||F||_\eta(\frac{1}{-(a-\eta)} + \frac{1}{b-\eta}).
\end{aligned}
$$

$\mathcal{K}_u F$ is continuous. The estimate shows that $\mathcal{K}_u$, viewed as a mapping from $BC^\eta(\mathbb{R}_-, X)$ into itself, satisfies the inequality

$$
||\mathcal{K}_u|| \le \frac{K}{-(a-\eta)} + \frac{K}{b-\eta}. \tag{3.4}
$$

The difference between two solutions of (3.1) is a solution of the homogeneous equation (2.1) with $0 \le s \le t < \infty$. The second part of the second statement then follows from Exercise 2.6. This proves (ii). □

Exercise 3.3. Prove (i) along the same lines.

Remark **3.4**. It is a consequence of (3.4) that for any compact interval J contained in (a, b), the norm of $\mathcal{K}_u$ is bounded uniformly for $\eta \in J$.

VIII.4 The unstable manifold

In this section we "generalize" the unstable subspace for equation (2.2) to the unstable manifold for (1.1). This is done in two steps. First, we prove the existence and uniqueness of a Lipschitz smooth unstable manifold in the space of bounded continuous functions. Next, we prove that solutions on the unstable manifold are not merely bounded but converge as $t \to -\infty$ to the origin. This second statement can be proven independently from the first one. We still want to include the first step because it yields a stronger uniqueness result.

A Lipschitz unstable manifold requires only a Lipschitz smooth nonlinearity. We conclude by proving that if the nonlinearity is of class C^k, then the same is true for the unstable manifold.

We recall equation (1.1), i.e., we consider for functions with values in X the AIE

$$
u(t) = T(t-s)u(s) + \int_s^t T^{\odot *}(t-\tau)R(u(\tau))\,d\tau, \quad -\infty < s \le t < \infty. \tag{4.1}
$$

We want R to represent the higher order terms and we need some smoothness. We therefore assume that

HR1 the mapping $u \mapsto R(u)$ from X into $X^{\odot *}$ is an element of $C^k(X, X^{\odot *})$ for some $k \geq 1$,

HR2 $R(0) = 0$ and $DR(0) = 0$.

Corollary 4.1. *There exists a continuous function $\alpha : \mathbb{R}_+ \to \mathbb{R}_+$ such that $\alpha(0) = 0$ and for all $\delta > 0$ and $\varphi \in \mathrm{B}_\delta(X)$, $||R(\varphi)|| \leq \alpha(\delta)||\varphi||$.*

With the mapping $R : X \to X^{\odot *}$ is associated the *substitution operator* (or Nemitsky operator) $\widetilde{R} : BC(\mathbb{R}_-, X) \to BC(\mathbb{R}_-, X^{\odot *})$ defined by $(\widetilde{R}(h))(t) = R(h(t))$.

Lemma 4.2. *The mapping $\widetilde{R}$ defined above is locally Lipschitz continuous. If $\alpha(\rho) = \sup_{\|u\| \leq \rho} \|DR(u)\|$, then $\alpha(\rho)$ is a Lipschitz constant for $\widetilde{R}$ on the ball in $BC(\mathbb{R}_-, X)$ of radius ρ.*

Exercise 4.3. Prove this lemma using the Lemma 1.1(ii) from Appendix IV.

We now introduce the mapping $\mathcal{G} : BC(\mathbb{R}_-, X) \times X_+ \to BC(\mathbb{R}_-, X)$ defined by

$$\mathcal{G}(u, \varphi)(t) = T(t)\varphi + \mathcal{K}_u(\widetilde{R}(u))(t).$$

This mapping is well defined, as one can infer from Lemmas 2.3 and 3.2. Note that the subscript u in the above equation has nothing to do with the function u and that fixed points of $\mathcal{G}(\,\cdot\,, \varphi)$ correspond to solutions of (4.1) which stay bounded for $t \to -\infty$.

Theorem 4.4. (*Unstable Lipschitz continuous manifold.*) *There exist R_1 and R_2 positive such that for each $\varphi \in \mathbf{B}_{R_1}(X_+)$ there is a unique element $u = u^*(\varphi) \in \mathbf{B}_{R_2}(BC(\mathbb{R}_-, X))$ such that $u = \mathcal{G}(u, \varphi)$. The mapping $\varphi \mapsto u^*(\varphi)$ is Lipschitz continuous.*

Proof. To apply Banach's Fixed Point Theorem, also called the contraction mapping principle, we show that a closed subset of $BC(\mathbb{R}_-, X)$ exists which is invariant under $\mathcal{G}(\,\cdot\,, \varphi)$ and on which $\mathcal{G}(\,\cdot\,, \varphi)$ is a strict contraction, provided φ is restricted to a suitable subset. If $\|u\| \leq \rho$, then

$$\|\mathcal{G}(u, \varphi)\| \leq \alpha(\rho)\, \|\mathcal{K}_u\|\, \|u\| + K\, \|\varphi\|.$$

We want to show that for some positive numbers R_1 and R_2 the closed ball $\overline{\mathbf{B}}_{R_2}(BC(\mathbb{R}_-, X))$ is invariant under $\mathcal{G}(\,\cdot\,, \varphi)$, provided $\varphi \in \overline{\mathbf{B}}_{R_1}(X_+)$ and that for φ in this ball, $\mathcal{G}(\cdot, \varphi)$ is a strict contraction on $\overline{\mathbf{B}}_{R_2}(BC(\mathbb{R}_-, X))$.

As $DR(0) = 0$, it follows that $\lim_{\rho \to 0} \alpha(\rho) = 0$. Hence, if we choose R_2 such that $\|\mathcal{K}_u\|\alpha(R_2) \leq \frac{1}{2}$ and if we choose $R_1 = R_2/2K$, then we meet both conditions and we can apply the Banach Fixed-Point Theorem. We obtain a mapping $\varphi \mapsto u^*(\varphi)$ such that $\mathcal{G}(u^*(\varphi), \varphi) = u^*(\varphi)$. Moreover, it follows from

$$\begin{aligned}\|u^*(\varphi_1) - u^*(\varphi_2)\| &= \|\mathcal{G}(u^*(\varphi_1), \varphi_1) - \mathcal{G}(u^*(\varphi_2), \varphi_2)\| \\ &\leq \|\mathcal{K}_u\|\alpha(\rho)\|u^*(\varphi_1) - u^*(\varphi_2)\| + K\|\varphi_1 - \varphi_2\|\end{aligned}$$

that $\varphi \mapsto u^*(\varphi)$ is Lipschitz continuous. □

Exercise 4.5. Verify that $P_+ u^*(\varphi)(0) = \varphi$.

Definition 4.6. The image of the mapping $\mathcal{U} : \mathbf{B}_{R_1}(X_+) \longrightarrow X$ defined by

$$\mathcal{U}(\varphi) = u^*(\varphi)(0)$$

is called a *local unstable manifold* and is denoted by $\mathcal{W}_u$.

Alternatively, but equivalently, we can define $\mathcal{W}_u$ as the graph of a Lipschitz map $U : \mathbf{B}_{R_1}(X_+) \to X_-$ given by

$$U(\varphi) = (I - P_+)(u^*(\varphi)(0)).$$

Theorem 4.7. *The unstable manifold has the following properties:*

(i) *$\mathcal{W}_u$ is invariant in the sense that for φ sufficiently small and $t \leq 0$, $u^*(\varphi)(t) \in \mathcal{W}_u$, i.e. $u^*(\varphi)(t) = \mathcal{U}(P_+ u^*(\varphi)(t))$.*

(ii) *There exists a positive constant L such that, for sufficiently small $\delta > 0$, $\mathcal{U}$ is a Lipschitzian homeomorphism of the ball of radius δ/L in X_+ onto the set*

$$\begin{aligned}\{\varphi \in X \mid \|P_+\varphi\| \leq \frac{\delta}{L},\ &\textit{and there exists a solution of } (4.1) \textit{ on} \\ \mathbb{R}_- \ \textit{ such that }\ &u(0) = \varphi,\ \|u(t)\| \leq \delta,\ t \leq 0\}.\end{aligned}$$

Proof. If $u^*(\varphi)$ satisfies (4.1), so does $u^*(\varphi)(t + \cdot)$. If $\|\varphi\|$ is small enough, then for $t \leq 0$, $P_+ u^*(\varphi)(t)$ will belong to the neighbourhood where $\mathcal{U}$ is defined. The first statement follows from the uniqueness assertion in Theorem 4.4. The second statement is essentially a reformulation of Theorem 4.4. Let R_1 and R_2 be as in Theorem 4.4. We choose

$$L = \frac{K}{1 - \|\mathcal{K}_u\|\alpha(R_2)},$$

which is the Lipschitz constant obtained in the proof of Theorem 4.4. The result now follows for $\delta \leq R_1 L$. □

Actually, as we shall show below, $\mathcal{U}$ inherits more smoothness from R. What kept us from directly applying the Implicit Function Theorem to the equation $\mathcal{F} = 0$, with $\mathcal{F}$ defined by

$$\mathcal{F}(u, \varphi) = u - \mathcal{G}(u, \varphi),$$

is the fact that the substitution operator $\widetilde{R}$ is not necessarily a C^k-mapping from $BC(\mathbb{R}_-, X)$ into itself. (The difficulty is that we need compactness to arrive at uniformity and that boundedness does not suffice. See Appendix IV.) However, solutions on the unstable manifold approach zero exponentially as $t \to -\infty$. This indicates that we can restrict ourselves to looking for the zero set of $\mathcal{F}$ in a space of exponentially decaying functions. This helps finding domain and range for $\mathcal{F}$ such that $\mathcal{F}$ is smooth.

Exercise 4.8. Show that for each ν positive, there exist δ and C positive such that for $\|\varphi\| \leq \delta$ and for $t \leq 0$, we have the estimate

$$\|u^*(\varphi)(t)\| \leq Ce^{(b-\nu)t}.$$

Hint: Derive an equation for $z(t) = P_+(u^*(\varphi)(t))$ and use the invariance of the unstable manifold. Apply Gronwall's Lemma.

Exercise 4.9. Repeat the construction of the local unstable manifold in the space $BC^\eta(\mathbb{R}_-, X)$.

If η is positive, then $\widetilde{R}$ viewed as a mapping

$$\widetilde{R} : BC^\eta(\mathbb{R}_-, X) \to BC^\eta(\mathbb{R}_-, X^{\odot *})$$

is k times continuously differentiable [see Appendix IV, Lemma 1.1(iv)]. This allows us to directly apply the Implicit Function Theorem.

Lemma 4.10. *The restriction of the mapping $\varphi \mapsto u^*(\varphi)$ to a small neighbourhood of 0 in X_+ is k times continuously differentiable.*

Proof. Choose $\eta \in (0, b)$. Then, from Lemma 1.1 of Appendix IV and Lemma 3.2(ii) it follows that $\mathcal{F}$ viewed as a mapping from $BC^\eta(\mathbb{R}_-, X) \times X_+$ into $BC^\eta(\mathbb{R}_-, X)$ is k times continuously differentiable, $\mathcal{F}(0,0) = 0$ and $D_1\mathcal{F}(0,0) = I$. The Implicit Function Theorem guarantees the existence of neighbourhoods $\mathcal{V}$ and $\mathcal{W}$ of the origin in respectively X_+ and $BC^\eta(\mathbb{R}_-, X)$ and the existence of a k-times continuously differentiable mapping $\varphi \mapsto \tilde{u}^*(\varphi)$ from $\mathcal{V}$ into $\mathcal{W}$ such that $\mathcal{F}(\tilde{u}^*(\varphi), \varphi) = 0$. The uniqueness assertion in Theorem 4.4 implies that actually $\tilde{u}^* = u^*$, provided we choose $\mathcal{W} \subset \mathbf{B}_{R_2}(BC^\eta(\mathbb{R}_+, X))$ small enough such that $\mathcal{V} \subset \mathbf{B}_{R_1}(X_+)$. $\square$

Corollary 4.11. *The local unstable manifold given by the restriction of $\mathcal{U}$ to a sufficiently small ball is k times continuously differentiable if R is k times continuously differentiable.*

Exercise 4.12. Prove that $\mathcal{W}_u$ is tangent to X_+ at zero: $(d\mathcal{U}/d\varphi)(0)\psi = \psi$.

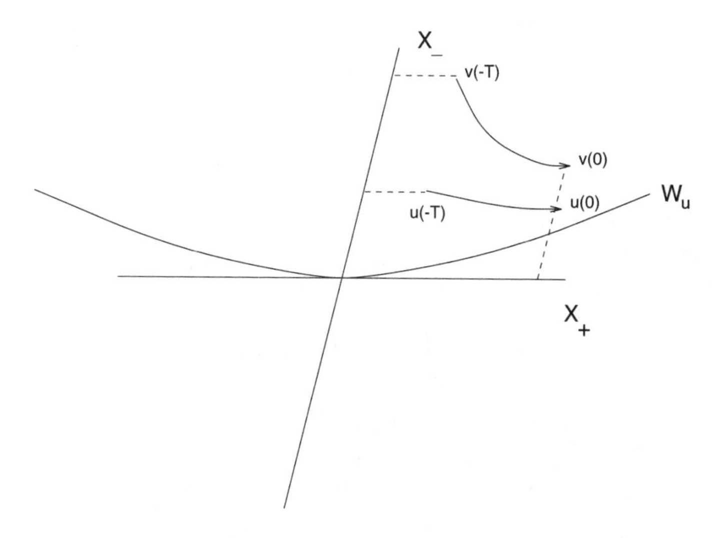

Fig. VIII.2. Attractivity of the local unstable manifold.

The next theorem, which is valid for $k = 1$ (even a Lipschitz condition is sufficient), describes that orbits which have stayed in a small neighbourhood of the equilibrium for a long time necessarily are very close to the unstable manifold.

Theorem 4.13. (*Conditional attractivity of the unstable manifold.*) *For every positive constant* ν*, there exist positive constants* C *and* δ *such that, if* u *and* v *are solutions of* (4.1) *on the interval* $I = [-T, 0]$, $T > 0$, *satisfying*

(i) $P_+u(0) = P_+v(0)$ *or* (i)′ $P_+u(-T) = P_+v(-T)$,

(ii) *for all* $t \in I$, $\|u(t)\| \leq \delta$ *and* $\|v(t)\| \leq \delta$,

then,

$$\|(I - P_+)(u(0) - v(0))\| \leq C\|(I - P_+)(u(-T) - v(-T))\|e^{(a+\nu)T}.$$

Proof. Assume (i) and (ii). Let $L = L(\delta)$ be a Lipschitz constant of R on the ball with radius δ. Let

$$z_+(t) = \|P_+(u(t) - v(t))\|, \qquad z_-(t) = \|(I - P_+)(u(t) - v(t))\|.$$

By taking the X_- component of (4.1) we get the estimate for $t \in I$:

$$\begin{aligned} z_-(0) &\le \|T(-t)(I-P_+)(u(t)-v(t))\| \\ &\quad + \|\int_t^0 T^{\odot *}(-\tau)(I-P_+)(R(u(\tau))-R(v(\tau)))\,d\tau\| \\ &\le Ke^{-at}z_-(t) + KL\int_t^0 e^{-a\tau}(z_-(\tau)+z_+(\tau))\,d\tau, \end{aligned}$$

i.e.,

$$e^{at}z_-(0) \le Kz_-(t) + KL\int_t^0 e^{a(t-\tau)}(z_-(\tau)+z_+(\tau))\,d\tau.$$

We need an estimate for the term $\int_t^0 e^{a(t-\tau)}z_+(\tau)\,d\tau$. Taking the X_+ component of (4.1) and using $z_+(0)=0$, we deduce that

$$z_+(t) \le -KL\int_0^t e^{b(t-\tau)}(z_-(\tau)+z_+(\tau))\,d\tau.$$

Applying Gronwall's inequality, we obtain

$$z_+(t) \le -KL\int_0^t e^{(-KL+b)(t-\tau)}z_-(\tau)\,d\tau.$$

If we let

$$K' = \frac{(KL)^2}{b-a-KL},$$

which is positive provided δ is sufficiently small, we obtain after some straightforward computations that for $t \in I$

$$e^{at}z_-(0) \le Kz_-(t) + (KL+K')\int_t^0 e^{a(t-\tau)}z_-(\tau)\,d\tau.$$

If we let $y(s) = z_-(t-s)e^{as}$, then

$$y(t) \le Ky(0) - (KL+K')\int_0^t y(t-\tau)\,d\tau,$$

and from Gronwall's lemma we deduce that

$$y(t) \le Ky(0)e^{-(KL+K')t}.$$

So, finally, we get that for $t \in I$

$$z_-(0) \le Kz_-(t)e^{-(a+KL+K')t}.$$

This proves the theorem. □

Exercise 4.14. Repeat the proof, this time assuming (i)′ and (ii).

VIII.5 Invariant wedges and instability

A set $\mathcal{O} \subset X$ has property C if $\lambda x \in \mathcal{O}$ for every $x \in \mathcal{O}$ and $0 < \lambda < \infty$ [i.e., $(0,\infty) \cdot \mathcal{O} \subset \mathcal{O}$]. If $\mathcal{O}$ has property C and, in addition,

$$\{x \mid x \in \mathcal{O} \text{ and } -x \in \mathcal{O}\} = \emptyset \text{ or } \{0\},$$

then $\mathcal{O}$ is called a *cone*. A set $\mathcal{O}$ with property C which is not a cone is called a *wedge*.

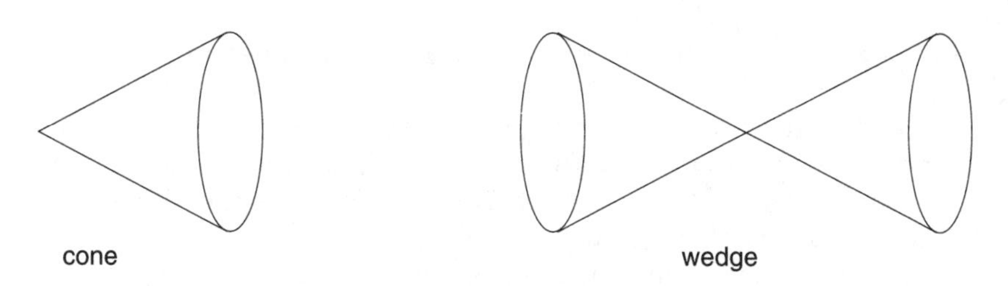

Fig. VIII.3. A cone and a wedge.

In this section we prove the existence of a family of *conditionally positively invariant wedges*, K_q, $q \in \mathbb{R}_+$. By conditionally invariant we mean that there exists a neighbourhood $\mathcal{V}$ of the origin in X such that for each initial condition $\varphi \in \mathrm{K}_q \cap \mathcal{V}$, the orbit through φ, $\Sigma(t,\varphi)$, remains for positive t in K_q as long as it stays in $\mathcal{V}$. If $0 \neq \varphi \in \mathrm{K}_q \cap \mathcal{V}$, then the orbit through φ leaves the neighbourhood $\mathcal{V}$ in finite time. Note that part of the material in this section is very much reminiscent of the proof of Proposition VII.5.10. To give a more detailed description of this unstable behaviour, we establish the existence of a quadratic Lyapunov functional. This Lyapunov functional is used in Chapter XV, in particular in Theorem XV.5.1.

We make the following assumptions:

H1 $X^{\odot *} = X^{\odot *}_{c-} \oplus X_{c+}$, where $X^{\odot *}_{c-}$ and X_{c+} are closed subspaces of $X^{\odot *}$ and X, respectively.

H2 $X^{\odot *}_{c-}$ and X_{c+} are invariant under $T^{\odot *}(t)$ and $T(t)$, respectively.

H3 X_{c+} is finite dimensional.

H4 The decomposition is an exponential dichotomy on $\mathbb{R}$ with respect to c, i.e., there exist real numbers a and b, $a < c < b$, and a positive constant K such that

$$(5.1)\qquad \begin{aligned} \|T(t)\varphi\| &\le K e^{bt}\|\varphi\|, && t \le 0,\ \varphi \in X_{c+},\\ \|T^{\odot *}(t)\varphi^{\odot *}\| &\le K e^{at}\|\varphi^{\odot *}\|, && t \ge 0,\ \varphi^{\odot *} \in X^{\odot *}_{c-}, \end{aligned}$$

H5 The number c, figuring in H4, is strictly positive.

Remark **5.1**. These assumptions certainly hold if $T(t)$ is eventually compact and the generator A has no eigenvalues on the line $\mathrm{Re}(z) = c$. In that case, X_{c+} is finite dimensional. More precisely, in that case we have

$$X_{c+} = \bigoplus_{\lambda \in \Lambda_{c+}} \mathcal{R}(P_\lambda^{\odot *}), \qquad X_{c-}^{\odot *} = \bigcap_{\lambda \in \Lambda_{c+}} \mathcal{N}(P_\lambda^{\odot *}),$$

where

$$\Lambda_{c+} = \{\lambda \in \sigma(A^{\odot *}) \mid \mathrm{Re}\,\lambda > c\}.$$

If we let

$$\begin{aligned} \gamma_{c+} &= \inf\{\mathrm{Re}\,\lambda \mid \lambda \in \Lambda_{c+}\}, \\ \gamma_{c-} &= \sup\{\mathrm{Re}\,\lambda \mid \lambda \in \sigma(A^{\odot *}) \text{ and } \mathrm{Re}\,\lambda < c\}, \end{aligned}$$

then for any positive number ϵ we can take $b = \gamma_{c+} - \epsilon$ and $a = \gamma_{c-} + \epsilon$.

Let $P_{c+}^{\odot *}$ denote the projection with range X_{c+} and kernel $X_{c-}^{\odot *}$. If we restrict the domain of $P_{c+}^{\odot *}$ to X_{c+}, then we shall write P_{c+} instead. The results in Sections 2–4 can be generalized to this more general splitting of the state space. For the case of ODE, see [49].

In fact, we may (and will) assume that in the above estimates $K = 1$ after introduction of an equivalent norm on $X^{\odot *}$. This norm is defined by

$$(5.2) \qquad |||\varphi^{\odot *}||| \stackrel{\text{def}}{=} \sup_{t \le 0} \frac{\|T^{\odot *}(t) P_{c+}^{\odot *} \varphi^{\odot *}\|}{e^{bt}} + \sup_{t \ge 0} \frac{\|T^{\odot *}(t)(I - P_{c+}^{\odot *})\varphi^{\odot *}\|}{e^{at}}.$$

Exercise 5.2. Show that there exists a positive constant γ such that for all $\varphi^{\odot *} \in X^{\odot *}$: $\|\varphi^{\odot *}\| \le |||\varphi^{\odot *}||| \le \gamma \|\varphi^{\odot *}\|$, i.e., $|||\cdot|||$ is indeed an equivalent norm on $X^{\odot *}$.

Exercise 5.3. Show that in this norm on $X^{\odot *}$, (5.1) holds with $K = 1$.

Definition 5.4. For $q > 0$ we define the wedge K_q in X by

$$\mathrm{K}_q = \{\varphi \in X \,:\, |||(I - P_{c+})\varphi||| \le q|||P_{c+}\varphi|||\}.$$

Exercise 5.5. Show that for $t \ge 0$, K_q is $T(t)$ invariant.

Theorem 5.6. *There exists $\delta_0 > 0$ such that $\mathrm{K}_q \cap \mathbf{B}_{\delta_0}(X)$ is conditionally positively invariant under the flow associated with* (1.1).

Proof. Fix $t_0 > 0$ and let $\delta > 0$ be given. There exist $\omega \ge 0$ such that $\|T(t)\| \le Me^{\omega t}$ for $t \ge 0$ [as $T(t)$ is a C_0-semigroup] and R is Lipschitz continuous. Therefore we can find C_1 positive such that if $|||\Sigma(t, \varphi)||| \le \delta$ for $0 \le t \le t_1 \le t_0$, then $|||\Sigma(t, \varphi)||| \le C_1|||\varphi||| \le (1 + q)C_1|||P_{c+}\varphi|||$ on the interval $I = [0, t_1]$ for some positive constant C_1. It then follows from equation (1.1) that

$$\begin{aligned}|||P_{c+}\Sigma(t,\varphi)||| &\geq e^{bt}|||P_{c+}\varphi||| - \alpha(\delta)(1+q)C_1t|||P_{c+}\varphi|||, \\ |||(I-P_{c+})\Sigma(t,\varphi)||| &\leq qe^{at}|||P_{c+}\varphi||| + \alpha(\delta)(1+q)C_1t|||P_{c+}\varphi|||.\end{aligned} \tag{5.3}$$

Define

$$Q(t) = \frac{|||(I-P_{c+})\Sigma(t,\varphi)|||}{|||P_{c+}\Sigma(t,\varphi)|||}$$

and

$$\rho(t) = \frac{qe^{at} + \alpha(\delta_0)C_1t}{e^{bt} - \alpha(\delta_0)C_1t}.$$

If φ is on the boundary of the wedge, then $Q(0) = q$. Moreover, from (5.3) it follows that $Q(t) \leq \rho(t)$ if $|||\Sigma(t,\varphi)|||$ is less than δ. It is a simple calculation to show that $\rho'(0) < 0$ if we choose δ_0 such that

$$\alpha(\delta_0) \leq \frac{q(b-a)}{C_1(1+q)^2}.$$

This shows that the boundary of the wedge inside the ball of radius δ_0 moves into the interior of the wedge under the flow and therefore proves the theorem. □

In the next theorem we establish an instability result by proving that an orbit through $0 \neq \varphi \in \mathrm{K}_q$ must leave in finite time a small neighbourhood of the origin in X. See also Propositions VII.5.10 and VII.5.11. In the proof we use an estimate, the proof of which we give as an exercise.

Exercise 5.7. Show that for $t \geq 0$ and $\varphi \in \mathrm{K}_q$,

$$|||T(t)\varphi||| \geq \frac{e^{bt}|||\varphi|||}{(1+q)}.$$

Theorem 5.8. *There exists $\delta_0 > 0$ such that if $0 \neq \varphi \in \mathrm{K}_q \cap \mathbf{B}_{\delta_0}(X)$ then $|||\Sigma(\varphi, t_1)||| = \delta_0$ for some positive $t_1 < \infty$.*

Proof. Choose t_0 such that $e^{bt_0}/(1+q) > 1$. There exists a positive constant C_1 such that if $\varphi \in \mathrm{K}_q$ and if $|||\Sigma(t,\varphi)||| \leq \delta$ on the interval $[0, t_0]$, then

$$\begin{aligned}|||\Sigma(t_0,\varphi)||| &\geq |||T(t_0)\varphi||| - C_1\alpha(\delta)|||\varphi||| \\ &\geq \Big(\frac{e^{bt_0}}{1+q} - C_1\alpha(\delta)\Big)|||\varphi|||.\end{aligned}$$

If we choose $\delta = \delta_0$ sufficiently small, then we achieve that

$$\Big(\frac{e^{bt_0}}{1+q} - C_1\alpha(\delta_0)\Big) > 1.$$

Therefore an orbit cannot stay in $\mathrm{K}_q \cap \mathbf{B}_{\delta_0}(X)$ for all positive time. Finally, by the previous result, if δ_1 is sufficiently small, then an orbit that starts in $V = \mathrm{K}_q \cap \mathrm{B}_{\delta_1}$ can only leave V through $\partial\mathrm{B}_{\delta_1}$; hence, choosing $\delta = \min\{\delta_0, \delta_1\}$, we arrive at the stated conclusion. □

We can even improve on this instability result by showing that there exists a continuous functional V on X, which is continuously differentiable (in a sense to be made precise later) and which has the following additional properties: if we choose δ_0 sufficiently small and if we let W be the restriction of V to $\mathrm{K}_q \cap \mathbf{B}_{\delta_0}(X)$, then W has a strict minimum at $\varphi = 0$ and the derivative along orbits is positive definite. Let Φ be a row vector of elements of X_+ which together form a (real) basis for X_+. Let M be the matrix such that $A\Phi = \Phi M$ and let $\varphi \mapsto \Gamma(\varphi)$ be the coordinate map on X_+ with respect to the basis Φ. We define $V : X \to \mathbb{R}$ by

$$V(\varphi) = \int_{-\infty}^{0} |e^{Ms} z|^2 \, ds$$

where z are the coordinates of $P_{c+}^{\odot *}\varphi$ with respect to the basis Φ, i.e., $P_{c+}^{\odot *}\varphi = \Phi\Gamma(\varphi) = \Phi z$. This integral is convergent because all the eigenvalues of M have real parts larger than or equal to $b > 0$. Note that V is a quadratic functional, i.e.,

$$V(s\varphi) = s^2 V(\varphi). \tag{5.4}$$

Lemma 5.9. *There exists $c_1 > 0$ such that $V(\varphi) \geq c_1 |||P_{c+}^{\odot *}\varphi|||^2$ for all $\varphi \in X$.*

Proof.

$$V(\varphi) = \int_{-\infty}^{0} |e^{Ms} z|^2 \, ds = z^T H z$$

where $H = \int_{-\infty}^{0} e^{M^T s} e^{Ms} \, ds$. H is nonsingular and positive. If we let

$$\beta = \min\{z^T H z \mid |z| = 1\},$$

then $\beta > 0$ and

$$V(\varphi) \geq \beta |z|^2$$

from which the result follows. □

Lemma 5.10. *V has within $\mathrm{K}_q \cap \mathbf{B}_{\delta_0}(X)$ a strict minimum at $\varphi = 0$. For $\varphi \in \mathrm{K}_q \cap \mathbf{B}_{\delta_0}(X)$ and provided δ_0 is sufficiently small,*

$$\dot{V}(\varphi) = \lim_{t \downarrow 0} \frac{1}{t} \big(V(\Sigma(t,\varphi)) - V(\varphi) \big) \tag{5.5}$$

exists and is positive, except at $\varphi = 0$.

Proof. The first assertion follows from Lemma 5.9. It is the content of Theorem VI.2.2 that $z(t) = \Gamma(P_{c+}^{\odot *}\Sigma(t,\varphi))$ satisfies the differential equation

$$\frac{d}{dt} z(t) = M z(t) + \Gamma(P_{c+}^{\odot *} R(\Sigma(t,\varphi))).$$

As $|||R(\varphi)||| = o(|||\varphi|||)$ for $|||\varphi||| \downarrow 0$, it follows that $z(t) = e^{Mt}z(0) + o(|t\, z(0)|)$. Hence,

$$\begin{aligned}
\dot{V}(\varphi) &= \lim_{t\downarrow 0} \frac{1}{t}\big(V(\Sigma(t,\varphi)) - V(\varphi)\big) \\
&= \lim_{t\downarrow 0} \frac{1}{t}\Big(\int_{-\infty}^{0} |e^{Ms}z(t)|^2\, ds - \int_{-\infty}^{0} |e^{Ms}z(0)|^2\, ds\Big) \\
&= \lim_{t\downarrow 0} \frac{1}{t}\Big(\int_{-\infty}^{0} |e^{Ms}(e^{Mt}z(0) + o(|t\, z(0)|))|^2\, ds - \int_{-\infty}^{0} |e^{Ms}z(0)|^2\, ds\Big) \\
&= \lim_{t\downarrow 0} \frac{1}{t}\Big(\int_{0}^{t} |e^{Ms}z(0)|^2\, ds + o(|t\, z(0)^2|)\Big) \\
&= |z(0)|^2 + o(|z(0)|^2).
\end{aligned}$$

Hence, if $\varphi \in \mathrm{K}_q$ and $|||\varphi|||$ is small enough, then there exists a positive constant C such that $\dot{V}(\varphi) \geq C|||\varphi|||^2$. This proves the second statement. □

VIII.6 The stable manifold

We now characterise the solutions of (1.1) which are defined and stay bounded for $t \to \infty$. Again we assume that there is no spectrum on the imaginary axis. The proof is based on arguments which are completely analogous to those employed in Section 4.

Theorem 6.1. (*Stable manifold of class C^k.*) *Assume that* H1–H4 *and* HR1 *and* HR2 *hold. There exists a C^k-mapping S of a neighbourhood of* 0 *in X_- into X_+ such that the graph $\mathcal{W}_s$ of S in $X = X_- \oplus X_+$ has the following properties:*

(i) *$\mathcal{W}_s$ is invariant in the sense that for $\varphi \in \mathcal{W}_s$ sufficiently small, $\Sigma(t;\varphi) \in \mathcal{W}_s$ for $t \geq 0$, i.e., $S((I - P_+)\Sigma(t;\varphi)) = P_+\Sigma(t;\varphi)$ for $t \geq 0$.*

(ii) *$\mathcal{W}_s$ is tangent to X_- at zero: $\frac{dS}{d\varphi}(0)\psi = \psi$.*

(iii) *There exists a positive constant L such that for sufficiently small and positive δ, S is a diffeomorphism of the ball with radius $\frac{\delta}{L}$ in X_- onto the set $\{\varphi \in X \mid \|(I - P_+)\varphi\| \leq \frac{\delta}{L}$, and there exists a solution of* (1.1) *on $\mathbb{R}_+$ such that $\Sigma(0,\varphi) = \varphi$ and $\|\Sigma(t,\varphi)\| \leq \delta$ for all positive t $\}$.*

(iv) *For each positive ν there exist δ and C positive such that for all $\varphi \in X_-$ such that $\|\varphi\| \leq \delta$ and for $t \geq 0$ we have the estimate*

$$\|\Sigma(t; S\varphi)\| \leq Ce^{(a+\nu)t}.$$

(v) *In addition, the stable manifold is attractive in the following sense: for every positive constant* ν *there exist positive constants* C *and* δ *such that if* u *and* v *are solutions of* (1.1) *on the interval* $I = [0, T]$, $T > 0$, *satisfying*

(a) $P_+ u(0) = P_+ v(0)$, *or* (a)$'$ $P_+ u(T) = P_+ v(T)$,

(b) *for all* $t \in I$, $\|u(t)\| \leq \delta$, $|v(t)\| \leq \delta$,

then

$$\|P_+(u(0) - v(0))\| \leq C \|P_+(u(T) - v(T))\| e^{(b-\nu)T}.$$

Exercise 6.2. Provide the details for a proof of this theorem.

Exercise 6.3. Give a verbal description of the meaning of statement (v).

VIII.7 Comments

Lemma 3.2 goes back to work of Perron [236]. See also Duistermaat [79]. The proof of Theorem 4.4 in the case of a system of ODE can be found in the works of Coddington and Levinson [49] and Hale [101]. The proof of Lemma 4.13 is borrowed from Ball [13]. Section 5 is partly based on results from Kirchgässner and Scheurle [150] and Iooss [134].

Chapter IX

The center manifold

IX.1 Introduction

In this chapter we analyse the behaviour of the nonlinear semiflow near a *nonhyperbolic* equilibrium; that is, we consider the situation where A does have spectrum on the imaginary axis. We use the decomposition of X as

$$X = X_- \oplus X_0 \oplus X_+.$$

X_0 is the subspace such that $\sigma(A_0) \subset i\mathbb{R}$, where A_0 is the map from $X_0 \to X_0$ induced by A. X_0 is called the *center subspace*, and $X_- \oplus X_+$ the *hyperbolic subspace*. The linear homogeneous equation

$$u(t) = T(t-s)u(s), \qquad -\infty < s \leq t < \infty, \tag{1.1}$$

restricted to the subspace X_0 is equivalent to a system of linear ODE with all the eigenvalues on the imaginary axis. If for one of the eigenvalues the algebraic multiplicity exceeds the geometric multiplicity, then this ODE has solutions with polynomial growth. However, all solutions are bounded by exponentials with arbitrary small positive exponents.

The center manifold is the nonlinear analogue of the space X_0 for the nonlinear equation

$$u(t) = T(t-s)u(s) + \int_s^t T^{\odot *}(t-\tau)R(u(\tau))\, d\tau, \quad -\infty < s \leq t < \infty. \tag{1.2}$$

We would like to play the same game as we did in the previous chapter. However, there are a few problems. First of all, the possibility for the linear equation to have unbounded solutions in X_0 shows that we cannot work in the space $BC(\mathbb{R}, X)$. Instead we must work in a function space that allows limited exponential growth both at plus and minus infinity. But, in general, such a space will not be mapped into itself by the substitution operator associated with R. So, we must modify R outside a small ball. (Recall that we are only describing the dynamics in a *neighbourhood* of an equilibrium.) Unfortunately, this cannot be done straightforwardly in a smooth manner. We work in an infinite dimensional space, and in these spaces, cutoff functions are not smooth in general. For RFDE we can exploit that the nonlinearity has finite dimensional range, cf. Exercise 4.6. But in the main text [cf. Section 4] we employ a different idea, viz. to use a priori estimates in the stable- and unstable directions. Finally, we must consider carefully how the substitution operator in BC^η spaces inherits smoothness from the defining function.

First, we will modify the nonlinearity suitably. Then, we will construct a Lipschitz continuous global center manifold. Finally, we state a general result on contractions on scales of Banach spaces. The general results obtained in this context are then used to get optimal smoothness of the center manifold.

IX.2 Spectral decomposition

Apart from $\odot$-reflexivity we make the following assumptions.

H1 $X^{\odot *} = X_-^{\odot *} \oplus X_0 \oplus X_+$, where $X_-^{\odot *}, X_0$ and X_+ are closed subspaces of $X^{\odot *}$, with X_0 and X_+ contained in X.

H2 $X_-^{\odot *}, X_0$ and X_+ are invariant under $T^{\odot *}(t)$.

H3 $T(t)$ can be extended to a one-parameter group on $X_0 \oplus X_+$.

H4 The decomposition is an exponential trichotomy on $\mathbb{R}$: there exist real numbers a and b, $a < 0 < b$, such that for each positive number ϵ there exists a positive constant K such that

$$
\begin{aligned}
\|T(t)\varphi\| &\le K e^{bt}\|\varphi\| && \text{for } t \le 0 \quad \text{and} \quad \varphi \in X_+,\\
\|T(t)\varphi\| &\le K e^{\epsilon|t|}\|\varphi\| && \text{for } t \in \mathbb{R} \text{ and } \varphi \in X_0,\\
\|T^{\odot *}(t)\varphi^{\odot *}\| &\le K e^{at}\|\varphi^{\odot *}\| && \text{for } t \ge 0 \text{ and } \varphi^{\odot *} \in X_-^{\odot *}.
\end{aligned}
$$

Remark **2.1**. These assumptions certainly hold if $T(t)$ is eventually compact. In that case, X_0 and X_+ are finite dimensional. More precisely, we have that

$$
\begin{aligned}
X_+ &= \bigoplus_{\lambda \in \Lambda_+} \mathcal{R}(P_\lambda^{\odot *}), \text{where } \Lambda_+ = \{\lambda \in \sigma(A^{\odot *}) \mid \operatorname{Re}\lambda > 0 \text{ and } \operatorname{Im}\lambda \ge 0\},\\
X_0 &= \bigoplus_{\lambda \in \Lambda_0} \mathcal{R}(P_\lambda^{\odot *}), \text{ where } \Lambda_0 = \{\lambda \in \sigma(A^{\odot *}) \mid \operatorname{Re}\lambda = 0 \text{ and } \operatorname{Im}\lambda \ge 0\},\\
X_-^{\odot *} &= \bigcap_{\lambda \in \Lambda_+ \cup \Lambda_0} \mathcal{N}(P_\lambda^{\odot *}).
\end{aligned}
$$

Here $P_\lambda^{\odot *}$ is the spectral projection operator on X (over $\mathbb{R}$) associated with $\lambda \cup \bar{\lambda}$. See Chapter IV Corollary 2.19 and Exercises 2.20-2.22. If we let

$$
\begin{aligned}
\gamma_+ &= \inf\{\operatorname{Re}\lambda \mid \lambda \in \Lambda_+\},\\
\gamma_- &= \sup\{\operatorname{Re}\lambda \mid \lambda \in \sigma(A^{\odot *}) \text{ and } \operatorname{Re}\lambda < 0\},
\end{aligned}
$$

then for any positive number ϵ less than both γ_+ and $-\gamma_-$, we can take $b = \gamma_+ - \epsilon$ and $a = \gamma_- + \epsilon$.

We will denote the projection operator of $X^{\odot *}$ with range X_+ and kernel $X_-^{\odot *} \oplus X_0$ by $P_+^{\odot *}$. Likewise, $P_0^{\odot *}$ is the projection operator with range X_0 and kernel $X_-^{\odot *} \oplus X_+$. When restricted to X, we suppress the superscript $^{\odot *}$.

For the description of the center subspace we supplement Definition 2.5 in Chapter VIII with:

Definition 2.2. Let E be a Banach space; then we define a one-parameter family of Banach spaces by

$$BC^{\eta}(\mathbb{R}, E) = \{F \in C(\mathbb{R}, E) \mid \sup_{t \in \mathbb{R}} e^{-\eta|t|}||F(t)|| < \infty\},$$

where the norm is the weighted supremum

$$||F||_{\eta} = \sup_{t \in \mathbb{R}} e^{-\eta|t|}||F(t)||.$$

Sometimes we suppress the subscript η for the norm if that does not cause confusion.

Remark that $\eta \leq \psi$ implies that $BC^{\eta}(\mathbb{R}, E) \subset BC^{\psi}(\mathbb{R}, E)$ and that

$$||u||_{\psi} \leq ||u||_{\eta} \quad \text{for all } u \in BC^{\eta}(\mathbb{R}, E). \tag{2.2}$$

These properties are often summarized by saying that $\left(BC^{\eta}(\mathbb{R}, E)\right)_{\eta \in \mathbb{R}}$ forms a *scale of Banach spaces* [in the general definition, a constant $C_{\psi,\eta}$ is allowed in the estimate (2.2)].

Definition 2.3. Let $\eta_1 \leq \eta_2$; then $\mathcal{J}_{\eta_2\eta_1}$ denotes the continuous embedding operator from $BC^{\eta_1}(\mathbb{R}, X)$ into $BC^{\eta_2}(\mathbb{R}, X)$.

The proof of the next lemma is left as an exercise.

Lemma 2.4. *Let* $\eta \in (0, \min\{-a, b\})$.

(i) $X_- = \{\varphi \in X \mid$ *there exists a solution of* (1.1) *on* $\mathbb{R}_+$ *through* φ *which belongs to* $BC^{a}(\mathbb{R}_+, X)\}$;

(ii) $X_+ = \{\varphi \in X \mid$ *there exists a solution of* (1.1) *on* $\mathbb{R}_-$ *through* φ *which belongs to* $BC^{b}(\mathbb{R}_-, X)\}$;

(iii) $X_0 = \{\varphi \in X \mid$ *there exists a solution of* (1.1) *on* $\mathbb{R}$ *through* φ *which belongs to* $BC^{\eta}(\mathbb{R}, X)\}$.

We conclude this section with a number of exercises which lead to a result that we shall use in the next chapter when computing the direction of Hopf bifurcation.

Exercise 2.5. Prove that each of the subspaces $X_-^{\odot*}$, X_0 and X_+ is invariant under the resolvent $(zI - A^{\odot*})^{-1}$.

Exercise 2.6. Show that in the vertical strip $\{z \mid a < \mathrm{Re} z < b\}$, singularities of the resolvent are necessarily located on the imaginary axis.

Exercise 2.7. Show that $(zI - A^{\odot*})^{-1} P_+^{\odot*}\psi = \int_0^{-\infty} e^{-z\tau} T^{\odot*}(\tau) P_+^{\odot*}\psi \, d\tau$ for $\mathrm{Re}\, z < b$.
Hint: Consider the semigroup $S(t) = T^{\odot*}(-t)P_+^{\odot*}$ on X_+ and apply Proposition 1.11 of Appendix II.

Exercise 2.8. Show that in the vertical strip $\{z \mid a < \mathrm{Re} z < b\}$ the following identity holds:

$$(zI - A^{\odot*})^{-1}(I - P_0^{\odot*})\psi = \int_0^{\infty} e^{-z\tau} T^{\odot*}(\tau) P_-^{\odot*}\psi \, d\tau + \int_0^{-\infty} e^{-z\tau} T^{\odot*}(\tau) P_+^{\odot*}\psi \, d\tau.$$

IX.3 Bounded solutions of the inhomogeneous linear equation

For the same reason as in the previous chapter, we need a pseudo-inverse for bounded solutions of the linear inhomogeneous equation

$$u(t) = T(t-s)u(s) + \int_s^t T^{\odot*}(t-\tau)F(\tau)\, d\tau. \tag{3.1}$$

This means that for a given forcing function F we characterise the set of all bounded solutions of (3.1). However, in this context, *bounded* means being an element of a function space that contains elements which have an unbounded range!

Definition 3.1. We define, formally at first, for $t \in \mathbb{R}$ and for F a mapping from $\mathbb{R}$ into $X^{\odot*}$,

$$(3.2)\qquad \begin{aligned}(\mathcal{K}F)(t) = \int_0^t T^{\odot *}(t-\tau)P_0^{\odot *}F(\tau)\,d\tau + \int_\infty^t T^{\odot *}(t-\tau)P_+^{\odot *}F(\tau)\,d\tau \\ + \int_{-\infty}^t T^{\odot *}(t-\tau)(I - P_0^{\odot *} - P_+^{\odot *})F(\tau)\,d\tau,\end{aligned}$$

Below we shall use the symbol $\mathcal{K}$ only in a context which guarantees that the integrals do indeed exist.

Lemma 3.2. *Let* $\eta \in (0, \min\{-a, b\})$.

(i) *Formula (3.2) defines a bounded linear mapping from* $BC^\eta(\mathbb{R}, X^{\odot *})$ *into* $BC^\eta(\mathbb{R}, X)$. $\mathcal{K}F$ *is the unique solution of* (3.1) *in* $BC^\eta(\mathbb{R}, X)$ *with vanishing* X_0 *component at* $t = 0$.

(ii) *If* $F \in BC(\mathbb{R}, X^{\odot *})$ *then* $(I - P_0)\mathcal{K}F \in BC(\mathbb{R}, X)$ *and the linear mapping* $F \mapsto (I - P_0)\mathcal{K}F$ *from* $BC(\mathbb{R}, X^{\odot *})$ *into* $BC(\mathbb{R}, X)$ *is bounded.*

Proof. Let $\epsilon < \min\{-a, b\} - \eta$ and $\epsilon < \eta$. Then a straightforward calculation shows that for each $t \in \mathbb{R}$ and $F \in BC^\eta(\mathbb{R}, X^{\odot *})$, we have the estimate

$$e^{-\eta|t|}\|\mathcal{K}F(t)\| \le K\|F\|_\eta\Big(\frac{1}{\eta - \epsilon} + \frac{1}{-a-\eta} + \frac{1}{b-\eta}\Big).$$

So,

$$(3.3)\qquad ||\mathcal{K}||_\eta \le K\Big(\frac{1}{\eta - \epsilon} + \frac{1}{-a-\eta} + \frac{1}{b-\eta}\Big).$$

As $\mathcal{K}F$ is continuous, this proves that indeed $\mathcal{K}F \in BC^\eta(\mathbb{R}, X)$ and that $\mathcal{K}$ is bounded. Straightforward inspection shows that $\mathcal{K}F$ is a solution of (3.1). Finally, as the difference of two solutions satisfying (3.1) is a solution of the homogeneous equation (1.1), the uniqueness part of assertion (i) follows from Lemma 2.3(iii). This proves (i). If $P_0^{\odot *}F = 0$ and $F \in BC(\mathbb{R}, X^{\odot *})$, it follows from H4 that

$$||\mathcal{K}F(t)|| \le K||F||_0\Big(\frac{1}{-a} + \frac{1}{b}\Big);$$

hence

$$(3.4)\qquad ||(I - P_0)\circ\mathcal{K}||_0 \le K\Big(\frac{1}{-a} + \frac{1}{b}\Big).$$

This proves (ii). □

Remark **3.3**. We shall write $\mathcal{K}_\eta$ if we want to emphasize that the domain and range of $\mathcal{K}$ are BC^η spaces.

Exercise 3.4. Carry out the above-mentioned straightforward calculations in detail.

IX.4 Modification of the nonlinearity

We recall the hypotheses on the nonlinearity from the previous chapter:

HR1 the mapping $u \mapsto R(u)$ from X into $X^{\odot *}$ is C^k for some $k \geq 1$,

HR2 $R(0) = 0$ and $DR(0) = 0$.

Let $\xi : \mathbb{R}_+ \to \mathbb{R}$ be a C^∞-smooth function such that

(i) $\xi(y) = 1$ for $0 \leq y \leq 1$,

(ii) $0 \leq \xi(y) \leq 1$ for $1 \leq y \leq 2$,

(iii) $\xi(y) = 0$ for $y \geq 2$.

We modify R in the center and in the hyperbolic directions separately; for δ positive, we let

$$R_\delta(x) = R(x)\xi\Big(\frac{\|P_0 x\|}{\delta}\Big)\xi\Big(\frac{\|(I-P_0)x\|}{\delta}\Big). \tag{4.1}$$

Lemma 4.1. *Let E and F be Banach spaces, let f be a Lipschitz continuous mapping from E into F with Lipschitz constant $L(\delta)$ on the ball of radius δ and let $f(0) = 0$. Let ξ_δ from E into $\mathbb{R}$ be defined by $\xi_\delta(x) = \xi(\frac{||x||}{\delta})$. Then there exists C such that for all $\delta > 0$ the mapping $x \mapsto \xi_\delta(x) f(x)$ is globally Lipschitz continuous from E into F with Lipschitz constant $(2C+1)L(2\delta)$.*

Proof. There exists a positive constant C such that ξ_δ is globally Lipschitz continuous with Lipschitz constant C/δ. The following estimates hold:

$$\begin{aligned}
&||f(x)\xi_\delta(x) - f(y)\xi_\delta(y)|| \\
&\quad \leq ||f(x) - f(y)||\xi(\frac{||y||}{\delta}) + ||\xi(\frac{||x||}{\delta}) - \xi(\frac{||y||}{\delta})||\,||f(x)|| \\
&\quad \leq \begin{cases} L(2\delta)||x-y|| + \frac{C}{\delta}L(2\delta)2\delta||x-y|| & \text{if } ||x||,\ ||y|| \leq 2\delta \\ 0 & \text{if } ||x||,\ ||y|| \geq 2\delta \\ \frac{C}{\delta}L(2\delta)2\delta||x-y|| & \text{if } ||x|| \leq 2\delta,\ ||y|| \geq 2\delta \end{cases} \\
&\quad \leq L(2\delta)(2C+1)||x-y||.
\end{aligned}$$

This proves the lemma. □

This lemma has the immediate

Corollary 4.2. *The mappings R_δ from X into $X^{\odot *}$ are globally Lipschitz continuous with Lipschitz constants L_{R_δ} that go to zero as δ goes to zero.*

Exercise 4.3. Show that $||R_\delta(x)|| \leq 4\delta L_{R_\delta}$ for all $x \in X$.

Exercise 4.4. Let E and F be Banach spaces and let $f : E \to F$ be a globally Lipschitz continuous function, with Lipschitz constant L. Let $\tilde{f}$ be the substitution operator from $BC^\eta(\mathbb{R}, E)$ into $BC^\eta(\mathbb{R}, F)$ defined by

$$(\tilde{f}(h))(s) = f(h(s)).$$

Show that $\tilde{f}$ is globally Lipschitz continuous with the same Lipschitz constant.

Corollary 4.5. *The substitution operators*

$$\widetilde{R}_\delta : BC^\eta(\mathbb{R}, X) \to BC^\eta(\mathbb{R}, X^{\odot *})$$

defined by $(\widetilde{R}_\delta(h))(t) = R_\delta(h(t))$ *are globally Lipschitz continuous with Lipschitz constants* L_{R_δ} *that go to zero as* δ *goes to zero.*

Exercise 4.6. Consider the special case of a RFDE. Derive a modified AIE from a modification of the right hand side of the RFDE. [For further motivation of this exercise, see Remark 7.11.]

IX.5 A Lipschitz center manifold

We choose the exponential weight η as in Lemma 3.2 and we define the mapping $\mathcal{G}$ from $BC^\eta(\mathbb{R}, X) \times X_0$ into $BC^\eta(\mathbb{R}, X)$ by

$$\mathcal{G}(u, \varphi) = T(\,\cdot\,)\varphi + \mathcal{K}_\eta \widetilde{R}_\delta(u). \tag{5.1}$$

We choose δ small enough to guarantee that

$$\mathrm{L}_{R_\delta} \|\mathcal{K}_\eta\| < \frac{1}{2}. \tag{5.2}$$

If

$$\|\varphi\| < \frac{r}{2K},$$

then $\mathcal{G}(\,\cdot\,, \varphi)$ leaves the ball with radius r in $BC^\eta(\mathbb{R}, X)$ invariant. Moreover, $\mathcal{G}(\,\cdot\,, \varphi)$ is Lipschitz continuous with Lipschitz constant $\frac{1}{2}$. Note that r is arbitrary. We thus obtain the following:

Theorem 5.1. *If* δ *is chosen as above, then there exists a globally Lipschitz continuous mapping* u^* *from* X_0 *into* $BC^\eta(\mathbb{R}, X)$ *such that* $u = u^*(\varphi)$ *is the unique solution in* $BC^\eta(\mathbb{R}, X)$ *of the equation*

$$u = \mathcal{G}(u, \varphi).$$

Definition 5.2. (Lipschitz center manifold.) The global center manifold, $\mathcal{W}_c$, is the image of the mapping $\mathcal{C} : X_0 \longrightarrow X$ defined by

$$\mathcal{C}(\varphi) = u^*(\varphi)(0). \tag{5.3}$$

Alternatively, but equivalently, we can define $\mathcal{W}_c$ as the graph of the Lipschitz-map $C : X_0 \to X_- \oplus X_+$ defined by

$$C(\varphi) = (I - P_0)(u^*(\varphi)(0)).$$

As $\mathcal{G}(0,0) = 0$, it follows that $u^*(0) = 0$ and, consequently, $\mathcal{C}(0) = 0$.

Theorem 5.3. (*Center manifold: invariance and relation to bounded orbits.*)

(i) *$\mathcal{W}_c$ is locally positively invariant in the sense that if $\varphi \in \mathcal{W}_c$ and the solution of* (1.2) *with $u(0) = \varphi$, $\Sigma(t,\varphi)$ satisfies $||\Sigma(t,\varphi)|| < \delta$ for $t \in [0,T]$, then $\Sigma(T,\varphi) \in \mathcal{W}_c$.*

(ii) *$\mathcal{W}_c$ contains all solutions of* (1.2) *which are defined on $\mathbb{R}$ and have their norm bounded above by δ.*

(iii) *The origin is contained in $\mathcal{W}_c$: $\mathcal{C}(0) = 0$.*

Exercise 5.4. Provide a proof of this theorem. Look at the proof of Theorem VIII.4.7 for inspiration.

Here we must comment on the dependence of the definition of the center manifold on our choice of η and δ. If δ is fixed and I is any subset of the interval $(0, \min\{-a, b\})$ on which (5.2) holds, then, due to (2.2) and the uniqueness of $u^*(\varphi)$ in any individual $BC^\eta(\mathbb{R}, X)$ space, the construction of the center manifold does not depend on the choice of η. However, it does depend on the choice of δ and the cut off function ξ. This is the celebrated nonuniqueness of the center manifold. However, if one computes the *Taylor series expansion* of the center manifold at the origin (up to order k), then this is *unique.* We will encounter an illustration of this fact in the next chapter on the Hopf bifurcation theorem. For a detailed proof of this statement, we refer to [251]. An easy but nevertheless important observation, concerning the dependence of u^* on δ, is the following:

Lemma 5.5. *Let δ be positive and sufficiently small. Then*

$$\|(I - P_0)u^*(\varphi)\|_0 < \delta.$$

Proof. As $u^*(\varphi) = \mathcal{G}(u^*, \varphi)$, it follows that

$$(I - P_0)u^*(\varphi) = (I - P_0)\mathcal{K}\widetilde{R_\delta}(u^*(\varphi)).$$

We observe that (see Exercise 4.3) $||R_\delta(u^*(\varphi))||_0 \le 4\delta L_{R_\delta}$. It follows from (3.3) that

$$||(I - P_0)u^*(\varphi)||_0 \le 4\delta K\Big(\frac{1}{-a} + \frac{1}{b}\Big)L_{R_\delta},$$

which is less than δ if

$$L_{R_\delta} < \Big(4K\big(\frac{1}{-a} + \frac{1}{b}\big)\Big)^{-1}.$$

□

It is a consequence of the previous lemma that the center manifold is located in the part of the space where the modification in the hyperbolic directions is trivial (provided δ has been chosen small enough). This observation will be helpful when proving the smoothness of the center manifold in section 8.

IX.6 Contractions on embedded Banach spaces

It has turned out that the introduction of function spaces with weighted norms is fruitful. A drawback, however, is that regularity considerations become cumbersome. We therefore need an appropriate version of the contraction mapping theorem. First, we state some regularity theorems for the substitution operator to motivate the formulation of the contraction theorem that we will use. To avoid cumbersome notation, we discuss the smoothness of the substitution operator for C^1 maps. In Appendix IV the corresponding results are formulated and proved for C^k-maps.

Let E and F be Banach spaces and let $g : E \to F$ be given. For a given mapping $h : \mathbb{R} \to E$, we let $G(h)$ be the mapping from $\mathbb{R}$ into F defined by

$$G(h)(t) = g(h(t)).$$

The smoothness of G depends on the smoothness of g, but also on the spaces that we choose as the range and the domain of G. If g is continuously differentiable then we expect G to be differentiable as well. The next lemmas and exercises address these matters.

Lemma 6.1. *If $g : E \to F$ is continuous and $\sup_{x\in E} ||g(x)|| < \infty$, $\eta_1 \in \mathbb{R}$ and $\eta_2 > 0$, then for each $h \in BC^{\eta_1}(\mathbb{R}, E)$, $G(h) \in BC^{\eta_2}(\mathbb{R}, F)$ and the mapping $G : BC^{\eta_1}(\mathbb{R}, E) \to BC^{\eta_2}(\mathbb{R}, F)$ is continuous.*

Exercise 6.2. Give a proof of this lemma in the same spirit as the proof of Lemma 1.1(v) of Appendix IV. Give a counterexample to show that we cannot replace the condition $\eta_2 > 0$ by the condition $\eta_2 = 0$.

Let now $g : E \to F$ be differentiable. For given mappings h and f from $\mathbb{R}$ into E, we let $G^{(1)}(h)(f)$ be the mapping from $\mathbb{R}$ into F defined by $G^{(1)}(h)(f)(t) = Dg(h(t))(f(t))$.

Lemma 6.3. *If $g : E \to F$ is continuously differentiable, $\eta_2 \geq \eta_1$ and $\sup_{x\in E} ||Dg(x)|| < \infty$, then $G^{(1)}(h) \in \mathcal{L}\Big(BC^{\eta_1}(\mathbb{R}, E), BC^{\eta_2}(\mathbb{R}, F) \Big)$ and the norm of this bounded linear mapping satisfies the estimate:*

$$||G^{(1)}(h)|| \leq \sup_{x\in E} ||Dg(x)||. \tag{6.2}$$

Proof. See Lemma 1.1(v) of Appendix IV. □

Exercise 6.4. If $\eta_2 > \eta_1$, then for $\eta \in \mathbb{R}$, the mapping $h \mapsto G^{(1)}(h)$ is continuous from $BC^{\eta}(\mathbb{R}, E)$ into $\mathcal{L}(BC^{\eta_1}(\mathbb{R}, E), BC^{\eta_2}(\mathbb{R}, F))$. Show this. Also show that if $\eta_1 = \eta_2$, then this mapping need *not* be continuous.

The only candidate for the derivative of the substitution operator G is the above-defined $G^{(1)}$. The previous exercise indicates that whether or not this is indeed the derivative depends in a subtle way on the spaces chosen for the domain and the range. We have the following result (again, see Appendix 4, Lemma 1.1(v)–(vi)).

Lemma 6.5. *Let g be as in Lemma 6.3 and let $\eta_2 > \eta_1$. G, considered as a mapping from $BC^{\eta_1}(\mathbb{R}, E)$ into $BC^{\eta_2}(\mathbb{R}, F)$, is continuously differentiable and $DG(h) = G^{(1)}(h)$.*

Note that, in general, $G : BC^{\eta_1}(\mathbb{R}, E) \to BC^{\eta_1}(\mathbb{R}, F)$ is not even differentiable, but as a mapping from $BC^{\eta_1}(\mathbb{R}, E)$ into $BC^{\eta_2}(\mathbb{R}, F)$, G is continuously differentiable.

After this lengthy introduction we are now ready to formulate the contraction theorem that we need later to prove the differentiability of the center manifold.

Let Y_0, Y, Y_1 and Λ be Banach spaces with norms denoted by, respectively,

$$\|\cdot\|_0, \quad \|\cdot\|, \quad \|\cdot\|_1 \quad \text{and} \quad |\cdot|,$$

and such that Y_0 is continuously embedded in Y, and Y is continuously embedded in Y_1. The embedding operators are denoted by J_0 and J respectively.

$$Y_0 \overset{J_0}{\hookrightarrow} Y \overset{J}{\hookrightarrow} Y_1$$

We will consider a fixed-point equation

$$y = f(y, \lambda) \tag{6.3}$$

where $f : Y \times \Lambda \to Y$ satisfies the following hypotheses:

Hf1 The function $g : Y_0 \times \Lambda \to Y_1$, $(y_0, \lambda) \mapsto g(y_0, \lambda) = Jf(J_0 y_0, \lambda)$ is of class C^1. There exist mappings

$$f^{(1)} : \ J_0 Y_0 \times \Lambda \to \mathcal{L}(Y),$$
$$f_1^{(1)} : \ J_0 Y_0 \times \Lambda \to \mathcal{L}(Y_1)$$

such that

$$D_1 g(y_0, \lambda)\xi = Jf^{(1)}(J_0 y_0, \lambda) J_0 \xi \quad \text{for all } (y_0, \lambda, \xi) \in Y_0 \times \Lambda \times Y_0$$

and

$$Jf^{(1)}(J_0y_0, \lambda)y = f_1^{(1)}(J_0y_0, \lambda)Jy \quad \text{for all } (y_0, \lambda, y) \in Y_0 \times \Lambda \times Y.$$

Hf2 There exists some $\kappa \in [0, 1)$ such that for all y, $\tilde{y} \in Y$ and for all $\lambda \in \Lambda$,

$$\|f(y, \lambda) - f(\tilde{y}, \lambda)\| \leq \kappa\|y - \tilde{y}\|$$

and

$$\|f^{(1)}(y, \lambda)\| \leq \kappa, \qquad \|f_1^{(1)}(y, \lambda)\| \leq \kappa.$$

It follows from Hf2 that (6.3) has for each $\lambda \in \Lambda$ a unique solution $\Psi = \Psi(\lambda)$. We assume that

Hf3 $\Psi = J_0 \circ \Phi$ for some continuous $\Phi : \Lambda \to Y_0$.

Hf4 $f_0 : Y_0 \times \Lambda \to Y$, $(y_0, \lambda) \mapsto f_0(y_0, \lambda) = f(J_0y_0, \lambda)$ has a continuous partial derivative

$$D_2f_0 : Y_0 \times \Lambda \to \mathcal{L}(\Lambda, Y).$$

Hf5 The mapping $(y, \lambda) \mapsto J \circ f^{(1)}(J_0y, \lambda)$ from $Y_0 \times \Lambda$ into $\mathcal{L}(Y, Y_1)$ is continuous.

Lemma 6.6. *Assume that* Hf1–Hf4 *hold. Then Ψ is locally Lipschitz continuous.*

Proof.

$$\begin{aligned}
\|\Psi(\lambda) - \Psi(\mu)\| &= \|f(\Psi(\lambda), \lambda) - f(\Psi(\mu), \mu)\| \\
&\leq \|f(\Psi(\lambda), \lambda) - f(\Psi(\mu), \lambda)\| \\
&\qquad + \|f_0(\Phi(\mu), \lambda) - f_0(\Phi(\mu), \mu)\| \\
&\leq \kappa\|\Psi(\lambda) - \Psi(\mu)\| \\
&\qquad + |\lambda - \mu| \sup_{s\in[0,1]} \|D_2f_0(\Phi(\mu), s\lambda + (1-s)\mu)\|.
\end{aligned} \tag{6.4}$$

Now fix some $\lambda \in \Lambda$ and let $C = C(\lambda) > ||D_2f_0(\Phi(\lambda), \lambda)||$. Since D_2f_0 and Φ are both continuous there exists some $\delta > 0$ such that

$$\sup_{s\in[0,1]} ||D_2f_0(\Phi(\mu), s\lambda + (1-s)\mu)|| \leq C$$

provided $|\mu - \lambda| \leq \delta$. For such μ the inequality (6.4) implies then

$$||\Psi(\lambda) - \Psi(\mu)|| \leq C(1-\kappa)^{-1}|\lambda - \mu|.$$

This proves the local Lipschitz continuity of Ψ. □

The hypotheses allow us to consider the following equation in $\mathcal{L}(\Lambda, Y)$,

$$A = f^{(1)}(\Psi(\lambda), \lambda)A + D_2f_0(\Phi(\lambda), \lambda).$$

Because of Hf2, this equation has for each $\lambda \in \Lambda$ a unique solution $A = \mathcal{A}(\lambda) \in \mathcal{L}(\Lambda, Y)$.

Lemma 6.7. *Assume that* Hf1–Hf5 *hold. Then the mapping $J \circ \Psi$ is of class C^1 and $D(J \circ \Psi)(\lambda) = J \circ \mathcal{A}(\lambda)$ for all $\lambda \in \Lambda$.*

Proof. Let $\lambda\ in \Lambda$ and $\mu \in \Lambda$ be given.

$$
\begin{aligned}
&J\Psi(\mu) - J\Psi(\lambda) - J\mathcal{A}(\lambda)(\mu - \lambda)\\
&\quad = Jf(\Psi(\mu), \mu) - Jf(\Psi(\lambda), \lambda) - Jf^{(1)}(\Psi(\lambda), \lambda)\mathcal{A}(\lambda)(\mu - \lambda)\\
&\qquad - JD_2 f_0(\Phi(\lambda), \lambda)(\mu - \lambda)\\
&\quad = g(\Phi(\mu), \mu) - g(\Phi(\lambda), \lambda) - Jf^{(1)}(\Psi(\lambda), \lambda)\mathcal{A}(\lambda)(\mu - \lambda)\\
&\qquad - D_2 g(\Phi(\lambda), \lambda)(\mu - \lambda)\\
&\quad = g(\Phi(\mu), \mu) - g(\Phi(\lambda), \mu) - Jf^{(1)}(\Psi(\lambda), \lambda)\mathcal{A}(\lambda)(\mu - \lambda)\\
&\qquad + g(\Phi(\lambda), \mu) - g(\Phi(\lambda), \lambda) - D_2 g(\Phi(\lambda), \lambda)(\mu - \lambda)\\
&\quad = Jf^{(1)}(\Psi(\lambda), \lambda)\left[\Psi(\mu) - \Psi(\lambda) - \mathcal{A}(\lambda)(\mu - \lambda)\right] + R(\lambda, \mu),\\
&\quad = f_1^{(1)}(\Psi(\lambda), \lambda)\left[J\Psi(\mu) - J\Psi(\lambda) - J\mathcal{A}(\lambda)(\mu - \lambda)\right] + R(\lambda, \mu),
\end{aligned}
\tag{6.5}
$$

where

$$
\begin{aligned}
R(\lambda, \mu) = &\int_0^1 \left[Jf^{(1)}(sJ_0\Phi(\mu) + (1 - s)J_0\Phi(\lambda), \mu) - Jf^{(1)}(J_0\Phi(\lambda), \lambda)\right]\\
&\qquad\qquad \times \left[\Psi(\mu) - \Psi(\lambda)\right] ds\\
&+ \int_0^1 \left[D_2 g(\Phi(\lambda), s\mu + (1 - s)\lambda) - D_2 g(\Phi(\lambda), \lambda)\right]\left[\mu - \lambda\right] ds.
\end{aligned}
$$

From Hf5 and the continuity of $D_2 g$ and Φ, we can find for each $\epsilon > 0$ some $\delta > 0$ such that

$$
\sup_{s \in [0,1]} ||Jf^{(1)}(sJ_0\Phi(\mu) + (1 - s)J_0\Phi(\lambda), \mu) - Jf^{(1)}(J_0\Phi(\lambda), \lambda)|| < \epsilon
$$

and

$$
\sup_{s \in [0,1]} ||D_2 g(\Phi(\lambda), s\mu + (1 - s)\lambda) - D_2 g(\Phi(\lambda), \lambda)|| < \epsilon,
$$

provided $|\mu - \lambda| < \delta$. Let C be a Lipschitz constant for Ψ in a neighbourhood of λ. Using (Hf2), we obtain

$$
||R(\lambda, \mu)|| \leq \epsilon(C + 1)|\mu - \lambda| \quad \text{for } |\mu - \lambda| < \delta.
$$

Substituting this in (6.5) and using (Hf2) we find for $|\mu - \lambda| < \delta$

$$
||J\Psi(\mu) - J\Psi(\lambda) - J\mathcal{A}(\lambda)(\mu - \lambda)||_1 \leq \epsilon((\frac{1 + C}{1 - \kappa})|\mu - \lambda|.
$$

This proves that $J \circ \Psi$ is differentiable at λ with $D(J \circ \Psi)(\lambda) = J \circ \mathcal{A}(\lambda)$. Next we show that $\lambda \mapsto J \circ \mathcal{A}(\lambda)$ is continuous. As

$$
\begin{aligned}
J\mathcal{A}(\lambda) - J\mathcal{A}(\mu) &= Jf^{(1)}(J_0\Phi(\lambda),\lambda)\mathcal{A}(\lambda) + D_2g(\Phi(\lambda),\lambda) \\
&\qquad - Jf^{(1)}(J_0\Phi(\mu),\mu)\mathcal{A}(\mu) - D_2g(\Phi(\mu),\mu) \\
&= f_1^{(1)}(J_0\Phi(\lambda),\lambda)(J\mathcal{A}(\lambda) - J\mathcal{A}(\mu)) \\
&\qquad + \big(Jf^{(1)}(J_0\Phi(\lambda),\lambda) - Jf^{(1)}(J_0\Phi(\lambda),\mu)\big)\mathcal{A}(\mu) \\
&\qquad + \big(Jf^{(1)}(J_0\Phi(\lambda),\mu) - Jf^{(1)}(J_0\Phi(\mu),\mu)\big)\mathcal{A}(\mu) \\
&\qquad + D_2g(\Phi(\lambda),\lambda) - D_2g(\Phi(\mu),\mu),
\end{aligned}
$$

it follows that

$$
\begin{aligned}
&||J\mathcal{A}(\lambda) - J\mathcal{A}(\mu)|| \\
&\quad \le \frac{1}{1-\kappa}\Big(||Jf^{(1)}(J_0\Phi(\lambda),\lambda) - Jf^{(1)}(J_0\Phi(\lambda),\mu)||\;||\mathcal{A}(\mu)|| \\
&\qquad + ||Jf^{(1)}(J_0\Phi(\lambda),\mu) - Jf^{(1)}(J_0\Phi(\mu),\mu)||\;||\mathcal{A}(\mu)|| \\
&\qquad + ||D_2g(\Phi(\lambda),\lambda) - D_2g(\Phi(\mu),\mu)||\Big)
\end{aligned}
$$

From the hypotheses we infer that indeed $\lambda \mapsto J \circ \mathcal{A}(\lambda)$ is continuous. This completes the proof of the lemma. □

IX.7 The center manifold is of class C^k

In this section we assume that

H5 X_0 is finite dimensional.

We introduce some notation that we specifically use in this section. We will write BC^η instead of $BC^\eta(\mathbb{R}, X)$. We define the set $BC^\infty = \cup_{\eta>0} BC^\eta$ without specifying any topology on it and we let

$$
V^\eta = \{u \in BC^\eta \mid ||(I - P_0)u||_0 < \infty\}. \tag{7.1}
$$

Exercise 7.1. Show that provided with the norm

$$
||u||_{V^\eta} = ||P_0 u||_\eta + ||(I - P_0)u||_0,
$$

(where $||\cdot||_\eta$ is the BC^η norm on $\mathbb{R}$ and $||\cdot||_0$ is the supremum norm on $\mathbb{R}$) the space V^η is a Banach space.

We define the set

$$
V_\delta^\infty = \{u \in BC^\infty \mid ||(I - P_0)u||_0 < \delta\}. \tag{7.2}
$$

If $p \ge 1$ and $Y, Y_1, \ldots, Y_p$ are Banach spaces, then we denote by $\mathcal{L}^{(p)}(Y_1 \times \ldots \times Y_p, Y)$ the Banach space of continuous p-linear mappings

$A : Y_1 \times \ldots \times Y_p \to Y$. If $Y_i = Y_1$, $i = 1, \ldots, p$, then we use the shorthand $\mathcal{L}^{(p)}(Y_1, Y)$. Recall the definition of R_δ given in formula (4.1). On the set $\{x \mid ||(I - P_0)x|| < \delta\}$ the function R_δ is k-times differentiable. It is here that we use H5. We denote its p-th derivative, $1 \leq p \leq k$, by $D^p R_\delta$. For $u \in V_\delta^\infty$ we define

$$(7.3) \qquad \big(\widetilde{R_\delta}^{(p)}(u)\big)(t) = D^p R_\delta(u(t)), \quad \text{for all } t \in \mathbb{R}.$$

Note that $\widetilde{R_\delta}^{(p)}$ maps V_δ^∞ into $\cap_{\zeta>0} BC^\zeta(\mathbb{R}, \mathcal{L}^{(p)}(X, X^{\odot *}))$. For a fixed $u \in V_\delta^\infty$ we can consider $\widetilde{R_\delta}^{(p)}$ as a p-linear operator on BC^∞ given by

$$\widetilde{R_\delta}^{(p)}(u)(v_1, \ldots, v_p)(t) = D^p R_\delta(u(t))(v_1(t), \ldots, v_p(t))$$
$$\text{for all } t \in \mathbb{R} \text{ and } v_1, \ldots, v_p \in BC^\infty.$$

So far we have proved that the center manifold is Lipschitz smooth. Recall that we have solved the fixed-point equation

$$(7.4) \qquad u = G(u, \varphi) = T(\cdot)\varphi + \mathcal{K} \circ \tilde{R}_\delta(u)$$

in the space BC^η. Choose an interval $[\tilde{\eta}, \overline{\eta}] \subset (0, \min\{-\gamma_-, \gamma_+\})$ such that $k\tilde{\eta} < \overline{\eta}$. Next, choose $\delta > 0$ small enough such that

$$(7.5) \qquad L_{R_\delta} ||\mathcal{K}||_\eta < \frac{1}{4} \qquad \text{for all } \eta \in [\tilde{\eta}, \overline{\eta}].$$

It is a consequence of (3.2) and Corollary (4.2) that this can be done.

Let $u_\eta^* : X_0 \mapsto BC^\eta$ denote the fixed point of (7.4). For all $\tilde{\eta} \leq \eta_1 \leq \eta_2 \leq \overline{\eta}$, we have $\mathcal{J}_{\eta_2\eta_1} \circ u_{\eta_1}^* = u_{\eta_2}^*$. These embedding operators, with suitable choice of η_1 and η_2, play the role of the embedding operators J_0 and J in section 6.

$\mathcal{R}_\delta$ is not smooth in general, so we cannot apply Lemma 6.5. Fortunately, it is a consequence of Lemma 5.5 that u_η^* takes values in an open subset of X where R_δ is smooth.

Exercise 7.2. Use Lemma 1.1.(v) of Appendix IV to prove the following result. Let $\eta \in (\tilde{\eta}, \overline{\eta}]$. The mapping $u \mapsto \mathcal{J}_{\eta\tilde{\eta}} \circ \mathcal{K} \circ \widetilde{R_\delta}(u)$ from $V_\delta^{\tilde{\eta}}$ into BC^η is of class C^1 with derivative $u \mapsto \mathcal{J}_{\eta\tilde{\eta}} \circ \mathcal{K} \circ \widetilde{R_\delta}^{(1)}(u) \in \mathcal{L}(V^{\tilde{\eta}}, BC^\eta)$.

There are four results that we will need. Having given the details of the proofs of Lemma 1.1.(v)-(vi) and Lemma 1.2 in Appendix IV we leave the proof as exercises. The reader may also consult [276].

Lemma 7.3. *Let $1 \leq p \leq k$, $\zeta_i > 0$ for $i = 1, \ldots, p$, $\zeta = \zeta_1 + \ldots + \zeta_p$ and $\eta \geq \zeta$. Then we have*

$$\widetilde{R_\delta}^{(p)}(u) \in \mathcal{L}^{(p)}(BC^{\zeta_1} \times \ldots \times BC^{\zeta_p}, BC^\eta(\mathbb{R}, X^{\odot *})) \qquad \textit{for all } u \in V_\delta^\infty,$$

with

$$||\widetilde{R_\delta}^{(p)}(u)||_{\mathcal{L}^{(p)}} \leq \sup_{t\in\mathbb{R}} e^{-(\eta-\zeta)|t|}||D^p R_\delta(u(t))||$$
$$= ||\widetilde{R_\delta}^{(p)}(u)||_{\eta-\zeta}.$$

The mapping $u \mapsto \widetilde{R_\delta}^{(p)}(u)$ *is continuous from* V_δ^σ, $\sigma > 0$, *into* $\mathcal{L}^{(p)}(BC^{\zeta_1} \times \ldots \times BC^{\zeta_p}, BC^\eta(\mathbb{R}, X^{\odot *}))$ *if* $\eta > \zeta$.

Lemma 7.4. *Let* $\eta_2 > k\eta_1 > 0$ *and* $1 \leq p \leq k$. *Then the mapping* $\widetilde{R_\delta} : V_\delta^{\eta_1} \to BC^{\eta_2}(\mathbb{R}, X^{\odot *})$ *is of class* C^k *with*

$$D^p\widetilde{R_\delta}(u) = \widetilde{R_\delta}^{(p)}(u) \in \mathcal{L}^{(p)}(BC^{\eta_1}, BC^{\eta_2}(\mathbb{R}, X^{\odot *})).$$

Proof for k=1. Let $||v||_{V_{\eta_1}} = 1$ and $u \in V_\delta^{\eta_1}$. As in the proof of Lemma 1.1.(v) one shows that

$$||\widetilde{R_\delta}(u+\epsilon v) - \widetilde{R_\delta}(u) - \epsilon\widetilde{R_\delta}^{(1)}(u)\, v||_{\eta_2} = o(\epsilon), \quad \epsilon \to 0,$$

and that the mapping $u \mapsto \widetilde{R_\delta}^{(1)}(u) \in \mathcal{L}(V^{\eta_1}, BC^{\eta_2})$ is continuous. The domain of $\widetilde{R_\delta}^{(1)}(u)$ extends naturally to BC^{η_1}. □

Lemma 7.5. *Under the conditions of Lemma 7.3 the mapping* $\widetilde{R_\delta}^{(p)} : V_\delta^\sigma \to \mathcal{L}^{(p)}(BC^{\zeta_1} \times \ldots \times BC^{\zeta_p}, BC^\eta(\mathbb{R}, X^{\odot *}))$ *is of class* C^{k-p} *provided* $\eta > \zeta + (k-p)\sigma$.

Lemma 7.6. *Let* $1 \leq p < k$, $\zeta_i > 0$ *for* $i = 1 \ldots p$, $\zeta = \zeta_1 + \ldots + \zeta_p$. *Let* Φ *be a mapping of class* C^1 *from* X_0 *into* $V_\delta^{\zeta_0}$. *Then the mapping* $\widetilde{R_\delta}^{(p)} \circ \Phi$ *from* X_0 *into* $\mathcal{L}^{(p)}(BC^{\zeta_1} \times \ldots \times BC^{\zeta_p}, BC^\eta)$ *is of class* C^1, *provided* $\eta > \zeta + \zeta_0$, *and then*

$$D(\widetilde{R_\delta}^{(p)} \circ \Phi(\phi))(v_1, \ldots, v_p)(s) = \widetilde{R_\delta}^{(p+1)}(\Phi(\phi))(\Phi'(\phi)(s), v_1(s), \ldots, v_p(s)).$$

Theorem 7.7. *For each* $p \in \mathbb{N}$ *with* $1 \leq p \leq k$ *and for each* $\eta \in (p\tilde{\eta}, \overline{\eta}]$, *the mapping* $J_{\eta\tilde{\eta}} \circ u_{\tilde{\eta}}^*$: $X_0 \to BC^\eta$ *is of class* C^p.

For each $\eta \in \mathbb{R}$, the evaluation map $ev : BC^\eta \to X$ defined by $ev(u) = u(0)$ is a bounded linear mapping. Therefore, we have as an immediate

Corollary 7.8. *The center manifold is of class* C^k.

Proof of Theorem 7.7, partly through exercises. $k = 1$. Let $\eta \in (\tilde{\eta}, \overline{\eta}]$. We apply Lemma 6.7 with

$$
\begin{aligned}
&Y_0 = V^{\tilde\eta}, \quad Y = BC^{\tilde\eta}, \quad Y_1 = BC^{\eta}, \quad \Lambda = X_0\\
&f(u,\varphi) = T(\cdot)\varphi + \mathcal{K}\circ \widetilde{R_\delta}(u), \quad \varphi\in X_0,\ u\in BC^{\tilde\eta}\\
&f^{(1)}(u,\varphi) = \mathcal{K}\circ \widetilde{R_\delta}^{(1)}(u) \in \mathcal{L}(BC^{\tilde\eta}), \quad \varphi\in X_0,\ u\in V_\delta^{\tilde\eta}\\
&f_1^{(1)}(u,\varphi) = \mathcal{K}\circ \widetilde{R_\delta}^{(1)}(u) \in \mathcal{L}(BC^{\eta}), \quad \varphi\in X_0,\ u\in V_\delta^{\tilde\eta}
\end{aligned}
$$

$J = \mathcal{J}_{\eta\tilde\eta}$, and J_0 likewise is defined by considering $V^{\tilde\eta}$ as a subset of $BC^{\tilde\eta}$.

Exercise 7.9. Verify the hypotheses Hf1–Hf5 of Lemma 6.7. Use Lemma 7.4 with $k=1$ for Hf1, formula (7.5) for Hf2, Lemma 5.5 for Hf3 and Exercise 7.2 to verify Hf5.

We conclude that $J_{\eta\tilde\eta}\circ u^*_{\tilde\eta}$ is of class C^1, and that $D(J_{\eta\tilde\eta}\circ u^*_{\tilde\eta}) \in \mathcal{L}(X_0, BC^\eta)$ is the unique solution of the equation

$$
u^{(1)} = \mathcal{K}\circ \widetilde{R_\delta}^{(1)}(u^*_{\tilde\eta}(\varphi))u^{(1)} + T(\cdot) = F_1(u^{(1)},\varphi) \tag{7.6}
$$

in $\mathcal{L}(X_0, BC^\eta)$. The mapping $F_1 : \mathcal{L}(X_0, BC^\eta)\times X_0 \to \mathcal{L}(X_0, BC^\eta)$ is a uniform contraction for each $\eta\in[\tilde\eta,\overline\eta]$; hence its fixed point $u^{*(1)}(\varphi)$ belongs in fact to $\mathcal{L}(X_0, BC^{\tilde\eta})$ which is embedded in $\mathcal{L}(X_0, BC^\eta)$ if $\eta\ge\tilde\eta$. The mapping $u^{*(1)} : X_0 \to \mathcal{L}(X_0, BC^\eta)$ is continuous if $\eta\in(\tilde\eta,\overline\eta]$.

If $k\ge 2$, we use induction on p. Let $1\le p\le k$ and suppose that for all q with $1\le q\le p$ and for all $\tilde\eta\in(q\tilde\eta,\overline\eta]$, the mapping $J_{\eta\tilde\eta}\circ u^*_{\tilde\eta} : X_0\to BC^\eta$ is of class C^q with $D^q(J_{\eta\tilde\eta}\circ u^*_{\tilde\eta}) = J_{\eta\tilde\eta}\circ u^{*(q)}_{\tilde\eta}$, $u^{*(q)}_{\tilde\eta}(\varphi)\in \mathcal{L}^{(q)}(X_0, BC^{q\tilde\eta})$ for each $\varphi\in X_0$ and $J_{\eta\tilde\eta}\circ u^{*(q)}_{\tilde\eta} : X_0 \to \mathcal{L}^{(q)}(X_0, BC^\eta)$ is continuous if $\eta\in(q\tilde\eta,\overline\eta]$. Suppose, moreover, that $u^{*(p)}_{\tilde\eta}(\varphi)$ is the unique solution of an equation of the form

$$
u^{(p)} = \mathcal{K}\circ \widetilde{R_\delta}^{(1)}(u^*_{\tilde\eta}(\varphi))u^{(p)} + H_p(\varphi) = F_p(u^{(p)},\varphi)
$$

with $H_1(\varphi) = T(\cdot)$ and for $p\ge 2$, $H_p(\varphi)$ is given as a finite sum of terms of the form

$$
\mathcal{K}\circ \widetilde{R_\delta}^{(q)}(u^*_{\tilde\eta}(\varphi))(u^{*(r_1)}_{\tilde\eta}(\varphi),\ldots,u^{*(r_q)}_{\tilde\eta}(\varphi))
$$

with $2\le q\le p$, $1\le r_i<p$ for $i=1\ldots q$ and $r_1+\ldots+r_q = p$. We remark that since $u^{*(r)}_{\tilde\eta}(\varphi)\in \mathcal{L}^{(r)}(X_0, BC^{r\tilde\eta})$ for $1\le r\le p$, we have $H_p(\varphi)\in \mathcal{L}^{(p)}(X_0, BC^{p\tilde\eta})$. Hence the mapping $F_p : \mathcal{L}^{(p)}(X_0, BC^\eta)\times X_0 \to \mathcal{L}^{(p)}(X_0, BC^\eta)$ is well defined and a uniform contraction for all $\eta\in[p\tilde\eta,\overline\eta]$. However, the first term of F_p (which is linear in $u^{(p)}$) is not continuously differentiable, neither in the variable $u^{(p)}$ nor in the parameter φ.

Fix some $\eta\in((p+1)\tilde\eta,\overline\eta]$ and choose $\sigma\in(\tilde\eta,\eta/(p+1))$ and $\zeta\in((p+1)\sigma,\eta)$. We will show that the hypotheses of Lemma 6.7 are satisfied for

$$Y_0 = \mathcal{L}^{(p)}(X_0, BC^{p\sigma}), \quad Y = \mathcal{L}^{(p)}(X_0, BC^{\zeta})$$
$$Y_1 = \mathcal{L}^{(p)}(X_0, BC^{\eta}), \quad \Lambda = X_0$$
$$f = F_p.$$

We have Hf2 because of (7.5), while Hf3 follows from the induction hypothesis and $p\sigma > p\tilde{\eta}$. In order to check Hf1, we first check that the mapping $(u^{(p)}, \varphi) \mapsto \mathcal{K} \circ \widetilde{R_\delta}^{(1)}(u^*_{\tilde{\eta}}(\varphi))u^{(p)}$ from $\mathcal{L}^{(p)}(X_0, B^{p\sigma}) \times X_0$ into $\mathcal{L}^{(p)}(X_0, BC^{\eta})$ is C^1 with respect to $u^{(p)}$. As this mapping is linear in $u^{(p)}$, it suffices to show that the mapping $\varphi \to \mathcal{K} \circ \widetilde{R_\delta}^{(1)}(u^*_{\tilde{\eta}}(\varphi))$ is continuous from X_0 into $\mathcal{L}(BC^{p\sigma}, BC^{\eta})$. This follows from Lemma 7.3, $\eta > p\sigma$ and the continuity of $J_{p\sigma\,\tilde{\eta}} \circ u^*_{\tilde{\eta}} : X_0 \to BC^{p\sigma}$. To verify Hf4 we show that the same mapping is of class C^1 from X_0 into $\mathcal{L}(BC^{p\sigma}, BC^{\zeta})$. This follows from Lemma 7.6 and $\zeta - p\sigma > \sigma$. In case $p \geq 2$ we also have to show that $H_p : X_0 \to BC^{\zeta}$ is of class C^1. This is indeed the case because the derivative of $\mathcal{K} \circ \widetilde{R_\delta}^{(q)}(u^*_{\tilde{\eta}}(\varphi))(u^{*(r_1)}(\varphi), \ldots, u^{*(r_q)}(\varphi))$ with respect to φ is given by (see Lemma 7.6)

$$\begin{aligned} D_\varphi \mathcal{K} \circ \widetilde{R_\delta}^{(q)} &(u^*_{\tilde{\eta}}(\varphi))(u^{*(r1)}(\varphi), \ldots, u^{*(r_q)}(\varphi)) \\ &= \mathcal{K} \circ \widetilde{R_\delta}^{(q+1)}(u^*_{\tilde{\eta}}(\varphi))(u^{*(1)}(\varphi), u^{*(r_1)}(\varphi), \ldots, u^{*(r_q)}(\varphi)) \\ &+ \sum_{k=1}^{q} \mathcal{K} \circ \widetilde{R_\delta}^{(q)}(u^*_{\tilde{\eta}}(\varphi))(u^{*(r_1)}(\varphi), \ldots, u^{*(r_k+1)}(\varphi), \ldots, u^{*(r_q)}(\varphi)). \end{aligned}$$

We conclude from Lemma 6.7 that $u^{*(p)} : X_0 \to \mathcal{L}^{(p)}(X_0, BC^{\eta})$ is of class C^1 with $u^{*(p+1)}(\varphi) = Du^{*(p)}(\varphi) \in \mathcal{L}^{(p+1)}(X_0, BC^{\eta})$ given by the unique solution of the equation

$$u^{(p+1)} = \mathcal{K} \circ \widetilde{R_\delta}^{(1)}(u^*_{\tilde{\eta}}(\varphi))u^{(p+1)} + H_{p+1}(\varphi)$$

with

$$H_{p+1} = \mathcal{K} \circ \widetilde{R_\delta}^{(2)}(u^*_{\tilde{\eta}}(\varphi))(u^{*(p)}(\varphi), u^{*(1)}(\varphi) + DH_p(\varphi) \quad \text{for all } \varphi \in X_0.$$

Hence $J_{\eta\tilde{\eta}} \circ u^*_{\tilde{\eta}} : X_0 \to BC^{\eta}$ is of class C^{p+1} if $\eta \in ((p+1)\tilde{\eta}, \overline{\eta}]$. □

Corollary 7.10. $DC(0)\psi = \psi$ *for all* $\psi \in X_0$

Proof. Let $\tilde{\eta}$, $\overline{\eta}$, $u^*_{\tilde{\eta}}$, $u^{*(1)}_{\tilde{\eta}}$ be as in the proof of Theorem 7.7 Fix $\eta \in (\tilde{\eta}, \overline{\eta}]$. Then

$$\mathcal{C}(\varphi) = ev(u^*_{\tilde{\eta}}(\varphi)) = ev(\mathcal{J}_{\eta\tilde{\eta}} u^*_{\tilde{\eta}}(\varphi)),$$

where $ev : BC^{\eta}(\mathbb{R}, X) \to X$ is defined by $ev(u) = u(0)$. Therefore,

$$D\mathcal{C}(0)\psi = ev(D(\mathcal{J}_{\eta\tilde{\eta}} \circ u^*_{\tilde{\eta}})(0)(\psi)) = ev(u^{*(1)}_{\eta}(0)\psi).$$

As $D\widetilde{R_\delta}(0) = 0$ and $u^*_{\tilde{\eta}}(0) = 0$ it follows from (7.6) that $u^{*(1)}_{\tilde{\eta}}(0) = T(\cdot)$. Hence $DC(0)\psi = ev(T(\cdot)\psi) = \psi$ □

Remark **7.11.**

(i) Notice that the proof of smoothness in this section works for nonlinearities in the AIE which are smooth and globally Lipschitz with a sufficiently small Lipschitz constant. Such nonlinearities were obtained from the originally given ones by the modification in Section 4. If the originally given nonlinearity has finite dimensional range, like in the case of RFDE, then one can also proceed as in Exercise 4.6 in order to obtain a locally equivalent AIE with suitable nonlinearity.

(ii) Suppose that $X_-^{\odot *} \oplus X_0$ is infinite dimensional and that we are interested in center stable manifolds, i.e., in graphs over $X_-^{\odot *} \oplus X_0$ which are tangent to this space at 0 and which consist of solutions in $BC^\eta(\mathbb{R}_+, X)$ with $\eta > 0$ small. Existence and Lipschitz continuity can be shown as in Sections 1–5. But more smoothness remains a problem, as far as general nonlinearities are concerned. In the special case of C^k-nonlinearities with finite dimensional range, one can use modifications as in Exercise 4.6 and arguments analogous to those in Sections 5–7, in order to obtain center stable manifolds of class C^k.

IX.8 Dynamics on and near the center manifold

We must keep in mind that we have constructed a *global center manifold* for the modified equation

$$(8.1) \quad u(t) = T(t-s)u(s) + \int_s^t T^{\odot *}(t-\tau) R_\delta(u(\tau))\, d\tau, \quad -\infty < s \le t < \infty.$$

In order to draw conclusions for the original equation (1.2), we must restrict to a neighbourhood of the equilibrium. If X_0 is finite dimensional, then the advantage of the center manifold is that it allows us to restrict the infinite dimensional dynamical system to finite dimensions. The system of ODE in finite dimensions is obtained by projection on the center directions as follows. If $u^*(\varphi)$ is the function defined in Theorem 5.1 and $y(t) = P_0(u^*(\varphi)(t))$, then $y(t)$ satisfies the equation

$$(8.2) \qquad y(t) = T(t)y(0) + \int_0^t T^{\odot *}(t-\tau) P_0^{\odot *} R_\delta(\mathcal{C}(y(\tau))\, d\tau$$

and, consequently,

$$(8.3) \qquad \dot{y} = Ay + P_0^{\odot *} R_\delta(\mathcal{C}(y)).$$

We conclude this section by stating the attractivity property of the center manifold. For the proof we refer to [13] (and Exercise 8.3 below). In particular, the center manifold is attractive in the absence of unstable directions. In this case, stability assertions for small solutions of (8.3) carry over to the full infinite dimensional dynamical system.

Theorem 8.1. (*Attraction of the center manifold.*) *For every positive constant ν there exist positive constants C and δ such that*

(i) *if u and v are solutions of* (1.2) *on the interval $I = [T, 0]$, $T < 0$, satisfying*

(a) $(P_+ + P_0)u(0) = (P_+ + P_0)v(0)$, *or* (a)′ $(P_+ + P_0)u(T) = (P_+ + P_0)v(T)$;

(b) *for all $t \in I$, $\|u(t)\| \leq \delta$ and $\|v(t)\| \leq \delta$,*

then

$$\|(I - P_+ - P_0)(u(0) - v(0))\| \leq C\|(I - P_+ - P_0)(u(T) - v(T))\|e^{-(a+\nu)T};$$

(ii) *if u and v are solutions of* (1.2) *on the interval $I = [0, T]$, $T > 0$, satisfying*

(a) $(I - P_+)u(0) = (I - P_+)v(0)$, *or (a)′* $(P_- + P_0)u(T) = (P_- + P_0)v(T)$,

(b) *for all $t \in I$, $\|u(t)\| \leq \delta$ and $\|v(t)\| \leq \delta$,*

then

$$\|P_+(u(0) - v(0))\| \leq C\|P_+ + (u(T) - v(T))\|e^{-(b-\nu)T}.$$

Exercise 8.2. Show that it is an immediate consequence of this theorem that all solutions which remain bounded and sufficiently small for all time lie on the center manifold.

In particular, the above exercise applies to small periodic solutions and small homoclinic and heteroclinic trajectories.

Exercise 8.3. Prove this theorem.
Hint: Convince yourself that the proof of Theorem VIII.4.13 still works when $b = 0$. Next, combine the results of Theorems VIII.4.13 and VIII.6.1.

IX.9 Parameter dependence

Quite often a system of equations will depend on parameters. It can happen that, for so-called critical values of the parameters, eigenvalues lie on the imaginary axis. As the parameters are varied in a full neighbourhood of the critical values, we may expect that these eigenvalues move around in both the left and the right half-planes. So the dimension of the local stable and unstable manifold varies with the parameters.

If a simple eigenvalue passes through zero, we expect a saddle node bifurcation of equilibria [159]. If a pair of complex conjugate eigenvalues passes through the imaginary axis, we expect the bifurcation of periodic solutions from the equilibrium (Hopf bifurcation; see [159] and the next chapter). In both cases, changes in the dynamics occur in a small neighbourhood of the equilibrium as parameters change. We have seen that the center manifold theorem captures solutions that exist and stay small for all time, but we must modify the theorem in order to include parameter dependence and bifurcation.

We consider the abstract integral equation

$$u(t) = T(t-s)u(s) + \int_s^t T^{\odot *}(t-\tau)R(u(\tau),\mu)\,d\tau \tag{9.1}$$

for $-\infty < s \leq t < \infty$, where

HRμ1 the mapping $(u,\mu) \mapsto R(u,\mu)$ from $X \times \mathbb{R}^p$ into $X^{\odot *}$ is C^k-smooth,

HRμ2 $R(0,\mu_0) = 0$, $D_1R(0,\mu_0) = 0$ and $D_2R(0,\mu_0) = 0$.

We shift the origin in parameter space to the critical value of the parameter, writing $\nu = \mu - \mu_0$. For $t \geq 0$ we define the family of bounded linear operators $\{\mathbf{T}(t)\}$ from $X \times \mathbb{R}^p$ into $X \times \mathbb{R}^p$ by

$$\mathbf{T}(t)(\varphi,\nu) = (T(t)\varphi,\nu).$$

In this way we include the parameters into our dynamical system. As the parameters have trivial dynamics (they are constant along an orbit), each of them contributes a simple eigenvalue at zero to the spectrum (see the first statement of the next lemma).

As the operator matrix for $\mathbf{T}(t)$ is diagonal and each component is $\odot$-reflexive, $\mathbf{T}(t)$ is $\odot$-reflexive as well; see Exercise II.6.1.. We denote the generator of $\{\mathbf{T}(t)\}$ by $\mathbf{A}$ and the generator of $\{T(t)\}$ by A. The proof of the following lemma is left as an exercise.

Lemma 9.1.

(i) *$X \times \mathbb{R}^p$ is $\odot$-reflexive with respect to $\{\mathbf{T}(t)\}$.*

(ii) *$\sigma(\mathbf{A}) = \sigma(A) \bigcup \{0\}$.*

(iii) *The space $(X \times \mathbb{R}^p)^{\odot *}$ is isometrically isomorphic to $X^{\odot *} \times \mathbb{R}^p$ and $\mathbf{T}^{\odot *}(t)(x^{\odot *},\nu) = (T^{\odot *}(t)x^{\odot *},\nu)$.*

We use the following notations. An element of $X \times \mathbb{R}^p$ is denoted by $w = (u,\nu)$. $P_0^{\odot *}$ is the projection of $X^{\odot *}$ on the center subspace of A. $\mathbf{P}^{\odot *}$ is the projection of $X^{\odot *} \times \mathbb{R}^p$ onto the center subspace of $\mathbf{A}$ given by $\mathbf{P}^{\odot *}w = (P_0^{\odot *}u,\nu) = (y,\nu)$. $\mathbf{R}$ denotes the mapping from $X \times \mathbb{R}^p$ into $X^{\odot *} \times \mathbb{R}^p$ defined by

$$\mathbf{R}(w) = (R(u,\mu_0+\nu),0). \tag{9.2}$$

We rewrite (9.1) as

$$(9.3)\qquad (u(t),\nu) = \mathbf{T}(t-s)(u(s),\nu) + \int_s^t \mathbf{T}^{\odot *}(t-\tau)\mathbf{R}(u(\tau),\mu_0+\nu)\,d\tau.$$

We are now in a position to apply the center manifold theorem to equation (9.3). This gives us a C^k-mapping $\mathcal{C}$ from $X_0 \times \mathbb{R}^p$ into $X \times \mathbb{R}^p$. Note that the (modified) equation has a vanishing nonlinearity in the second component. Hence, the mapping $\mathcal{C}$ takes the form $\mathcal{C}(\varphi,\nu) = (\mathcal{C}_1(\varphi,\nu),\nu)$. We project the center manifold flow on the center subspace $X_0 \times \mathbb{R}^p$. This gives the differential equation

$$\frac{dz}{dt} = \mathbf{A}z + \mathbf{P}^{\odot *}(\mathbf{R}(\mathcal{C}(z))),$$

where $z = (y,\nu)$ and which is equivalent to the system of equations

$$(9.4)\qquad \begin{cases} \dot{y} = Ay + P_0^{\odot *}R(\mathcal{C}_1(y,\nu),\mu_0+\nu), \\ \dot{\nu} = 0. \end{cases}$$

IX.9.1 Concretization in the case of RFDE

We will apply this construction to the RFDE

$$(9.5)\qquad \begin{cases} \dot{x}(t) = \int_0^h d\zeta(\theta,\mu)x(t-\theta) + g(x_t,\mu), & t \geq 0, \\ x(\sigma) = \varphi(\sigma), & -h \leq \sigma \leq 0, \end{cases}$$

and we will show to what function this leads for R in (9.2). Here, we suppose that g is a C^k-mapping, $k \geq 1$, from $X \times \mathbb{R}$ into $\mathbb{R}^n$, $g(0,\mu) = 0$ and $D_1 g(0,\mu) = 0$. For $z \in \mathbb{C}$ and $\mu \in \mathbb{R}^p$ we define $\Delta(z,\mu)$ by

$$(9.6)\qquad \Delta(z,\mu) = zI - \int_0^h e^{-z\theta}\,d\zeta(\theta,\mu).$$

If $\det \Delta(z,\mu_0)$ vanishes for purely imaginary values of z, we say that μ_0 is a critical value. If this is the case, we rewrite (9.5) as

$$(9.7)\qquad \begin{aligned} \dot{x}(t) = &\int_0^h d\zeta(\theta,\mu_0)x(t-\theta) + \int_0^h (d\zeta(\theta,\mu) - d\zeta(\theta,\mu_0))x(t-\theta) \\ &+ g(x_t,\mu), \qquad t \geq 0, \end{aligned}$$

with initial condition $x_0 = \varphi$. This equation is of the form (9.1) if we let $\{T(t)\}$ be the linear semigroup corresponding to the kernel $\zeta(\,\cdot\,,\mu_0)$, $u(t) = x_t$ and

$$(9.8)\qquad R(\varphi,\mu) = r^{\odot *}\Big(g(\varphi,\mu) + \int_0^h (d\zeta(\theta,\mu) - d\zeta(\theta,\mu_0))\varphi(-\theta)\Big)$$

Remark **9.2**. When $\theta \mapsto \zeta(\theta, \mu)$ has a jump at a μ-dependent position, the differentiability assumption HRμ1 is *not* satisfied. This is a smoothness problem that occurs whenever one takes one of the "delays" as a parameter. Yet it is possible to prove a Hopf bifurcation theorem and to compute the direction of bifurcation [see Hale [103] and Diekmann and van Gils [66]]. Somewhat unprecisely one can say that the reason is that solutions defined for all time are necessarily smooth and that therefore one can differentiate with respect to a translation parameter. Thus one expects that, provided one can use a contraction argument to construct a center manifold, the equation describing the dynamics on it should depend smoothly on such "delay-like" parameters. As far as we know a proof of this smoothness property is not yet elaborated in the literature.

IX.10 A double eigenvalue at zero

In this section we consider a RFDE, depending on two parameters, which has the origin as an equilibrium and is such that for special (critical) values of the parameters, the linearization has a double eigenvalue at zero and no spectrum in the right half-plane. The center manifold reduction allows us to reduce the dynamics to an ODE in the plane. This ODE in the plane has an equilibrium, and the linearization at this equilibrium has a double eigenvalue 0 at the critical parameter values. If the geometric multiplicity of this eigenvalue is 1, then this is called a Takens-Bogdanov point. As such a vector field is nonhyperbolic, the higher order terms in the Taylor expansion near the origin determine the local phase portrait. It is known that for the Takens-Bogdanov singularity the quadratic terms are needed, but higher order terms are not; see [271, 18, 19, 98]. After application of a normal form reduction and neglect of third order terms, the vector field has at criticality, the form

$$\begin{aligned} \dot{x} &= y, \\ \dot{y} &= ax^2 + bxy. \end{aligned} \tag{10.1}$$

If both a and b are nonzero, then the vector field (10.1) has a universal unfolding given by

$$\begin{aligned} \dot{x} &= y, \\ \dot{y} &= \mu_1 + \mu_2 y + ax^2 + bxy. \end{aligned} \tag{10.2}$$

In the RFDE, the origin is an equilibrium for all values of the parameter. Hence, the dynamics on the center manifold cannot be described by (10.2). Unfolding (10.1) within the class of vector fields having the origin as an equilibrium yields (see [26]):

$$\begin{aligned} \dot{x} &= y, \\ \dot{y} &= \mu_1 x + \mu_2 y + a x^2 + bxy. \end{aligned} \tag{10.3}$$

The objective here is to show that for a certain RFDE involving a scalar nonlinear function f, the dynamics on the center manifold is indeed described by (10.3). In particular, we show that if $f''(0) \neq 0$, then a and b are both nonzero. Through a series of exercises we will study the dynamics of (10.3).

Consider the RFDE

$$\dot{x}(t) = \alpha x(t) + f(x(t-1)), \tag{10.4}$$

where $f \in C^2(\mathbb{R}, \mathbb{R})$ and $f(0) = 0$. Linearization around the zero equilibrium yields

$$\dot{x}(t) = \alpha x(t) + f'(0) x(t-1). \tag{10.5}$$

This equation has a double eigenvalue zero at $(\alpha, f'(0)) = (1, -1)$ (see, e.g., Exercises IV.3.18 end IV.3.19). We investigate (10.5) in a neighbourhood of this point. Therefore, we introduce the small parameters λ and μ by

$$\begin{aligned} \alpha &= 1 + \lambda, \\ f'(0) &= -1 + \mu. \end{aligned} \tag{10.6}$$

Writing

$$g(\varphi, \lambda, \mu) = \lambda \varphi(0) + \mu \varphi(-1) + f(\varphi(-1)) - (\mu - 1)\varphi(-1), \tag{10.7}$$

we have

$$\dot{x}(t) = x(t) - x(t-1) + g(x_t, \lambda, \mu), \tag{10.8}$$

and g satisfies

$$g(\varphi, \lambda, \mu) = \mathcal{O}(|\varphi| + |\lambda| + |\mu|)^2. \tag{10.9}$$

The corresponding abstract integral equation is

$$u(t) = T(t)u(0) + \int_0^t T^{\odot *}(t-\tau) r^{\odot *} g(u(\tau), \lambda, \mu)\, d\tau, \tag{10.10}$$

where $\{T(t)\}$ is the semigroup of the linear part of (10.8) at $\lambda = \mu = 0$. The infinitesimal generator has a double eigenvalue at zero. All the other eigenvalues lie in the left half-plane (see Section XI.2, in particular Figure XI.1). Therefore, the center manifold is two dimensional and locally attracting. We compute, to lowest order, the vector field on X_0 describing the flow on the center manifold.

According to Exercise IV.3.19, with $\zeta(\tau) = 1$ for $0 < \tau < 1$ and 0 otherwise, we have

$$(10.11)\qquad \begin{aligned} P_0\varphi(\theta) &= \operatorname*{Res}_{z=0}\,((zI-A)^{-1}\varphi)(\theta) \\ &= (2\theta+\frac{2}{3})\varphi(0) - \int_0^1 (2(\theta-\sigma)+\frac{2}{3})\varphi(\sigma-1)\,d\sigma. \end{aligned}$$

Remark **10.1**. This spectral projection operator, first defined on X, can be extended to the spectral projection operator on $X^{\odot *}$ by a small adaptation of the formula; compare Corollary IV.5.2 with Theorem IV.3.1.

The projection of equation (10.8) on the center manifold is given by

$$(10.12)\qquad \dot{u} = Au + P_0^{\odot *} r^{\odot *} g(\mathcal{C}(u,\lambda,\mu),\lambda,\mu), \quad u \in X_0.$$

Remark **10.2**. By $\mathcal{C}(u,\lambda,\mu)$ we mean the two-dimensional slice, of the actually four-dimensional invariant manifold, that is obtained by keeping λ and μ constant. See Section 8. Here our notation is sloppy.

Recall that $\mathcal{C}(u,\lambda,\mu) = u + \mathcal{O}(||u||^2)$. Hence, (10.12) can be written as

$$(10.13)\qquad \begin{aligned} \dot{u}(t) = Au(t) + P_0^{\odot *} r^{\odot *}(&\frac{f''(0)}{2}u(t)(-1)^2 \\ &+ \lambda u(t)(0) + \mu u(t)(-1)) + \mathcal{O}((|\lambda|+|\mu|)||u||^2 + ||u||^3). \end{aligned}$$

Introducing coordinates in X_0:

$$(10.14)\qquad u = x\varphi_0 + y\varphi_1,$$

where

$$(10.15)\qquad \varphi_0(\theta) = 1, \quad \varphi_1(\theta) = \theta, \quad -1 \le \theta \le 0$$

[see (IV.5.8)], we obtain the ODE in the plane

$$(10.16)\qquad \begin{aligned} \dot{x} &= y + \frac{2}{3}h(x,y,\lambda,\mu), \\ \dot{y} &= 2h(x,y,\lambda,\mu) \end{aligned}$$

with

$$(10.17)\qquad \begin{aligned} h(x,y,\lambda,\mu) = &\frac{f''(0)}{2}(x-y)^2 + \lambda x + \mu(x-y) \\ &+ \mathcal{O}((|x|+|y|)^3 + (|x|+|y|)^2(|\lambda|+|\mu|)). \end{aligned}$$

Putting

$$(10.18)\qquad \eta = \frac{dx}{dt},$$

we obtain the system of equations

$$\begin{aligned} \dot{x} &= \eta, \\ \dot{\eta} &= \frac{f''(0)}{2}(x^2 - \frac{4}{3}x\eta + \frac{1}{3}\eta^2) + (2\lambda + 2\mu)x + (\frac{2}{3}\lambda - \frac{4}{3}\mu)y \\ &\quad + \mathcal{O}((|x| + |\eta|)^3 + (|x| + |\eta|)^2(|\lambda| + |\mu|)). \end{aligned} \tag{10.19}$$

Exercise 10.3. Show that the nonlinear transformation

$$x = u + \frac{1}{6}f''(0)u^2, \qquad \eta = v + \frac{1}{3}f''(0)uv$$

brings (10.19) into the form

$$\begin{aligned} \dot{u} &= v + \text{hot}, \\ \dot{v} &= f''(0)(u^2 - \frac{4}{3}uv) + 2(\lambda + \mu) + \frac{2}{3}(\lambda - 2\mu) + \text{hot}, \end{aligned} \tag{10.20}$$

where

$$\text{hot} = \mathcal{O}((|\lambda| + |\mu|)(|u| + |v|)^2 + (|u| + |v|)^3). \tag{10.21}$$

Exercise 10.4. Determine a rescaling of u, v and the time (not changing the direction of the time) that transforms (10.20) to

$$\begin{aligned} \dot{u} &= v + \text{hot}, \\ \dot{v} &= u^2 - uv + \frac{32}{9}(\lambda + \mu) + \frac{8}{9}(\lambda - 2\mu) + \text{hot}. \end{aligned} \tag{10.22}$$

Hence, we have shown that the equation on the center manifold takes precisely the most general form that one can get for this singularity. The fact that this equation comes from a RFDE after reduction does not impose additional constraints.

In the next exercises we analyse in some detail the system

$$\begin{aligned} \dot{u} &= v, \\ \dot{v} &= u^2 - uv + \beta_1 u + \beta_2 v. \end{aligned} \tag{10.23}$$

Apart from the higher order terms, (10.22) is of the form (10.23). It is known that adding higher order terms does not change the phase portraits

Exercise 10.5.

(i) Determine the fixed points, the linearization at the fixed points and their nature, for system (10.23).

(ii) Determine the locus in the $\beta_1 - \beta_2$ plane of Hopf bifurcation points and transcritical bifurcation points.

(iii) Determine the direction of bifurcation and the stability of the bifurcating periodic orbits at the Hopf bifurcation points.

The periodic orbits that appear at a Hopf bifurcation disappear at a homoclinic orbit. To find the locus of homoclinic points, we first apply the rescaling

$$
\begin{aligned}
& u \to \epsilon^2 \overline{u}, \quad v \to \epsilon^3 \overline{v}, \quad t \to \epsilon \overline{t}, \\
& \beta_1 \to \begin{cases} \epsilon^2 \text{ if } \beta_1 > 0 \\ -\epsilon^2 \text{ if } \beta_1 < 0 \end{cases}, \quad \beta_2 \to \epsilon^2 \overline{\beta_2}.
\end{aligned} \tag{10.24}
$$

Dropping the bars, we obtain the system

$$
\begin{aligned}
\dot{u} &= v, \\
\dot{v} &= u^2 \pm u - \epsilon u v + \epsilon \beta_2 v.
\end{aligned} \tag{10.25}
$$

If $\epsilon = 0$, then this system is Hamiltonian with Hamiltonian function

$$
H(u, v) = \frac{v^2}{2} - \frac{u^3}{3} \mp \frac{u^2}{2} = h. \tag{10.26}
$$

The choice of the sign in front of u^2 depends on the choice of the sign of β_1, see (10.24).

Exercise 10.6. Draw the phase portraits of these two Hamiltonian systems. In particular, show that if $\beta_1 > 0$, then there is a homoclinic orbit at the level $h = 0$ and v is then given by $v = \frac{1}{3}\sqrt{3}u\sqrt{2u+3}$; if $\beta_1 < 0$, then there is a homoclinic orbit at the level $h = \frac{1}{6}$, and in this case, v is given by $v = \frac{1}{3}\sqrt{3}(u-1)\sqrt{2u+1}$.

Let l be the part of the u-axis between the two fixed points. Denote by $\widetilde{\gamma_h}$ the part of the orbit of (10.25) starting at l at the level $H = h$ at $t = 0$ until it returns to l for the first time at $t = T$. We denote by ΔH the difference between the value of H at $t = 0$ and $t = T$. Continuous dependence on the vector field implies that $\widetilde{\gamma_h} = \gamma_h + \mathcal{O}(\epsilon)$, uniformly for values of h in a compact subset of l, where γ_h is the compact connected part of the level set $H = h$.

The perturbed system (10.25) has a periodic orbit if $\Delta H = 0$, i.e.,

$$
\begin{aligned}
0 = \Delta H &= \int dH \\
&= \int_0^T \frac{\partial H}{\partial u} v \, dt + \int_0^T \frac{\partial H}{\partial v} (u^2 \pm u - \epsilon u v + \epsilon \beta_2 v) \, dt \\
&= \epsilon \Big(\oint_{\gamma_h} -uv \, du + \oint_{\gamma_h} v \, du \Big) + \mathcal{O}(\epsilon^2) \\
&= \epsilon \oint_{\gamma_h} v \, du \Big(\frac{\oint_{\gamma_h} -uv \, du}{\oint_{\gamma_h} v \, du} + \beta_2 \Big) + \mathcal{O}(\epsilon^2).
\end{aligned} \tag{10.27}
$$

Exercise 10.7. Show that

$$\oint_{\gamma_h} uv\, du \,(\oint_{\gamma_h} v\, du)^{-1} = \begin{cases} -\dfrac{6}{7} & \text{if } \beta_1 > 0 \text{ and } h = 0, \\ \dfrac{1}{7} & \text{if } \beta_1 < 0 \text{ and } h = \dfrac{1}{6}. \end{cases}$$

With all the information gathered in the exercises we obtain the following bifurcation diagram:

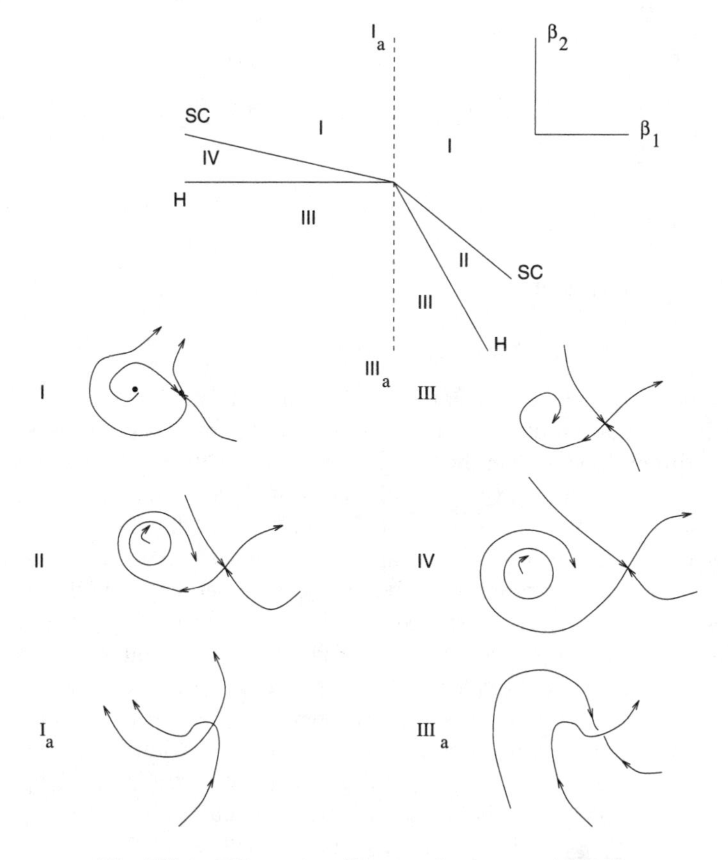

Fig. IX.1. Bifurcation diagram for system (10.23).

It can be shown that the limit cycle in regions II and V is unique. This is a consequence of the fact that the mapping

$$h \mapsto \oint_{\gamma_h} -uv\, du \,(\oint_{\gamma_h} v)^{-1}$$

is strictly monotonic. See [19].

Exercise 10.8. In this exercise, we follow the route of duality to obtain the projection operator on X_0 instead of residue calculus.

(i) In X^*, let

$$\chi_0(t) = 2(1-t), \quad \chi_1(t) = \frac{2}{3} + \frac{4}{3}t - t^2, \quad 0 < t \le 1. \tag{10.24}$$

With u_0 and u_1 as in (10.16) show that $A^*\chi_0 = 0$, $A^*\chi_1 = \chi_0$ and

$$\langle \varphi_0, \chi_0 \rangle = 1 - \langle \varphi_1, \chi_0 \rangle = 1 - \langle \varphi_0, \chi_1 \rangle = \langle \varphi_1, \chi_1 \rangle = 0.$$

(ii) Define $P : X \to X$ by

$$P\varphi = \langle \varphi, \chi_0 \rangle \varphi_1 + \langle \varphi, \chi_1 \rangle \varphi_0. \tag{10.25}$$

Show that P defined by (10.25) is the same as the projection operator defined in (10.10).

IX.11 Comments

Since the introduction of the center manifold some twenty years ago by Pliss [240] and Kelley [149] many papers have been published which consider the reduction process in different contexts. A readable presentation for an ODE in finite dimensions is given by Vanderbauwhede [277]. For Volterra integral equations of convolution type, see the work of Diekmann and van Gils [66]. For equations in abstract spaces, see the works of Chow and Lu [39] and Vanderbauwhede and Iooss [278], and the references given there.

There are two different methods to prove a center manifold theorem. The method of graph transforms, see, for instance, the book by Hirsch, Pugh and Shub [126], is a geometric construction. The other method, which we use here, is more analytical and uses the variation-of-constants formula. This goes at least back to the work of Perron [236].

Lemma 6.7 is a slight modification of a result in Vanderbauwhede and van Gils [276]. Corollary 7.7 has been proven for Volterra integral equations of convolution type by Diekmann and van Gils [66] and for RFDE by van Gils [88]. The proof of Theorem 8.1 is due to Ball [13].

An elementary derivation of the parameter-dependent normal form for the Takens-Bogdanov singularity can be found in the book by Kuznetsov [159], including a proof of the monotonicity of the quotient of the integrals.

Chapter X
Hopf bifurcation

X.1 Introduction

If the center subspace is finite dimensional, the center manifold theorem enables us to reduce an infinite dimensional dynamical system to a finite dimensional one. As a consequence, results on bifurcation from equilibria (e.g., saddle node, Hopf, Takens-Bogdanov bifurcation [159])) can be "lifted" from the theory of ODE to the theory of RFDE. The alternative is to study such bifurcations directly in the infinite dimensional setting and to "imitate" the ODE proof. However, this usually entails a number of technical complications (which are, in one way or another, connected with lack of smoothness) and, therefore, the alternative is, in our opinion, less attractive.

In this chapter we will derive the Hopf bifurcation theorem for RFDE, taking as our starting point the ODE version. We give a formula for the direction of bifurcation and we will derive the following stability result: if the center manifold is attractive and the periodic orbit is asymptotically stable within the center manifold, then it is asymptotically stable.

X.2 The Hopf bifurcation theorem

First, we recall the Hopf bifurcation theorem in finite dimensions (see Appendix VIII and, for instance, the book by Golubitsky and Schaeffer [92] for a detailed proof). Consider the system of ODEs

$$\dot{x} = f(x, \mu), \tag{2.1}$$

where $x \in \mathbb{R}^n$ and $\mu \in \mathbb{R}$. We assume that

Hf1 $f(0, \mu) = 0$, $f \in C^k(\mathbb{R}^n \times \mathbb{R}, \mathbb{R})$, $k \geq 2$,

Hf2 $D_1 f(0, \mu_0)$ has simple (i.e., of algebraic multiplicity one) eigenvalues at $\pm i\omega_0$, $\omega_0 > 0$, and no other eigenvalue of $D_1 f(0, \mu_0)$ belongs to $i\omega_0\mathbb{Z}$,

Hf3 $\mathrm{Re}\,(D\sigma(\mu_0)) \neq 0$, where $\sigma(\mu)$ is the branch of eigenvalues of $D_1 f(0, \mu_0)$ through $i\omega_0$ at $\mu = \mu_0$.

Theorem 2.1. (*Hopf bifurcation for* ODE.) *Let the above hypotheses be satisfied. Then there exist C^{k-1}-functions $\epsilon \mapsto \mu^*(\epsilon)$, $\epsilon \mapsto \omega^*(\epsilon)$ and $\epsilon \mapsto x^*(\epsilon)$, defined for ϵ sufficiently small, taking values in $\mathbb{R}$, $\mathbb{R}$ and $C(\mathbb{R};\mathbb{R}^n)$, respectively, such that at $\mu = \mu^*(\epsilon)$, $x^*(\epsilon)(t)$ is a $\frac{2\pi}{\omega^*(\epsilon)}$ periodic solution of (2.1). Moreover, μ^* and ω^* are even, $\mu(0) = \mu_0$, $\omega(0) = \omega_0$, $x^*(-\epsilon)(t) = x^*(\epsilon)(t + \frac{\pi}{\omega^*(\epsilon)})$ and $x^*(\epsilon)(t) = \epsilon \mathrm{Re}\,(e^{i\omega_0 t}p) + o(\epsilon)$ for $\epsilon \downarrow 0$, where p is an eigenvector of $D_1 f(0, \mu_0)$ at $i\omega_0$. In addition, if x is a small periodic solution of this equation with μ close to μ_0 and minimal period close to $\frac{2\pi}{\omega_0}$, then, $x(t) = x^*(\epsilon)(t + \theta)$ and $\mu = \mu^*(\epsilon)$ for some ϵ and some $\theta \in [0, 2\pi/\omega^*(\epsilon))$ (with ϵ unique modulo its sign).*

Exercise 2.2. Let f satisfy H1–H3. Show that $q \cdot D_1 D_2 f(0, \mu_0)p = D_1\sigma(\mu_0)$, where $\sigma(\mu)$ is the branch of eigenvalues through $i\omega_0$ at $\mu = \mu_0$. Here, p and q are eigenvectors of $D_1 f(0, \mu_0)$ and $D_1 f(0, \mu_0)^T$ respectively at $i\omega_0$, normalized such that $q \cdot p = \sum_1^n \bar{q}_i\, p_i = 1$.

Next, for $\mu \in \mathbb{R}$ and $-\infty < s \le t < \infty$ we consider the AIE

$$u(t) = T(t-s)u(s) + \int_s^t T^{\odot *}(t-\tau)\, R(u(\tau), \mu)\, d\tau. \tag{2.2}$$

We assume that H1–H4 from Section IX.2 hold and that X_0 is finite dimensional (H5 in Section IX.7). In addition, we assume that

HR1 the mapping $(u, \mu) \mapsto R(u, \mu)$ from $X \times \mathbb{R}$ into $X^{\odot *}$ is C^k-smooth, $k \ge 2$,

HR2 $R(0, \mu) = 0$ for all $\mu \in \mathbb{R}$, $D_1 R(0, \mu_0) = 0$.

Note that these hypotheses are slightly different from the corresponding ones in Section IX.8 on parameter dependence. HR2 guarantees that $u = 0$ is a solution of (2.2) for all $\mu \in \mathbb{R}$. This makes certain bifurcation formulas simpler and there is no loss of generality compared to the situation of Section IX.8, as we can always achieve this by applying the Implicit Function Theorem and a change of variables. Also, we require more smoothness ($k \ge 2$) and the parameter space be one dimensional.

We must relate assumptions and quantities connected with the ODE on the center manifold to corresponding assumptions and quantities associated with T and R. We start to assume that

HT A has simple eigenvalues at $\pm i\omega_0$, $\omega_0 > 0$, and no other eigenvalue belongs to $i\omega_0\mathbb{Z}$.

We decompose X according to the spectrum of A as in Section IX.2. The flow on the center manifold is described by the differential equation for $y \in X_0$ [see (IX.8.3)]:

$$\dot{y} = Ay + P_0^{\odot *} R(\mathcal{C}_1(y, \nu), \mu_0 + \nu), \tag{2.3}$$

where, as before, $\nu = \mu - \mu_0$.

Let $\phi \in X_0$ be an eigenvector of A at $i\omega_0$. Let $\phi^\odot \in X^\odot$ be an eigenvector of A^* at $i\omega_0$ which we assume to be normalized such that $\langle \phi^\odot, \phi \rangle = 1$. Note that this can be achieved, since we assume the eigenvalue to be simple. As before, $P_0^{\odot *}$ is the real spectral projection operator on $X^{\odot *}$ with range X_0. To describe the subspace of X_0 corresponding to the eigenvalues $\pm i\omega_0$, we introduce a basis and coordinates. Here, it is convenient to combine one complex coordinate and two complex conjugate basis vectors to describe a two-dimensional real subspace (see Exercises III.7.26-27 and IV.2.17):

$$X_{\omega_0} = \{\, z\phi + \overline{z}\overline{\phi} \mid z \in \mathbb{C} \,\}.$$

Combining Theorem IV.2.5(vii) and Exercise IV.2.17, we find that

$$P_{\omega_0}^{\odot *}\psi = \langle \phi^\odot, \psi \rangle \phi + \overline{\langle \phi^\odot, \psi \rangle}\,\overline{\phi} \tag{2.4}$$

or, in other words, that the coordinate of ψ is given by $z = \langle \phi^\odot, \psi \rangle$. Let

$$X_0 = X_{\omega_0} \oplus W$$

be the decomposition corresponding to $P_{\omega_0}^{\odot *}$.

The ODE (2.3) implies that, when $y = z\phi + \overline{z}\overline{\phi} + w$,

$$\frac{dz}{dt} = i\omega_0 z + \langle\, \phi^\odot, R(\mathcal{C}_1(z\phi + \overline{z}\overline{\phi} + w, \nu), \mu_0 + \nu)\,\rangle.$$

Since $\mathcal{C}_1(y, \nu) = y + \mathcal{O}(y^2)$, we find by Taylor series expansion that

$$\begin{aligned}\frac{dz}{dt} =& \big(i\omega_0 + \langle\, \phi^\odot, D_1R(0, \mu_0 + \nu)\phi\,\rangle\big)z + \langle\, \phi^\odot, D_1R(0, \mu_0 + \nu)\overline{\phi}\,\rangle \overline{z} \\ &+ \mathcal{O}(|z|^2) + \text{terms related to } w.\end{aligned}$$

We claim that from this it follows that

$$\operatorname{Re}(D\sigma(0)) = \operatorname{Re}\langle\, \phi^\odot, D_2D_1R(0, \mu_0)\phi\,\rangle,$$

where $\sigma(\nu)$ is the branch of eigenvalues of the linearization (at $y = 0$) of (2.3) through $i\omega_0$ at $\nu = 0$. The claim is easily substantiated. Details are provided by the next three exercises.

Exercise 2.3. Show that the transformation

$$z = y_1 + iy_2$$

relates the two-dimensional real ODE

$$\begin{pmatrix} \dot{y_1} \\ \dot{y_2} \end{pmatrix} = \begin{pmatrix} a_{11} & a_{12} \\ a_{21} & a_{22} \end{pmatrix} \begin{pmatrix} y_1 \\ y_2 \end{pmatrix} \tag{2.5}$$

to the "quasi-complex" ODE

$$\dot{z} = \frac{1}{2}(a_{11} + a_{22} + i(a_{21} - a_{12}))z + \frac{1}{2}(a_{11} + a_{22} + i(a_{12} - a_{21}))\overline{z}.$$

Exercise 2.4. With the same setting and notation as in the foregoing exercise, let the a_{ij} be differentiable functions of a parameter ν such that a_{11} and a_{22} are $\mathcal{O}(\nu)$ and $a_{12} = -\omega_0 + \mathcal{O}(\nu)$ and $a_{21} = \omega_0 + \mathcal{O}(\nu)$. Let $\sigma(\nu)$ denote the branch of eigenvalues with $\sigma(0) = i\omega_0$. Prove that $\operatorname{Re} D\sigma(0) = Dm(0)$, where $m(\nu)$ is the real part of the coefficient of z in the equation for $\dot{z}$.

Exercise 2.5. Extend the analysis of the last two exercises to the situation where y has more components which, however, are decoupled at $\nu = 0$.

Hence, to satisfy the transversality condition Hf3, we assume

HR3 $\operatorname{Re} \langle \phi^{\odot}, D_1 D_2 R(0, \mu_0)\phi \rangle \neq 0$.

We can now state the Hopf bifurcation theorem for (semiflow Σ associated with the) AIE (2.2).

Theorem 2.6. *Assume that* HR1–3 *hold. Then there exist C^{k-1} functions $\epsilon \mapsto \mu^*(\epsilon)$, $\epsilon \mapsto \varphi^*(\epsilon)$ and $\epsilon \mapsto \omega^*(\epsilon)$ with values in $\mathbb{R}$, X_0 and $\mathbb{R}$ respectively, defined for ϵ sufficiently small, such that the solution of* (2.2) *with initial condition $\varphi(\epsilon) = \mathcal{C}_1(\varphi^*(\epsilon), \mu^*(\epsilon))$ is $\frac{2\pi}{\omega^*(\epsilon)}$-periodic. Moreover, $\mu^*(\epsilon)$ and $\omega^*(\epsilon)$ are even, $\mu^*(0) = \mu_0$, $\omega^*(0) = \omega_0$, $\varphi(-\epsilon) = \Sigma(\frac{\pi}{\omega^*(\epsilon)}, \varphi(\epsilon))$ and if $u(t)$ is any small periodic solution of this equation with μ close to μ_0 and period close to $\frac{2\pi}{\omega_0}$, then $\mu = \mu^*(\epsilon)$ and there exists $\sigma \in [0, 2\pi/\omega^*(\epsilon))$ such that $u(\sigma) = \mathcal{C}_1\big(\varphi^*(\epsilon), \mu^*(\epsilon)\big)$.*

X.2.1 Application to RFDE

We specialize to RFDE.

$$
(2.6) \qquad \begin{cases} \dot{x}(t) = \int_0^h d\zeta(\theta, \mu) x(t-\theta) + g(x_t, \mu), & t \geq 0, \\ x(\sigma) = \varphi(\sigma), & -h \leq \sigma \leq 0. \end{cases}
$$

The equivalence between the RFDE and the AIE becomes clear (compare IX.9.1) if we let $\{T(t)\}$ be the linear semigroup corresponding to the kernel $\zeta(\,\cdot\,, \mu_0)$, $u(t) = x_t$ and

$$
(2.7) \qquad R(u, \mu) = \sum_{j=1}^{n} r_j^{\odot *} \Big(g(u, \mu) + \int_0^h \big(d\zeta(\theta, \mu) - d\zeta(\theta, \mu_0)\big)\varphi(-\theta) \Big)_j .
$$

We satisfy HR1 and HR2 if we assume that

Hg1 g is a C^k-smooth mapping, $k \geq 2$, from $X \times \mathbb{R}$ into $\mathbb{R}^n$,

Hg2 $g(0, \mu) = 0$ and $D_1 g(0, \mu) = 0$ for all $\mu \in \mathbb{R}$,

Hζ1 the mapping $(\mu, \varphi) \mapsto \int_0^h d\zeta(\theta, \mu)\varphi(-\theta)$ from $\mathbb{R} \times X$ into $\mathbb{R}^n$ is C^k-smooth, $k \geq 2$.

The assumptions on the smoothness with respect to the parameter can be weakened. This becomes important if, for instance, the delay is a parameter. We do not pursue this here, but refer to [66] and [103].

We recall some facts from Chapter IV, in particular from Sections IV.3 and IV.5. For $z \in \mathbb{C}$ and $\mu \in \mathbb{R}$, we define $\Delta(z,\mu)$ by

$$\Delta(z,\mu) = zI - \int_0^h e^{-z\theta} d\zeta(\theta,\mu). \tag{2.8}$$

A has $i\omega_0$ as an eigenvalue if there exists $0 \neq p \in \mathbb{C}^n$ such that $\Delta(i\omega_0,\mu_0)p = 0$. An eigenvector is given by $\phi(\theta) = e^{i\omega_0\theta}p$. Let $0 \neq q \in \mathbb{C}^n$ satisfy $\Delta(i\omega_0,\mu_0)^T q = 0$. If

$$\begin{aligned} \phi^{\odot}(t) &= q + \int_0^t k(\tau)\, d\tau, \qquad t \geq 0, \\ k(t) &= \int_t^h q e^{i\omega_0(t-\tau)} d\zeta(\tau,\mu_0), \qquad t \geq 0, \end{aligned} \tag{2.9}$$

then $\phi^{\odot}$ is an eigenvector of A^* at the eigenvalue $i\omega_0$,

$$A^*\phi^{\odot} = i\omega_0\phi^{\odot}, \tag{2.10}$$

and from (2.8) and (2.9), it follows that

$$\begin{aligned} \langle \phi^{\odot}, \phi \rangle &= \int_0^h d\phi^{\odot}(\tau)\phi(-\tau) \\ &= q \cdot D_1\Delta(i\omega_0,\mu_0)p, \end{aligned}$$

where we use the notation $a \cdot b = \sum_1^n a_i b_i$, for $a, b \in \mathbb{C}^n$. If $i\omega_0$ is a *simple* eigenvalue, we can normalize $\langle \phi^{\odot}, \phi \rangle$ to 1 (see Exercise IV.3.10 and Corollary IV.5.12). Moreover, from (2.7), (2.8) and (2.9) it follows that

$$\operatorname{Re}\langle \phi^{\odot}, D_1D_2R(0,\mu_0)\phi \rangle = -\operatorname{Re}(q \cdot D_2\Delta(i\omega_0,\mu_0)p).$$

In order to satisfy HT and HR3, we assume that

Hζ2 $z = \pm i\omega_0$ are simple roots of $\det \Delta(z,\mu_0) = 0$ and no other root belongs to $i\omega_0\mathbb{Z}$,

Hζ3 $\operatorname{Re}(q \cdot D_2\Delta(i\omega_0,\mu_0)p) \neq 0$.

The Hopf bifurcation theorem for RFDE can now be stated.

Theorem 2.7. (*Hopf bifurcation for RFDE.*) *Assume* (Hζ1-Hζ3) and (Hg1-Hg2). *Then there exist C^{k-1}-functions $\epsilon \mapsto \mu^*(\epsilon)$, $\epsilon \mapsto \rho^*(\epsilon)$ and $\epsilon \mapsto \omega^*(\epsilon)$, with values in $\mathbb{R}$ and a mapping $\epsilon \mapsto \psi(\epsilon)$ taking values in W $(\subset X_0)$, defined for ϵ sufficiently small, such that the solution of* (2.6) *with initial condition $\varphi(\epsilon) = \mathcal{C}_1(\rho^*(\epsilon)(\phi + \overline{\phi} + \psi(\epsilon)), \mu^*(\epsilon))$ is $\frac{2\pi}{\omega^*(\epsilon)}$-periodic. Moreover, $\mu^*(\epsilon)$ and $\omega^*(\epsilon)$ are even in ϵ, $\mu^*(0) = \mu_0$, $\omega^*(0) = \omega_0$ and if $x(t)$ is any small periodic solution of this equation with μ close to μ_0 and period* *close to* $\frac{2\pi}{\omega_0}$, *then* $\mu = \mu^*(\epsilon)$ *for some* ϵ *and there exists* $\sigma \in [0, 2\pi/\omega^*(\epsilon))$ *such that* $x(\theta+\sigma) = C_1(\rho^*(\epsilon)(\phi+\overline{\phi}+\psi(\epsilon)), \mu^*(\epsilon))(\theta), \theta \in [-h, 0]$. *Finally,* $\rho^*(\epsilon) = \epsilon + o(\epsilon)$ *and* $\psi(\epsilon) = o(1)$ *as* $\epsilon \to 0$.

X.3 The direction of bifurcation

Our next goal is to give a formula for the third term in the Taylor expansion of $\mu^*(\epsilon)$ (as μ is even, the second term necessarily vanishes). If this term is positive, then the periodic solutions exist for values of μ larger than the critical value. The bifurcation is called *supercritical.* In addition, the sign determines whether or not the critical Floquet multiplier of the bifurcating periodic solution of the ODE exceeds 1. If the equilibrium is stable for $\mu < \mu_0$ and the third term in the Taylor expansion is positive, then the critical multiplier moves *into* the unit circle. The bifurcating solution is stable. The stability of the equilibrium for $\mu < \mu_0$ is exchanged with the stability of the periodic orbit for $\mu > \mu_0$. Conversely, if the sign is negative, then the periodic solutions exist for values of μ smaller than the critical value and the critical Floquet multiplier moves outside the unit circle. The bifurcation is then called *subcritical* and the periodic solution is necessarily unstable. For ODE, this result is already in the paper of Hopf [128]; see also [134] and Appendix VIII.

In infinite dimensions, the Hopf bifurcation theorem has been proved by Crandall and Rabinowitz [53]. They also show the relation between the direction of bifurcation and the position of the critical multiplier relative to the unit circle. From this result we cannot immediately conclude that a supercritical bifurcation gives rise to stable solutions. We have to do some more work.

There are two ways to come to a stability result. One way is to develop Floquet theory in an infinite dimensional setting. This is done in Chapter XIV.

The alternative, which we adopt in this chapter, is to exploit the fact that we have reduced the dynamics to a finite dimensional center manifold. If at the bifurcation point no spectrum of A is in the right half-plane, then the manifold is attracting; see Theorem IX.7.4. If, moreover, the periodic solution is asymptotically stable within the center manifold, then, as we will show, we know enough to conclude that the periodic orbit is asymptotically stable. The first result summarizes the outcome of Exercise VIII.1.10.

Theorem 3.1. (*Direction of Hopf bifurcation for* ODE.) *Let* Hf1–Hf3 *be satisfied and let* $k \geq 3$. *Let* p *be as in Theorem* 2.1. *Let* q *be the vector in* $\mathbb{C}^n$ *satisfying*

(i) $D_1 f(0, \mu_0)^T q = i\omega_0 q$,

(ii) $q \cdot p = 1$.
If we write $\mu^*(\epsilon) = \mu_2\epsilon^2 + o(\epsilon^2)$, *then* μ_2 *is given by the formula*

$$\mu_2 = \frac{-\operatorname{Re} c}{\operatorname{Re}\,(q \cdot D_1 D_2 f(0,\mu_0)p)}, \tag{3.1}$$

where c *is given by*

$$\begin{aligned} c = {} & \frac{1}{2} q \cdot (D_1)^3 f(0,\mu_0)(p^2,\overline{p}) \\ & + q \cdot (D_1)^2 f(0,\mu_0)(-D_1 f(0,\mu_0)^{-1}(D_1)^2 f(0,\mu_0)(p,\overline{p}),p) \\ & + \frac{1}{2}\, q \cdot (D_1)^2 f(0,\mu_0)((2i\omega_0 - D_1 f(0,\mu_0))^{-1} \\ & \quad \times ((D_1)^2 f(0,\mu_0)(p,p),\overline{p})). \end{aligned} \tag{3.2}$$

Exercise 3.2. Show that the expression for c given in [114, formula III.3.5], reduces to the expression given above. Exploit the fact that u_1 and v_1 are eigenvectors of $D_1 f(x^*,\nu_c)$ to elaborate the resolvents, and then show that several terms cancel each other.

We want to apply this to (2.2) after reduction to (2.3). We need to calculate the Taylor series expansion of $R(\mathcal{C}_1(y,\mu_0),\mu_0)$ with respect to y up to and including order 3. It turns out that it suffices to calculate the Taylor series expansion of $\mathcal{C}_1(y,\mu_0)$ up to and including order 2. So far, we know that

$$\mathcal{C}_1(0,\mu_0) = 0; \qquad D_1\mathcal{C}_1(0,\mu_0)\varphi = \varphi, \quad \varphi \in X_0;$$

see Theorem IX.5.3 and Corollary IX.7.10. We compute the second order derivative.

Lemma 3.3.

$$\begin{aligned} & D_1^2\mathcal{C}(0,\mu_0)(\psi_1,\psi_2) \\ & = \int_0^\infty T^{\odot *}(\tau)P_-^{\odot *} D_1^2\, R(0,\mu_0)(T(-\tau)\psi_1, T(-\tau)\psi_2)\, d\tau \\ & \quad + \int_0^{-\infty} T^{\odot *}(\tau)P_+^{\odot *} D_1^2\, R(0,\mu_0)(T(-\tau)\psi_1, T(-\tau)\psi_2)\, d\tau. \end{aligned}$$

Proof. Recall [see (IX.5.1)] that $u^*(\varphi,\mu_0)$ satisfies the integral equation

$$\begin{aligned} u^*(\varphi,\mu_0)(t) = T(t)\varphi & + \int_{-\infty}^t T^{\odot *}(t-\tau)\, P_-^{\odot *}\, R(u^*(\varphi,\mu_0)(\tau),\mu_0)\, d\tau \\ & + \int_\infty^t T^{\odot *}(t-\tau)\, P_+^{\odot *}\, R(u^*(\varphi,\mu_0)(\tau),\mu_0)\, d\tau \\ & + \int_0^t T^{\odot *}(t-\tau)\, P_0^{\odot *}\, R(u^*(\varphi,\mu_0)(\tau),\mu_0)\, d\tau. \end{aligned}$$

As u^* depends continuously differentiably on φ, $D_1 u^*(\varphi, \mu_0)\psi_1$ satisfies the integral equation

$$\begin{aligned}(D_1 u^*(\varphi, \mu_0)\psi_1)(t) &= T(t)\psi_1 \\ &+ \int_{-\infty}^{t} T^{\odot *}(t-\tau)\, P_-^{\odot *}\, D_1 R(u^*(\varphi, \mu_0)(\tau), \mu_0)(D_1 u^*(\varphi, \mu_0)\psi_1)(\tau)\, d\tau \\ &+ \int_{\infty}^{t} T^{\odot *}(t-\tau)\, P_+^{\odot *}\, D_1 R(u^*(\varphi, \mu_0)(\tau), \mu_0)(D_1 u^*(\varphi, \mu_0)\psi_1)(\tau)\, d\tau \\ &+ \int_{0}^{t} T^{\odot *}(t-\tau)\, P_0^{\odot *}\, D_1 R(u^*(\varphi, \mu_0)(\tau), \mu_0)(D_1 u^*(\varphi, \mu_0)\psi_1)(\tau)\, d\tau\end{aligned}$$

for $t \in \mathbb{R}$. Using that $u^*(0, \mu_0)(\tau) = 0$ for all $\tau \in \mathbb{R}$, evaluation at $\varphi = 0$ yields

$$(D_1 u^*(0, \mu_0)\psi_1)(t) = T(t)\psi_1.$$

Computing the second derivative with respect to φ at $\varphi = 0$, we find

$$\begin{aligned}& D_1^2 u^*(\varphi, \mu_0)(\psi_1, \psi_2)(t)\big|_{\varphi=0} \\ &= \int_{-\infty}^{t} T^{\odot *}(t-\tau)\, P_-^{\odot *}\, D_1^2 R(0, \mu_0)(T(\tau)\psi_1, T(\tau)\psi_2)\, d\tau \\ &+ \int_{\infty}^{t} T^{\odot *}(t-\tau)\, P_+^{\odot *}\, D_1^2 R(0, \mu_0)(T(\tau)\psi_1, T(\tau)\psi_2)\, d\tau \\ &+ \int_{0}^{t} T^{\odot *}(t-\tau)\, P_0^{\odot *}\, D_1^2 R(0, \mu_0)(T(\tau)\psi_1, T(\tau)\psi_2)\, d\tau.\end{aligned}$$

Evaluation at $t = 0$ gives the result. □

Lemma 3.4. *Let ϕ be an eigenvector of A at the eigenvalue $i\omega_0$; then the following relations hold:*

$$D_1^2 C_1(0, \mu_0)(\phi, \phi) = (2i\omega_0 I - A^{\odot *})^{-1}(I - P_0^{\odot *})\, D_1^2 R(0, \mu_0)(\phi, \phi),$$

$$D_1^2 C_1(0, \mu_0)(\phi, \overline{\phi}) = -(A^{\odot *})^{-1}(I - P_0^{\odot *})\, D_1^2 R(0, \mu_0)(\phi, \overline{\phi}).$$

Proof. If we use that $T(t)\phi = e^{i\omega t}\phi$ for all $t \in \mathbb{R}$, it follows from Lemma 3.3 that

$$\begin{aligned}D_1^2 C_1(0, \mu_0)(\phi, \phi) &= \int_{0}^{\infty} e^{-2i\omega_0\tau}\, T^{\odot *}(\tau)\, P_-^{\odot *} D_1^2\, R(0, \mu_0)(\phi, \phi)\, d\tau \\ &+ \int_{0}^{-\infty} e^{-2i\omega_0\tau}\, T^{\odot *}(\tau)\, P_+^{\odot *} D_1^2\, R(0, \mu_0)(\phi, \phi)\, d\tau.\end{aligned}$$

For z on the imaginary axis and for all $\varphi^{\odot *} \in X^{\odot *}$, we have the identity (compare Exercises 2.5–2.8 at the end of Section IX.2)

$$(zI - A_{\mu_0}^{\odot *})^{-1}\varphi^{\odot *} - (zI - A_{\mu_0}^{\odot *})^{-1}P_0^{\odot *}\varphi^{\odot *}$$
$$= \int_0^{\infty} e^{-z\tau}\, T^{\odot *}(\tau) P_-^{\odot *}\varphi^{\odot *}\, d\tau + \int_0^{-\infty} e^{-z\tau}\, T^{\odot *}(\tau) P_+^{\odot *}\varphi^{\odot *}\, d\tau.$$

Combination of these two identities yields the first assertion. □

Exercise 3.5. Convince yourself that the second assertion in the above lemma is correct.

Exercise 3.6. Define H from X_0 into $X^{\odot *}$ by $H(\varphi) = R(\mathcal{C}_1(\varphi, \mu_0), \mu_0)$. Show that

$$\begin{aligned} D^3 H(0)(\psi_1, \psi_2, \psi_3) = {} & D_1^3 R(0, \mu_0)(\psi_1, \psi_2, \psi_3) \\ & + D_1^2 R(0, \mu_0)((D_1^2 \mathcal{C}_1(0, \mu_0), \mu_0)(\psi_1, \psi_2), \psi_3) \\ & + D_1^2 R(0, \mu_0)((D_1^2 \mathcal{C}_1(0, \mu_0), \mu_0)(\psi_2, \psi_3), \psi_1) \\ & + D_1^2 R(0, \mu_0)((D_1^2 \mathcal{C}_1(0, \mu_0), \mu_0)(\psi_3, \psi_1), \psi_2). \end{aligned}$$

Theorem 3.7. *Let* $\mathrm{HB}_\mu 1$, $\mathrm{HB}_\mu 2$, $\mathrm{HR}_\mu 1$, $\mathrm{HR}_\mu 2$ *and* HT *be satisfied for some* $k \geq 3$. *Let* ϕ, $\phi^\odot$ *be eigenvectors of* A *and* A^* *at* $i\omega_0$, *normalized so that* $\langle \phi^\odot, \phi \rangle = 1$. *Let* $\mu^*(\epsilon)$ *be as defined in Theorem 2.2. The coefficient* μ_2 *in the expansion* $\mu^*(\epsilon) = \mu_2 \epsilon^2 + o(\epsilon^2)$ *is given by*

$$\mu_2 = \frac{-\operatorname{Re} c}{\operatorname{Re} \langle \phi^\odot, D_1 D_2 R(0, \mu_0)\phi \rangle},$$

where c *is explicitly given by*

$$\begin{aligned} c = {} & \frac{1}{2}\langle \phi^\odot, D_1^3 R(0, \mu_0)(\phi, \phi, \overline{\phi}) \rangle \\ & + \langle \phi^\odot, D_1^2 R(0, \mu_0)(-A^{\odot *^{-1}} D_1^2 R(0, \mu_0)(\phi, \overline{\phi}), \phi) \rangle \\ & + \frac{1}{2}\langle \phi^\odot, D_1^2 R(0, \mu_0)((2i\omega_0 - A^{\odot *})^{-1} D_1^2 R(0, \mu_0)(\phi, \phi), \overline{\phi}) \rangle. \end{aligned}$$

Proof. The identity for c follows from the combination of Theorem 3.1 with Lemma 3.3 and Exercise 3.6. The denominator in the expression of μ_2 is the real part of the derivative of the critical eigenvalue at criticality, i.e., $\operatorname{Re} D\sigma(\mu_0)$, where $\sigma(\mu)$ is the branch of eigenvalues through $i\omega_0$ at $\mu = \mu_0$ (compare Section 2 and Exercise 2.2) and equals the denominator in the expression for μ_2 in Theorem 3.1. □

Theorem 3.8. *Let* $\gamma(t)$ *be a* T*-periodic solution of (2.2) which is asymptotically stable within the center manifold. If there is no spectrum of* A *in the right half-plane and* $\sup_{t \in [0,T]} ||\gamma(t)||$ *is sufficiently small, then* γ *is asymptotically stable.*

Proof. We write an element u in X as

$$u = (P_0 u, (I - P_0)u) = (y, z). \tag{3.3}$$

As $X_+ = \{0\}$, we can write equation (2.2) as the system of equations

$$\begin{aligned} \frac{dy}{dt} &= Ay + P_0^{\odot *} R(y + z, \mu), \\ z(t) &= T(t-s)z(s) \\ &\quad + \int_s^t T^{\odot *}(t-\tau)\,(I - P_0^{\odot *})\,R\big(y(\tau) + z(\tau), \mu\big)\,d\tau. \end{aligned} \tag{3.4}$$

From now on we shall suppress the dependence of the center manifold on the parameter in the notation, and we write for $\varphi \in X_0$,

$$\mathcal{C}(\varphi) = \varphi + (I - P_0)\mathcal{C}(\varphi) \stackrel{\text{def}}{=} \varphi + h(\varphi).$$

We apply the change of coordinates

$$Y = y, \qquad Z = z - h(y). \tag{3.5}$$

In these coordinates, the center manifold coincides with the subspace $\{(Y, Z) \mid Z = 0\}$ and (3.4) takes the form

$$\begin{aligned} \frac{dY}{dt} &= AY + P_0^{\odot *} R(Y + h(Y) + Z, \mu), \\ Z(t) &= z(t) - h(y(t)), \end{aligned} \tag{3.6}$$

where

$$\begin{aligned} Z(t) &= T(t-s)z(s) - h(y(t)) \\ &\quad + \int_s^t T^{\odot *}(t-\tau)\,(I - P_0^{\odot *})\,R(Y(\tau) + h(Y(\tau)) + Z(\tau), \mu)\,d\tau \\ &= T(t-s)Z(s) + T(t-s)h(y(s)) - h(y(t)) \\ &\quad + \int_s^t T^{\odot *}(t-\tau)\,(I - P_0^{\odot *})\,R(Y(\tau) + h(Y(\tau)) + Z(\tau), \mu)\,d\tau \\ &= T(t-s)Z(s) + \int_s^t T^{\odot *}(t-\tau)\,(I - P_0^{\odot *})\,\widetilde{R}(Y(\tau), Z(\tau), \mu)\,d\tau \end{aligned}$$

and

$$\widetilde{R}(Y, Z, \mu) = R(Y + h(Y) + Z, \mu) - R(Y + h(Y), \mu).$$

In the last step, we have used the invariance of the center manifold. Let $\gamma(\theta)$, $\theta \in [0, 2\pi)$ be a parameterization of the closed orbit on the center manifold. We know that it is asymptotically stable. It is possible to introduce a moving orthonormal coordinate system $\{v(\theta),\ \xi_2(\theta), \ldots,\ \xi_n(\theta)\}$, 2π-periodic in θ, with $n = \dim X_0$, and such that $v(\theta) = (|\frac{d\gamma(\theta)}{d\theta}|)^{-1}\frac{d\gamma(\theta)}{d\theta}$ (see, for instance, [101, Section VI.1]). The transformation

$$Y = \gamma(\theta) + Q(\theta)\eta, \qquad Q = (\,\xi_2 \dots \xi_n\,) \tag{3.7}$$

leads to a system of equations for θ and η

$$\begin{aligned} \frac{d\theta}{dt} &= \Theta(\eta, \theta), \\ \frac{d\eta}{dt} &= \tilde{g}(\eta, \theta), \end{aligned} \tag{3.8}$$

where Θ and $\tilde{g}$ are 2π-periodic in t and $\eta = 0$ is an asymptotically stable solution of (3.8). As $\Theta(0, t) > 0$, we can reparameterize the time and consider, instead of (3.8),

$$\frac{d\eta}{d\theta} = \frac{\tilde{g}(\eta, \theta)}{\Theta(\eta, \theta)} = g(\eta, \theta). \tag{3.9}$$

A theorem due to Massera [191, Theorem 8] states that if (3.8) is periodic and has $\eta = 0$ as an asymptotically stable solution, then a positive definite function $\widetilde{V}$ exists such that $d\widetilde{V}/d\theta$ [the derivative of $\widetilde{V} = \widetilde{V}(\eta, \theta)$ along orbits of (3.9)] is negative definite (also see [101, Theorem X.4.2 and [168, Theorem VI.21.2]). We denote the solution of (3.8) with initial values $\theta = 0$, $\eta = \eta_0$ at $t = 0$ by $\theta = \theta(t; \eta_0)$, $\eta = \eta(t; \eta_0)$. For $|\eta_0|$ small enough, we can obtain t as a function of θ and η_0 by the implicit function theorem. Note that $t(0, \eta_0) = 0$. We now define $V = V(\eta, t)$ by

$$\widetilde{V}(\eta, \theta) = \widetilde{V}(\eta(t; \eta_0), \theta(t; \eta_0)) = \widetilde{V}(\eta, t(\theta, \eta_0)) = V(\eta, t).$$

V has the same properties as $\widetilde{V}$, i.e., V is positive definite with a negative definite derivative along the orbits of (3.8) in a neighbourhood of the periodic orbit.

We extend the domain of definition of V (first defined on the center manifold) and define W on $\mathrm{B}_\delta(\eta) \times Z \times \mathbb{R}$ by

$$W(\eta, Z, t) = V(\eta, t). \tag{3.10}$$

We let

$$\mathcal{V}_{\delta,\sigma} = \big\{(\eta, Z, t) \mid V(\eta, t) < \delta,\ ||Z|| < \sigma\big\}. \tag{3.11}$$

As R is continuous and $\frac{dV}{dt} < 0$ on $\mathcal{V}_{\delta,\sigma} \cap \{(\eta, Z, t) \mid |Z| = 0\}$, it follows that there exists a continuous mapping $\delta \mapsto \sigma^*(\delta)$, which is positive for δ positive, such that

$$\frac{dW}{dt}(\eta, Z, t) < 0 \quad \text{for all } (\eta, Z, t) \in \partial\mathcal{V}_{\delta,\sigma^*(\delta)} \text{ such that } ||Z|| < \delta. \tag{3.12}$$

As the center manifold is attracting, we know that there exists positive constants K and ν such that

$$||Z(t)|| \le Ke^{-\nu t}||Z(0)||$$

for t positive and provided that $||Z(t)|| + ||Y(t)||$ is less than some fixed constant. Using an equivalent norm (see Section IX.8), we may take $K = 1$. If $\sup_{\theta\in[0,2\pi)} ||\gamma(\theta)||$ is small enough, then there exists a positive number ρ such that

$$(3.13) \qquad \text{for all } t > 0, \quad ||\eta(t)|| + ||Z(t)|| \le \rho \quad \text{and} \quad |Z(t)|| \le e^{-\nu t}||Z(0)||.$$

Hence, we conclude that $\mathcal{V}_{\delta,\sigma^*(\delta)}$ is invariant under the flow. This shows that γ is stable. It also follows from (3.10) and (3.11) that an orbit which starts in $\mathcal{V}_{\delta_0,\sigma^*(\delta_0)}$ enters $\mathcal{V}_{\delta,\sigma^*(\delta)}$ for any δ positive. So, γ is asymptotically stable. □

X.3.1 Application to RFDE

We recall (2.6):

$$\begin{cases} \dot{x}(t) = \int_0^h d\zeta(\theta,\mu)x(t-\theta) + g(x_t,\mu), & t \ge 0, \\ x(\theta) = \varphi(\theta), & -h \le \theta \le 0. \end{cases}$$

We apply the result from Section 3 to this RFDE.

Theorem 3.9. *Let ϕ, q, p and $\Delta(z,\mu)$ be as in Section* 2.1 *and assume that* Hg1, Hg2, Hζ1, Hζ2 *and* Hζ3 *hold for some $k \ge 3$. Let $\mu^*(\epsilon)$ be as defined in Theorem* 2.3. *If we write $\mu^*(\epsilon) = \mu_2\epsilon^2 + o(\epsilon^2)$, then*

$$\mu_2 = \frac{\operatorname{Re} c}{\operatorname{Re}(q \cdot D_2\Delta(i\omega_0,\mu_0)p)},$$

where

$$\begin{aligned} c = \frac{1}{2} q \cdot D_1^3 g(0,\mu_0)(\phi,\phi,\overline{\phi}) \\ + q \cdot D_1^2 g(0,\mu_0)(e^{0\cdot}\,\Delta(0,\mu_0)^{-1}D_1^2 g(0,\mu_0)(\phi,\overline{\phi}),\phi) \\ + \frac{1}{2} q \cdot D_1^2 g(0,\mu_0)(e^{2i\omega_0\cdot}\,\Delta(2i\omega_0,\mu_0)^{-1}D_1^2 g(0,\mu_0)(\phi,\phi),\overline{\phi}). \end{aligned}$$

Proof. We elaborate the general formula for the direction of bifurcation in this special case. As $R(u,\mu) = r^{\odot *}g(u,\mu)$, we use the identity

$$(A^{\odot *} - zI)^{-1}r_i^{\odot *} = -e^{z\cdot}\,\Delta(z)^{-1}e_i,$$

where e_i is the ith unit vector in $\mathbb{C}^n$ and $z \in \rho(A^{\odot *})$; see Corollary IV.5.4. This yields the formula for c. Finally, $\operatorname{Re}\langle\phi^{\odot}, D_1D_2R(0,\mu_0)\phi\rangle$ equals $\operatorname{Re}(q \cdot \Delta(i\omega_0,0)p)$ in this case. □

Example 3.10. Consider the equation

$$\dot{x}(t) = -\mu(1 + x(t))x(t-1), \tag{3.14}$$

which has an equilibrium at $x \equiv 0$. The linearized equation

$$\dot{y}(t) = -\mu y(t-1) \tag{3.15}$$

has as its the characteristic function

$$\Delta(z, \mu) = z + \mu e^{-z}. \tag{3.16}$$

$\Delta(\,\cdot\,, \mu)$ has a pair of complex conjugate purely imaginary roots $\pm i\frac{\pi}{2}$ at $\mu = \mu_0 = \frac{\pi}{2}$, whereas all the other roots (at this value of μ) have a negative real part (see Section XI.2, in particular Figure XI.1). Let A be the generator of the linear semigroup corresponding to (3.15) at $\mu = \mu_0$. An eigenvector of A at the eigenvalue $\frac{\pi}{2}$ is given by $p(t) = e^{i\frac{\pi}{2}t}$. One verifies easily that an eigenvector of A^* at the eigenvalue $\frac{\pi}{2}$ is given by $q(t) = qe^{i\frac{\pi}{2}t}$, $q \neq 0$. We put $q = 1 - \frac{\pi i}{2}/(1 + \frac{\pi^2}{4})$ so that $\langle q, p\rangle = 1$ (the reader is invited to verify this). The nonlinearity in this case is given by $g(\varphi, \mu) = -\mu(1+\varphi(0))\varphi(-1)$; hence

$$D_1^2 g(0, \frac{\pi}{2})(\varphi_1, \varphi_2) = -\frac{\pi}{2}(\varphi_1(0)\varphi_2(-1) + \varphi_1(-1)\varphi_2(0)).$$

So, $D_1^2 g(0, \frac{\pi}{2})(p, \overline{p}) = 0$, and $D_1^2 g(0, \frac{\pi}{2})(p, p) = i\pi$. A short computation leads to

$$c = \frac{\pi}{5(1 + \frac{\pi^2}{4})}(\frac{1}{2} - \frac{3}{4}\pi - i(\frac{\pi}{4} - \frac{3}{2})).$$

Thus

$$\operatorname{Re} c = \frac{\pi}{20(1 + \frac{\pi^2}{4})}(2 - 3\pi) < 0.$$

Finally,

$$-\operatorname{Re} q D_2 \Delta(i\frac{\pi}{2}, \frac{\pi}{2})p = \frac{2\pi}{1 + 4\pi^2} > 0.$$

We conclude that the bifurcation is *supercritical*, i.e., a branch of stable periodic solutions bifurcates from the equilibrium for values of μ larger than $\frac{\pi}{2}$.

Exercise 3.11. Consider the scalar integral equation

$$x(t) = \int_0^h p(x(t-\theta))\, k(\theta)\, d\theta,$$

where $p : \mathbb{R} \to \mathbb{R}$ and $p(0) = 0$. Show along the lines indicated below that the formula for c in this case is given by

$$\begin{aligned} c =& \frac{1}{2} q p'''(0)\, \mathcal{L}\{k\}(i\omega_0) + (p''(0))^2\, \Delta(0)^{-1}\, \mathcal{L}\{k\}(0)\, \mathcal{L}\{k\}(i\omega_0) \\ &+ \frac{1}{2}\, (p''(0))^2\, \Delta(2i\omega_0)^{-1}\, \mathcal{L}\{k\}(2i\omega_0)\, \mathcal{L}\{k\}(i\omega_0), \end{aligned}$$

where $\mathcal{L}$ denotes the Laplace transform: $\mathcal{L}\{f\}(z) = \int_0^\infty e^{-zt} f(t)\, dt$. The notation for the Laplace transform we adopt here is different from the one used previously, to avoid confusion with the notation for the complex conjugate.

(i) Rewrite (3.10) in the form of the abstract integral equation

$$u(t) = T(t-s)u(s) + \int_s^t T^{\odot *}(t-\tau)R(u(\tau))\, d\tau,$$

where the spaces and the semigroups are the same as in Exercise II.5.6 and Exercise III.4.7 and formula (1.4). Verify that R is the operator from $L_1([0,h];\mathrm{I\!R})$ into $\mathrm{NBV}([0,h);\mathrm{I\!R})$ defined by

$$R(\varphi) = \int_0^h (p(\varphi(\theta)) - p'(0)\varphi(\theta))\, k(\theta)\, d\theta \cdot H,$$

where H is the element of $\mathrm{NBV}([0,h);\mathrm{I\!R})$ defined by $H(a) = 1$ for $a > 0$.

(ii) Show that

(a) $DR(0)(\varphi) = 0$,

(b) $D^2R(0)(\varphi_1, \varphi_2) = \int_0^h p''(0)\,(\varphi_1(\theta), \varphi_2(\theta))\, k(\theta)\, d\theta \cdot H$. Let

$$\Delta(z) = I - \int_0^h p'(0) e^{-z\theta}\, k(\theta)\, d\theta.$$

(iii) Verify that $A\varphi = i\omega_0\varphi$ and $\varphi(0) = 1$ is equivalent to $\varphi(t) = e^{i\omega_0 t}$ and $\Delta(i\omega_0) = 0$, respectively. Next show that $A^\odot\psi = i\omega_0\psi$ is equivalent to

$$\psi(\theta) = \int_\theta^h e^{i\omega_0(\theta-\tau)} p'(0) k(\tau)\, d\tau\ \psi(0), \qquad \Delta(i\omega_0)\psi(0) = 0.$$

(iv) Conclude from (iii) that $A\varphi = i\omega_0\varphi$, $A^\odot\psi = i\omega_0\psi$ and $\langle\psi, \varphi\rangle = 1$ implies that

$$\psi^\odot(0) = q = \frac{1}{\Delta'(i\omega_0)}.$$

Exercise 3.12. Again we consider the scalar integral equation

$$x(t) = \int_0^h p(x(t-\theta))\, k(\theta)\, d\theta,$$

where $p : \mathrm{I\!R} \to \mathrm{I\!R}$. But now we start from a nontrivial equilibrium, i.e., we give up the condition that $p(0) = 0$, but here we assume that the kernel is normalized to 1, i.e., $\int_0^h k(a)\, da = 1$ and that $\overline{x} = p(\overline{x})$ for some $\overline{x}$. Reduce this case to the previous exercise by shifting this equilibrium to the origin and show that the formula for c reduces to

$$c = -\frac{1}{(p'(\overline{x}))^2\, \mathcal{L}\{k\}'(i\omega_0)} \times \Big(\frac{1}{2} p'''(\overline{x}) + \frac{(p''(\overline{x}))^2}{1 - p'(\overline{x})} + \frac{(p''(\overline{x}))^2\, \mathcal{L}\{k\}(2i\omega_0)}{2(1 - p'(\overline{x}))\, \mathcal{L}\{k\}(2i\omega_0)}\Big). \tag{3.17}$$

X.4 Comments

There exists a vast literature on the Hopf Bifurcation Theorem. We do not even attempt to give credit to all those who have contributed to the applicability under varying hypotheses.

For ordinary differential equations, the analysis of bifurcation of periodic solutions from an equilibrium goes back to Andronov and Chaikin [6] and Hopf [128]. The Crandall and Rabinowitz proof [53] covers the theorem in the context of analytic semigroups.

For RFDE, the first results are obtained by Chafee [28]; see also Hale [102]. Averaging and normal form calculations were developed by Chow and Mallet-Paret [41].

A formula for determining the direction of Hopf bifurcation for scalar one-delay equations is given by Claeyssen [43]. More general results for RFDE were given by Stech [262] (allowing infinite delays) and van Gils [88]. For Volterra integral equations of convolution type, see Diekmann and van Gils [66].

Infinite delays, including bifurcation formulas, can be found in the work of Staffans [260].

The relation between stability on the center manifold and stability in the full state space has been investigated by Negrini and Tesei [205] (in a Banach space setting) and by Hassard, Kazarinoff and Wan [114] (in a Hilbert space setting). The proof of Theorem 3.8 is adopted from [114].

Chapter XI

Characteristic equations

XI.1 Introduction: an impressionistic sketch

As illustrated by the saddle point property and the Hopf bifurcation theorem, an analysis of the qualitative behaviour of solutions near an equilibrium starts with an analysis of solutions of the linearized equation. For delay equations, the latter, in turn, reduces to an analysis of the characteristic equation. Particularly relevant questions are: how are the roots of the characteristic equation located relative to the imaginary axis, and when do roots cross the imaginary axis as parameters are varied? (cf. Chapters I, IV and X).

In general, no necessary and sufficient conditions are known for all roots to be in the left half-plane. (Note the contrast with the ODE case where the characteristic equation reduces to a polynomial and the Routh-Hurwitz theorem provides exactly such conditions.)

In principle, roots can cross at any point of the imaginary axis as parameters are varied and sometimes this can be exploited as follows. Let the characteristic equation be

$$F(p, z) = 0, \tag{1.1}$$

where p denotes the parameter vector. We can try to solve the equation

$$F(p, i\nu) = 0 \tag{1.2}$$

for p as a function of $\nu \in \mathbb{R}$ to find "surfaces" in parameter space such that for p on such a surface, the characteristic equation has a root exactly on the imaginary axis. These surfaces divide the parameter space into regions and sometimes it is possible to conclude, by additional arguments, for which region the steady state is stable. The boundary of that region in parameter space is called the *stability boundary* and its shape may allow us to conclude whether the mechanism corresponding to some component of the parameter vector is stabilizing or destabilizing.

The idea just described requires a parameter which is at least two dimensional [the equation $F = 0$ consists of two real equations and ν is one dimensional, so, generically, the surfaces are $(\dim p)-1$ dimensional]. When $\dim p = 2$, the method is most effective, since the parameterized curves in the plane can be displayed graphically and human beings are particularly good at absorbing two-dimensional visual information. In this chapter we restrict our attention to a two-dimensional parameter and we shall write $p = (\alpha, \beta)$, with $\alpha, \beta \in \mathbb{R}$, and rewrite (1.2) as

$$F(\alpha, \beta, i\nu) = 0. \tag{1.3}$$

As a rule, solving (1.3) requires a numerical implementation of the implicit function theorem, but in exceptional cases, notably when α and β appear linearly, one can find explicit expressions for α and β as a function of ν. Even then it may require a computer to draw the corresponding curves in the (α, β)-plane parameterized by ν. In most examples, there are denumerably many curves, since the function $\nu \mapsto \big(\alpha(\nu), \beta(\nu)\big)$ has singularities at points $\pm\nu_k$, $k = 1, 2, \ldots$ (alternatively, one may think of one curve which "visits" the point at infinity in the parameter plane infinitely often; see Fig XI.1). When two such curves intersect, there are simultaneously two pairs of roots on the imaginary axis for the corresponding parameter values. In easy situations, the curves are nested and do not intersect. Still one has, in order to decide in what region of the parameter space the steady state is stable, to determine on which side of a curve the corresponding roots are in the left, respectively right, half-plane (note that the stability region is not necessarily connected, since roots may return to the left half-plane after a journey into the right half-plane).

In applications (see, e.g., Chapters XV and XVI) we frequently encounter equations of the form

$$\dot{x}(t) = \alpha x(t) + f(x(t-1)) \tag{1.4}$$

where $f(0) = 0$. Linearization around the zero equilibrium yields the delay equation

$$\dot{x}(t) = \alpha x(t) + \beta x(t-1) \tag{1.5}$$

with characteristic equation

$$z = \alpha + \beta e^{-z}. \tag{1.6}$$

For (1.6), one can analyse completely how the location of the roots in the complex plane depends on (α, β). The main ingredient for this analysis is the division of the complex plane into horizontal strips such that the number of roots in a given strip is independent of the parameter (α, β). To find out what the number of roots in a given strip is, one can proceed as follows: find a half-line in the parameter plane, say $\{(\alpha, \beta) = (s\alpha_0, s\beta_0),\ s \geq 0\}$, such that in the strip under scrutiny all roots in the right half-plane are known for small s (e.g., there are no roots in the right half-plane) and such that there are no roots in the left half-plane for s sufficiently large; next, analyse how many roots cross the imaginary axis as s increases.

Part of the arguments work for the more general equation

$$z = \alpha \int_0^1 e^{-z\theta}\, d\eta_0(\theta) + \beta e^{-z}, \tag{1.7}$$

where $\eta_0 \in \mathrm{NBV}$ and η_0 is continuous at $\theta = 1$.

Note that (1.7) is obtained from

$$z = \alpha \int_0^1 e^{-z\theta}\, d\eta(\theta), \tag{1.8}$$

where $\eta \in \mathrm{NBV}$, by splitting of the jump at $\theta = 1$ and introducing the parameters α and β accordingly.

Exercise 1.1. Use a scaling of the time variable to bring the equation

$$\dot{x}(t) = \gamma x(t) + h(x(t-\delta))$$

into the form (1.4).

The organization of this chapter is as follows. In Section 2 we determine the stability boundary in (α, β)-parameter space for the prototype equation (1.6). We analyse relevant properties of the curves C_k and compute on which side of such a curve the critical roots are in the right half-plane. Using the fact that the roots depend continuously on parameters and are confined to a finite portion of the right half-plane, we are able to give a complete partitioning of the parameter plane into regions defined by the number of roots in the right half-plane (see Fig. XI.1). In Section 3 we continue the analysis of the prototype equation, but now we also consider the positioning of the roots relative to each other, not just relative to the imaginary axis. Using the invariance of horizontal strips in the complex plane, we show that the ordering of the imaginary parts of the roots coincides with the ordering of the real parts. This is a special feature which does not carry over to a wide class of problems. However, it plays an important role in obtaining more detailed information about the flow of certain nonlinear equations (cf. Chapter XV), in particular in the proof that a so-called Morse decomposition exists (see Mallet-Paret [176]).

Section 4 is devoted to case studies. We analyse three examples from population dynamics and show how the techniques of Section 2 or ad hoc techniques can be used to find Hopf bifurcation points in parameter space. Biological interpretations of the results are presented. For two of the three examples, we go on to determine the direction of Hopf bifurcation from the formula derived in Chapter X. Together, these examples should give a fair idea of what tools (or tricks) are available, but, unfortunately, they are not really representative for characteristic equations in general. The examples are carefully chosen such that strong conclusions can be obtained, but for characteristic equations in general, the rule seems to be that they are not at all amenable to analysis.

XI.2 The region of stability in a parameter plane

First we concentrate on the prototype equation

$$z - \alpha - \beta e^{-z} = 0. \tag{2.1}$$

Writing $z = \mu + i\nu$, we find two real equations

$$\begin{aligned} \mu - \alpha - \beta e^{-\mu}\cos\nu &= 0, \\ \nu + \beta e^{-\mu}\sin\nu &= 0 \end{aligned} \tag{2.2}$$

and solving for α and β, we obtain, for $\sin\nu \neq 0$,

$$\begin{aligned} \alpha &= \mu + \frac{\nu}{\sin\nu}\cos\nu, \\ \beta &= -\frac{\nu}{\sin\nu}e^{\mu}. \end{aligned} \tag{2.3}$$

If we now fix μ at the value zero, we get

$$\alpha = \frac{\nu\cos\nu}{\sin\nu} \quad \text{and} \quad \beta = -\frac{\nu}{\sin\nu}.$$

Since these functions are even in ν (as they should), we restrict our attention to $\nu \geq 0$. Motivated by the observation that these expressions have singularities at $\nu = k\pi$, $k = 1, 2, \ldots$, we introduce some notation.

The intervals I_k are for $k = 0, 1, 2, \ldots$ defined by

$$I_k = \big((2k-1)\pi, (2k+1)\pi\big). \tag{2.5}$$

Each I_k is divided into the point $2k\pi$ and two intervals:

$$\begin{aligned} I_k^- &= \big((2k-1)\pi, 2k\pi\big), \\ I_k^+ &= \big(2k\pi, (2k+1)\pi\big). \end{aligned} \tag{2.6}$$

Note that the sign indicates the sign of the sine-function on the interval. Finally, we define curves $C_k^{\pm}$ in the (α, β)-plane parameterized by ν as follows:

$$C_k^{\pm} = \big\{(\alpha, \beta) = \big(\frac{\nu\cos\nu}{\sin\nu}, -\frac{\nu}{\sin\nu}\big) \mid \nu \in I_k^{\pm}\big\}. \tag{2.7}$$

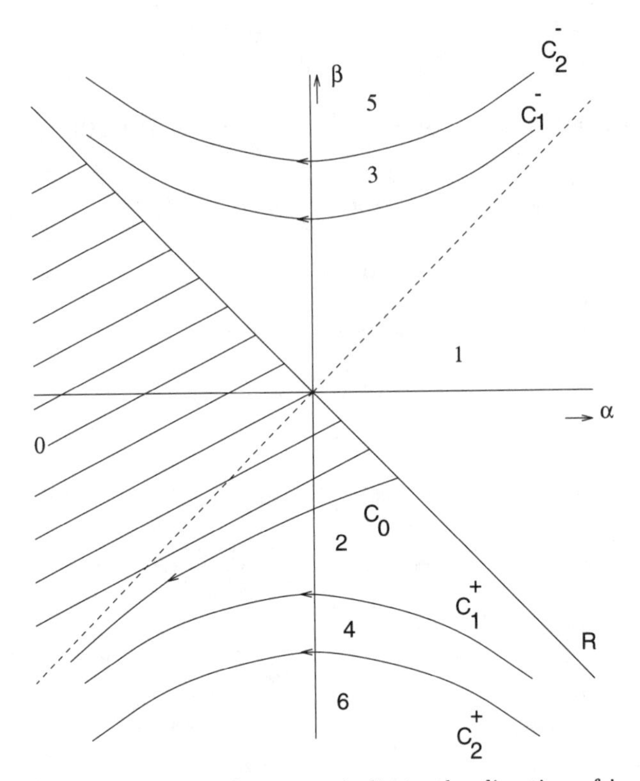

Fig. XI.1. The arrows along the curves indicate the direction of increasing ν. The numbers in the regions refer to the number of roots in the right half-plane. For the shaded region, this number is zero, so this is the stability region. The stability boundary consists of two parts: a piece of the line R and the curve C_0.

For $k = 0$ we shall omit the sign index, since C_0^+ coincides with C_0^-. In Figure XI.1 the curves $C_k^{\pm}$ are drawn for $k = 0, 1, 2$ together with the line

$$R = \{(\alpha, \beta) \mid \alpha = -\beta\} \tag{2.8}$$

at which (2.1) has $z = 0$ as a root.

Along R we are entitled to expect a steady state bifurcation for the nonlinear problem, and along C_0 we are dealing with a Hopf bifurcation. At the intersection of R and C_0 we have a double eigenvalue zero (see Section IX.10 for a derivation of the bifurcation diagram for this codimension two bifurcation).

The next set of elementary exercises concerns analytical proofs of some of the features of the curves $C_k^{\pm}$. Subsequently, we shall explain how the information about the number of roots in the right half-plane is obtained.

Exercise 2.1. Verify that the curves $C_k^{\pm}$ cannot intersect.

Exercise 2.2. Verify that the curves C_k^- lie in the region $\beta > \max\{-\alpha, \alpha\}$ and the curves C_0, C_k^+ in the region $\beta < \min\{-\alpha, \alpha\}$.

Exercise 2.3. Verify that the curves $C_k^{\pm}$ are asymptotic to the lines $\beta = \pm\alpha$ as shown in Figure XI.1.

Exercise 2.4. Verify that the curves are ordered as shown in Figure XI.1 by analysing the ordering of their intersection with the line $\alpha = 0$ and using Exercise 2.1.

Exercise 2.5. Show that along the curves $C_k^{\pm}$ α is decreasing and β has at most one extremum.

At a curve $C_k^{\pm}$, a pair of roots is exactly on the imaginary axis [recall that if (μ, ν) is a root, then so is $(\mu, -\nu)$]. Next, we want to know on which side of the curve these roots are in the right half-plane. In order to determine this we shall consider auxiliary curves along which the imaginary part is fixed and the real part varies (this is just a convenient choice out of an infinity of curves transversal to the $C_k^{\pm}$ under consideration at the point under consideration).

Fixing ν at the value ν_0 in (2.3), we find

$$\alpha = \mu + \frac{\nu_0}{\sin \nu_0} \cos \nu_0 \quad \text{and} \quad \beta = -\frac{\nu_0}{\sin \nu_0} e^{\mu} \tag{2.9}$$

which has tangent vector

$$v_2 = \Big(1, -\frac{\nu_0}{\sin \nu_0}\Big). \tag{2.10}$$

We have to compute the inner product of v_2 and a designated normal vector for $C_k^{\pm}$. The arrows along $C_k^{\pm}$ indicate the direction of increasing ν. They are used to fix what we mean by "to the left" and "to the right" of $C_k^{\pm}$. If $v_1 = (p, q)$ is a tangent vector to $C_k^{\pm}$ and $v_1^{\perp} := (-q, p)$ is the designated normal vector, then a vector $w = (r, s)$ points to the left iff $wv_1^{\perp} = -rq + ps > 0$. In the present situation,

$$v_1 = \Big(\frac{\cos \nu_0 \sin \nu_0 - \nu_0}{\sin^2 \nu_0}, \frac{\nu_0 \cos \nu_0 - \sin \nu_0}{\sin^2 \nu_0}\Big) \tag{2.11}$$

and hence

$$\begin{aligned} v_2 v_1^{\perp} &= \frac{\sin \nu_0 - \nu_0 \cos \nu_0}{\sin^2 \nu_0} - \frac{\nu_0}{\sin \nu_0} \frac{\cos \nu_0 \sin \nu_0 - \nu_0}{\sin^2 \nu_0} \\ &= \frac{\sin^2 \nu_0 - 2\nu_0 \cos \nu_0 \sin \nu_0 + \nu_0^2}{\sin^3 \nu_0}. \end{aligned} \tag{2.12}$$

Exercise 2.6. Let the function h be defined by

$$h(\nu) = \nu^2 - 2\nu \cos\nu \sin\nu + \sin^2\nu.$$

Show that $h(\nu) > 0$ for $\nu > 0$. Can you find an "abstract" (topological) argument to show that necessarily the sign of $v_2 v_1^{\perp}$ does not change along a curve $C_k^{\pm}$? For a "concrete" analytical argument, see Proposition 2.13 below.

We conclude from (2.12) and Exercise 2.6 that the sign of $\sin\nu_0$ determines on which side of the curve $C_k^{\pm}$ the roots are in the right half-plane. In fact, we can draw the following precise conclusion:

> when moving away from C_0 or C_k^+ to the left or from C_k^- to the right, the critical roots move into the right half-plane.

Exercise 2.7. Show that, when moving from the line R into the region $\{(\alpha, \beta) \mid \alpha > -\beta\}$, the real root at zero moves to the right when $\beta > -1$ and to the left when $\beta < -1$.

Observing that for $\beta = 0$ there is just one root, viz. $z = \alpha$, we can now conclude that the number of roots in the right half-plane for the various regions of parameter space is as indicated in Figure XI.1 by using, essentially, Rouché's theorem (see Dieudonné [70, Theorem 9.17.3]). The following lemma is an application of that theorem. (It is a special case, in the sense that we restrict to a two dimensional parameter, of Theorem 9.17.4 of [70], to which we refer for the proof.)

Lemma 2.8. (*Continuity of the roots of an equation as a function of parameters*). *Let Ω be an open set in $\mathbb{C}$, F a continuous complex-valued function on $\mathbb{R} \times \mathbb{R} \times \Omega$ such that, for each (α, β), $z \mapsto F(\alpha, \beta, z)$ is analytic in Ω. Let ω be an open subset of Ω whose closure $\overline{\omega}$ in $\mathbb{C}$ is compact and contained in Ω. Let (α_0, β_0) be such that no zero of $F(\alpha_0, \beta_0, z)$ is on the boundary of ω. Then there exists a neighbourhood U of (α_0, β_0) in $\mathbb{R} \times \mathbb{R}$ such that*

(i) *for any $(\alpha, \beta) \in U$, $F(\alpha, \beta, z)$ has no zeros on the boundary of ω;*
(ii) *the number of zeros of $F(\alpha, \beta, z)$ in ω, taking multiplicities into account, is constant for $(\alpha, \beta) \in U$.*

Since the right half-plane does not have compact closure, we first have to derive a priori bounds for the roots in that half-plane, which show that roots can enter or leave the right half-plane only through the imaginary axis, when α and β are varying in a compact set. Such bounds follow immediately from (2.2). Indeed, for $\mu > 0$ we deduce from (2.2) the inequalities

$$\mu < \alpha + |\beta| \quad \text{and} \quad \nu < |\beta|. \tag{2.13}$$

So now we know that the number of roots in the right half-plane is constant in the regions bounded by the curves $C_k^{\pm}$ and the line R. Consider a point (α_0, β_0) on $C_k^{\pm}$ or R. Let $\epsilon > 0$ be such that for the parameter point (α_0, β_0), there are no roots with $-\epsilon < \mu < 0$ and no roots with $0 < \mu < \epsilon$ [such an ϵ can be found since the roots of an analytic function cannot have a finite accumulation point and, moreover, (2.13) excludes that a sequence of roots exists with ν tending to infinity]. Then the number of roots in the half-plane $\mu > -\epsilon$ is constant in a neighbourhood of (α_0, β_0), and so is the number of roots in the half-plane $\mu > \epsilon$ and in the vertical strip $-\epsilon < \mu < \epsilon$. For this vertical strip, we know when the roots have $\mu > 0$ and when they have $\mu < 0$. Hence we know how the number of roots in the right half-plane changes when crossing $C_k^{\pm}$ or R. We have rigorously justified the numbers given in Figure XI.1.

Exercise 2.9. (About multiple delays.) Equations of the form

$$\dot{x}(t) = g(x(t-\gamma)) + f(x(t-1)) \tag{2.14}$$

with $f(0) = g(0) = 0$, lead, after linearization, to the characteristic equation

$$z - \alpha e^{-\gamma z} - \beta e^{-z} = 0. \tag{2.15}$$

This characteristic equation is relatively easy for integer values of γ.

(i) Analyse the case $\gamma = 2$. In particular, draw the analogue of Figure XI.1 for this case (see Chow and Mallet-Paret [41]);
(ii) Do the same for $\gamma = 4$ (see Nussbaum and Potter [229]).

Exercise 2.10. Determine the region of stability for

$$z^2 = \alpha z e^{-z} + \beta. \tag{2.16}$$

Exercise 2.11. (The delay as a bifurcation parameter.) Consider the equation

$$\dot{x}(t) = \gamma x(t) + h(x(t-\delta)) \tag{2.17}$$

with $h(0) = 0$. Use Exercise 1.1 to conclude that varying δ amounts to moving along the half-line in the (α, β)-plane through the origin and the point $(\gamma, h'(0))$ and conclude that, consequently, all stability and bifurcation information can be read off directly from Figure XI.1.

To conclude this section, we formulate some local results for the case where one cannot find explicit expressions for (α, β) but has to use the implicit function theorem instead. Given a characteristic equation

$$F(\alpha, \beta, z) = 0, \tag{2.18}$$

we define

$$\begin{aligned} G_1(\alpha, \beta, \mu, \nu) &= \operatorname{Re} F(\alpha, \beta, \mu + i\nu), \\ G_2(\alpha, \beta, \mu, \nu) &= \operatorname{Im} F(\alpha, \beta, \mu + i\nu) \end{aligned} \tag{2.19}$$

and note that the Cauchy-Riemann equations take the form

$$\frac{\partial G_1}{\partial \mu} = \frac{\partial G_2}{\partial \nu}, \qquad \frac{\partial G_1}{\partial \nu} = -\frac{\partial G_2}{\partial \mu}. \tag{2.20}$$

We assume that not all these partial derivatives vanish. Suppose we have found, in one way or another, a point $(\alpha_0, \beta_0, 0, \nu_0)$ at which G vanishes [in other words, a parameter combination (α_0, β_0) for which the characteristic equation has a root at position ν_0 on the imaginary axis]. Let M denote the matrix of partial derivatives with respect to (α, β) evaluated at that point, i.e.,

$$M = \begin{pmatrix} \frac{\partial G_1}{\partial \alpha} & \frac{\partial G_1}{\partial \beta} \\ \frac{\partial G_2}{\partial \alpha} & \frac{\partial G_2}{\partial \beta} \end{pmatrix}_{(\alpha_0,\beta_0,0,\nu_0)}. \tag{2.21}$$

Assume that M is nonsingular. Then the equation

$$G(\alpha, \beta, 0, \nu) = 0 \tag{2.22}$$

has a locally unique solution curve $\big(\alpha(\nu), \beta(\nu)\big)$ with tangent vector

$$v_1 = -M^{-1} w_1, \tag{2.23}$$

where, by definition,

$$w_1 = \begin{pmatrix} \frac{\partial G_1}{\partial \nu} \\ \frac{\partial G_2}{\partial \nu} \end{pmatrix}_{(\alpha_0,\beta_0,0,\nu_0)}. \tag{2.24}$$

Let N denote the matrix

$$\begin{pmatrix} 0 & -1 \\ 1 & 0 \end{pmatrix}; \tag{2.25}$$

then $N v_1$ is a left-pointing normal vector to the curve $\big(\alpha(\nu), \beta(\nu)\big)$.

Similarly, we can define a locally unique solution curve for the equation

$$G(\alpha, \beta, \mu, \nu_0) = 0 \tag{2.26}$$

with tangent vector

$$v_2 = -M^{-1} w_2 \tag{2.27}$$

where, by definition,

$$w_2 = \begin{pmatrix} \frac{\partial G_1}{\partial \mu} \\ \frac{\partial G_2}{\partial \mu} \end{pmatrix}_{(\alpha_0,\beta_0,0,\nu_0)}. \tag{2.28}$$

Note that the Cauchy-Riemann equations imply that $w_2 = -N w_1$.

So the critical roots are in the right half-plane in the parameter region to the left of the curve $(\alpha(\nu), \beta(\nu))$, when we follow this curve in the direction of increasing ν, whenever

$$-NM^{-1}w_1 \cdot M^{-1}Nw_1 > 0 \tag{2.29}$$

and to the right whenever this quantity is negative. So the next exercise yields a proof of the subsequent proposition, which summarizes our conclusion.

Exercise 2.12. Verify by an explicit computation that the right hand side of (2.29) is equal to $-|w_1|^2(\det M)^{-1}$.

Proposition 2.13. *The critical roots are in the right half-plane in the parameter region to the left of the curve* $(\alpha(\nu), \beta(\nu))$*, when we follow this curve in the direction of increasing* ν*, whenever* $\det M < 0$ *and to the right when* $\det M > 0$*. Here* M *is the matrix defined in* (2.21).

Exercise 2.14. In order to give an alternative, but essentially equivalent proof of Proposition 2.13, consider

$$G(\alpha(s), \beta(s), \mu, \nu) = 0$$

where

$$(\alpha(s), \beta(s)) = (\alpha_0, \beta_0) + sNv_1$$

[i.e., $s \mapsto (\alpha(s), \beta(s))$ follows the left-pointing normal vector at (α_0, β_0) for $s > 0$]. Show that the Cauchy-Riemann equations guarantee that there is a unique solution $(\mu(s), \nu(s))$ with $(\mu(0), \nu(0)) = (0, \nu_0)$. Compute $\frac{\partial \mu}{\partial s}(0)$ and show that it is negative when $\det M > 0$ and positive when $\det M < 0$.

Exercise 2.15. Bounds like (2.13) can be derived for a more general class of characteristic equations. For instance, consider the equation

$$z = \alpha \int_0^1 e^{-z\theta}\, d\eta_0(\theta) + \beta e^{-z}, \tag{2.30}$$

where $\eta_0 \in \mathrm{NBV}$ has norm equal to one and η_0 is continuous at $\theta = 1$. Show that for any root with $z = \mu + i\nu$ with $\mu > 0$, the inequalities

$$\mu < |\alpha| + |\beta|, \qquad \nu < |\alpha| + |\beta| \tag{2.31}$$

must hold.

XI.3 Strips

In this section we are more ambitious, in the sense that we want to derive as much information as we can about the location of the roots in the complex plane. Again we concentrate on the prototype equation

$$z - \alpha - \beta e^{-z} = 0. \tag{3.1}$$

We extend the intervals $I_k^{\pm}$ [recall (2.5)-(2.6) and recall in particular that the sign indicates the sign of the sine function] into horizontal strips $\Sigma_k^{\pm}$ in the complex plane:

$$\begin{aligned} \Sigma_k^+ &= \{\mu + i\nu \mid \nu \in I_k^+ = (2k\pi, (2k+1)\pi)\}, \\ \Sigma_k &= \{\mu + i\nu \mid \nu \in I_k = ((2k-1)\pi, (2k+1)\pi)\}, \\ \Sigma_k^- &= \{\mu + i\nu \mid \nu \in I_k^- = ((2k-1)\pi, 2k\pi)\}. \end{aligned} \tag{3.2}$$

The reason for considering these strips is that, as one can conclude right away from the imaginary part of (3.1) [see (2.2)], a root can never lie on the boundary of a strip, no matter what the values of α and β are (exclude in this statement the boundary $\nu = 0$ of $\Sigma_0^{\pm}$). We are now ready to formulate the first two main results.

Theorem 3.1. *For $\beta > 0$, equation (3.1) has a unique and simple root λ_k in the strip Σ_k for $k = 0, 1, \ldots$ and no other roots. For $k = 1, 2, \ldots$, the root in Σ_k is contained Σ_k^-.*

Theorem 3.2. *For $\beta < 0$, equation (3.1) has a unique and simple root λ_k in the strip Σ_k^+ for $k = 1, 2, \ldots$. There are two roots in Σ_0 (which are real and simple for $-e^{\alpha-1} < \beta < 0$ and complex conjugate for $\beta < -e^{\alpha-1}$). There are no other roots.*

As a first step to proving these theorems we consider the case $\alpha = 0$.

Lemma 3.3. *The equation $z = \beta e^{-z}$ has a simple purely imaginary root*

$$\begin{cases} z = i(\pi/2 + 2m\pi) & \text{for } \beta = -(\pi/2 + 2m\pi), \\ z = 0 & \text{for } \beta = 0, \\ z = i(\pi/2 + (2m+1)\pi) & \text{for } \beta = (\pi/2 + (2m+1)\pi), \end{cases}$$

where $m = 0, 1, 2, \ldots$ and there are no other purely imaginary roots.

Proof. The equations for the real and imaginary part read

$$0 = \beta \cos \nu, \tag{3.3}$$

$$\nu = -\beta \sin \nu. \tag{3.4}$$

From (3.3) we find $\nu = \pi/2 + k\pi$. Subsequently, (3.4) yields that $\beta = \pi/2 + k\pi$ when k is odd and $\beta = -(\pi/2 + k\pi)$ when k is even. This proves the lemma. □

Lemma 3.4. *For $|\beta|$ close to zero, all roots of the equation $z = \beta e^{-z}$ with nonzero imaginary part are in the left half-plane.*

Proof. The equations for the real and imaginary part read

$$\mu = \beta e^{-\mu} \cos \nu, \tag{3.5}$$

$$\nu = -\beta e^{-\mu} \sin \nu. \tag{3.6}$$

So we find

$$(\mu e^{\mu})^2 + (\nu e^{\mu})^2 = \beta^2. \tag{3.7}$$

Assume that $\mu \geq 0$; then (3.7) implies that $\nu^2 \leq \beta^2$. Consequently, $|\frac{\sin \nu}{\nu} \beta|$ is smaller than one for β sufficiently small. This, however, contradicts

$$|\frac{\sin \nu}{\nu} \beta| = e^{\mu} > 1$$

[see (3.6)]. □

Exercise 3.5. Show that the conclusion of Lemma 3.4 actually holds for $0 < |\beta| < \frac{\pi}{2}$.

Lemma 3.6. *If $|\beta|$ is close to zero, the equation $z = \beta e^{-z}$ has no positive real roots for $\beta < 0$ and precisely one if $\beta > 0$.*

Proof. If $\nu = 0$, then (3.5) becomes $\beta = \mu e^{\mu}$ which has a unique positive solution when $\beta > 0$ and no positive solutions when $\beta < 0$. □

Lemma 3.7. *For any given strip $\{\mu + i\nu \mid m_1 < \nu < m_2,\ \mu < 0\}$ in the left half-plane, there exists a β sufficiently large such that the equation $z = \beta e^{-z}$ has no roots in this strip.*

Proof. Since $\mu \mapsto \mu e^{\mu}$ is bounded for negative μ and ν is bounded by assumption, the lemma follows from (3.7). □

Of course, the idea is again to exploit the continuity of the roots as a function of (α, β) (see Lemma 2.8). So now we must find a priori bounds for the roots in a horizontal strip. It is at this point that we have to avoid $\beta = 0$.

Lemma 3.8. *Given a horizontal strip, there are upper and lower bounds for the real parts of the roots (3.1), uniformly for (α, β) in compact subsets of $\{(\alpha, \beta) \mid \beta < 0\} \cup \{(\alpha, \beta) \mid \beta > 0\}$.*

Proof. For the upper bound, see (2.13). Combining the equations for the real and the imaginary part, we find [cf. (2.2)]

$$\beta^2 = (\mu - \alpha)^2 e^{2\mu} + \nu^2 e^{2\mu}. \tag{3.8}$$

The right hand side of (3.8) tends to zero as $\mu \to -\infty$, uniformly for ν in a given bounded set and α in a given bounded set, whereas the left hand side is clearly bounded away from zero, uniformly for such α, ν and β bounded away from zero. This then gives the lower bound for μ. □

Proof of Theorem 3.1. As noted earlier, there cannot be a root on the boundary of Σ_0 and $\Sigma_k^{\pm}$, $k = 1, 2, \ldots$. Together with Lemma 3.8, this information allows us to conclude from Lemma 2.8 that, within Σ_0 and $\Sigma_k^{\pm}$, the number of roots is finite and independent of (α, β) in $\{(\alpha, \beta) \mid \beta > 0\}$. So it suffices to analyse equation (3.1) for a particular choice of α and β. Take $\alpha = 0$.

We first consider $\Sigma_0 = 0$. When $\beta > 0$ and $-\pi < \nu < \pi$, the second equation of (2.2) requires that $\nu = 0$. Subsequently, the first equation of (2.2) implies that $\mu > 0$. From Lemma 3.6, it now follows that (3.1) has a unique root in Σ_0.

Next consider a strip Σ_k^+ with $k \geq 1$. According to Lemma 3.3, there are no roots on the imaginary axis within this strip for positive values of β. So Lemma 2.8 yields that the number of roots in the (half-) strips $\Sigma_k^+ \cap \{\mu + i\nu \mid \mu < 0\}$ and $\Sigma_k^+ \cap \{\mu + i\nu \mid \mu > 0\}$ is independent of β for $\beta \in (0, \infty)$. Lemma 3.7 implies that the first is zero, and Lemma 3.4 that the second is zero.

Finally, consider a strip Σ_k^- with $k \geq 1$. According to Lemma 3.3, there is a unique $\beta^* > 0$ such that there is a root on the imaginary axis. Reasoning as in the case of Σ_k^+, we find that there cannot be any roots in $\Sigma_k^- \cap \{\mu + i\nu \mid \mu < 0\}$ for $\beta > \beta^*$ and that there cannot be any roots in $\Sigma_k^- \cap \{\mu + i\nu \mid \mu > 0\}$ for $0 < \beta < \beta^*$.

There exists $\epsilon > 0$ such that, for $\beta = \beta^*$, any root other than the one on the imaginary axis has a real part bigger than ϵ or smaller than $-\epsilon$. So for $\beta = \beta^*$, the number of roots in Σ_k^- equals

$$1 + \#\text{roots in } \Sigma_k^- \cap \{\mu + i\nu \mid \mu < -\epsilon\} + \#\text{roots in } \Sigma_k^- \cap \{\mu + i\nu \mid \mu > \epsilon\}.$$

The last two terms are necessarily zero, as one can conclude from another application of Lemma 2.8 and our knowledge about the value for $\beta > \beta^*$ and $0 < \beta < \beta^*$, respectively. So the number of roots in Σ_k^- equals one for $\beta = \beta^*$ and hence for all (α, β) with $\beta > 0$. □

Exercise 3.9. Consider the case $\alpha = 0$, $\beta < 0$ and the strip Σ_0. Prove that

(i) for $-\frac{1}{e} < \beta < 0$, there are precisely two simple real roots $\lambda_0 < \lambda_{00} < 0$;

(ii) for $\beta \uparrow 0$, $\lambda_0 \to -\infty$ and $\lambda_{00} \to 0$;

(iii) for $\beta \downarrow -\frac{1}{e}$, both λ_0 and λ_{00} tend to -1;

(iv) for $\beta < -\frac{1}{e}$, there is a pair of conjugate roots $\mu_0 \pm i\nu_0$;

(v) for $-\frac{\pi}{2} < \beta < -\frac{1}{e}$, $\mu_0 < 0$ and for $\beta < -\frac{\pi}{2}$, $\mu_0 > 0$.

Exercise 3.10. Prove Theorem 3.2.

Exercise 3.11. Consider the case $\alpha = 0$. Prove that for $\beta \neq 0$ and any root $z = \mu + i\nu$ of (3.1), we have

$$\frac{\partial \mu}{\partial \beta} = \frac{1}{\beta}\{(1+\mu)\mu + \nu^2\}, \tag{3.9}$$

$$\frac{\partial \nu}{\partial \beta} = \frac{\nu}{\beta}. \tag{3.10}$$

Conclude that for the roots in the strips Σ_k, $k = 1, 2, 3, \ldots$, $\operatorname{sign} \frac{\partial \mu}{\partial \beta} = \operatorname{sign} \beta$.

We are now ready to state and prove that the ordering of the imaginary parts of the roots induces an ordering of the real parts of the roots.

Theorem 3.12. *Using the notation of Theorems* 3.1 *and* 3.2, *we have the inequalities*

$$(\lambda_0 \text{ or } \mu_0) > \mu_1 > \mu_2 > \cdots \tag{3.11}$$

or, in other words, the real parts of the roots of (3.1) *are ordered.*

Proof. The set of parameters such that $\mu_k > \mu_{k+1}$ is open, since the roots depend continuously on the parameters (cf. Lemma 2.8). For a point in the closure of this set, we clearly have $\mu_k = \mu_{k+1}$. But then necessarily

$$\cos \nu_k = \cos \nu_{k+1},$$
$$\nu_{k+1} \sin \nu_k = \nu_k \sin \nu_{k+1},$$

which, in turn, implies that

$$\left(\frac{\nu_k}{\nu_{k+1}}\right)^2 = \frac{\sin^2 \nu_k}{\sin^2 \nu_{k+1}} = \frac{1-\cos^2 \nu_k}{1-\cos^2 \nu_{k+1}} = 1.$$

However, the definition of the strips excludes the possibility that $\nu_k = \nu_{k+1}$ (and we consider positive ν, so $\nu_k = -\nu_{k+1}$ is excluded as well). So the set is both open and closed. It remains to exclude the possibility that it is empty. Take $\alpha = 0$ and $\beta = -(\frac{\pi}{2} + 2k\pi)$; then $\mu_k = 0$. We claim that $\mu_{k+1} < 0$ for this value of β. Indeed, $\mu_{k+1} = 0$ for $\beta = -(\frac{\pi}{2} + 2(k+1)\pi)$ and $\frac{\partial \mu_{k+1}}{\partial \beta} < 0$ for $\beta < 0$ (cf. Exercise 3.11), so our claim must be correct. □

Exercise 3.13. Consider the equation

$$\dot{x}(t) = -x(t-\tau). \tag{3.12}$$

Describe, e.g., in a picture, what happens with the roots of the characteristic equation when τ increases from zero to infinity. Explain the meaning of the statement: delayed negative feedback is a destabilizing mechanism.

Exercise 3.14. Show that one can remove the parameter α from the equation (3.1) by the transformation $y = z - \alpha$ and that, consequently, it suffices to study the case $\alpha = 0$.

Although the case $\alpha = 0$ is representative for the general case as far as the number of roots in a given strip is concerned, one has to be careful in "translating" results concerning the location of the roots with respect to the imaginary axis. Here Figure XI.1 can be extremely helpful.

Exercise 3.15. Consider the half-line $\{(\alpha, \beta) = (s\alpha_0, s\beta_0), s \geq 0\}$ in the parameter space with $|\frac{\beta_0}{\alpha_0}| < 1$. Show that no root of (3.1) can cross the imaginary axis as we move along this half-line. Show that all roots are in the left half-plane when $\alpha_0 < 0$, whereas exactly one root is in the right half-plane when $\alpha_0 > 0$.

Exercise 3.16. Consider the characteristic equation

$$z = s - se^{-z}, \qquad s \geq 0. \tag{3.13}$$

Describe the motion of the roots in the complex plane as $s \to \infty$.

Exercise 3.17. Show that for (α, β) such that $\alpha + \beta > 0$ and $\beta < 0$, we have $\lambda_0 < 0 < \lambda_{00}$, whereas for (α, β) such that $\alpha + \beta < 0$, both roots in Σ_0 lie on the same side of the imaginary axis (or precisely on the imaginary axis).

Exercise 3.18. (A continuation of Exercise 2.9 about multiple delays.) When analysing $z - \alpha e^{-\gamma z} - \beta e^{-z} = 0$, with $\gamma > 1$ and integer, the case $\alpha = 0$ is no longer representative (the continuation argument breaks down since one has no analogue of Lemma 3.8) and one should study the case $\beta = 0$ instead. Prove the following results:

(i) When γ is even, there are $\frac{\gamma}{2}$ roots in any strip $\Sigma_k^{\pm}$, $k = 1, 2, \ldots$. If $\alpha < 0$, there are γ roots in Σ_0, whereas there are $\gamma + 1$ roots in Σ_0 if $\alpha > 0$.

(ii) When γ is odd and $\alpha < 0$, there are $\gamma - 1$ roots in Σ_0, $\frac{1}{2}(\gamma - 1)$ roots in Σ_k^- and $\frac{1}{2}(\gamma + 1)$ roots in Σ_k^+, $k = 1, 2, \ldots$. When γ is odd and $\alpha > 0$, there are γ roots in Σ_0, $\frac{1}{2}(\gamma + 1)$ roots in Σ_k^- and $\frac{1}{2}(\gamma - 1)$ roots in Σ_k^+, $k = 1, 2, \ldots$.

Exercise 3.19. (A continuation of Exercise 2.12.) Consider the characteristic equation

$$z - \alpha \int_0^1 e^{-z\theta} \, d\eta_0(\theta) - \beta e^{-z} = 0. \tag{3.14}$$

Show that there are upper and lower bounds for the real parts of the roots in a given horizontal strip, uniformly for (α, β) in a compact subset of either of the two half-planes $\beta > 0$ and $\beta < 0$.
Hint: Derive, first, the identity

$$\beta^2 = \big(\mu e^{\mu} - e^{\mu}\alpha \int_0^1 e^{-\mu\theta} \cos\nu\theta \, d\eta_0(\theta)\big)^2$$
$$+ \big(\nu e^{\mu} + e^{\mu}\alpha \int_0^1 e^{-\mu\theta} \sin\nu\theta \, d\eta_0(\theta)\big)^2 \tag{3.15}$$

and then repeat the argument used below equation (3.8). In this second step, use that since η_0 is continuous at 1, the variation of η_0 in a neighbourhood of 1 is small, and so for every $\epsilon > 0$, there is a $\delta > 0$ such that, uniformly for $\mu < 0$,

$$\Big|\int_{1-\delta}^1 e^{-\mu\theta} \cos\nu\theta \, d\eta_0(\theta)\Big| < \epsilon e^{-\mu}.$$

Corollary 3.20. *Equation* (3.13) *has at most finitely many roots in a given horizontal strip.*

XI.4 Case studies

In each of the following three subsections we discuss a simple model from population dynamics, to a large extent by means of a series of exercises. After introducing the equations, we find the steady states (constant solutions) and linearize around them to derive a characteristic equation. For a convenient choice of parameters, we try to determine the stability region in parameter space. As a next step, we use, in Subsections 4.1 and 4.3, the formulas for the direction of Hopf bifurcation from Chapter X to find out whether we are dealing with a supercritical or a subcritical bifurcation.

XI.4.1 A competition equation

Consider the equation

$$x(t) = \frac{\gamma}{2\epsilon} \int_{1-\epsilon}^{1+\epsilon} x(t-\tau) e^{-x(t-\tau)} \, d\tau, \tag{4.1}$$

where $x(t)$ is the population birth rate at time t. Here, $\gamma > 0$ can be thought of as the expected number of offspring produced by a newborn individual during its entire life span in the absence of density effects. The reproductivity is concentrated in a "window" of width 2ϵ, $0 < \epsilon < 1$, centered at 1 (so time is scaled such that the midpoint of the reproductivity period, which can be thought of as a generation time, is reached exactly one time unit after birth). Finally, density dependence is incorporated in the exponential correction factor. Such a factor can be derived from a model for nursery competition by a time scale argument (see Metz and Diekmann [193] Example 6.2.4, p. 121 and Section VI.3.2, p. 222).

The steady states are

$$\overline{x}_1 = 0 \quad \text{and} \quad \overline{x}_2 = \log\gamma \tag{4.2}$$

(note that $\overline{x}_2$ is only then biologically meaningful when $\gamma > 1$). If we linearize the equation about such a steady state and if we, subsequently, substitute an elementary solution $t \mapsto e^{zt}$, we obtain the characteristic equation for the linearized equation.

Define $K : \mathbb{R} \to \mathbb{R}$,

$$K(\tau) = \begin{cases} 1/2\epsilon & \text{for } 1-\epsilon \le \tau \le 1+\epsilon, \\ 0 & \text{elsewhere.} \end{cases} \tag{4.3}$$

The characteristic equations corresponding to $\overline{x}_1$ and $\overline{x}_2$ are, respectively,

$$\gamma\overline{K}(z) = 1, \tag{4.4}$$

$$(1 - \log\gamma)\overline{K}(z) = 1, \tag{4.5}$$

where $\overline{K}(z)$ denotes the Laplace transform of K.

Exercise 4.1. Let K be a measurable nonnegative function with compact support, defined on $\mathbb{R}_+$ and such that $\int_0^\infty K(\tau)\,d\tau = 1$. Consider the equation

$$\alpha\overline{K}(z) = 1. \tag{4.6}$$

Prove that

(i) for $\alpha > 1$, there is a dominant positive real root;

(ii) for $0 < \alpha < 1$, there is a dominant negative real root;

(iii) for $\alpha \to 0$, all roots will tend to infinity in the left half-plane (more precisely, their real part tends to minus infinity);

(iv) for $-1 \le \alpha < 0$, all roots have nonzero imaginary part and negative real part;

(v) for $\alpha < -1$, all roots have nonzero imaginary part (and for some, the real part may be positive).

Exercise 4.2. Show that $\overline{x}_1$ is stable for $0 < \gamma < 1$ and unstable for $\gamma > 1$ and that $\overline{x}_2$ is stable for $1 < \gamma < e^2$.

Exercise 4.3. Let K be given by (4.3). Show that

$$\overline{K}(i\nu) = \frac{\sin\epsilon\nu}{\epsilon\nu}(\cos\nu - i\sin\nu) \tag{4.7}$$

and conclude that the roots of the equation

$$\operatorname{Im}\overline{K}(i\nu) = 0$$

are precisely the points $\nu = k\pi$, $k \in \mathbb{Z}$.

Exercise 4.4. Show that the characteristic equation (4.5) has a root $z = ik\pi$, $k \in \mathbb{Z}$, provided

$$\log \gamma = 1 + (-1)^{k+1} \frac{\epsilon k\pi}{\sin \epsilon k\pi} \tag{4.8}$$

and that this is a complete list of the roots on the imaginary axis.

Exercise 4.5. For the steady state $\overline{x}_2$, the stability boundary in the $(\log \gamma, \epsilon)$-plane is given by (4.8), with $k = 1$, and the line $\gamma = 1$. Verify this.
Hint: Show that $|\frac{k\xi}{\sin k\xi}| > |\frac{\xi}{\sin \xi}|$ for $\xi \neq 0$ and $k = 2, 3, 4, \ldots$.

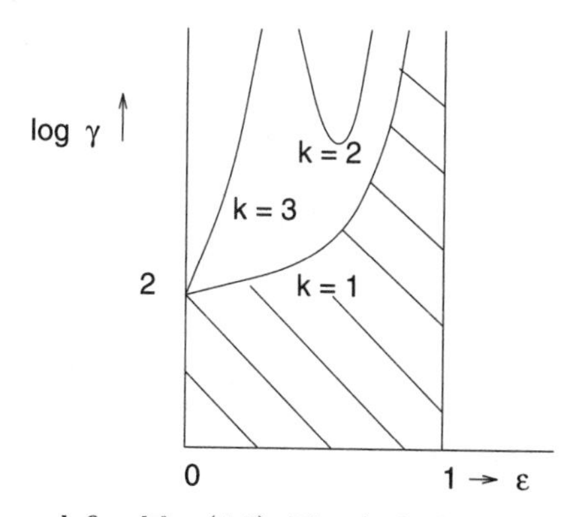

Fig. XI.2. The curves defined by (4.8). The dashed area is the stability domain of the steady state $\overline{x}_2$. Note that one can pass the upper stability boundary either by shortening the reproductive period (decreasing ϵ) or by increasing the fertility (increasing γ) or by a combination of these two effects. At this upper boundary, a Hopf bifurcation occurs. The period of the periodic solution is ± 2 (i.e., approximately twice the generation time).

Exercise 4.6. Next, we want to determine the direction of bifurcation by specifying the ingredients for formula (X.3.15) of Exercise X.3.12. So define $p : \mathbb{R} \to \mathbb{R}$ by

$$p(x) = \gamma x e^{-x}. \tag{4.9}$$

Show that

(i) $Dp(\overline{x}_2) = 1 - \log \gamma$;

(ii) $D^2(\overline{x}_2) = \log \gamma - 2$;

(iii) $D^3(\overline{x}_2) = 3 - \log \gamma$.

Also show that

(iv) $\overline{K}(2\pi i) = \frac{\sin 2\epsilon\pi}{2\epsilon\pi}$;

(v) $\operatorname{Re} D\overline{K}(\pi i) = \frac{\sin \epsilon\pi}{\epsilon\pi} > 0$ for $0 \leq \epsilon < 1$.

Substitute these results into (X.3.15) and conclude that

$$\begin{aligned}\operatorname{sign}\operatorname{Re} c = \operatorname{sign}\Big\{ &\frac{\epsilon\pi}{2\sin\epsilon\pi} - 1 - \frac{(\epsilon\pi(\sin\epsilon\pi)^{-1}-1)^2}{\epsilon\pi(\sin\epsilon\pi)^{-1}+1} \\ &- \frac{(\epsilon\pi(\sin\epsilon\pi)^{-1}-1)^2\sin 2\epsilon\pi}{4\epsilon\pi(1+\cos\epsilon\pi)}\Big\}.\end{aligned} \tag{4.10}$$

For $\epsilon = 0$ and $\epsilon = 1/2$, one finds $\operatorname{Re} c < 0$, i.e., a supercritical bifurcation to a stable periodic solution, whereas for $\epsilon \uparrow 1$, $\operatorname{Re} c > 0$, i.e., a subcritical bifurcation to an unstable periodic solution. Numerical evaluation of the right hand side of (4.10) yields the conclusion that $\operatorname{Re} c < 0$ for $\epsilon < \epsilon_d$ and $\operatorname{Re} c > 0$ for $\epsilon > \epsilon_d$, where $\epsilon_d = 0.66081$. So the direction of bifurcation changes exactly once along the upper stability boundary.

In conclusion of this subsection we refer to Swick [269] for a description of numerical experiments which indicate that the periodic solution may undergo a sequence of period doublings culminating in chaotic behaviour when γ is further increased (for some fixed small value of ϵ). Such in close analogy with the situation for the difference equation $x(t) = \gamma x(t-1)e^{-x(t-1)}$. See Mallet-Paret and Nussbaum [180] for a general methodology to study such phenomena.

XI.4.2 A simple deterministic epidemic model

Next, we discuss a simple model for the spread, in a closed population, of an infectious disease which confers only temporary immunity. Consider a population divided into two classes S and I. The class S consists of those individuals who are susceptible to a certain infectious disease and the class I of those who experience the consequences of an infection. We distinguish the members of I according to the time elapsed since they were infected. In particular, let $i(t,\tau)$ denote the density, at time t, of those members of I which have class-age τ. We assume that:

(i) The population is demographically closed and all changes are due to the infection mechanism. In other words,

$$S(t) + I(t) = N, \tag{4.11}$$

where N denotes the population size.

(ii) The interaction of infectives and susceptibles is of "mass-action" type, with a weighted average over the age-structured class of infectives. More precisely, there exists a nonnegative function $A(\tau)$, describing the infective "force" of an individual who was infected τ units of time ago, such that

$$i(t,0) = S(t)\int_0^\infty A(\tau)i(t,\tau)\,d\tau. \tag{4.12}$$

(iii) The infective "force" reduces to zero after a finite time: there exists a least positive number $\tau_1 < \infty$ such that the support of A is contained in $[0, \tau_1]$.
(iv) The disease confers only temporary immunity: there exists a number $\tau_2 \geq \tau_1$ such that every infected individual becomes susceptible again exactly τ_2 units of time after its contagion.

Because of (iv) we can rewrite (4.11) as

$$S(t) + \int_0^{\tau_2} i(t, \tau)\, d\tau = N. \tag{4.13}$$

Noting that $i(t, \tau) = i(t - \tau, 0)$ and eliminating $S(t)$ from (4.12) and (4.13), we obtain

$$i(t, 0) = \left(N - \int_0^{\tau_2} i(t - \tau, 0)\, d\tau\right) \int_0^{\tau_1} A(\tau) i(t - \tau, 0)\, d\tau, \tag{4.14}$$

which upon the transformation of variables

$$\begin{aligned} x(t) &= \frac{\tau_2}{N} i(\tau_2 t, 0), \\ b(t) &= \tau_2 A(\tau_2 t)\Big(\int_0^{\tau_2} A(\tau)\, d\tau\Big)^{-1}, \\ \gamma &= N \int_0^{\tau_2} A(\tau)\, d\tau \end{aligned}$$

leads to a nonlinear Volterra integral equation

$$x(t) = \gamma\left(1 - \int_0^1 x(t - \tau)\, d\tau\right) \int_0^1 b(\tau) x(t - \tau)\, d\tau. \tag{4.15}$$

The steady states are given by

$$\overline{x}_1 = 0 \quad \text{and} \quad \overline{x}_2 = 1 - \gamma^{-1} \tag{4.16}$$

(note that $\overline{x}_2$ is only then biologically meaningful when $\gamma > 1$). The characteristic equations corresponding to $\overline{x}_1$ and $\overline{x}_2$ are, respectively,

$$\gamma \bar{b}(z) = 1, \tag{4.17}$$

$$\bar{b}(z) + (1 - \gamma) \frac{1 - e^{-z}}{z} = 1. \tag{4.18}$$

Exercise 4.7. Show that $\overline{x}_1$ is stable for $0 < \gamma < 1$ and unstable for $\gamma > 1$. Show that $\overline{x}_2$ is stable for γ close to but larger than 1.

As formulated, the problem contains only one parameter, viz. γ. Of course, we can introduce further parameters in the function b, but since the integral of b is normalized to be one, they will necessarily show up in a complicated nonlinear way. So we shall stick to one parameter.

A special feature of the present problem is that the limiting situation for $\gamma \to \infty$ is rather easy. Indeed, for $\gamma \to \infty$, (4.18) reduces to

$$1 - e^{-z} = 0, \tag{4.19}$$

which has roots $\pm 2k\pi i$ exactly on the imaginary axis.

Exercise 4.8. Use the implicit function theorem to prove that equation (4.18) has roots which converge to $\pm 2k\pi i$ for $\gamma \to \infty$ and that they do so from the right half-plane when $b_k > 0$ and from the left half-plane when $b_k < 0$, where, by definition,

$$b_k = 2 \int_0^1 b(\tau) \sin(2\pi\tau)\, d\tau. \tag{4.20}$$

From the last exercise, we can conclude that $\overline{x}_2$ is unstable for large values of γ when at least some $b_k > 0$, and stable when all $b_k < 0$. Actually, one can prove a lot more. When looking for roots on the imaginary axis, say at position ν, we obtain two equations in two unknowns ν and γ. Since the equations are linear in γ, we can eliminate γ to obtain one equation in the unknown ν. By a miracle, this nonlinear equation can be analysed completely in this special case (see the exercises below or Diekmann and Montijn [69]). This analysis then yields the following result.

Theorem 4.9. *As γ increases from 1 to ∞, exactly as many pairs of complex conjugated roots of (4.18) cross the imaginary axis as there are $k \in \mathbb{N}$ for which $b_k > 0$. These are simple and go from left to right with a positive speed. The one on the positive imaginary axis passes through $I_k^- = \big((2k-1)\pi, 2k\pi\big)$.*

Exercise 4.10. Define $f : \mathbb{R} \to \mathbb{R}$ by

$$\begin{aligned} f(\nu) = -1 &+ \frac{\int_0^1 b(\tau)\cos(\nu\tau)\,d\tau \int_0^1 \sin(\nu\tau)\,d\tau}{\int_0^1 \sin(\nu\tau)\,d\tau} \\ &- \frac{\int_0^1 b(\tau)\sin(\nu\tau)\,d\tau \int_0^1 \cos(\nu\tau)\,d\tau}{\int_0^1 \sin(\nu\tau)\,d\tau}. \end{aligned} \tag{4.21}$$

Show that equation (4.18) has a root $z = i\nu$ on the imaginary axis iff $f(\nu) = 0$ and

$$\gamma = 1 + \frac{\int_0^1 b(\tau)\sin(\nu\tau)\,d\tau}{\int_0^1 \sin(\nu\tau)\,d\tau}. \tag{4.22}$$

Exercise 4.11. Let $k \in \mathbb{N}$. Prove the following statements:

(i) If $b_k = 0$, then f has no zero in $I_k = \big((2k-1)\pi, (2k+1)\pi\big)$.

(ii) If $b_k \neq 0$, then f has precisely one zero in I_k, say ν_k, and ν_k is simple.

(iii) If $b_k > 0$, then $\nu_k \in I_k^-$ and $f'(\nu_k) > 0$.

(iv) If $b_k < 0$, then $\nu_k \in I_k^+$ and $f'(\nu_k) < 0$.

Hint: Rewrite, after some manipulation with trigonometric identities, the equation $f(\nu) = 0$ as the fixed-point problem

$$\nu = 2k\pi + (-1)^{k+1} 2 \arcsin\left(\int_0^1 b(\tau) \sin(\nu(\tau - 1/2))\, d\tau\right);$$

next, use the Cauchy-Schwarz inequality with respect to the measure $b(\tau)\, d\tau$ to deduce that we are dealing with a contraction mapping on I_k.

Exercise 4.12. Show that only roots in I_k^- yield a value of γ greater than one.

Exercise 4.13. Use the implicit function theorem together with the Cauchy-Riemann equations to deduce that

$$\operatorname{sign} \frac{\partial \mu}{\partial \gamma}(\gamma^*) = \operatorname{sign} f'(\nu), \tag{4.23}$$

where μ is the real part of the root which is, for $\gamma = \gamma^*$, at position ν on the imaginary axis.

Exercise 4.14. Show that $b_1 > 0$ if the support of b is contained in $[0, 1/2]$. Next, derive the following conclusion: if the period of immunity is long compared to the period of infectivity, then enlargement of the population density leads to a destabilization of the steady endemic state and one can expect to see oscillations. Hint: Recall the transformation below formula (4.14).

Only the first bifurcating periodic solution can possibly be stable. Numerically, one finds that the first usually corresponds to the root with the smallest ν (i.e., the smallest k). Here, "usually" refers to choices for the kernel b. Note that the smallest frequency corresponds to the largest period. However, there are exceptional cases in which a smaller period comes first and the observed trend is not a valid general principle!

XI.4.3 A simple predator-prey-patch model

Many natural populations have a geographical distribution which is far from uniform. In fact, a population consists sometimes of an ensemble of local subpopulations, or colonies, which are connected by occasional migrations. In this context, we also speak of patches. New patches are "created" by individuals dispersing from existing patches. Likewise, when we consider prey-predator interaction, a prey patch may be invaded by a predator. In many situations, notably in spider mite–predatory mite interaction, the invasion by a predator leads, after a while, to the extermination of the prey and subsequently, by lack of food, to the extinction of the local predator population. Despite the local extinction, one can have overall population stability. The following caricatural model is concerned with just such a situation. In it, patches are considered as a kind of superindividuals. In particular, we shall analyse the stabilizing influence of prey dispersal from predator invaded patches. For background information, see Diekmann, Metz and Sabelis [68], Hastings [115], Jansen and Sabelis [139] and Sabelis and Diekmann [247].

Consider the following system of equations

$$\begin{aligned} \dot{x}(t) &= a\big(x(t) + ey(t)\big) - bx(t)y(t), \\ y(t) &= b\int_{t-1}^{t} x(\tau)y(\tau)\,d\tau. \end{aligned} \tag{4.24}$$

Here, x denotes the number of prey patches and a the rate at which dispersing prey from an arbitrary prey patch found new colonies in patches which were still empty. The rate at which predators invade an arbitrary prey patch is given by by, where y denotes the number of predator patches (a shorthand for prey-predator patches) and b a reaction coefficient. A predator patch exists exactly one unit of time (so we have already scaled the time variable such that the time between predator invasion and local extinction is the new unit). Finally, the rate at which prey dispersing from an arbitrary predator patch found new colonies in empty patches is given by ae. Hence, e is the contribution of a predator patch to the creation of new prey patches relative to the contribution of a prey patch. The integral equation for y states that at time t, the number of predator patches equals the sum of those patches which were invaded by a predator during the preceding time interval of one unit.

The steady states of (4.24) are given by

$$\overline{x}_1 = \overline{y}_1 = 0 \quad \text{and} \quad \overline{x}_2 = b^{-1}, \quad \overline{y}_2 = \frac{a}{b(1-ea)}. \tag{4.25}$$

Note that the steady state with index 1 is unstable for all $a > 0$ (prey patches will increase exponentially in number in the absence of predators!) and that $\overline{y}_2$ is biologically meaningful only when $ea < 1$, which we assume from now on to hold true (it is amusing to think about the case $ea > 1$ and it helps to think about it in terms of the biological interpretation, but this is completely outside the scope of this chapter).

Exercise 4.15. Show that the characteristic equation corresponding to the nontrivial steady state $(\overline{x}_2, \overline{y}_2)$ is given by

$$z - 1 + e^{-z} + \alpha + \beta\frac{1-e^{-z}}{z} = 0, \tag{4.26}$$

where

$$\alpha = \frac{ea^2}{1 - ea}, \qquad \beta = \frac{a(1 - 2ea)}{1 - ea}. \tag{4.27}$$

Exercise 4.16. Explain the fact that the parameter b does not occur in (4.26)–(4.27) by showing that it can be eliminated from (4.24) by a scaling of both x and y with a factor b.

Exercise 4.17. Show that the transformation $(e, a) \to (\alpha, \beta)$ given by (4.27) is invertible with inverse

$$a = \alpha + \beta, \qquad e = \frac{\alpha}{2\alpha^2 + 3\alpha\beta + \beta^2}. \tag{4.28}$$

Exercise 4.18. Verify that (4.26) has a root $z = 0$ if and only if $\alpha + \beta = 0$.

Exercise 4.19. Put $z = \mu + i\nu$ in (4.26) and solve for α and β to obtain

$$\begin{aligned} \alpha &= 1 - \mu - e^{-\mu}\cos\nu \\ &\quad + \frac{(\nu - e^{-\mu}\sin\nu)\big(\mu(1 - e^{-\mu}\cos\nu) + \nu e^{-\mu}\sin\nu\big)}{\mu e^{-\mu}\sin\nu + \nu(e^{-\mu}\cos\nu - 1)} \end{aligned} \tag{4.29}$$

and

$$\beta = (\mu^2 + \nu^2)\frac{e^{-\mu}\sin\nu - \nu}{\mu e^{-\mu}\sin\nu + \nu(e^{-\mu}\cos\nu - 1)}. \tag{4.30}$$

When $\mu = 0$, i.e., when we concentrate on roots which lie exactly on the imaginary axis, (4.29) and (4.30) simplify to

$$\alpha = 2 + \frac{\nu\sin\nu}{\cos\nu - 1} \tag{4.31}$$

and

$$\beta = \nu\frac{\sin\nu - \nu}{\cos\nu - 1}. \tag{4.32}$$

Since these functions have singularities for $\nu = \pm 2k\pi$, we now introduce intervals

$$J_k = (2k\pi, 2(k+1)\pi) \tag{4.33}$$

and denote by C_k the curve in the (α, β)-plane parameterized by ν as in (4.31)–(4.32) for $\nu \in J_k$, with the convention that C_0 also contains the limit point $(0, 0)$ for $\nu \to 0$:

$$\big(\alpha(0), \beta(0)\big) = (0, 0). \tag{4.34}$$

Note once again that the symmetry allows us to restrict our attention to $k \geq 0$, i.e., $\nu \geq 0$.

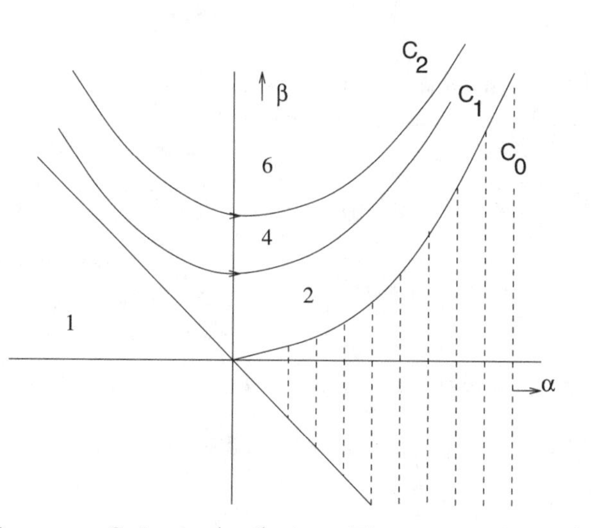

Fig. XI.3. The curves C_k in the (α, β)-plane. The arrows refer to the direction of increasing ν. The numbers refer to the roots of (4.26) in the right half-plane. The stability domain is the dashed region.

Exercise 4.20.

(i) Prove that C_k and C_l, with $k \neq l$, cannot intersect.

(ii) Prove that C_k cannot intersect the line $\alpha + \beta = 0$ for $k \geq 1$ and for $k = 0$, $\nu > 0$.

Hint for (ii): Consider, with α and β given by (4.31)–(4.32),

$$h(\nu) = \big(\alpha(\nu) + \beta(\nu)\big)(\cos\nu - 1)$$

and prove that $h(\nu) < 0$ for $\nu \neq 0$.

Exercise 4.21.

(i) Verify that each of the curves C_k intersects the line $\alpha = 2$ exactly once.

(ii) Show that $\beta_{k+1} > \beta_k$, where β_k is, by definition, such that C_k passes through $(2, \beta_k)$ or, in other words, β_k is the β-coordinate of the intersection of C_k and the line $\alpha = 2$.

(iii) Combine the information of this and the foregoing exercise to conclude that the curves C_k are ordered according to k.

We now have an analytical proof of the most relevant features of the following figure, which was produced by computer on the basis of (4.31)–(4.32).

It remains to determine the number of roots in the right half-plane in the various regions defined by the curves C_k. In order to get started, we need a reference point.

Exercise 4.22. Take $(\alpha, \beta) = (1, 0)$. Prove that all roots of (4.26) are in the left half-plane.

Second, we verify that the roots cannot "escape" or "pop up" at infinity in the right half-plane.

Exercise 4.23. Let z be a root of (4.26) with $\operatorname{Re} z > 0$ and $|z| > 0$. Show that necessarily

$$|z| < 1 + |1 + \alpha - \beta| + |\beta|.$$

Third, we have to determine which way the roots move in the complex plane if we move away from a curve C_k in the parameter plane.

Exercise 4.24. Identify F from (2.18) with the left hand side of (4.26). Show that the matrix M defined in (2.21) is then given by

$$\begin{pmatrix} 1 & \frac{\sin \nu}{\nu} \\ 0 & \frac{\cos \nu - 1}{\nu} \end{pmatrix}$$

and that, consequently, $\det M < 0$.

Applying Proposition 2.13, we conclude that the critical roots move from the imaginary axis into the right half-plane if we move from a curve C_k into the region in parameter space to the left of that curve. We can now copy the arguments of Section 2 to conclude that the number of roots in the right half-plane is, for the various regions in parameter space, exactly as indicated in Figure XI.3.

Remark **4.25**. As formulated, the left hand side of (4.26) has a singularity for $z = 0$. But this is a removable singularity, and for $z = 0$, one should replace $\beta \frac{1-e^{-z}}{z}$ by β. So we can indeed invoke Lemma 2.8.

Exercise 4.26. Show that $z = 0$ is a simple root of (4.26) for $\alpha + \beta = 0$ and $\alpha \neq 0$. Let μ be the real part of the continuation of this root when we follow the line $\{(\alpha_0 + \gamma, -\alpha_0 + \gamma) \mid \gamma \in \mathbb{R}\}$ in parameter space. Compute that

$$\frac{d\mu}{d\gamma}(0) = -\frac{4}{\alpha_0}$$

and verify that this result is consistent with the numbers in Figure XI.3.

Exercise 4.27. (Some more detailed information on the curves C_k.)

(i) Compute, for $\epsilon \to 0$,

$$\alpha(2k\pi + \epsilon) \sim -\frac{4k\pi}{\epsilon},$$

$$\beta(2k\pi + \epsilon) \sim \frac{8k^2\pi^2}{\epsilon^2}.$$

(ii) Use computer algebra to verify that β is increasing along C_0 and has a unique minimum along any C_k with $k \geq 1$.

Exercise 4.28. (Strips again, or an alternative argument to deduce that roots do not return to the left half-plane). Show that $z = \mu + 2k\pi i$ cannot be a root of (4.26) for $\alpha > 0$ and $\mu > 0$. Conclude that for $\alpha > 0$, the root which enters the right half-plane when passing C_0 is caught in the strip $0 < \nu < 2\pi$, $\mu > 0$. So, when we subsequently pass C_1 (while remaining in the region with $\alpha > 0$), the root which is now on the imaginary axis cannot be a continuation of the first one, hence must come from the left half-plane and enter the right half-plane.

Before we proceed with the calculation of the direction of bifurcation along C_0, we use (4.28) to translate our results so far into the natural parameters a and e. The outcome is Fig. XI.4.

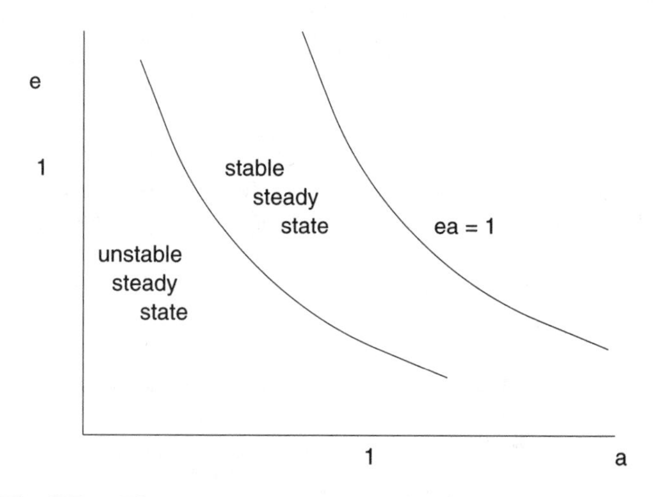

Fig. XI.4. The stability region in the (a, e)- parameter plane.

We note that, for a given value of a, one can stabilize the steady state by increasing e. It is this observation which leads to the conclusion that prey dispersal from predator invaded patches is a stabilizing mechanism.

In Chapter X, we derived a formula for the direction of Hopf bifurcation starting from an abstract integral equation. In order to apply the results to the present example, we have to show how equations (4.24) can be rewritten as an abstract integral equation. Here, we like to emphasize that it is the biological interpretation which suggests how to do it and that it may be rather hard to find a convenient framework on the basis of mathematical arguments only.

The verbal description of the model in the beginning of this subsection suggests considering

$$
\begin{aligned}
&\dot{x} = a(x+ey) - bxy,\\
&n(t,0) = bx(t)y(t),\\
&\frac{\partial n}{\partial t} + \frac{\partial n}{\partial \tau} = 0,\\
&y(t) = \int_0^1 n(t,\tau)\,d\tau,
\end{aligned}
\tag{4.35}
$$

where n is the density function of predator patches with respect to their "age", i.e., the time elapsed since predator invasion took place. So, as state space we take $\mathbb{R} \times L^1([0,1],\mathbb{R})$, where the first component refers to the variable $x(t)$ and the second to $n(t,\cdot)$.

Exercise 4.29. Derive (4.24) from (4.35) by integrating the first order PDE for n along the characteristics $t - \tau =$ constant. Show that any solution of (4.35) defines for $t > 1$, a solution of (4.24) and that, conversely, one can associate with any solution of (4.24) for $t > 1$, a solution of (4.35) by putting $n(t,\tau) = bx(t-\tau)y(t-\tau)$. So the only difference between (4.24) and (4.35) is the way an initial condition is prescribed.

The unperturbed semigroup $\{T_0(t)\}$ acts trivially on the first and by translation on the second component:

$$
T_0(t)(y,\varphi) = (y,\psi)
\tag{4.36}
$$

with

$$
\psi(a) = \begin{cases} \varphi(a-t), & a \geq t,\\ 0, & a < t. \end{cases}
\tag{4.37}
$$

So $A_0(z,\varphi) = (0,-\varphi')$, with $\mathcal{D}(A_0) = \{(y,\varphi) \mid \varphi \in \mathrm{AC},\ \varphi(0) = 0\}$. The duality diagram for the state spaces will therefore be

$$
\begin{array}{ccc}
\mathbb{R} \times L^1([0,1],\mathbb{R}) & \longrightarrow & \mathbb{R} \times L^\infty([0,1],\mathbb{R})\\
\uparrow & & \downarrow\\
\mathbb{R} \times \mathrm{NBV}([0,1),\mathbb{R}) & \longleftarrow & \mathbb{R} \times C_0([0,1),\mathbb{R}).
\end{array}
$$

This is, apart from an additional $\mathbb{R}$-component, the same as in Exercises II.2.1, II.5.6 and III.4.7 and Appendix VII. Note that when doing spectral analysis, we work with the complexified space or, equivalently, replace everywhere $\mathbb{R}$ by $\mathbb{C}$ in the spaces above.

The embedding operator $j : X \to X^{\odot\odot} \subset X^{\odot *}$ is defined by

$$
j(\varphi)(\theta) = \int_0^\theta \varphi(\sigma)\,d\sigma, \quad 0 \leq \theta \leq 1.
\tag{4.38}
$$

Next, we define $B : X \to X^{\odot *}$ by

$$B(y,\varphi) = (a_{11}y + a_{12}\int_0^1 \varphi(\theta)d\theta\ ,\ (a_{21}y + \int_0^1 \varphi(\theta)d\theta)\cdot H), \tag{4.39}$$

where

$$a_{11} = -\frac{ea^2}{1-ea}, \qquad a_{12} = ea - 1, \qquad a_{21} = \frac{a}{1-ea} \tag{4.40}$$

and where $\tau \mapsto H(\tau)$ is the NBV function which is identically one for $\tau > 0$.

Likewise, we introduce a mapping R from X into $X^{\odot *}$ by defining

$$R(y,\varphi) = (-y\int_0^1 \varphi(\theta)d\theta\ ,\ y\int_0^1 \varphi(\theta)d\theta \cdot H). \tag{4.41}$$

The nonlinear system of equations (4.35) can now be written in the form of an abstract integral equation as

$$u(t) = T(t)(y,\varphi) + j^{-1}\int_0^t T^{\odot *}(t-\tau)R(u(\tau))\,d\tau, \tag{4.42}$$

where $T(t)$ is the perturbed linear semigroup defined by the solutions of

$$T(t)(y,\varphi) = T_0(t)(y,\varphi) + j^{-1}\int_0^t T_0^{\odot *}(t-\tau)BT(\tau)(y,\varphi)\,d\tau. \tag{4.43}$$

Exercise 4.30. Convince yourself that $A_0^{\odot *} + Bj^{-1}$ corresponds to the linearization at the nontrivial steady state.

Exercise 4.31. Check that solutions of (4.42) correspond to solutions of (4.35).

To calculate in this case the direction of Hopf bifurcation, we need to gather all the ingredients that are needed to evaluate the formula for c in Theorem X.3.7. In particular, we need a formula for the resolvent and for the eigenfunctions of $A^{\odot *}$. First, we state the result. At the end of this section we shall explain how the representation was derived.

In $\mathrm{NBV}([0,1),\mathbb{C})$ we use the splitting

$$\phi = wH + \psi,$$

with ψ continuous at 0 (cf. the text between Exercises IV.5.17 and IV.5.18 and note that here we use the notation H instead of δ). An element in $X^{\odot *}$ is denoted by the triple (y, w, ψ) and the decomposition by $X^{\odot *} = \mathbb{C}\times\mathbb{C}\times Y$.

Lemma 4.32.

(i) *The resolvent of $A^{\odot *}$ has the following representation:*

(4.44)

$$(zI - A^{\odot*})^{-1} = \Delta(z)^{-1} \begin{pmatrix} 1 - G(z;1) & a_{12}(1 - G(z;1)) & a_{12} \int_0^1 de_{-z}* \\ 0 & 0 & 0 \\ a_{21}(1 - G(z;\,\cdot\,)) & (z - a_{11})G(z;\,\cdot\,) & \Delta(z)F \end{pmatrix}$$

where

(4.45)

$$\begin{aligned} \Delta(z) &= (z - a_{11})(1 - G(z;1)) - a_{12}a_{21}G(z;1), \\ G(z;\theta) &= \frac{1 - e^{-z\theta}}{z}, \quad 0 \le \theta \le 1, \\ F(\psi)(\theta) &= \Delta(z)^{-1}\big(a_{12}a_{21} + z - a_{11}\big)G(z;\theta) \\ &\quad \times \int_0^1 d_\sigma\,((e_{-z} * \psi)(\sigma)) + (e_{-z} * \psi)(\theta). \end{aligned}$$

(ii) $(y^{\odot}, \psi^{\odot})$ *is an eigenfunction of* $A^{\odot}$ *with eigenvalue* λ *if and only if*

$$\begin{pmatrix} y^{\odot} \\ \psi^{\odot} \end{pmatrix} = \begin{pmatrix} c_1 \\ c_2 G(\lambda;\,\cdot\, - 1) \end{pmatrix}$$

with

$$c_2 = \frac{\lambda a_{12}}{\lambda - 1 + e^{-\lambda}} c_1.$$

(iii) $(y, 0, \phi)$ *is an eigenfunction of* $A^{\odot*}$ *with eigenvalue* λ *if and only if*

$$\begin{pmatrix} y \\ \phi \end{pmatrix} = \begin{pmatrix} c_1 \\ c_2(1 - e^{-\lambda\,\cdot}) \end{pmatrix}$$

with

$$c_2 = \frac{\lambda a_{21}}{\lambda - 1 + e^{-\lambda}} c_1.$$

(iv) *If* $(y^{\odot}, \psi^{\odot})$ *is an eigenfunction of* $A^{\odot}$ *at* $i\omega$ *and* (y, ϕ) *is an eigenfunction of* $A^{\odot*}$ *at* $i\omega$, *then*

$$\langle (y^{\odot}, \psi^{\odot}), (y, \phi) \rangle = (\, y^{\odot} \quad \psi^{\odot}(0)\,)\, E'(i\omega) \begin{pmatrix} y \\ \phi'(0) \end{pmatrix}.$$

Exercise 4.33. Show that it is a consequence of Lemma 4.32(i) that

$$(zI - A^{\odot*})^{-1}(y, w, 0) = (u, 0, \psi)$$

with

$$\begin{aligned} u &= \Delta(z)^{-1}((1 - G(z;1))y + a_{12}G(z;1)w), \\ \psi'(a) &= \Delta(z)^{-1}(a_{21}e^{-za}y + (z - a_{11})e^{-za}w). \end{aligned}$$

We have now gathered all the ingredients to compute the direction of bifurcation at a Hopf bifurcation point. Unfortunately, it turns out to be a prohibitive amount of work to do this with fountainpen and paper. Therefore, we have written a MAPLE program, which is reproduced at the very end of this section, to compute the stability of periodic solutions.

There are two parameters in the problem, α and β. The condition for Hopf bifurcation is that $\Delta(z)$ has a purely imaginary root, say at $i\omega$. We find a curve in parameter space where Hopf bifurcation occurs which we parameterize by ω. We compute the coefficient c as a function of ω. If $\operatorname{Re} c$ is negative, then the bifurcating periodic solution is stable; if $\operatorname{Re} c$ is positive, it is unstable.

Unfortunately, the algebraic expression for c obtained with the aid of Maple is not such that one can decide about the sign of $\operatorname{Re} c$ by looking at it. Therefore, the best we could do at this point was to compute the value of $\operatorname{Re} c$ along the curve numerically. We have computed 1000 points on the curve C_0, equidistant in ν in the interval $(0, 2\pi)$. The result is plotted in Fig. XI.5. It appears that along the curve C_0, there is, at $\nu \sim \frac{2\pi}{3}$, a change from a subcritical bifurcation to an unstable periodic solution to a supercritical bifurcation to a stable periodic solution. For ν close to 2π [which is close to infinity in the (α, β)-parameter plane] the reverse change takes place.

From the fact that the bifurcation is subcritical for low values of ν, we infer that most likely this model exhibits bistable behaviour for certain parameter values, since the unstable bifurcating periodic solution, together with its stable manifold, may serve as a separatrix between the stable steady state and another attractor (possibly another periodic solution lying on the same branch, when the branch bends backward and then gains stability). We have not investigated this by numerical experiments or any other means.

Proof of Lemma 4.32 (i). We compute the resolvent in $X^{\odot *}$ using the identity

$$(zI - A^{\odot *}) = (I - B\,j^{-1}\,(zI - A_0^{\odot *})^{-1})(zI - A_0^{\odot *}).$$

Exercise 4.34. Verify that

$$(zI - A_0^{\odot *})^{-1} = \frac{1}{z}\begin{pmatrix} 1 & 0 & 0 \\ 0 & 0 & 0 \\ 0 & 1 - e^{-z\cdot} & ze_{-z} * \end{pmatrix}.$$

Exercise 4.35. Verify that $B\,j^{-1}$ can be represented by the matrix

$$B\,j^{-1} = \begin{pmatrix} a_{11} & 0 & a_{12}\int_0^1 d \\ a_{21} & 0 & \int_0^1 d \\ 0 & 0 & 0 \end{pmatrix}.$$

Here we use the notation $(\int_0^1 d)(\psi) = \int_0^1 d\psi(\tau) = \psi(1)$.

A combination of these two facts yields that

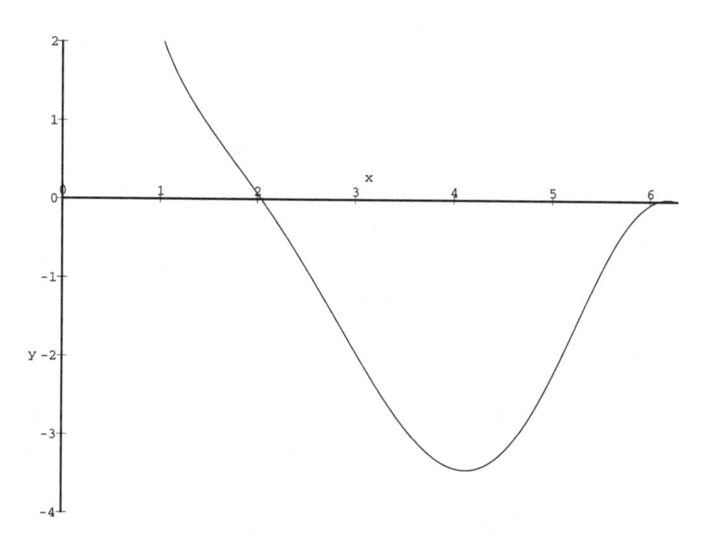

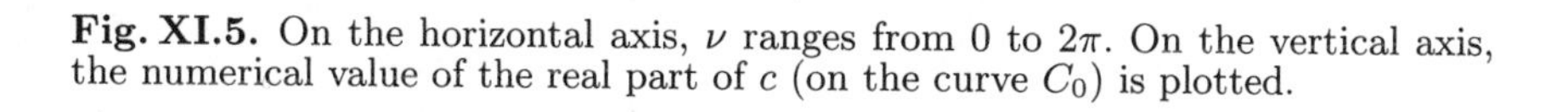
Fig. XI.5. On the horizontal axis, ν ranges from 0 to 2π. On the vertical axis, the numerical value of the real part of c (on the curve C_0) is plotted.

$$(4.46)\qquad B\,j^{-1}(zI - A_0^{\odot *})^{-1} = \begin{pmatrix} \frac{a_{11}}{z} & a_{12}\int_0^1 e^{-z\tau}\,d\tau & a_{12}\int_0^1 d\,e_{-z} * \\ \frac{a_{21}}{z} & \int_0^1 e^{-z\tau}\,d\tau & \int_0^1 d\,e_{-z} * \\ 0 & 0 & 0 \end{pmatrix}.$$

Exercise 4.36. Show that

$$\int_0^1 de_{-z} * \psi = \int_0^1 \frac{1 - e^{z(\sigma-1)}}{z} \psi'(\sigma)\,d\sigma.$$

Define $P(z) : \mathbb{R}^2 \to \mathbb{R}^2$ by

$$(4.47)\qquad P(z) = \begin{pmatrix} 1 - \frac{a_{11}}{z} & -a_{12}\frac{1-e^{-z}}{z} \\ -\frac{a_{21}}{z} & 1 - \frac{1-e^{-z}}{z} \end{pmatrix},$$

and $Q(z) : Y \to \mathbb{R}^2$ by

$$(4.48)\qquad Q(z) = \begin{pmatrix} -a_{12}\int_0^1 d\,e_{-z} * \\ -\int_0^1 d\,e_{-z} * \end{pmatrix}.$$

Then

$$\det P(z) = \frac{\Delta(z)}{z},$$

with Δ defined by (4.45), and

$$I - Bj^{-1}(zI - A_0^{\odot *})^{-1} = \begin{pmatrix} P & Q \\ 0 & I \end{pmatrix}.$$

Hence

$$(I - Bj^{-1}(zI - A_0^{\odot *})^{-1})^{-1} = \begin{pmatrix} P^{-1} & 0 \\ 0 & I \end{pmatrix} \begin{pmatrix} I_2 & -Q \\ 0 & I \end{pmatrix}. \tag{4.49}$$

Combining (4.46)–(4.48) and (4.49), it follows that formula (4.44) holds. □

Exercise 4.37. Let $z = \lambda$ be a simple zero of $\Delta(z)$. Let P be the corresponding projection operator. Show by residue calculus that

$$P^{\odot *} \begin{pmatrix} y \\ w \\ \psi \end{pmatrix} = \Delta'(\lambda)^{-1} \begin{pmatrix} g_0 \\ 0 \\ g(\cdot) \end{pmatrix},$$

where

$$g_0 = (1 - \frac{1-e^{-\lambda}}{\lambda})y + \frac{a_{12}\,(1-e^{-\lambda})}{\lambda} w + a_{12} \int_0^1 (\frac{1-e^{-\lambda(1-u)}}{\lambda})\, d\psi(u)$$

and $g(\cdot)$ is the function defined by

$$g(t) = \frac{a_{21}\,(1-e^{-\lambda t})y}{\lambda} + \frac{(\lambda - a_{11})(1-e^{-\lambda t})w}{\lambda} + \frac{(a_{12}\,a_{21} + \lambda - a_{11})(1-e^{-\lambda t}) \int_0^1 (\frac{1-e^{-\lambda(1-u)}}{\lambda}) d\psi(u)}{\lambda}.$$

Exercise 4.38. Show that this result implies that

$$\mathcal{R}(P^{\odot *}) = \{ \begin{pmatrix} c_1 \\ 0 \\ c_2(1-e^{-\lambda \cdot}) \end{pmatrix} \mid c_2 = \frac{\lambda a_{21}}{\lambda - 1 + e^{-\lambda}} c_1 \}. \tag{4.50}$$

From Exercise 4.38, Lemma 4.32(ii) follows.

Exercise 4.39. Verify by direct computation that $(P^{\odot *})^2 = P^{\odot *}$.

Exercise 4.40. Verify by direct computation that the nonzero elements of the range of $P^{\odot *}$ are eigenvectors of $A^{\odot *}$ corresponding to the eigenvalue λ.

Exercise 4.41. Verify that

$$P^{\odot}\begin{pmatrix} c \\ g \end{pmatrix} = \Delta'(\lambda)^{-1}\begin{pmatrix} \frac{\lambda-1-e^{-\lambda}}{\lambda}c + a_{21}\int_0^1 e^{-\lambda\theta}g(\theta)d\theta \\ (a_{12}c + (a_{12}a_{21} + \lambda - a_{11})\int_0^1 e^{-\lambda\theta}g(\theta)d\theta)\frac{1-e^{\lambda(\cdot-1)}}{\lambda} \end{pmatrix}$$

and conclude that

$$\mathcal{R}(P^{\odot}) = \{\begin{pmatrix} c_1 \\ c_2(\frac{1-e^{\lambda(\cdot-1)}}{\lambda}) \end{pmatrix} \mid c_2 = \frac{\lambda a_{12}}{\lambda - 1 + e^{-\lambda}}c_1\}. \tag{4.51}$$

This proves Lemma 4.32(iii).

Exercise 4.42. Use (4.50) and (4.51) to verify Lemma 4.32(iv) by direct computation.

Exercise 4.43. Show that in X the explicit representation of the resolvent takes the form

$$(zI - A)^{-1}\begin{pmatrix} \eta \\ \psi \end{pmatrix} = \begin{pmatrix} \xi \\ \phi \end{pmatrix}$$

with

$$\xi = \Delta(z)^{-1}\{\frac{z-1+e^{-z}}{z}\eta + a_{12}\int_0^1 \frac{1-e^{z(\sigma-1)}}{z}\psi(\sigma)d\sigma\},$$

$$\phi(\theta) = \Delta(z)^{-1}\{a_{21}\eta + (a_{12}a_{21} + z - a_{11})\int_0^1 \frac{1-e^{z(\sigma-1)}}{z}\psi(\sigma)\,d\sigma\}e^{-z\theta} + \int_0^\theta e^{-z(\theta-\sigma)}\psi(\sigma)\,d\sigma.$$

In conclusion of this section, we list the Maple program that we used to derive the an expression for the coefficient c.

```
with(linalg):
RootOf(_Z^2+1):=sqrt(-1);

_M:=proc(lambda)
        array([[lambda-a11,-a12*(int(exp(-lambda*s),s=0..1))],
               [-a21,1-(int(exp(-lambda*s),s=0..1))]]);
end;

_dM:=proc(lambda)
        array([[1,-a12*(int(-s*exp(-lambda*s),s=0..1))],
               [0,-int(-s*exp(-lambda*s),s=0..1)]]);
end;

resolvente:=proc(lambda,mapping)
```

```
#
# the mapping is assumed to be of the type (y,wH)
# where H is the Heaviside mapping.
# mapping[1]=y and mapping[2]=w.
# The result is the pair (u, psi(tau)), see (4.57)-(4.58).
#
array([1/det(_M(lambda))
          *(1-1*(int(exp(-lambda*s),s=0..1))*mapping[1]+
          a12*(int(exp(-lambda*s),s=0..1))*mapping[2]),
       unapply(1/det(_M(lambda))*
          (a21*exp(-lambda*tau)*mapping[1]
          +(lambda-a11)*exp(-lambda*tau)*mapping[2])
       ,tau)]);
end;

ddR:=proc(f1,f2)
#
# fk, (k=1,2) is the pair (fk[1],fk[2]) with fk[1] a real number
# and fk[2] a mapping (compare(4.44)).
# The result takes the form (y,wH), where H is the Heaviside
# function and is represented as the pair (y,w).
#
        array([(-f1[1]*int(f2[2](tau),tau=0..1)
                -f2[1]*int(f1[2](tau),tau=0..1)),
               (+f1[1]*int(f2[2](tau),tau=0..1)
                +f2[1]*int(f1[2](tau),tau=0..1))]);
end;

#
# Phi is the eigenfunction in X at the eigenvalue iw
#

Phi_0:=array([1-int(exp(-I*w*s),s=0..1),a21]);

Phi:=array([Phi_0[1],unapply(Phi_0[2]*exp(-I*w*tau),tau)]);

Phibar_0:=array([1-int(exp(I*w*s),s=0..1),a21]);

Phibar:=array([Phibar_0[1],unapply(Phibar_0[2]
                                    *exp(I*w*tau),tau)]);

#
# Psi is the eigenfunction in X-star at the eigenvalue -iw.
#
Psibar_0:=array([a21,I*w-a11]);
```

```
inproduct:=proc(a,b) a[1]*b[1]+a[2]*b[2] end;

#
# We need to scale Psi such that <psi,phi>=1,
# This is done using (4.67)
#

rbar:=1/inproduct(Psibar_0,evalm( _dM(I*w)  Phi_0)):

c:=rbar*(inproduct(Psibar_0,ddR(resolvente(0, ddR(Phi,Phibar))
   ,Phi))+ 1/2*inproduct(Psibar_0, ddR(resolvente(2*I*w,
    ddR(Phi,Phi)),Phibar))):

_s:=solve({alpha=_e*a^2/(1-_e*a),
          beta=a*(1-2*_e*a)/(1-_e*a)},{_e,a});

assign(_s);
#
# see formula(4.26)
#

a11:=simplify(-_e*a^2/(1-_e*a));   #(4.43)
a21:=simplify(a/(1-_e*a));        #(4.43)
a12:=simplify(_e*a-1);            #(4.43)

_s1:=solve({evalc(Re(det(_M(I*w)))),
           evalc(Im(det(_M(I*w))))},{alpha,beta});
assign(_s1);

rec:=evalc(Re(c)):

a:=array(1..1000);

b:=array(1..1000);

for i from 1 to 1000 do
      a[i]:=subs(w=0.002*i*Pi,rec);
      hulp:=a[i];
      b[i]:=evalhf(hulp);
od:
```

XI.5 Comments

The monograph [268] by Stépán gives an up-to-date systematic overview of theorems and tools for the stability analysis of RFDE in terms of roots of the characteristic equation. The books by Kolmanovskii and Nosov [151] and MacDonald [172] contain both a rich variety of examples and a discussion of various analytical techniques.

Chapter XII

Time-dependent linear systems

XII.1 Introduction

In this chapter we shall deal with the nonautonomous linear RFDE

$$\dot{x}(t) = \int_0^h d_\theta[\zeta(t,\theta)]x(t-\theta), \qquad t \geq s, \tag{1.1}$$

with initial condition

$$x(s+\theta) = \phi(\theta), \qquad -h \leq \theta \leq 0, \tag{1.2}$$

and with abstract integral equations (AIE)

$$u(t) = T_0(t-s)\phi + \int_s^t T_0^{\odot *}(t-\tau)B(\tau)u(\tau)\,d\tau, \quad t \geq s, \tag{1.3}$$

where $B(t) : X \to X^{\odot *}$ is a family of bounded linear operators. The connection between the two problems is provided by the specification of $X, \{T_0(t)\}, \{B(t)\}$ and the formula

$$u(t) = x_t. \tag{1.4}$$

In Section 2 we give the definitions of forward and backward evolutionary systems (which are related to each other by duality) and we show that the solution operators associated with (1.3) constitute a forward evolutionary system. In Section 3 we specialise to the setting of RFDE and show that (1.4) yields a one-to-one relation between (1.1) and (1.2), on the one hand, and (1.3) on the other. In Section 4 we show, by means of an example involving a moving point delay, that strong continuity of $t \mapsto B(t)$ is not enough to have a duality framework exactly as in the autonomous case. Subsequently, we prove that continuity of $t \mapsto B(t)$ with respect to the operator norm is a sufficient condition. In Section 5, finally, we concentrate on operators $B(t)$ with finite dimensional range, the associated finite dimensional Volterra integral equations and the interpretation of the adjoint evolutionary system associated with a RFDE.

XII.2 Evolutionary systems

A nonautonomous system is characterised by the fact that not just the time difference between the initial time and the present time matters but also the initial time itself. Hence we have to work with two-parameter families of operators $U(t, s)$, where s corresponds to the initial time and t to the current time. The set $\triangle \subset \mathbb{R}^2$ on which U is defined is of the form

$$\triangle = \{(t, s) \mid \alpha \leq s \leq t \leq \omega\} \tag{2.1}$$

where $\alpha, \omega \in \mathbb{R} \cup \{-\infty, +\infty\}$ with $\alpha < \omega$ and where, here and in the following, one should read $\leq$ as $<$ whenever the left hand side equals $-\infty$ or the right hand side equals $+\infty$.

Definition 2.1. A two-parameter family $U = \{U(t, s)\}_{(t,s)\in\triangle}$ of bounded linear operators on a Banach space X is called a *forward evolutionary system* on X whenever

$$U(s, s) = I \quad \text{(the identity)}, \qquad \alpha \leq s \leq \omega, \tag{2.2}$$

$$U(t, r)U(r, s) = U(t, s), \qquad \alpha \leq s \leq r \leq t \leq \omega. \tag{2.3}$$

One can read (2.3) as: following the state of the system as it evolves from time s to time r and then from time r to time t amounts to the same as following the state as it evolves from time s to time t. Property (2.2) states that the operators act on the initial state prescribed at the time given by the second argument.

The adjoint operators of a forward evolutionary system do *not* (necessarily) constitute a forward evolutionary system since $(AB)^* = B^*A^*$, which may be different from A^*B^*.

Exercise 2.2. Let U be a forward evolutionary system on X. Define $V(s, t) = (U(t, s))^*$ and note the interchanging of the arguments! Rewrite (2.3) in terms of V and try to give an interpretation.

Definition 2.3. A two-parameter family $V = \{V(s, t)\}_{(t,s)\in\triangle}$ of bounded linear operators on a Banach space Y is called a *backward evolutionary system* on Y whenever

$$V(t, t) = I, \qquad \alpha \leq t \leq \omega, \tag{2.4}$$

$$V(s, r)V(r, t) = V(s, t), \qquad \alpha \leq s \leq r \leq t \leq \omega. \tag{2.5}$$

One can read (2.5) as: following the state of the system backward from the final state at time t to the state at time s, it does not matter whether or not we "stop" at a time r in between. Property (2.4) states that the operators act on the final state prescribed at the time given by the second argument.

Exercise 2.4. Verify that the adjoint of a forward evolutionary system is a backward evolutionary system and that the adjoint of a backward evolutionary system is a forward evolutionary system.

Definition 2.5. The forward evolutionary system U is said to be *strongly continuous* if for every $x \in X$ the mapping $(t,s) \mapsto U(t,s)x$ is continuous from $\triangle$ to X. Strong continuity of a backward evolutionary system is defined similarly.

The formal calculations of the next exercise are intended to provide insight in the relationship of the differential equations that would be satisfied by, respectively, a forward evolutionary system and the adjoint backward evolutionary system, if only we would have sufficient regularity of orbits.

Exercise 2.6. Verify formally that the differential equation

$$\frac{\partial U}{\partial t}(t,s) = A(t)U(t,s) \tag{2.6}$$

implies that $V(s,t) = (U(t,s))^*$ satisfies

$$\frac{\partial V}{\partial s}(s,t) = -A(s)^*V(s,t), \tag{2.7}$$

which is a differential equation with respect to the "initial" time variable s. Hint: Use that $U(t,r)U(r,s) = U(t,s)$ is independent of r and that, consequently, differentiation with respect to r yields zero.

Unlike the case of C_0-semigroups, one does not have a well-defined notion of a generating family $A(t)$ for a given evolutionary system $U(t,s)$. In particular, there is no analogue of the Hille-Yosida theorem. For the kind of application we are dealing with in this book, this does not really matter, as one can avoid talking about abstract differential equations and, instead, show directly the equivalence of an abstract integral equation, based on the variation-of-constants formula, and the corresponding functional differential equation. This equivalence will be proven in the next section. In the remainder of this section, we concentrate on the constructive definition of an evolutionary system in terms of solutions of an abstract integral equation. Again, we adopt a perturbation approach.

Let the Banach space X be $\odot$-reflexive with respect to a C_0-semigroup $\{T_0(t)\}$. The perturbation is now given in the form of a family $\{B(t)\}_{\alpha \le t \le \omega}$ of bounded linear operators from X into $X^{\odot *}$. We assume that this family is strongly continuous, i.e., for every $x \in X$, the mapping $t \mapsto B(t)x$ is continuous from $[\alpha, \omega]$ to $X^{\odot *}$.

Theorem 2.7. *The variation-of-constants equation*

$$U(t,s)\phi = T_0(t-s)\phi + \int_s^t T_0^{\odot *}(t-\tau)B(\tau)U(\tau,s)\phi\, d\tau, \tag{2.8}$$

where $(t,s) \in \Delta$ *and* $\phi \in X$, *uniquely defines a strongly continuous forward evolutionary system* U. *The expansion*

$$U(t,s) = \sum_{n=0}^{\infty} U_n(t,s), \tag{2.9}$$

where

$$U_n(t,s)\phi = \int_s^t T_0^{\odot *}(t-\tau)B(\tau)U_{n-1}(\tau,s)\phi\, d\tau, \tag{2.10}$$

and

$$U_0(t,s) = T_0(t-s) \tag{2.11}$$

converges in the uniform (operator) topology, uniformly on Δ. *Furthermore,*

$$||U(t,s)|| \leq M e^{(\omega_0 + MK(t,s))(t-s)}, \tag{2.12}$$

where

$$K(t,s) = \sup_{s \leq \tau \leq t} ||B(\tau)|| \tag{2.13}$$

and M *and* ω_0 *are such that* $||T_0(t)|| \leq M e^{\omega_0 t}$.

The proof of Theorem 2.7 employs standard arguments similar to those used to prove Theorem II.2.4. For instance, one needs the following lemma to replace Lemma III.2.1.

Lemma 2.8. *Let* $f : \Delta \to X^{\odot *}$ *be continuous. Define* $v : \Delta \to X^{\odot *}$ *by*

$$v(t,s) = \int_s^t T_0^{\odot *}(t-\tau)f(\tau,s)\, d\tau. \tag{2.14}$$

Then v *is continuous and takes values in* $j(X)$.

Exercise 2.9.

(i) Work out the details of the proofs of Lemma 2.8 and Theorem 2.7.
(ii) Prove that one can sharpen the estimate (2.12) to

$$||U(t,s)|| \leq M e^{\int_s^t (\omega_0 + M||B(\tau)||)\, d\tau} \tag{2.15}$$

when $t \mapsto ||B(t)||$ is measurable.

Exercise 2.10. State and prove a result concerning the continuous dependence of the evolutionary system $U(t,s)$ on the perturbing family $\{B(t)\}$.

Exercise 2.11. Let B be periodic with period p. Show that $U(t+p,s+p) = U(t,s)$ for all t and s.

XII.3 Time-dependent linear RFDE

In the setting of RFDE we have

$$X = C\big([-h,0],\mathbb{R}^n\big) \simeq \big\{(\alpha,\phi) \in \mathbb{R}^n \times C([-h,0],\mathbb{R}^n) \mid \phi(0) = \alpha\big\}$$

and

$$X^{\odot *} = \mathbb{R}^n \times L^\infty\big([-h,0],\mathbb{R}^n\big)$$

or, these spaces with $\mathbb{R}^n$ replaced by $\mathbb{C}^n$ (in this context, recall Exercise III.7.9). The perturbations $B(t) : X \to X^{\odot *}$ are defined by

$$B(t)\phi = (\langle \zeta(t,\cdot),\phi\rangle_n, 0), \tag{3.1}$$

where $\zeta(t,\cdot)$ is a family of NBV matrices such that for all $\phi \in X$, the mapping $t \mapsto \langle \zeta(t,\cdot),\phi\rangle_n$ is continuous from (a subset of) $\mathbb{R}$ into $\mathbb{R}^n$. Our aim is to show that, with $T_0(t)$ and $T_0^{\odot *}(t)$ as defined by II.2.4 and II.5.7 respectively, solutions of the AIE (2.8) are in one-to-one correspondence with solutions of the initial-value problem for the RFDE

$$\begin{aligned} \dot{x}(t) &= \langle \zeta(t,\cdot), x_t\rangle_n \\ &= \int_0^h d_\theta[\zeta(t,\theta)]x(t-\theta), \qquad t \geq s. \end{aligned} \tag{3.2}$$

Theorem 3.1. *Let, for X, $T_0(t)$ and $B(t)$ as described above, $U(t,s)$ denote the evolutionary system defined by the* AIE (2.8). *Then $x(t)$ defined by*

$$x(s+\theta) = \varphi(\theta), \qquad -h \leq \theta \leq 0, \tag{3.3}$$

$$x(t) = (U(t,s)\varphi)(0), \qquad t \geq s, \tag{3.4}$$

satisfies (3.2) *and, conversely, if x is a solution of the RFDE* (3.2) *satisfying the initial condition* (3.3), *then for $t \geq s$ and $-h \leq \theta \leq 0$, we have*

$$\big(U(t,s)\varphi\big)(\theta) = \begin{cases} \varphi(t-s+\theta), & t+\theta \leq s, \\ x(t+\theta), & t+\theta \geq s. \end{cases} \tag{3.5}$$

Corollary 3.2. *The initial-value problem* (3.2)–(3.3) *admits a unique solution. (Here the solution concept is that of* Definition I.2.1, *with the trivial* *modifications required to take account of the initial time s and the time dependence of ζ.)*

For the proof of Theorem 3.1 we need the following variant of Lemma III.4.3.

Lemma 3.3. *Let e_i be the ith unit vector in $\mathbb{R}^n$ (or $\mathbb{C}^n$) and let $r_i^{\odot *} = (e_i, 0)$. For any $\eta \in L^1_{loc}$, we have*

$$\int_s^t T_0^{\odot *}(t-\tau)\eta(\tau) r_i^{\odot *}\, d\tau = e_i \int_s^{\max(s,t+\cdot)} \eta(\sigma)\, d\sigma. \tag{3.6}$$

Proof of Theorem 3.1. Fix $s \in \mathbb{R}$ and $\varphi \in X$. Define continuous functions y and x by

$$y(t) = \langle \zeta(t,\cdot), U(t,s)\varphi \rangle_n, \qquad t \geq s, \tag{3.7}$$

$$x(t) = (U(t,s)\varphi)(0), \qquad t \geq s. \tag{3.8}$$

Combining equation (2.8), evaluated at $\theta = 0$, with the identity (3.6) and the definitions of x and y, we find

$$x(t) = \varphi(0) + \int_s^t y(\tau)\, d\tau \tag{3.9}$$

from which we conclude that x is continuously differentiable for $t \geq s$ and

$$\dot{x}(t) = y(t), \qquad t \geq s. \tag{3.10}$$

On the other hand, (2.8), (3.6) and (3.9) together imply that for $t \geq s$ and $-h \leq \theta \leq 0$

$$(U(t,s)\varphi)(\theta) = \begin{cases} \varphi(t-s+\theta), & t+\theta \leq s, \\ x(t+\theta), & t+\theta \geq s. \end{cases} \tag{3.11}$$

Provided we extend x to the interval $[s-h, s]$ by (3.3), we can rewrite this equality as

$$U(t,s)\varphi = x_t. \tag{3.12}$$

Combining now (3.10), (3.7) and (3.11), we find

$$\dot{x}(t) = \langle \zeta(t,\cdot), x_t \rangle_n = \int_0^h d_\theta[\zeta(t,\theta)]\, x(t-\theta), \qquad t \geq s, \tag{3.13}$$

which is precisely equation (3.2).

Conversely, if x is a solution of the RFDE (3.2) with initial condition (3.3), then integration of (3.2) yields

$$x_t(\theta) = \big(T_0(t-s)\varphi\big)(\theta) + \int_s^{\max(s,t+\theta)} \langle \zeta(\tau,\cdot), x_\tau \rangle_n \, d\tau. \tag{3.14}$$

Using Lemma 3.3 in the other direction, we rewrite this equality in the form

$$x_t = T_0(t-s)\varphi + \int_s^t T_0^{\odot *}(t-\tau)B(\tau)x_\tau \, d\tau. \tag{3.15}$$

The uniqueness of solutions of equation (2.8) then implies at once that

$$U(t,s)\varphi = x_t. \tag{3.16}$$

□

Corollary 3.4. *For $t \geq s+h$, the operators $U(t,s)$ map X into $j^{-1}\mathcal{D}(A_0^{\odot *})$ and are compact.*

The proof of this corollary is essentially the same as the proof of Corollaries III.4.7 and III.4.8. Also see Proposition VII.6.5.

XII.4 Invariance of $X^{\odot}$: a counterexample and a sufficient condition

At the end of Section I.2 we have argued that one might specify an initial condition for a RFDE in the form of a point value at zero and a L^∞-history. Recently, it has become popular to study RFDE with so-called state-dependent delays such as

$$\dot{x}(t) = f\big(x(t-\sigma(x_t))\big) \tag{4.1}$$

where $\sigma : X \to \mathbb{R}_+$ is (at least) continuous. Of course, it may then happen that $\frac{d}{dt}\sigma(x_t) = 1$, in which case $t - \sigma(x_t)$ is constant and one clearly has to know a precise value for x in the corresponding point. The same conclusion pertains to the linear equation

$$\dot{x}(t) = x(t-\tau(t)) \tag{4.2}$$

when $\frac{d\tau}{dt}(t) = 1$ for some interval of time. So, if we try to "lift" the operators $U(t,s)$ to the space $X^{\odot *}$, by taking adjoints, restrictions and adjoints once more, something must go wrong. We shall now show that $X^{\odot}$ need not be invariant, so that it may be impossible to take restrictions.

Within the setting of RFDE we take $\alpha = 0, \omega = 1$, $n = 1, h = 1$ and for $-1 \leq \theta \leq 0$ and $0 \leq t \leq 1$,

$$\zeta(t,\theta) = H(\theta - t) \tag{4.3}$$

where, as before, H denotes the Heaviside function. Then

(4.4) $$B(t)\varphi = (\varphi(-t), 0)$$

and $\{B(t)\}_{t\in[0,1]}$ is indeed strongly continuous. The RFDE (3.2) with initial condition (3.3) then reduces to

(4.5) $$\dot{x}(t) = x_t(-t) = x(0) = \varphi(-s)$$

and so the solution is given by

(4.6) $$x(t) = \begin{cases} \varphi(t-s), & s-1 \le t \le s, \\ \varphi(0) + (t-s)\varphi(-s), & s \le t \le 1. \end{cases}$$

Therefore

(4.7) $$(U(t,s)\varphi)(\theta) = \begin{cases} \varphi(t+\theta-s), & -1 \le \theta \le -(t-s), \\ \varphi(0) + (t+\theta-s)\varphi(-s), & -(t-s) \le \theta \le 0. \end{cases}$$

Now let $V(s,t) = (U(t,s))^*$; then for $f \in X^* = \mathrm{NBV}$, we have

(4.8) $$\begin{aligned}\big(V(s,t)f\big)(\theta) = f(t-s)H(\theta) + \int_0^{t-s} f(\tau)\, d\tau H(\theta - s) \\ + f(t-s+\theta) - f(t-s).\end{aligned}$$

Exercise 4.1. Verify (4.8).

We see that, in general, $V(s,t)f$ is not continuous at $\theta = s$ and, consequently (cf. Theorem II.5.2), $V(s,t)f \notin X^\odot$ even if $f \in X^\odot$. We conclude from this example that under the present conditions on the family $\{B(t)\}$, the space $X^\odot$ is not necessarily invariant under the adjoint evolutionary system. This observation then raises the question of whether or not we can "force" $X^\odot$ to be invariant by sharpening the conditions on $\{B(t)\}$. In the remainder of this section we show that the answer is yes.

Assumption 4.2. The mapping $t \mapsto B(t)$ is continuous from $[\alpha, \omega]$ to $\mathcal{L}(X, X^{\odot *})$, equipped with the operator norm.

Convention **4.3.** Functions of the variables s and t are continuously extended as constants outside the interval $[\alpha, \omega]$.

Lemma 4.4. *Under Assumption* 4.2 *we have that uniformly for* $(t,s) \in \triangle$

(4.9) $$\lim_{h\to 0} ||U(t+h, s+h) - U(t,s)|| = 0.$$

Proof. The AIE (2.8) implies that

$$\begin{aligned}U(t+h,&s+h)\varphi - U(t,s)\varphi \\ &= \int_s^t T_0^{\odot *}(t-\tau)\big[B(\tau+h)-B(\tau)\big]U(\tau+h,s+h)\varphi\,d\tau \\ &\quad + \int_s^t T_0^{\odot *}(t-\tau)B(\tau)\big[U(\tau+h,s+h)\varphi - U(\tau,s)\varphi\big]\,d\tau.\end{aligned}$$

So, there exists a constant $M<\infty$ such that

$$\begin{aligned}\|U(t+h,s+h)\varphi - U(t,s)\varphi\| \le M\|\varphi\| &\sup_{\alpha\le\tau\le\omega}\|B(\tau+h)-B(\tau)\| \\ &+ M\int_s^t \|U(\tau+h,s+h)\varphi - U(\tau,s)\varphi\|\,d\tau.\end{aligned}$$

The assertion now follows from Gronwall's Lemma ([101], Section I.6). □

Theorem 4.5. *Assumption* 4.2 *guarantees that* $X^{\odot}$ *is invariant under*

$$V(s,t) = (U(t,s))^*.$$

Proof. Straightforward estimates show that

$$\|\int_s^t T_0^{\odot *}(t-\tau)B(\tau)U(\tau,s)\,d\tau\| \le C(t-s)$$

for some constant $C<\infty$. Hence

$$V(s,t)\varphi^* - \varphi^* = T_0^*(t-s)\varphi^* - \varphi^* + O(t-s)\|\varphi^*\|$$

and we conclude that $\varphi^*\in X^{\odot}$ if and only if $\|V(s,t)\varphi^* - \varphi^*\|\to 0$ as $s\uparrow t$.

Next observe that

$$V(s-h,t)\varphi^{\odot} - V(s,t)\varphi^{\odot} = V(s-h,s)V(s,t)\varphi^{\odot} - V(s,t)\varphi^{\odot}$$

and that, therefore, $X^{\odot}$ is invariant if and only if the mapping $s\mapsto V(s,t)\varphi^{\odot}$ is left continuous from $[\alpha,\omega]$ to X^* for every $\varphi^{\odot}\in X^{\odot}$ and every $t\in(\alpha,\omega]$.

Finally, the identity

$$V(s-h,t)\varphi^{\odot} = V(s-h,t-h)V(t-h,t)\varphi^{\odot}$$

and Lemma 4.4 [recall that $V(s,t)=(U(t,s))^*$ and that norm continuity of operator valued functions is preserved under taking adjoints] imply that under Assumption 4.2 the mapping $s\mapsto V(s,t)\varphi^{\odot}$ is indeed left continuous. □

When $X^{\odot}$ is invariant, we define

$$V^{\odot}(s,t) = V(s,t)\big|_{X^{\odot}}, \tag{4.10}$$

$$U^{\odot *}(t,s) = \left(V^{\odot}(s,t)\right)^{*}. \tag{4.11}$$

Exercise 4.6. Prove that $U^{\odot *}$ is a forward evolutionary system on $X^{\odot *}$ which extends jUj^{-1}.

Exercise 4.7. Prove that $V^{\odot}$ is strongly continuous.

Exercise 4.8. Verify the following versions of the variation-of-constants identity:

$$V(s,t)\varphi^{\odot} = T_0^{\odot}(t-s)\varphi^{\odot} + \int_s^t V^{*}(s,\tau)B^{*}(\tau)T_0^{\odot}(t-\tau)\varphi^{\odot}\,d\tau, \tag{4.12}$$

$$U(t,s)\varphi = T_0(t-s)\varphi + \int_s^t U^{\odot *}(t,\tau)B(\tau)T_0(\tau-s)\varphi\,d\tau, \tag{4.13}$$

$$V(s,t)\varphi^{\odot} = T_0^{\odot}(t-s)\varphi^{\odot} + \int_s^t T_0^{*}(\tau-s)B^{*}(\tau)V^{\odot}(\tau,t)\varphi^{\odot}\,d\tau, \tag{4.14}$$

where $\varphi \in X$ and $\varphi^{\odot} \in X^{\odot}$.

XII.5 Perturbations with finite dimensional range

In Section 3 we established the one-to-one correspondence between the AIE (2.8) and the RFDE (3.2) without talking, as we did in the autonomous case, about Volterra integral equations for $\mathbb{R}^n$ ($\mathbb{C}^n$)-valued functions. In this section we show that actually most of Section III.3 carries over to the time-dependent situation at the expense of increased notational complication. Many of the derivations amount to formula manipulation and are therefore cast in the form of exercises. We are not sure how useful the results of this section really are, but refer to [51], where closely related ideas are elaborated with control theory as the motivating driving force.

So, let us assume that the range of $B(t)$ lies in a fixed finite dimensional subspace of $X^{\odot *}$ or, more precisely, that

$$B(t)\varphi = \sum_{j=1}^{n} \langle r_j^{*}(t), \varphi\rangle r_j^{\odot *} \tag{5.1}$$

where $t \mapsto r_j^{*}(t)$ is continuous from $[\alpha,\omega]$ to X^{*} for $j = 1,\dots,n$. Note that the range of $B^{*}(t)$ is finite dimensional too, but does depend on t. Also note that both RFDE and age-dependent population models lead to perturbations of this form, but that the continuity of $t \mapsto r_j^{*}(t)$ imposes a certain restriction (cf. Section 4).

To motivate subsequent definitions and arguments, we start with a formal and, in fact, unjustifiable calculation. Suppose the AIE (2.8) would make sense on $X^{\odot *}$ or, in other words, suppose that we could write

$$U^{\odot *}(t,s)\varphi^{\odot *} = T_0^{\odot *}(t-s)\varphi^{\odot *} + \sum_{k=1}^{n} \int_s^t T_0^{\odot *}(t-\sigma)\langle r_k^*(\sigma), U^{\odot *}(\sigma,s)\varphi^{\odot *}\rangle r_k^{\odot *}\, d\sigma.$$

Choosing $\varphi^{\odot *} = r_j^{\odot *}$ and applying $r_i^*(t)$ to both sides of this equation, we obtain the finite dimensional integral equation

$$R(t,t-s) = K(t,t-s) + \int_s^t K(t,t-\sigma)R(\sigma,\sigma-s)\, d\sigma, \tag{5.3}$$

where R and K are "defined" by

$$K_{ij}(t,t-s) = \langle r_i^*(t), T_0^{\odot *}(t-s)r_j^{\odot *}\rangle, \tag{5.4}$$

$$R_{ij}(t,t-s) = \langle r_i^*(t), U^{\odot *}(t,s)r_j^{\odot *}\rangle. \tag{5.5}$$

Expressed in words, this means that the kernel R is the resolvent of the kernel K.

Exercise 5.1. In the same spirit, "define" for given $\varphi^{\odot *} \in X^{\odot *}$, the n-vectors $f(t,s)$ and $y(t,s)$ by

$$f_i(t,s) = \langle r_i^*(t), T_0^{\odot *}(t-s)\varphi^{\odot *}\rangle, \tag{5.6}$$

$$y_i(t,s) = \langle r_i^*(t), U^{\odot *}(t,s)\varphi^{\odot *}\rangle. \tag{5.7}$$

Show that y satisfies the equation

$$y(t,s) = f(t,s) + \int_s^t K(t,t-\sigma)y(\sigma,s)\, d\sigma \tag{5.8}$$

and that, consequently, y can be expressed explicitly in terms of the forcing function f and the resolvent R by the formula

$$y(t,s) = f(t,s) + \int_s^t R(t,t-\tau)f(\tau,s)\, d\tau. \tag{5.9}$$

Exercise 5.2. Show that $U^{\odot *}(t,s)$ can be expressed in terms of the unperturbed semigroup and the function y defined by (5.7) as follows:

$$U^{\odot *}(t,s)\varphi^{\odot *} = T_0^{\odot *}(t-s)\varphi^{\odot *} + \sum_{k=1}^{n} \int_s^t T_0^{\odot *}(t-\sigma)y_k(\sigma,s)r_k^{\odot *}\, d\sigma. \tag{5.10}$$

(Note that y depends on $\varphi^{\odot *}$.)

In the calculations above, undefined quantities abound. To avoid these, we shall now follow a roundabout way. First, we introduce K as the L^∞-derivative of a Lipschitz function, just as we did in Section III.3. We then derive R as the solution of the equation (5.3). For initial data in X and $X^\odot$, we derive integral equations in the spirit of (5.8) and then represent the solution as in (5.9). Finally, duality is used to extend these representations to, respectively, X^* and $X^{\odot *}$.

Lemma 5.3. *There exists a $n \times n$ matrix-valued function $K(t,\tau)$, $\alpha \le t \le \omega$, $\tau \ge 0$, with $K(t,\cdot) \in L^\infty_{loc}(\mathbb{R}_+)$, such that*

$$\int_0^u k_{ij}(t,\tau)\,d\tau = \langle r_i^*(t), \int_0^u T_0^{\odot *}(\tau) r_j^{\odot *}\,d\tau\rangle \tag{5.11}$$

for $\alpha \le t \le \omega$ and $u \ge 0$, where k_{ij} are the entries of K.

The proof of this lemma is, apart from notational differences, identical to the proof of Lemma III.3.1–Corollary III.3.2.

Exercise 5.4. Prove that $t \mapsto K(t,\cdot)$ is continuous from $[\alpha,\omega]$ to $L^\infty([0,t_{\max}])$ for any $t_{\max} < \infty$.
Hint: Show that

$$\begin{aligned}|\langle r_i^*(t_1), \int_{u_1}^{u_2} T_0^{\odot *}(\tau) r_j^{\odot *}\,d\tau\rangle - \langle r_i^*(t_2), \int_{u_1}^{u_2} T_0^{\odot *}(\tau) r_j^{\odot *}\,d\tau\rangle| \\ \le M\frac{e^{\omega u_2} - e^{\omega u_1}}{\omega}\|r_j^{\odot *}\|\|r_i^*(t_1) - r_i^*(t_2)\|.\end{aligned}$$

Exercise 5.5. Show that for any $\eta \in L^1$, the identity

$$\langle r_i^*(t), \int_0^u T_0^{\odot *}(u-\tau)\eta(\tau) r_j^{\odot *}\,d\tau\rangle = \int_0^u k_{ij}(t,u-\tau)\eta(\tau)\,d\tau$$

holds.
Hint: See the proof of Lemma III.3.3.

Lemma 5.6. *For given $\varphi \in X$, define the n-vectors $f(t,s)$ and $y(t,s)$ by*

$$f_i(t,s) = \langle r_i^*(t), T_0(t-s)\varphi\rangle, \tag{5.13}$$

$$y_i(t,s) = \langle r_i^*(t), U(t,s)\varphi\rangle. \tag{5.14}$$

Then the Volterra integral equation (5.8) *holds.*

We leave the proof of this lemma, which is based on an application of $r_i^*(t)$ to the AIE (2.8) and some formula manipulation, to the reader.

The dual version of Lemma 5.6 reads as follows.

Lemma 5.7. *For given $\varphi^{\odot} \in X^{\odot}$, define the n-vectors $g(s,t)$ and $z(s,t)$ by*

$$g_i(t,s) = \langle r_i^{\odot *}(t), T_0^{\odot}(t-s)\varphi^{\odot}\rangle, \tag{5.15}$$

$$z_i(t,s) = \langle r_i^{\odot *}(t), V^{\odot}(s,t)\varphi^{\odot}\rangle. \tag{5.16}$$

Then

$$z(s,t) = g(s,t) + \int_s^t K^T(\tau, \tau - s) z(\tau, t)\, d\tau. \tag{5.17}$$

The derivation of (5.17) starts from (4.14), which we here repeat, while inserting the specific form (5.1) for $B(t)$, as

$$\begin{aligned} V^{\odot}(s,t)\varphi^{\odot} &= T_0^{\odot}(t-s)\varphi^{\odot} \\ &\quad + \sum_{k=1}^{n} \int_s^t T_0^*(\tau - s)\langle r_k^{\odot *}, V^{\odot}(\tau, t)\varphi^{\odot}\rangle r_k^*(\tau)\, d\tau. \end{aligned} \tag{5.18}$$

Since r_k^* depends on time (such in contrast to $r_k^{\odot *}$), the derivation of (5.17) is slightly more involved than the derivation of (5.8). In fact, we need the following variant of Lemma III.3.3 and Exercise 5.5.

Lemma 5.8. *For any $\eta \in L^1$, the identity*

$$\langle r_j^{\odot *}, \int_s^t T_0^*(\tau - s) r_i^*(\tau)\eta(\tau)\, d\tau\rangle = \int_s^t k_{ij}(\tau, \tau - s)\eta(\tau)\, d\tau \tag{5.19}$$

holds.

Exercise 5.9. Prove Lemma 5.8.
Hint: Integrate both sides of (5.19) with respect to s from u to t. Next, combine the following three observations:
(i) $\int_u^t \int_s^t k_{ij}(\tau, \tau - s)\eta(\tau)\, d\tau\, ds = \int_u^t \langle r_i^*(\tau), \int_u^\tau T_0^{\odot *}(\tau - s) r_j^{\odot *}\, ds\rangle \eta(\tau)\, d\tau$,
(ii) $\int_u^t \langle r_j^{\odot *}, \int_s^t T_0^*(\tau - s) r_i^*(\tau)\eta(\tau)\, d\tau\rangle\, ds = \int_u^t \langle r_j^{\odot *}, \int_u^\tau T_0^*(\tau - s) r_i^*(\tau)\, ds\rangle \eta(\tau)\, d\tau$,
(iii) $\langle \varphi^*, \int_0^t T_0^{\odot *}(\tau)\varphi^{\odot *}\, d\tau\rangle = \langle \varphi^{\odot *}, \int_0^t T_0^*(\tau)\varphi^*\, d\tau\rangle$,
for all $\varphi^* \in X^*$, $\varphi^{\odot *} \in X^{\odot *}$ (cf. Lemma III.2.16).

Exercise 5.10. Prove Lemma 5.7.

When considering initial data in one of the "big" spaces $X^{\odot *}$ or X^*, one can no longer define the forcing function for the Volterra integral equation by the explicit expression (5.13) or (5.15). But one can use the same trick that led to the definition of K! We now summarize the results of this sections in two theorems.

Theorem 5.11. *Let, for given $\varphi^{\odot *} \in X^{\odot *}$, $y(t,s)$ be the unique L^∞-solution of the Volterra integral equation* (5.8) *with f defined by*

$$f_i(t,s) = \frac{d}{du}\langle r_i^*(t), \int_0^u T_0^{\odot *}(\tau)\varphi^{\odot *}\,d\tau\rangle\big|_{u=t-s} \tag{5.20}$$

where the derivative maps a Lipschitz function to an element of L^∞. Then, for $\alpha \le s \le t \le \omega$

$$U^{\odot *}(t,s)\varphi^{\odot *} = T_0^{\odot *}(t-s)\varphi^{\odot *} + \sum_{k=1}^{n}\int_s^t T_0^{\odot *}(t-\tau)r_k^{\odot *}y_k(\tau,s)\,d\tau. \tag{5.21}$$

Theorem 5.12. *Let, for given $\varphi^* \in X^*$, $z(s,t)$ be the unique L^∞-solution of the Volterra integral equation* (5.17) *with g defined by*

$$g_i(s,t) = \frac{d}{du}\langle \int_0^u T_0^*(\tau)\varphi^*\,d\tau, r_i^{\odot *}\rangle\big|_{u=t-s}, \tag{5.22}$$

where the derivative maps a Lipschitz function to an element of L^∞. Then, for $\alpha \le s \le t \le \omega$

$$V(s,t)\varphi^* = T_0^*(t-s)\varphi^* + \sum_{k=1}^{n}\int_s^t T_0^*(\sigma - s)r_k^*(\sigma)z_k(\sigma,t)\,d\sigma. \tag{5.23}$$

Proof of Theorem 5.11. We first note that the solution z of (5.17) is given explicitly by

$$z(s,t) = g(s,t) + \int_s^t R^T(\tau,\tau-s)g(\tau,t)\,d\tau, \tag{5.24}$$

which follows more or less directly from the following alternative way of writing the resolvent equation (5.3):

$$R(t,t-s) = K(t,t-s) + \int_s^t K^T(\tau,\tau-s)R^T(t,t-\tau)\,d\tau. \tag{5.25}$$

Now let $\varphi^\odot$ be an arbitrary element of $X^\odot$. Then, with $g(s,t)$ defined by (5.15), we have

$$\begin{aligned}\langle \int_s^t T_0^{\odot *}(t-\tau)r_k^{\odot *}y_k(\tau,s)\,d\tau, \varphi^\odot\rangle &= \int_s^t g_k(\tau,t)y_k(\tau,s)\,d\tau\\ = \int_s^t g_k(\tau,t)\big\{f_k(\tau,s) &+ \sum_{j=1}^{n}\int_s^\tau R_{kj}(\tau,\tau-\sigma)f_j(\sigma,s)\,d\sigma\big\}\,d\tau\\ = \int_s^t \{g_k(\sigma,t)f_k(\sigma,s) &+ \sum_{j=1}^{n}\int_\sigma^t g_k(\tau,t)R_{kj}(\tau,\tau-\sigma)\,d\tau\, f_j(\sigma,s)\big\}\,d\sigma\end{aligned}$$

and hence

$$\sum_{k=1}^{n} \langle \int_s^t T_0^{\odot *}(t-\tau) r_k^{\odot *} y_k(\tau, s)\, d\tau, \varphi^{\odot} \rangle$$
$$= \sum_{j=1}^{n} \int_s^t \{ g_j(\sigma, t) + \sum_{k=1}^{n} \int_\sigma^t g_k(\tau, t) R_{kj}(\tau, \tau - \sigma)\, d\tau \} f_j(\sigma, s)\, d\sigma$$
$$= \sum_{j=1}^{n} \int_s^t z_j(\sigma, t) f_j(\sigma, s)\, d\sigma.$$

On the other hand, we can use (5.18) and (5.16) to write

$$\langle U^{\odot *}(t,s)\varphi^{\odot *} - T_0^{\odot *}(t-s)\varphi^{\odot *}, \varphi^{\odot} \rangle$$
$$= \langle \varphi^{\odot *}, \sum_{k=1}^{n} \int_s^t T_0^{*}(\sigma - s) z_k(\sigma, t) r_k^{*}(\sigma)\, d\sigma \rangle$$
$$= \lim_{h \downarrow 0} \sum_{k=1}^{n} \langle \int_s^t T_0^{*}(\sigma - s) z_k(\sigma, t) r_k^{*}(\sigma)\, d\sigma, \frac{1}{h} \int_0^h T_0^{\odot *}(\tau) \varphi^{\odot *}\, d\tau \rangle$$
$$= \lim_{h \downarrow 0} \sum_{k=1}^{n} \int_s^t z_k(\sigma, t) f_k^h(\sigma, s)\, d\sigma,$$

where

$$f_i^h(t,s) = \langle r_i^{*}(t), T_0(t-s) \frac{1}{h} \int_0^h T_0^{\odot *}(\tau) \varphi^{\odot *}\, d\tau \rangle.$$

We claim that

$$\lim_{h \downarrow 0} \int_s^t z_k(\sigma, t) f_k^h(\sigma, s)\, d\sigma = \int_s^t z_k(\sigma, t) f_k(\sigma, s)\, d\sigma.$$

Combination of our two identities then shows the correctness of representation (5.21).

To prove the claim, we first note that

$$f_i^h(\sigma, s) = \frac{1}{h} \langle r_i^{*}(\sigma), \int_0^h T_0^{\odot *}(\tau + \sigma - s) \varphi^{\odot *}\, d\tau \rangle \to f(\sigma, s),$$

as $h \downarrow 0$, for almost all σ. Hence the claim follows from the dominated convergence theorem. □

Exercise 5.13. Provide the details for a proof of Theorem 5.12 along exactly the same lines.

Exercise 5.14. Show that with the choice of $\varphi^{\odot *} = r_i^{\odot *}$, we have that $y(t,s)$ is the ith column of $R(t, t-s)$.

Exercise 5.15. Argue as in Section III.4 to derive that for RFDE,

$$K(t,\tau) = \zeta(t,\tau),$$

and that the corresponding forcing function g in the Volterra integral equation (5.17) is obtained from the initial state φ^*, by the identification

$$g(s,t) = \varphi^*(t-s), \qquad t \geq s.$$

Exercise 5.16. Let H be a Banach space whose elements are functions defined on $\mathbb{R}_+$. Let the kernel $L(t,\sigma)$, $\alpha \leq t \leq \omega$, $\sigma \geq 0$, be such that

(i) for any $h \in H$, the equation

$$z(s,t) = h(t-s) + \int_s^t L(\tau,\tau-s)z(\tau,t)\,d\tau, \qquad \alpha \leq s \leq t \leq \omega,$$

admits a unique solution;

(ii) the formula

$$(V(s,t)h)(\sigma) = h(t-s+\sigma) + \int_s^t L(\tau,\tau-s+\sigma)z(\tau,t)\,d\tau$$

defines a two-parameter family of bounded linear operators on H.

Show that V is a backward evolutionary system on H.
Hint: Write $z(s,t;h)$ to emphasize the dependence on h and verify that for $\eta \geq 0$, such that $s-\eta \geq \alpha$,

$$z(s-\eta, s; V(s,t)h) = z(s-\eta, t; h).$$

Exercise 5.17. Combine the results of the last two exercises to derive an interpretation for the adjoint of the forward evolutionary system corresponding to translation along solutions of RFDE.

XII.6 Comments

Resolvent kernels and their properties are studied by Gripenberg et al. [97]. In Section III.6 we briefly touched upon the so-called structural operators F and G. The analogue theory for time-dependent RFDE is developed by Colonius, Manitius and Salamon [51]. For equations with state-dependent delays, we refer to the work of Mallet-Paret and Nussbaum [181, 182] and the references given there.

Chapter XIII
Floquet theory

XIII.1 Introduction

Floquet theory deals with periodic linear systems. Let a strongly continuous semigroup $\{T_0(t)\}_{t\geq 0}$ on a complex Banach space X be given. Consider a strongly continuous family of operators

$$B(t) \in \mathcal{L}(X, X^{\odot *}), \qquad t \in \mathbb{R},$$

which is periodic: there exists $p > 0$ such that

$$B(t+p) = B(t) \qquad \text{for all } t \in \mathbb{R}.$$

We consider the evolutionary system of operators

$$U(t,s), \quad t,s \in \mathbb{R} \quad \text{with } t \geq s,$$

associated with

$$u(t) = T_0(t-s)u(s) + \int_s^t T_0^{\odot *}(t-\tau)B(\tau)u(\tau)\,d\tau. \tag{1.1}$$

The asymptotic properties are determined by the period maps (or "monodromy operators")

$$U(t+p,t), \quad t \in \mathbb{R}$$

and, in particular, by their spectra.

The main result of the classical theory for ODE is a reduction of the periodic linear system to an autonomous linear system by a periodic transformation. We shall derive a related result, Theorem 4.5.

An important application of Floquet theory concerns periodic orbits of nonlinear systems: the local behaviour of a semiflow Σ close to a periodic orbit ϑ is, to the first order, determined by the derivatives $D_2\Sigma(t,\varphi)$, where $\varphi \in \vartheta$. According to Section VII.4, these maps are given by solutions of a periodic equation of the form (1.1). Upon complexification, the results of the present chapter become applicable. Details of this are contained in Chapter XIV.

Exercise 1.1. Set $\mathcal{C} = C([-h,0], \mathbb{R}^n)$. Consider a periodic solution $x : \mathbb{R} \to \mathbb{R}^n$ of a RFDE

$$\dot{x}(t) = g(x_t),$$

where $g : \mathcal{C} \to \mathbb{R}^n$ is of class C^1. Let Σ denote the associated semiflow. Show that the derivatives $D_2\Sigma(t, x_0)$ are given by the solutions of a periodic linear RFDE. The maps $U_x(t,s)$ of the corresponding evolutionary system are compact for $t \geq h + s$.

XIII.2 Preliminaries on periodicity and a stability result

Proposition 2.1. *Suppose* $u : [s, \infty) \to X$ *is the solution of* (1.1) *with* $u(s) = \varphi$. *Then* $v : [s+p, \infty) \to X$, $v(t) = u(t-p)$, *is a solution of* (1.1) *with* $v(s+p) = \varphi$.

Proof. For all $t \geq s + p$,

$$\begin{aligned} v(t) = u(t-p) &= T_0(t-p-s)\varphi + \int_s^{t-p} T_0^{\odot *}(t-p-\tau)B(\tau)u(\tau)\, d\tau \\ &= T_0(t-(s+p))\varphi + \int_s^{t-p} T_0^{\odot *}(t-(\tau+p))B(\tau)u(\tau)\, d\tau \\ &= T_0(t-(s+p))\varphi + \int_{s+p}^{t} T_0^{\odot *}(t-\tau)B(\tau-p)u(\tau-p)\, d\tau \\ &= T_0(t-(s+p))\varphi + \int_{s+p}^{t} T_0^{\odot *}(t-\tau)B(\tau)v(\tau)\, d\tau, \end{aligned}$$

where, in the last step, we used the periodicity of B. □

Corollary 2.2.
(i) *For* $t \geq s$, $U(t+p, s+p) = U(t,s)$.
(ii) *For* $t \geq 0$ *and* $j \in \mathbb{Z}$, $U(t+jp, jp) = U(t,0)$.
(iii) *For* $t \in \mathbb{R}$ *and* $j \in \mathbb{N}$, $U(t+jp, t) = U(t+p,t)^j$.

Proof. To prove (i), let $t \geq s$, $\varphi \in X$ and $U(t,s)\varphi = u(t)$, where $u : [s,\infty) \to X$ is the solution of (1.1) with $u(s) = \varphi$. We have

$$U(t+p, s+p)\varphi = v(t+p),$$

where $v : [s+p, \infty) \to X$ is the solution of (1.1) with $v(s+p) = \varphi$. Proposition 2.1 and uniqueness for the initial-value problems associated with (1.1) now imply the assertion. The identity in (ii) follows from (i) by induction. To prove (iii), let $t \in \mathbb{R}$ be given. In case $j = 0, 1$, there is nothing to prove. Suppose the assertion holds true for some $j \in \mathbb{N}$. Then, using (i) twice,

$$\begin{aligned} U(t+(j+1)p,t) &= U(t+jp+p,t) = U(t+jp,t-p) \\ &= U(t+jp,t)U(t,t-p) = U(t+jp,t)U(t+p,t) \\ &= U(t+p,t)^j U(t+p,t) = U(t+p,t)^{j+1}. \end{aligned}$$

□

Exercise 2.3. Consider a continuous map $\zeta : \mathbb{R} \times \mathcal{C} \to \mathbb{C}^n$ [where $\mathcal{C} = C([-h,0],\mathbb{C}^n)$] such that every $\zeta(t) : \mathcal{C} \to \mathbb{C}^n$, $\zeta(t)\varphi := \zeta(t,\varphi)$, is linear. Assume that for some $p > 0$,

$$\zeta(t+p) = \zeta(t).$$

Show that all iterates $U(t+p,t)^j$, $jp \geq h$, of maps from the evolutionary system associated with the periodic linear RFDE

$$\dot{x}(t) = \langle \zeta(t), x_t \rangle_n$$

are compact.

Theorem 2.4. (*Stability theorem.*) *Let $s \in \mathbb{R}$ be given. Assume*

$$\sigma(U(s+p,s)) \subset B_1(0).$$

Then there exist constants $C \geq 0$, $\epsilon > 0$, such that

$$||U(t,s)|| \leq Ce^{-\epsilon(t-s)} \qquad \textit{for all } t \geq s.$$

Proof. The spectral radius formula

$$\max_{\lambda \in \sigma(U(s+p,s))} |\lambda| = \lim_{j\to\infty} ||U(s+p,s)^j||^{\frac{1}{j}}$$

implies that for some $k \in \mathbb{N}$, $||U(s+p,s)^k|| < 1$. Since the curves

$$t \mapsto U(t,s)\varphi, \quad \varphi \in X,$$

are continuous, the family of maps $U(t,s)$, $s \leq t \leq s+p$, is pointwise bounded. So the Principle of Uniform Boundedness yields a constant $C_1 > 0$ such that

$$||U(t,s)|| \leq C_1 \quad s \leq t \leq s+p.$$

Set

$$C := C_1 \max_{j=0,\ldots,k-1} ||U(s+p,s)^j||$$

and let $t \geq s$ be given. Consider the largest integer m_t so that $s + m_t kp \leq t$, and the largest integer $m_t^* \in \{0,\ldots,k-1\}$ with $s + m_t kp + m_t^* p \leq t$. We have

$$\begin{aligned}
U(t,s)& \\
&= U(t, s+m_t kp + m_t^* p)U(s+m_t kp + m_t^* p, s+m_t kp)U(s+m_t kp, s) \\
&= U(t - m_t kp - m_t^* p, s)U(s+m_t^* p, s)U(s+m_t kp, s) \\
&= U(t - m_t kp - m_t^* p, s)U(s+p, s)^{m_t^*} U(s+p,s)^{km_t}.
\end{aligned}$$

Therefore

$$||U(t,s)|| \leq C||U(s+p,s)^k||^{m_t}$$

and it becomes obvious how to finish the proof. □

XIII.3 Floquet multipliers

Define the family of *period maps* V_t by

$$V_t := U(t+p, t).$$

According to Corollary 2.2(ii), $V_{t+p} = V_t$ for all $t \in \mathbb{R}$. We would like to compare the spectra σ_t of the operators V_t and derive relations between their generalized eigenspaces. We start with a proposition.

Proposition 3.1. *For t, s in $\mathbb{R}$ with $t \geq s$,*

$$V_t = A_t A_s \quad \textit{and} \quad V_s = A_s A_t,$$

where

$$A_s = U(s+jp, t), \quad A_t = U(t+p, s+jp)$$

and $j \in \mathbb{N}$ is chosen such that $s + (j-1)p \leq t < s + jp$.

Proof. We only prove the identity $V_s = A_s A_t$:

$$\begin{aligned}
V_s = V_{s+jp} &= U(s+(j+1)p, s+jp) \\
&= U(s+(j+1)p, t+p)U(t+p, s+jp) \\
&= U(s+jp, t)U(t+p, s+jp),
\end{aligned}$$

by Corollary 2.2(i). □

A bounded linear operator on X is said to have the *spectral isolation property* if each nonzero point in its spectrum is an isolated point of the spectrum. A family of bounded linear operators on X is said to have the spectral isolation property if each of its members has it. From now on, we assume

Hypothesis 3.2. The family $\{V_t\}_{t\in\mathbb{R}}$ has the spectral isolation property.

This is satisfied if, for example, iterates of V_t are compact. The latter is true for periodic linear RFDE [see Exercise 2.3 and Corollary 2.2(iii)].

For $t \in \mathbb{R}$ and $0 \neq \lambda \in \sigma_t$, let $\mathcal{M}_{\lambda,t}$ denote the generalized eigenspace of V_t.

Theorem 3.3. *Let t, s in $\mathbb{R}$ be given with $t \geq s$ and let $\lambda \in \mathbb{C}\backslash\{0\}$. Then*
(i) $\lambda \in \sigma_t$ *if and only if* $\lambda \in \sigma_s$
and, in this case,
(ii) $U(t,s)$ *reduces to a topological isomorphism of* $\mathcal{M}_{\lambda,s}$ *onto* $\mathcal{M}_{\lambda,t}$.

Proof. 1. Let $t \geq s$ be given.
2. The spectral projection $P_{\lambda,t}$ of V_t with respect to a point $\lambda \in \sigma_t\backslash\{0\}$ maps X onto $\mathcal{M}_{\lambda,t}$ and is given by the contour integral

$$P_{\lambda,t} = \frac{1}{2\pi i}\int_{\Gamma}(zI - V_t)^{-1}\,dz, \tag{3.1}$$

where Γ is a small circle enclosing only the isolated point λ of σ_t. Due to the holomorphy of the resolvent $(zI - V_t)^{-1}$ in a neighbourhood of the points $\lambda \in \mathbb{C}\backslash(\sigma_t \cup \{0\})$, we can extend the definition (3.1) and find

$$P_{\lambda,t} = 0 \quad \text{for } 0 \neq \lambda \in \mathbb{C}\backslash\sigma_t.$$

Analogously, we consider $P_{\lambda,s}$ and $\mathcal{M}_{\lambda,s}$ for all $\lambda \in \mathbb{C}\backslash\{0\}$.
3. We first prove that

$$A_t P_{\lambda,s} = P_{\lambda,t} A_t \quad \text{for } 0 \neq \lambda \in \mathbb{C}. \tag{3.2}$$

Choose $\epsilon > 0$ so small that $z \notin \sigma_t \cup \sigma_s$ for $\lambda \neq z \in \overline{B_\epsilon(\lambda)}$. The equation

$$A_t(zI - A_s A_t) = (zI - A_t A_s)A_t, \quad z \in \mathbb{C},$$

and Proposition 3.1 imply

$$(zI - V_t)^{-1}A_t = A_t(zI - V_s)^{-1}, \quad |z - \lambda| < \epsilon,$$

so that integration along Γ yields the assertion (3.2).
4. As above, $A_s P_{\lambda,t} = P_{\lambda,s} A_s$ for $0 \neq \lambda \in \mathbb{C}$.
5. Let $\lambda \in \mathbb{C}\backslash\{0\}$ be given. The map $\mathcal{M}_{\lambda,t} \to \mathcal{M}_{\lambda,s}$, $\varphi \mapsto A_s\varphi$, is an isomorphism: from Part 3, we have $A_t\mathcal{M}_{\lambda,s} \subset \mathcal{M}_{\lambda,t}$. Therefore Proposition 3.1 yields

$$\mathcal{M}_{\lambda,s} = V_s(\mathcal{M}_{\lambda,s}) = A_s A_t(X_{\lambda,s}) \subset A_s(\mathcal{M}_{\lambda,t}) \subset \mathcal{M}_{\lambda,s},$$

where the last inclusion follows from Part 4. So A_s, in fact, maps $\mathcal{M}_{\lambda,t}$ onto $\mathcal{M}_{\lambda,s}$. Injectivity is a consequence of $V_t = A_t A_s$ and the fact that V_t defines an isomorphism of $\mathcal{M}_{\lambda,t}$.

6. As in Part 5. one proves that $\mathcal{M}_{\lambda,t} \to \mathcal{M}_{\lambda,s}$, $\varphi \mapsto A_t\varphi$, is an isomorphism.
7. The spaces $\mathcal{M}_{\lambda,s}$, $\mathcal{M}_{\lambda,t}$, $0 \neq \lambda \in \mathbb{C}$, are closed. The Open Mapping Theorem implies that the continuous isomorphisms of Part 5 and 6 are topological.
8. Proof of assertion (i). Let $\lambda \in \mathbb{C}\backslash\{0\}$ be given. We have $\lambda \in \sigma_t$ if and only if $\mathcal{M}_{\lambda,t} \neq \{0\}$; the latter is equivalent to $\mathcal{M}_{\lambda,s} \neq \{0\}$ (see Part 5) which, in turn, is equivalent to $\lambda \in \sigma_s$.
9. Proof of assertion (ii). For $0 \neq \lambda \in \sigma_s$,

$$\begin{aligned} U(t,s) &= U(t, s+(j-1)p)U(s+(j-1)p, s) \\ &= U(t+p, s+jp)V_s^{j-1} \quad \text{(Corollary 2.2)} \\ &= A_t V_s^{j-1}, \end{aligned}$$

where j is as in Proposition 3.1. V_s defines a topological isomorphism of $\mathcal{M}_{\lambda,s}$, and A_t defines a topological isomorphism of $\mathcal{M}_{\lambda,s}$ onto $\mathcal{M}_{\lambda,t}$, according to Part 6. Now (ii) becomes obvious. □

Exercise 3.4. Prove that the complementary spaces

$$Q_{\lambda,t} := \mathcal{N}(P_{\lambda,t}), \quad t \in \mathbb{R},\ 0 \neq \lambda \in \sigma_t,$$

are invariant under the evolutionary system, i.e., $U(t,s)Q_{\lambda,s} \subset Q_{\lambda,t}$.

Definition 3.5. The spectral points $\lambda \in \sigma_t\backslash\{0\}$, $t \in \mathbb{R}$ arbitrary, are called the *Floquet multipliers* (of the evolutionary system).

XIII.4 Floquet representation on eigenspaces

Let a Floquet multiplier λ be given, and consider the p-periodic family of generalized eigenspaces $\mathcal{M}_{\lambda,t}$. The operators

$$U_t : \mathcal{M}_{\lambda,0} \to \mathcal{M}_{\lambda,t}, \qquad U_t\varphi := U(t,0)\varphi, \quad t \geq 0,$$

and

$$\widehat{U}_t : \mathcal{M}_{\lambda,t} \to \mathcal{M}_{\lambda,0}, \qquad U_t\varphi := U(0,t)\varphi, \quad t \leq 0,$$

are topological isomorphisms (see Theorem 3.3). For $t < 0$, we define

$$U_t := (\hat{U}_t)^{-1}.$$

Exercise 4.1. Prove that there exist solutions $u : \mathbb{R} \to X$ of equation (1.1) such that

$$u(t) \in \mathcal{M}_{\lambda,t} \quad \text{for all } t \in \mathbb{R}. \tag{4.1}$$

We wish to represent solutions as in Exercise 4.1, which are defined on all of $\mathbb{R}$, by the flow of a continuous linear vector field. This requires, first, a state space. For this, we take $\mathcal{M}_{\lambda,0}$. Observe that U_p maps $\mathcal{M}_{\lambda,0}$ onto $\mathcal{M}_{\lambda,0}(=\mathcal{M}_{\lambda,p})$.

Proposition 4.2. (*Embedding into a flow.*) *There exists a continuous linear vector field* $W \in \mathcal{L}(\mathcal{M}_{\lambda,0})$ *such that*

$$U_p = e^{pW}.$$

Proof. Recall that $U_p\varphi = V_0\varphi$ on $\mathcal{M}_{\lambda,0}$. Since $\lambda \neq 0$, we can choose a branch $\log_\lambda$ of the logarithm which is analytic in a neighbourhood of $\sigma(U_p) = \{\lambda\}$. Then we define $W := \frac{1}{p}\log_\lambda U_p$. □

In general, the map $U(p,0)$ on the full space X cannot be embedded into the flow of a continuous linear vector field on X.

Exercise 4.3. Give an example!

In order to relate solutions of equation (1.1) through the spaces $\mathcal{M}_{\lambda,t}$ to the flow $(e^{tW})_{t\in\mathbb{R}}$ on $\mathcal{M}_{\lambda,0}$, we consider the maps

$$R_t : \mathcal{M}_{\lambda,0} \to X, \qquad R_t\varphi := U_t e^{-tW}\varphi, \quad t \in \mathbb{R}.$$

Proposition 4.4. *For all* $t \in \mathbb{R}$, $R_{t+p} = R_t$.

Proof. 1. First, we show that $U_{t+p} = U_tU_p$. For $t \geq 0$ and $\varphi \in \mathcal{M}_{\lambda,0}$,

$$\begin{aligned} U_{t+p}\varphi &= U(t+p,0)\varphi = U(t+p,p)U(p,0)\varphi \\ &= U(t,0)U(p,0)\varphi, \quad \text{by Corollary 2.2(i).} \end{aligned}$$

For $t < 0 \leq t+p$ and $\varphi \in \mathcal{M}_{\lambda,0}$,

$$\begin{aligned} U_p\varphi &= U(p,0)\varphi = U(0,-p)\varphi = U(0,t)U(t,-p)\varphi \\ &= U(0,t)U(t+p,0)\varphi = U_t^{-1}U_{t+p}\varphi. \end{aligned}$$

For $t+p < 0$ and $\varphi \in \mathcal{M}_{\lambda,0}$,

$$\begin{aligned} U_pU_{t+p}^{-1}\varphi &= U(p,0)U(0,t+p)\varphi = U(p,t+p)\varphi \\ &= U(0,t)\varphi = U_t^{-1}\varphi. \end{aligned}$$

2. For $t \in \mathbb{R}$ and $\varphi \in \mathcal{M}_{\lambda,0}$,

$$\begin{aligned} R_{t+p}\varphi &= U_{t+p}e^{-(t+p)W}\varphi = U_tU_pe^{-pW}e^{-tW}\varphi \\ &= U_t\big(U_pU_p^{-1}\big)e^{-tW}\varphi = R_t\varphi. \end{aligned}$$

□

Theorem 4.5. *For every solution $u : \mathbb{R} \to X$ of equation* (1.1) *such that $u(t) \in \mathcal{M}_{\lambda,t}$ for all t, we have*

$$u(t) = R_t e^{tW} u(0) \qquad \text{for all } t \in \mathbb{R}.$$

Proof.

$$u(t) = U_t u(0) = U_t e^{-tW} e^{tW} u(0) = R_t e^{tW} u(0).$$

□

Exercise 4.6. Let a finite set Λ of Floquet multipliers be given. Let $\mathcal{M}_{\Lambda,t}$, for $t \in \mathbb{R}$, denote the associated generalized eigenspace of V_t. Prove that there exist a map $W \in \mathcal{L}(\mathcal{M}_{\Lambda,0})$ and a periodic family of transformations $R_t : \mathcal{M}_{\Lambda,0} \to X$ such that for every solution $u : \mathbb{R} \to X$ of equation (1.1) which satisfies

$$u(t) \in \mathcal{M}_{\Lambda,t} \quad \text{for all } t \in \mathbb{R},$$

one has

$$u(t) = R_t e^{tW} u(0), \quad t \in \mathbb{R}.$$

Remark **4.7**. The last statement contains the main result of the classical Floquet theory: in case of a periodic linear ODE on $X := \mathbb{C}^n$, one has

$$\mathbb{C}^n = \mathcal{M}_{\Lambda,t} \quad \text{for all } t \in \mathbb{R},$$

where Λ is the finite set of Floquet multipliers, and it follows that all solutions are given by the flow of a linear vector field.

Let us finally consider periodic linear RFDE

$$\dot{x}(t) = \langle \zeta(t), x_t \rangle_n \tag{4.2}$$

as in Exercise 2.3. It is sometimes convenient to have a flow representation for $\mathbb{C}^n$-valued solutions, in terms of coordinates. So fix a Floquet multiplier λ. Recall that now $\mathcal{M}_{\lambda,0}$ has finite dimension, say d. Choose a basis $\varphi_1, \ldots, \varphi_d$ of $\mathcal{M}_{\lambda,0}$. The linear vector field W on $\mathcal{M}_{\lambda,0}$ is then given by multiplication of coordinate vectors with a matrix $\widetilde{W} \in \mathbb{C}^{d,d}$. Set $\delta\varphi := \varphi(0)$, for $\varphi \in C$. The periodic family of maps $\delta R_t : \mathcal{M}_{\lambda,0} \to \mathbb{C}^n$, $t \in \mathbb{R}$, is given by matrices $\widetilde{R}_t \in \mathbb{C}^{n,d}$, with respect to the basis $\varphi_1, \ldots, \varphi_d$ and the standard basis of $\mathbb{C}^n$. For a solution $x : \mathbb{R} \to \mathbb{C}^n$ of equation (4.2) which satisfies $x_t \in \mathcal{M}_{\lambda,t}$ for all $t \in \mathbb{R}$, we obtain

$$x(t) = \delta x_t = \delta R_t e^{tW} x_0 = \widetilde{R}_t e^{t\widetilde{W}} z,$$

where $z = (z_1, \ldots, z_d)^T \in \mathbb{C}^d$ and $x_0 = \sum_{k=1}^d z_k \varphi_k$.

XIII.5 Comments

We refer to the works of Abraham and Robbin [1], Amann [5], Coddington and Levinson [49], Hartman [113] and Hale [101] for the classical Floquet theory, to Hale [102] for the case of RFDE and to Henry [121] for time-periodic linear perturbations of analytic semigroups. Also see the book by Kuchment [156] for PDE and Huang and Mallet-Paret [131] for a class of delay differential equations. Huang and Mallet-Paret [130] developed a homotopy method to locate Floquet multipliers for certain time-periodic linear delay differential equations.

In may be useful to look also for generalizations of the classical theory which aim at a reduction of the full evolutionary system associated with the periodic AIE (1.1), not only for solutions through eigenspaces, as in [102, 121] and in our Theorem 4.5. Of course, one cannot expect a reduction to a flow, under reasonably general assumptions. However, what are conditions which permit to reduce to a strongly continuous semiflow?

Chapter XIV
Periodic orbits

XIV.1 Introduction

Let a strongly continuous semigroup of operators $\{T_0(t)\}_{t\geq 0}$, on a real Banach space X be given. Assume that X is $\odot$-reflexive with respect to the semigroup. Consider a C^1-map $G : \mathcal{O} \to X^{\odot *}$ on some open set $\mathcal{O} \subset X$. The maximal solutions $u : I_\varphi \to X$ of the equation

$$u(t) = T_0(t)\varphi + \int_0^t T_0^{\odot *}(t-s)G(u(s))\,ds, \quad \varphi \in X, \tag{1.1}$$

define a continuous semiflow Σ whose partial derivatives $D_2\Sigma$ exist and are continuous on all of its domain

$$\Delta = \{(t,\varphi) \in [0,\infty) \times X : t \in I_\varphi\}.$$

In Chapters VIII–X we discussed the behaviour of Σ close to stationary points. These are the simplest positively invariant sets. The present chapter is devoted to the next order of complication. We provide the basic tools for the description of the behaviour of the semiflow close to the orbit $\vartheta \subset X$ of a periodic solution $u : \mathbb{R} \to X$ of equation (1.1).

The first step is linearization. Fix a point $\varphi = u(t_v) \in \vartheta$ and set $v = u(\,\cdot\, + t_v)$. The derivatives $D_2\Sigma(t,\varphi)$, $t \geq 0$ coincide with the evolution maps $U_v(t,0)$ determined by the solutions of the linear variational equation along v,

$$w(t) = T_0(t-s)w(s) + \int_s^t T_0^{\odot *}(t-\tau)DG\big(v(\tau)\big)w(\tau)\,d\tau. \tag{1.2}$$

Complexification yields an equation as in Chapter XIII. We assume the spectral isolation property (Hypothesis XIII.3.2) and obtain Floquet multipliers. Their relation to the translates of u and to the orbit ϑ is discussed in Section 2.

In Section 3 we introduce Poincaré maps Π on transversals to ϑ through some point $\varphi \in \vartheta$.

In order to be able to speak of transversals - and to construct a Poincaré map - we need derivatives of Σ with respect to the time variable t. So we have to make additional assumptions. Only then can properties of Σ close to ϑ be expressed in terms of the behaviour of Π close to its fixed point φ. The conceptual framework of Poincaré maps is most useful in dealing with many situations, from the existence of periodic solutions to the description of bifurcation and highly complicated phenomena, like chaos.

In Section 4 we establish the relation between Floquet multipliers and the spectra of $D\Pi(\varphi)$, for different choices of φ and the transversal.

So, there exists a route from the Floquet multipliers to the local behaviour of Σ close to ϑ. It should not be concealed that it is often difficult to get estimates of Floquet multipliers, as this requires a priori knowledge of the periodic orbit. Nevertheless, such estimates have been obtained in several problems; Section 5 contains references.

XIV.2 The Floquet multipliers of a periodic orbit

Suppose $u : \mathbb{R} \to X$ is a nonconstant periodic solution of (1.1), i.e., $\#u(\mathbb{R}) \geq 2$, and there exists $q > 0$ such that $u(\cdot + q) = u$. The number q is called a period of u, and $\vartheta = u(\mathbb{R})$ is called the orbit of u.

Proposition 2.1. *Suppose $u : \mathbb{R} \to X$ is a nonconstant periodic solution of equation* (1.1). *Then*

(i) *each $t > 0$ with $u(t) = u(0)$ is a period of u;*

(ii) *there exists a minimal period $p > 0$ of u;*

(iii) *the set of periods coincides with $p\mathbb{N}$;*

(iv) *for every periodic solution v of* (1.1) *with $v(\mathbb{R}) = \vartheta$, there exists a $t_v \in \mathbb{R}$ such that $v = u(\cdot + t_v)$.*

Proof. 1. There is a period $q > 0$ of u. Let $t > 0$ be given with $u(t) = u(0)$. Note

$$\begin{aligned} u(s) &= \Sigma(s, u(0)) = \Sigma(s, u(t)) = \Sigma(s, \Sigma(t, u(0)) \\ &= \Sigma(s + t, u(0)) = u(t + s), \quad s \geq 0. \end{aligned}$$

For $s < 0$, choose $j \in \mathbb{N}$ with $s + jq > 0$. Then

$$u(s) = u(s + jq) = u(s + jq + t) = u(s + t).$$

This proves (i).
2. Define $p = \inf\{q > 0 : u(q) = u(0)\}$. Assume $p = 0$. Using (i) we infer that $\{t \in \mathbb{R} : u(t) = u(0)\}$ is dense. By continuity, $u(\mathbb{R}) = \{u(0)\}$, a contradiction to our assumptions. It follows that $p > 0$ and continuity implies $u(p) = u(0)$ so that p is a period, due to (i). Smaller periods are not possible.

3. Proof of (iii). Each jp, $j \in \mathbb{N}$, is obviously a period. Suppose, on the other hand, $q > 0$ is a period. By minimality of p, we have $q \geq p$. Let j denote the largest integer so that $q \geq jp$. Then $u(0) = u(q) = u(q - jp)$. Either $q - jp = 0$, or $q - jp > 0$ is a period, due to (i). In the last case, we obtain $q - jp \geq p$, a contradiction to the choice of j.
4. Proof of (iv). For a periodic solution v such that $v(\mathbb{R}) = \vartheta$, we can choose t_v with $u(t_v) = v(0)$. The semiflow properties imply $v(t) = u(t+t_v)$ for all $t \geq 0$. In particular, $v(p) = v(0)$. According to (i), p is a period of v. This also leads to $v(t) = u(t + t_v)$ for $t < 0$. □

Consider a translate $v = u(\cdot + t_v)$ of u. We complexify the values of the p-periodic map

$$DG(v(\cdot)) : \mathbb{R} \to \mathcal{L}(X, X^{\odot *})$$

and apply the canonical isomorphism from $(X^{\odot *})_{\mathbb{C}}$ onto $(X_{\mathbb{C}})^{\odot *}$ to each $DG(v(t))_{\mathbb{C}}$, $t \in \mathbb{R}$ [see Section III.7]. This yields a p-periodic map

$$B_v : \mathbb{R} \to \mathcal{L}(X_{\mathbb{C}}, (X_{\mathbb{C}})^{\odot *}),$$

which is again strongly continuous in the sense of Chapters XII and XIII. The evolutionary system defined by the solutions of the equations

$$\widehat{w}(t) = T_0(t-s)_{\mathbb{C}}\widehat{w}(s) + \int_s^t (T_0(t-\tau)_{\mathbb{C}})^{\odot *} B_v(\tau)\widehat{w}(\tau)\, d\tau, \quad t \geq s,$$

$$\widehat{w}(s) \in X_{\mathbb{C}},$$

consists of the complexifications $U_v(t,s)_{\mathbb{C}}$ of the maps $U_v(t,s)$ given by the solutions to equation (1.2).

Proposition 2.2. *Let periodic solutions $v = u(\cdot + t_v)$ and $\widetilde{v} = u(\cdot + t_{\widetilde{v}})$ be given. For all $t, s \in \mathbb{R}$ with $t \geq s$, we then have*

$$U_v\big((t_{\widetilde{v}} - t_v) + t, (t_{\widetilde{v}} - t_v) + s\big)_{\mathbb{C}} = U_{\widetilde{v}}(t,s)_{\mathbb{C}}.$$

Proof. It is sufficient to prove this for the systems on X. Set $\delta = t_{\widetilde{v}} - t_v$. Then $v = \widetilde{v}(\cdot - \delta)$. Let $\varphi \in X$ be given. For reals t and s with $t \geq s$, we find

$$U_{\widetilde{v}}(t,s)\varphi = \widetilde{w}(t),$$

where $\widetilde{w} : [s, \infty) \to X$ is the (unique) solution to

$$\widetilde{w}(t) = T_0(t-s)\varphi + \int_s^t T_0^{\odot *}(t-\tau) DG(\widetilde{v}(\tau))\widetilde{w}(\tau)\, d\tau, \quad t > s.$$

The curve

$$\widehat{w} = \widetilde{w}(\cdot - \delta) : [s + \delta, \infty) \to X$$

satisfies $\widehat{w}(s+\delta) = \widetilde{w}(s) = \varphi$ and, for all $t \geq s+\delta$,

$$\begin{aligned}
\widehat{w}(t) &= \widetilde{w}(t-\delta) \\
&= T_0(t-\delta-s)\varphi + \int_s^{t-\delta} T_0^{\odot *}(t-\delta-\tau) DG(\widetilde{v}(\tau))\widetilde{w}(\tau)\, d\tau \\
&= T_0(t-(s+\delta))\varphi + \int_{s+\delta}^{t} T_0^{\odot *}(t-\tau) DG(\widetilde{v}(\tau-\delta))\widetilde{w}(\tau-\delta)\, d\tau \\
&= T_0(t-(s+\delta))\varphi + \int_{s+\delta}^{t} T_0^{\odot *}(t-\tau) DG(v(\tau))\widehat{w}(\tau)\, d\tau.
\end{aligned}$$

So

$$U_{\tilde{v}}(t,s)\varphi = \widetilde{w}(t) = \widehat{w}(t+\delta) = U_v(t+\delta, s+\delta)\varphi.$$

□

Corollary 2.3. $U_v(p + (t_{\tilde{v}} - t_v) + t, (t_{\tilde{v}} - t_v) + t) = U_{\tilde{v}}(t+p, t), \quad t \in \mathbb{R}.$

From now on, we assume

Hypothesis 2.4. There exists a $t_v \in \mathbb{R}$ such that the family of maps $U_v(t+p,t)_{\mathbb{C}}$, $t \in \mathbb{R}$, has the spectral isolation property.

Corollary 2.3 then implies that for every $t_{\tilde{v}}$, the family of maps

$$U_{\tilde{v}}(t+p,t)_{\mathbb{C}}, \quad t \in \mathbb{R},$$

has the spectral isolation property. Moreover, Proposition 2.2 and results from Chapter XIII yield that all sets

$$\sigma\big(U_{\tilde{v}}(t+p,t)_{\mathbb{C}}\big) \setminus \{0\},$$

where $t_{\tilde{v}}$ and t are arbitrary in $\mathbb{R}$, coincide. This justifies calling the nonzero spectral points of any of these maps "the Floquet multipliers of the periodic orbit ϑ". Observe that all generalized eigenspaces associated with one Floquet multiplier are isomorphic.

Exercise 2.5. Let a Floquet multiplier λ of ϑ be given. Describe the relations between the translates of u and the associated generalized eigenspaces precisely.

Proposition 2.6. *Suppose u is differentiable. Then, $1 \in \mathbb{C}$ is a Floquet multiplier, and there exists a translate $v = u(\cdot + t_v)$ such that $\dot{v}(0) = Dv(0)1$ satisfies*

$$U_v(p,0)\dot{v}(0) = \dot{v}(0) \neq 0,$$

i.e., $(\dot{v}(0), 0) \in X_{\mathbb{C}}$ is an eigenvector of the eigenvalue 1 of the map $U_v(p,0)_{\mathbb{C}}$.

Proof. The condition $\#u(\mathbb{R}) \geq 2$ implies $\dot{u}(t_v) \neq 0$ for some t_v. For $v = u(\,\cdot\, + t_v)$, $\dot{v}(0) \neq 0$. Differentiation of

$$v(t) = v(t+p) = \Sigma\big(p, v(t)\big)$$

at $t = 0$ yields $\dot{v}(0) = D_2\Sigma\big(p, v(0)\big)\dot{v}(0) = U_v(p,0)\dot{v}(0)$. □

Exercise 2.7. Consider a nonconstant periodic solution $x : \mathbb{R} \to \mathbb{R}^n$ of the RFDE

$$\dot{x}(t) = g(x_t), \tag{2.1}$$

where g is C^1 from some open subset of $C([-h,0],\mathbb{R}^n)$ into $\mathbb{R}^n$. Show that 1 is a Floquet multiplier of the orbit $\{x_t : t \in \mathbb{R}\}$ and that $\dot{x}_0$, $\dot{x}_0(s) = \dot{x}(s)$ for $-h \leq s \leq 0$ defines an eigenvector of the eigenvalue 1 of the map $D_2\Sigma(p, x_0)$ [where Σ denotes the semiflow of equation (2.1)].

Definition 2.8. The periodic orbit ϑ is called *hyperbolic* if 1 is a Floquet multiplier with one dimensional generalized eigenspace and if no other Floquet multipliers belong to the unit circle.

XIV.3 Poincaré maps

The present section requires basic facts from calculus on manifolds. All we need is collected in Appendix V. Let a solution u of equation (1.1) be given, defined on some interval $I \ni 0$ and a hyperplane $H \subset X$ so that $u(t) \in H$ for some $t > 0$. We may write $H = u(t) + Y$, where Y is a closed subspace of X of codimension 1, or, the nullspace of some $\varphi^* \in X^*$. The global chart $\varphi \mapsto \varphi - u(t)$ makes H a C^1-submanifold of X. Note

$$T_\varphi H = Y, \quad \text{for all } \varphi \in H.$$

In order to speak of transversality of u and H, and to make use of this, we assume that the semiflow Σ is C^1 on some neighbourhood of $(t, u(0))$. Now suppose that

$$D_1\Sigma\big(t, u(0)\big)1 = \dot{u}(t) \notin Y.$$

Proposition 3.1. *There exists a C^1-map*

$$\sigma : B_\epsilon\big(u(0)\big) \to (0,\infty), \quad \epsilon > 0,$$

such that $\sigma\big(u(0)\big) = t$ and

$$\Sigma\big(\sigma(\varphi), \varphi\big) \in H, \qquad \textit{for all } \varphi \in B_\epsilon\big(u(0)\big),$$

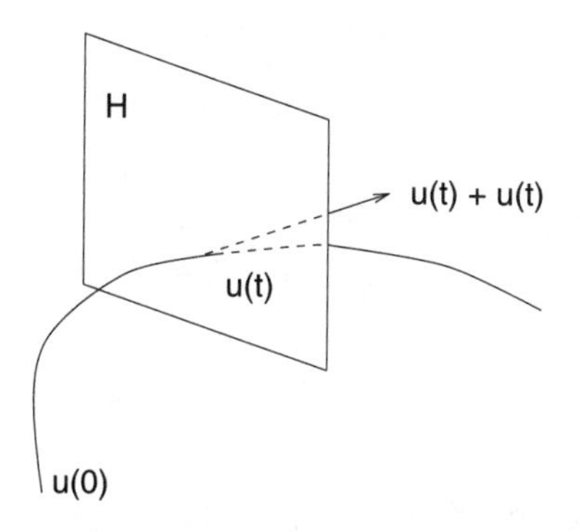

Fig. XIV.1. The hyperplane.

and $D_1\Sigma\big(\sigma(\varphi),\varphi\big)1 \notin Y$ *on* $B_\epsilon\big(u(0)\big)$.

Proof. Solve the equation

$$\varphi^*\big(\Sigma(s,\varphi)-u(t)\big)=0$$

close to the solution $(t,u(0))$. The transversality hypothesis yields

$$0\neq\varphi^*(\dot{u}(t))=D_1\big(\varphi^*\circ(\Sigma(\,\cdot\,,\,\cdot\,)-u(t))(t,u(0)\big)1$$

so that the Implicit Function Theorem is applicable. □

The C^1-map $\Pi_\sigma : B_\epsilon\big(u(0)\big)\to X$, $\Pi_\sigma(\varphi)=\Sigma\big(\sigma(\varphi),\varphi\big)$ has range in H. It maps initial values onto intersections of trajectories with H. It is not hard to obtain a geometric description of its derivatives.

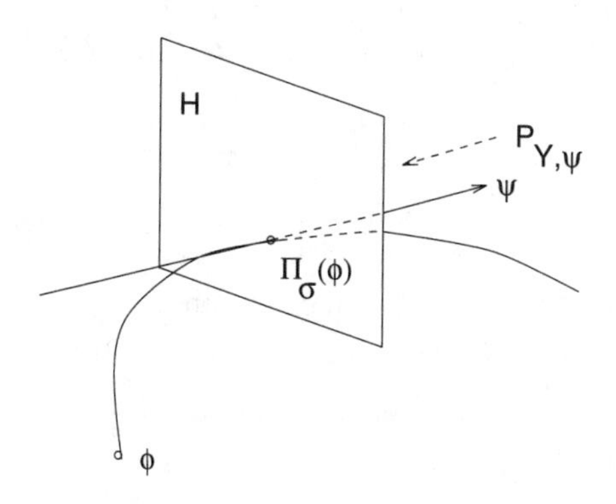

Fig. XIV.2. The projection for $\Pi_\sigma(\varphi)=0$.

Proposition 3.2.
$$D\Pi_\sigma(\varphi) = P_{Y,\psi} \circ D_2\Sigma\big(\sigma(\varphi),\varphi\big),$$
where $P_{Y,\psi}: X \to X$ is the projection onto Y parallel to the tangent vector
$$\psi = D_1\Sigma\big(\sigma(\varphi),\varphi\big)1.$$

Note that $P_{Y,\psi}$ in Proposition 3.2 is defined as
$$D_1\Sigma\big(\sigma(\varphi),\varphi\big)1 \notin Y$$
and that we have
$$P_{Y,\psi}\xi = \xi - \frac{\varphi^*(\xi)}{\varphi^*(\psi)}\psi, \quad \text{for } \xi \in X.$$

Proof of Proposition 3.2. Fix $\varphi \in B_\epsilon\big(u(0)\big)$ and $\chi \in X$. Differentiation of
$$\varphi^*\big(\Sigma(\sigma(\,\cdot\,),\,\cdot\,) - u(t)\big) = 0$$
at φ yields
$$\begin{aligned} 0 &= \varphi^*\big(D_1\Sigma(\sigma(\varphi),\varphi)D\sigma(\varphi)\chi + D_2\Sigma(\sigma(\varphi),\varphi)\chi\big) \\ &= \varphi^*\big(D\sigma(\varphi)\chi D_1\Sigma(\sigma(\varphi),\varphi)1 + D_2\Sigma(\sigma(\varphi),\varphi)\chi\big) \\ &= D\sigma(\varphi)\chi\varphi^*(\psi) + \varphi^*(D_2\Sigma(\sigma(\varphi),\varphi)\chi). \end{aligned}$$
It follows that
$$\begin{aligned} D\Pi_\sigma(\varphi)\chi &= D\sigma(\varphi)\chi D_1\Sigma(\sigma(\varphi),\varphi) + D_2\Sigma(\sigma(\varphi),\varphi)\chi \\ &= D\sigma(\varphi)\chi\psi + D_2\Sigma(\sigma(\varphi),\varphi)\chi \\ &= -\frac{\varphi^*(D_2(\sigma(\varphi),\varphi)\chi)}{\varphi^*(\psi)}\psi + D_2\Sigma(\sigma(\varphi),\varphi)\chi \\ &= P_{Y,\psi}\big(D_2\Sigma(\sigma(\varphi),\varphi)\chi\big). \end{aligned}$$
□

We apply the previous construction to a nonconstant periodic solution $u: \mathbb{R} \to X$ and to $t = q$, where q is a period of u.

Observe that the assumptions $u(q) \in H$, $\dot{u}(q) \notin Y$ are now equivalent to
$$u(0) \in H, \quad \dot{u}(0) \notin Y.$$
The map $\Pi : H \cap B_\epsilon\big(u(0)\big) \to H$,
$$\Pi(\varphi) = \Sigma\big(\sigma(\varphi),\varphi\big)\big),$$
is called a *Poincaré map*. It maps an open subset of the submanifold H into H; $u(0)$ is a fixed point. Its derivatives are given by
$$D\Pi(\varphi)\chi = P_{Y,\psi}D_2\Sigma\big(\sigma(\varphi),\varphi\big)\chi, \quad \psi = D_1\Sigma\big(\sigma(\varphi),\varphi\big)1$$
for $\chi \in Y = T_\varphi H$. At the fixed point,
$$D\Pi(u(0))\chi = P_{Y,\dot{u}(0)}D_2\Sigma\big(q,u(0)\big)\chi,$$
since $\dot{u}(0) = \dot{u}(q) = D_1\Sigma(q,u(0))1$ and $q = \sigma(u(0))$.

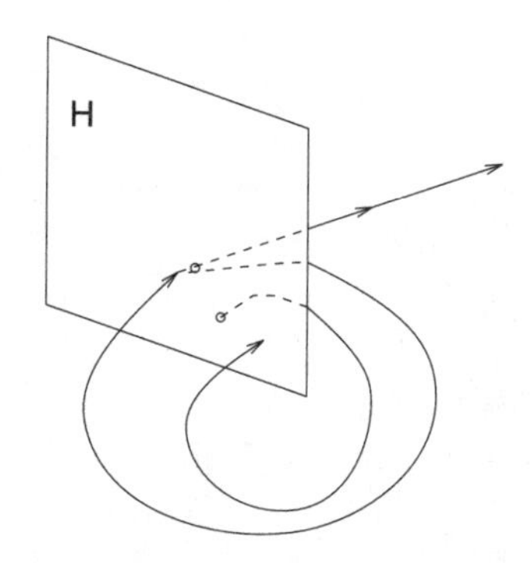

Fig. XIV.3. The Poincaré map.

Theorem 3.3. *In case $|\lambda| < 1$ for all spectral points λ of $D\Pi(u(0))$, the orbit $u(\mathbb{R})$ is stable and exponentially attractive with asymptotic phase. If there is a spectral point λ with $|\lambda| > 1$, then $u(\mathbb{R})$ is unstable.*

Exercise 3.4. Prove the previous theorem.
Hint: The assertion on the asymptotic phase means that for φ sufficiently close to $u(\mathbb{R})$ there exists a t_φ such that

$$\Sigma(t, \varphi) - u(t - t_\varphi) \to 0 \quad \text{as } t \to \infty.$$

The proof of the instability result is not hard if one uses the local unstable manifold of Π at $u(0)$.

If Σ is defined by the solutions of a RFDE

$$\dot{x}(t) = g(x_t), \tag{3.1}$$

one can say a bit more about the Poincaré maps. Let us assume for the remaining part of this section that g is a C^1-map from an open subset of $\mathcal{C} = C\big([-h, 0], \mathbb{R}^n\big)$ into $\mathbb{R}^n$ and that $x : \mathbb{R} \to \mathbb{R}^n$ is a periodic solution. It follows that x is differentiable, and so is the periodic trajectory

$$t \to x_t \quad [=: u(t)]$$

of the semiflow Σ on the domain of g;

$$D\big(s \mapsto x_s\big)(t)1 = \dot{x}_t,$$

where $\dot{x}_t(\theta) = \dot{x}(t + \theta)$ for $-h \leq \theta \leq 0$. Suppose H is a hyperplane in $\mathcal{C}$ which is transversal to $t \mapsto x_t$ at $t = 0$. Choose a period $q > h$. It follows that Σ is C^1 on a neighbourhood of $(q, u(0))$, and we have an associated Poincaré map

$$\Pi_g : B_\epsilon(x_0) \cap H \to H.$$

An important fact is that restrictions of Π_g, and $D\Pi_g(x_0)$, are compact maps:

Proposition 3.5.

(i) *There exists a* $\delta > 0$ *such that* $\overline{\Pi_g\bigl(B_\delta(x_0) \cap H\bigr)}$ *is a compact subset of* H*;*

(ii) $D\Pi_g(x_0) \in \mathcal{L}\bigl(T_{x_0}H\bigr)$ *is a compact map.*

Proof. 1. Choose a bounded open neighbourhood ϑ_b of the compact set $\vartheta = \{x_t : t \in \mathbb{R}\}$ on which g is bounded. Choose $\delta > 0$ so small that on $B_\delta(x_0) \cap H$,

$$h < \sigma(\varphi) < q + h, \quad \text{and} \quad \Sigma(t, \varphi) \in \vartheta_b \quad \text{for } 0 \le t \le q + h.$$

Apply Proposition VII.6.5 (to the semiflow associated with $g|\vartheta_b$) and deduce that

$$\Sigma\bigl([h, h+q] \times \bigl(B_\delta(x_0) \cap H\bigr)\bigr)$$

has compact closure in $\mathcal{C}$. The last set contains $\Pi_g\bigl(B_\delta(x_0) \cap H\bigr)$.
2. Let j denote the inclusion map from $Y = T_{x_0}H$ into $\mathcal{C}$. We have

$$D\Pi_g(x_0) = P_{Y,\dot{x}_0} \circ D_2\Sigma(q, x_0) \circ j$$

with $D_2\Sigma(q, x_0) = U(q, 0)$, where $U(q, 0)$ is given by the solutions of the linear variational equation

$$\dot{y}(t) = Dg(x_t)y_t.$$

Now, note that $U(q, 0)$ is compact since $q > h$. □

The last proposition shows that Π_g is not a diffeomorphism, not even a homeomorphism: existence of a continuous inverse I_g on some neighbourhood of x_0 in H would imply, via

$$\Pi_g\bigl(I_g(\varphi)\bigr) = \varphi,$$

that a neighbourhood of x_0 in H has compact closure, which is impossible since $\dim H = \infty$.

XIV.4 Poincaré maps and Floquet multipliers

Let a nonconstant periodic solution $u : \mathbb{R} \to X$ of equation (1.1) be given, with minimal period p. We make the following assumptions.

(F) There exists a translate $v = u(\cdot + t_v)$ such that the Spectral Isolation Hypothesis XIII.3.2 is satisfied for all maps $U_v(t+p,t)_{\mathbb{C}}$, $t \in \mathbb{R}$, from the evolutionary system given by equation $(1.2)_{\mathbb{C}}$.

(H) $H \subset X$ is a hyperplane which contains $u(0)$.

(Ξ) Σ is C^1 in a neighbourhood of $(p, u(0))$, and $\dot{u}(0) = \dot{u}(p) \notin T_{u(0)}H$.

Under condition (F), we have Floquet multipliers for the periodic orbit

$$\vartheta = u(\mathbb{R}).$$

Conditions (H) and (Ξ) guarantee that a Poincaré map

$$\Pi : B_\epsilon\big(u(0)\big) \cap H \to H$$

can be constructed. Let $\varphi^* \in X^*$ denote an element whose nullspace equals

$$Y = H - u(0) = T_\varphi H, \quad \text{for all } \varphi \in H.$$

We want to explore the relations between the spectrum of $D\Pi\big(u(0)\big)$ which is given by

$$P_{Y,\dot{u}(0)} \circ D_2\Sigma\big(p, u(0)\big)\big|Y$$

and the Floquet multipliers of ϑ, i.e., the nonzero spectral points of the complexification U of $D_2\Sigma\big(p, u(0)\big) = U_u(p, 0)$, in the notation of Section 2. It is convenient to introduce the inclusion map $j : Y_{\mathbb{C}} \to X_{\mathbb{C}}$ and the abbreviations

$$P = \big(P_{Y,\dot{u}(0)}\big)_{\mathbb{C}}, \qquad L = D\Pi\big(u(0)\big)_{\mathbb{C}}.$$

Then

$$L\varphi = \big(P \circ U \circ j\big)(\varphi)$$

on the complexified space $Y_{\mathbb{C}}$. Set $\psi = \big(\dot{u}(0), 0\big) \in X_{\mathbb{C}}$.

Exercise 4.1. Show that $Y_{\mathbb{C}}$ is the nullspace of $\varphi^*_{\mathbb{C}} : X_{\mathbb{C}} \to \mathbb{C}$ and that

$$P\varphi = \varphi - \frac{\varphi^*_{\mathbb{C}}(\varphi)}{\varphi^*_{\mathbb{C}}(\psi)}\psi, \quad \text{for all } \varphi \in X_{\mathbb{C}}.$$

Arguing as in the proof of Proposition 2.6, we find

Proposition 4.2.

$$U\psi = \psi \neq 0$$

(1 *is a Floquet multiplier*).

The key to the desired result is

Proposition 4.3. *For $\lambda \in \mathbb{C} \setminus \sigma(U)$, we have $\lambda \notin \sigma(L)$ and*

$$(\lambda I - L)^{-1}(\varphi) = P \circ (\lambda I - U)^{-1} \circ j(\varphi) \quad \text{on } Y_{\mathbb{C}}.$$

Proof. 1. Let $\lambda \in \mathbb{C} \setminus \sigma(U)$ be given. By Proposition 4.2, we find $\lambda \neq 1$ and

$$(\lambda I - U)^{-1}(\psi) = \frac{1}{\lambda - 1}\psi.$$

Since $P\psi = 0$ (see Exercise 4.1)

$$P(\lambda I - U)^{-1}(\psi) = \frac{1}{\lambda - 1}P\psi = 0. \tag{4.1}$$

2. For all $\varphi \in Y_{\mathbb{C}}$,

$$\begin{aligned}
&(\lambda I - L)\big(P \circ (\lambda I - U)^{-1} \circ j(\varphi)\big) \\
&\quad = (\lambda I - P \circ U \circ j)\big(P \circ (\lambda I - U)^{-1} \circ j(\varphi)\big) \\
&\quad = \big(P \circ (\lambda I - U) \circ j\big)\big(P(\lambda I - U)^{-1}(\varphi)\big) \\
&\quad = \big(P \circ (\lambda I - U)\big)\Big((\lambda I - U)^{-1}(\varphi) - \frac{\varphi^*_{\mathbb{C}}\big((\lambda I - U)^{-1}(\varphi)\big)}{\varphi^*_{\mathbb{C}}(\psi)}\psi\Big) \\
&\quad = P\varphi - \frac{\varphi^*_{\mathbb{C}}\big((\lambda I - U)^{-1}(\varphi)\big)}{\varphi^*_{\mathbb{C}}(\psi)} P\big((\lambda I - U)(\varphi)\big) = \varphi - 0 = \varphi.
\end{aligned}$$

Here we used $P\varphi = \varphi$ on $Y_{\mathbb{C}}$ and $U\psi = \psi$, $P\psi = 0$.
3. For all $\varphi \in Y_{\mathbb{C}}$,

$$\begin{aligned}
&\big(P \circ (\lambda I - U)^{-1} \circ j\big) \circ (\lambda I - L)(\varphi) \\
&\quad = \big(P \circ (\lambda I - U)^{-1}\big)\big(P \circ (\lambda I - U) \circ j(\varphi)\big) \\
&\quad = \big(P \circ (\lambda I - U)^{-1}\big)\Big((\lambda I - U)(\varphi) - \frac{\varphi^*_{\mathbb{C}}\big((\lambda I - U)(\varphi)\big)}{\varphi^*_{\mathbb{C}}(\psi)}\psi\Big) \\
&\quad = P\Big(\varphi - \frac{\varphi^*_{\mathbb{C}}\big((\lambda I - U)(\varphi)\big)}{\varphi^*_{\mathbb{C}}(\psi)} P\big((\lambda I - U)^{-1}(\varphi)\big)\Big) = \varphi - 0 = \varphi,
\end{aligned}$$

due to $P\varphi = \varphi$ and (4.1). □

Since all nonzero points of $\sigma(U)$ are isolated, the same result holds true for the subset $\sigma(L)$, and with each $\lambda \in \mathbb{C}\setminus\{0\}$ we can associate a projection operator for U as well as for L. Indeed, as in the proof of Theorem XIII.3.3, choose $\delta = \delta(\lambda)$ so that

$$\big(\overline{B_\delta(\lambda)} \setminus \{\lambda\}\big) \cap \sigma(U) = \emptyset = \big(\overline{B_\delta(\lambda)} \setminus \{\lambda\}\big) \cap \sigma(L).$$

Set $\Gamma(\theta) = \lambda + \delta e^{i\theta}$, for $0 \le \theta \le 2\pi$, and define

$$P_{\lambda,U} = \frac{1}{2\pi i}\int_{\Gamma}(zI - U)^{-1}\,dz \in \mathcal{L}(X_{\mathbb{C}}),$$
$$P_{\lambda,L} = \frac{1}{2\pi i}\int_{\Gamma}(zI - L)^{-1}\,dz \in \mathcal{L}(Y_{\mathbb{C}}).$$

Then, if λ belongs to the spectrum, the maps are the eigenprojections; for other λ, they are zero:

$$\lambda \in \sigma(U) \text{ is equivalent to } P_{\lambda,U} \neq 0,$$
$$\lambda \in \sigma(L) \text{ is equivalent to } P_{\lambda,L} \neq 0.$$

From Proposition 4.3, we deduce

Corollary 4.4. *For $0 \neq \lambda \in \mathbb{C}$ and $\varphi \in Y_{\mathbb{C}}$,*

$$P_{\lambda,L}(\varphi) = P \circ P_{\lambda,U} \circ j(\varphi).$$

Theorem 4.5.
(i) *We have $\sigma(U) \setminus \{0,1\} = \sigma(L) \setminus \{0,1\}$.*
(ii) *For $\lambda \in \sigma(U) \setminus \{0,1\}$, P defines a topological isomorphism from the generalized eigenspace $X_{\mathbb{C},\lambda} = P_{\lambda,U}X_{\mathbb{C}}$ onto $Y_{\mathbb{C},\lambda} = P_{\lambda,L}Y_{\mathbb{C}}$.*
(iii) *In case $X_{\mathbb{C},1} = \mathbb{C}\psi$ (i.e., 1 is a simple Floquet multiplier),*

$$1 \notin \sigma(L).$$

(iv) *In case $X_{\mathbb{C},1} \supset \mathbb{C}\psi$, $1 \in \sigma(L)$ and*

$$Y_{\mathbb{C},1} = X_{\mathbb{C},1} \cap Y_{\mathbb{C}}, \quad X_{\mathbb{C},1} = Y_{\mathbb{C},1} \oplus \mathbb{C}\psi.$$

Proof. 1. Let $\lambda \in \sigma(U)\setminus\{0,1\}$ be given. We show that P maps $P_{\lambda,U}X_{\mathbb{C}}$ one-to-one onto $P_{\lambda,L}Y_{\mathbb{C}}$: First, note that $P^{-1}(0) \cap P_{\lambda,U}X_{\mathbb{C}} = \{0\}$ follows from $P^{-1}(0) = \mathbb{C}\psi \subset P_{1,U}X_{\mathbb{C}}$ and from $P_{1,U}X_{\mathbb{C}} \cap P_{\lambda,U}X_{\mathbb{C}} = \emptyset$ (since $\lambda \neq 1$). Corollary 4.4 yields

$$P\big(P_{\lambda,U}Y_{\mathbb{C}}\big) = P_{\lambda,L}Y_{\mathbb{C}}.$$

It remains to prove $P_{\lambda,L}Y_{\mathbb{C}} \supset P_{\lambda,U}X_{\mathbb{C}}$. For $\varphi \in P_{\lambda,U}X_{\mathbb{C}}$, $\varphi = P_{\lambda,U}(\chi + \xi\psi)$ with $\chi \in Y_{\mathbb{C}}$ and $\xi \in \mathbb{C}$, since

$$X_{\mathbb{C}} = Y_{\mathbb{C}} \oplus \mathbb{C}\psi.$$

But $P_{\lambda,U}$ annihilates $P_{1,U}X_{\mathbb{C}} \supset \mathbb{C}\psi$. Therefore $\varphi = P_{\lambda,U}\chi \in P_{\lambda,U}Y_{\mathbb{C}}$.
2. Proof of (i). Let $\lambda \in \sigma(U) \setminus \{0,1\}$. Then $P_{\lambda,U}X_{\mathbb{C}} \neq \{0\}$. According to Part 1., $P_{\lambda,L}Y_{\mathbb{C}} \neq \{0\}$. Hence $\lambda \in \sigma(L)$.
3. Proof of (ii). Part 1 and the Open Mapping Theorem imply that P defines a topological isomorphism from the closed subspace $P_{\lambda,U}X_{\mathbb{C}}$ onto the closed subspace $P_{\lambda,L}Y_{\mathbb{C}}$.
4. In case $P_{1,U}X_{\mathbb{C}} = \mathbb{C}\psi (= P^{-1}(0))$, we find

$$P_{1,L}Y_{\mathbb{C}} = P \circ P_{1,U}(Y_{\mathbb{C}}) = \{0\},$$

or $1 \notin \sigma(L)$.
5. Suppose $P_{1,U}X_{\mathbb{C}} \supset \mathbb{C}\psi$. Proof that the closed space $X_{\mathbb{C},1} \cap Y_{\mathbb{C}}$ is a complement of $\mathbb{C}\psi$ in $X_{\mathbb{C},1}$: $\psi \notin Y_{\mathbb{C}}$ shows that the intersection of these spaces is zero. For every $\varphi \in X_{\mathbb{C},1}$, $\varphi = \chi + \xi\psi$ with some $\chi \in Y_{\mathbb{C}}$ and $\xi \in \mathbb{C}$. Hence $\chi = \varphi = \xi\psi \in X_{\mathbb{C},1} - \mathbb{C}\psi = X_{\mathbb{C},1}$, or $\chi \in X_{\mathbb{C},1} \cap Y_{\mathbb{C}}$.
6. Proof of (iv). It remains to show that, for $P_{1,U} \supset \mathbb{C}\psi$, we have

$$X_{\mathbb{C},1} \cap Y_{\mathbb{C}} = Y_{\mathbb{C},1}.$$

With $P\varphi = \varphi$ on $Y_{\mathbb{C}}$ and with $P\psi = 0$,

$$\begin{aligned} Y_{\mathbb{C},1} = P_{1,L}Y_{\mathbb{C}} &= P \circ P_{1,U} \circ j(Y_{\mathbb{C}}) \subset P(X_{\mathbb{C},1}) \\ &= P((X_{\mathbb{C},1} \cap Y_{\mathbb{C}}) \oplus \mathbb{C}\psi) = P(X_{\mathbb{C},1} \cap Y_{\mathbb{C}}) \\ &= X_{\mathbb{C},1} \cap Y_{\mathbb{C}}. \end{aligned}$$

For $\varphi \in X_{\mathbb{C},1} \cap Y_{\mathbb{C}}$, we note $\varphi = P\varphi$, $\varphi = P_{1,U}\varphi$, $\varphi = j(\varphi)$. Therefore

$$\varphi = P \circ P_{1,U} \circ j(\varphi) = P_{1,L}(\varphi) \in Y_{\mathbb{C},1}.$$

□

XIV.5 Comments

In case of vector fields on finite dimensional spaces, Floquet multipliers of periodic orbits are discussed in, e.g., the works of Abraham and Robbin [1], Amann [5] and Hale [101]. For Poincaré maps, see Hirsch and Smale [127] and Palis and de Melo [232].

Results for infinite dimensional systems are obtained by Hale [102] (RFDE) and by Henry [121] (nonlinear perturbations of analytic semigroups, semilinear parabolic PDE). Both books go beyond the scope of the previous sections and describe the semiflows close to periodic orbits in their respective cases.

We promised references to results where Floquet multipliers were analyzed and used to understand the dynamics of a given system. For autonomous differential delay and integral equations, see, e.g., the works of Chow, Diekmann and Mallet-Paret [34], Chow and Walther [42], Dormayer [73, 74, 75], Ivanov, Lani-Wayda and Walther [137], Mallet-Paret and Sell [185], Walther [291, 297, 299, 300, 302] and Xie [310, 311, 312].

In Sections 2–4, on the way to Theorem 4.5, we made a series of assumptions on the semiflow. One may ask to which extent such assumptions can be weakened. For example, are there less restrictive conditions than the Spectral Isolation Hypothesis XIII.3.2 so that one can define Floquet multipliers of a periodic orbit?

Another question concerns different Poincaré maps associated with one periodic orbit. For flows of vector fields, all such maps are conjugates of each other; conjugating diffeomorphisms are defined by going from one transversal hyperplane to the other as depicted in Fig. XIV.4.

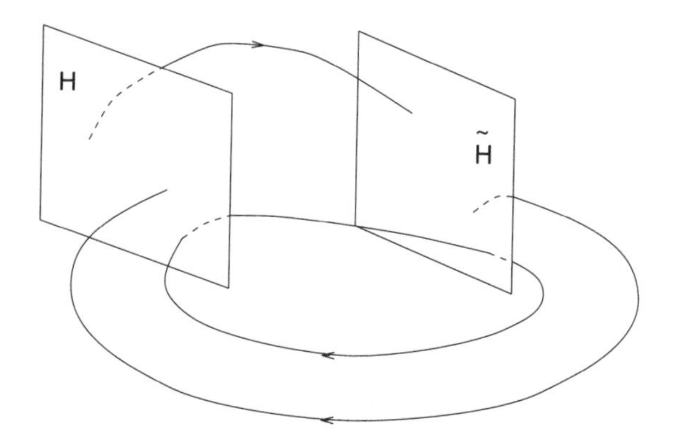

Fig. XIV.4.

In our more general situation, we only obtained that spectra coincide (since they are given by Floquet multipliers) and that generalized eigenspaces are isomorphic. What more can be said? Obvious difficulties are the lack of smoothness of the semiflow, and compactness (as in case of RFDE) which prevents candidates for conjugating maps, if defined, from being homeomorphisms.

The nonlinear analysis of a Poincaré map close to its fixed point on the periodic orbit begins with its local stable, center and unstable manifolds. The construction of these manifolds for diffeomorphisms has a long history. We saw, however, that the Poincaré maps discussed in this chapter are not diffeomorphisms in general. There exist examples of RFDE where Poincaré maps are not even injective; see Walther [290], an der Heiden and Walther [118]. Local invariant manifolds for general C^1-maps are constructed by Hale and Lin [109], Iooss [134] and Neugebauer [207].

Important for the investigation of global properties of smooth maps are hyperbolic invariant sets. For diffeomorphisms, they were defined by Smale [253]. A wider notion of hyperbolic sets, designed for applications to C^1-maps, is introduced and discussed by Steinlein and Walther [266, 267] and Lani-Wayda [162]. Notably, the Shadowing Lemma, which is the main tool for the description of the dynamics close to hyperbolic sets, remains valid when diffeomorphisms are replaced by C^1-maps (see [266, 163]). Applications to Poincaré maps for RFDE are given by Steinlein and Walther [267] and by Lani-Wayda [164]. See also Section XVI.3. For Inclination Lemmas in case of C^1-maps, we refer to Hale and Lin [109] and to Walther [293].

Chapter XV

The prototype equation for delayed negative feedback: periodic solutions

XV.1 Delayed feedback

In this chapter we study the nonlinear equation

$$\dot{x}(t) = f(x(t-\alpha)) \tag{1.1}$$

where $f : \mathbb{R} \to \mathbb{R}$ is a continuous function. The positive parameter α is called the time lag. Equation (1.1) is the prototype for nonlinear delayed feedback. We shall see that the lag α can cause much more complex behaviour of solutions than in the ODE case $\alpha = 0$ where only monotone solutions converging to equilibria or infinity are possible. In particular, we shall prove the existence of periodic solutions (Section 5).

In Sections 6–8 we shall employ a method developed in [285, 287, 289] to prove a theorem of Nussbaum [214] on global bifurcation of periodic solutions which is basic for understanding the dynamics of (1.1).

It is convenient to reparametrize the equation and to consider

$$\dot{x}(t) = \alpha f(x(t-1)), \tag{1.2}$$

which is obtained from (1.1) by scaling the time variable, i.e., by working with $x(\alpha t)$ instead of $x(t)$. The advantage is that we can now work in a fixed state space when α is varied.

In order to concentrate on the simplest nontrivial situation, we make the following assumption:

(NF) $f : \mathbb{R} \to \mathbb{R}$ is continuous with $f(0) = 0$, differentiable at $x = 0$ with $f'(0) = -1$, and for all $x \neq 0$, $xf(x) < 0$.

The sign condition means that we study negative feedback with respect to the zero solution $x \equiv 0$: a deviation $x(t) < 0$ (> 0) is followed by a motion $\dot{x}(t+1) > 0$ (< 0) into the opposite direction. A famous example is given by the delayed logistic equation of Hutchinson [132],

$$\dot{n}(t) = r[1 - \frac{n(t-\tau)}{K}]n(t), \tag{1.3}$$

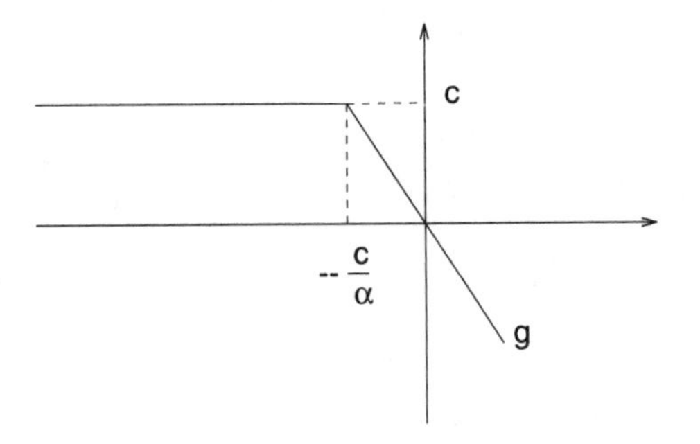

Fig. XV.1.

for the density of a single species population with a growth rate that is controlled by the simplest feedback mechanism but with reaction lag τ. [See also [54, 55, 56, 21]. Section 7.2 of [208] contains a derivation based on clear population dynamical assumptions. This leads to delay equations which are slightly different from (1.3).] The solutions $n > 0$ transform into solutions x of (1.2) with $\alpha = r\tau$ and $f(x) = 1 - e^x$ if one sets

$$x(t) = \log \frac{n(\tau t)}{K} \, .$$

Thus the logistic equation yields an example of a negative feedback equation where, in addition,

(B) f is bounded from above.

Examples from applications typically satisfy certain boundedness conditions. An equivalent form of (1.3) is

$$\dot{x}(t) = -\alpha x(t-1)[1 + x(t)]. \tag{1.4}$$

This is one of the first nonlinear differential delay equations which were intensely studied, beginning with the papers of Wright [308] and Kakutani and Markus [144].

The proof of the existence of periodic solutions of equation (1.2) in the general case, for $\alpha > \frac{\pi}{2}$ and arbitrary nonlinearities satisfying conditions (NF) and (B), is long. In special cases, however, periodic solutions can be found by an explicit construction. We end this section with an example of this kind.

Let $\alpha > 4$ and $c > 0$. Consider

$$\dot{x}(t) = g(x(t-1)), \tag{1.5}$$

where

$$g(x) = \begin{cases} c & \text{for } x \leq -\frac{c}{\alpha}, \\ -\alpha x & \text{for } x > -\frac{c}{\alpha}. \end{cases}$$

Every continuous $\varphi : [-1, 0] \to \mathbb{R}$ defines a solution $x : [-1, \infty) \to \mathbb{R}$ of equation (1.5) with $x_0 = \varphi$ (see, e.g., the argument in Section VII.2). For every φ which satisfies

$$\varphi \le -\frac{c}{\alpha} \quad \text{on } [-1, -\frac{1}{\alpha}], \qquad \varphi(t) = ct \quad \text{for } -\frac{1}{\alpha} < t \le 0,$$

we obtain

$$\dot{x}(t) = c \quad \text{for } 0 < t < 1 - \frac{1}{\alpha}$$

and in the second step

$$x(t) = c[1 - \frac{1}{2\alpha} - \frac{\alpha}{2}(t-1)^2] \quad \text{for } 1 - \frac{1}{\alpha} \le t \le 2 - \frac{1}{\alpha}.$$

The last formula follows from integrating equation (1.5) between $1 - \frac{1}{\alpha}$ and t, and from

$$-\frac{c}{\alpha} \le x(s) = cs \quad \text{for } -\frac{1}{\alpha} \le s \le 1 - \frac{1}{\alpha}.$$

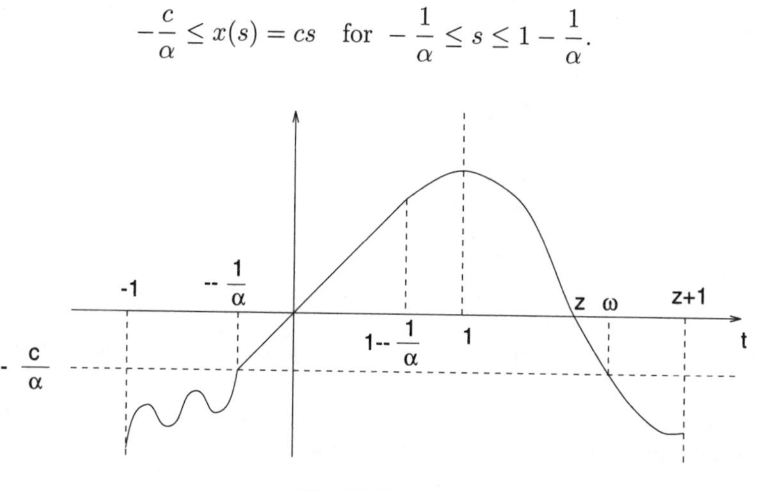

Fig. XV.2.

Beginning with

$$\varphi < 0 \quad \text{on } [-1, 0),$$

we infer from condition (NF) that

$$0 < \dot{x} \quad \text{and} \quad 0 < x \quad \text{on } (0, 1].$$

Consequently,

$$\dot{x} < 0 \quad \text{on } (1, 2].$$

Define $\omega = 1 + \frac{1}{\alpha}\sqrt{2\alpha + 1}$. Since $\alpha > 4$, we find $1 < \omega < 2 - \frac{1}{\alpha}$. Also

$$x(\omega) = -\frac{c}{\alpha}$$

so that there exists a first zero z of x on $(0,\infty)$, located in $(1,\omega)$. Arguing as above, we conclude that

$$\dot{x} < 0 \quad \text{and} \quad x < 0 \quad \text{on } (z, z+1)$$

and

$$0 < \dot{x} \quad \text{on } (z+1, z+2].$$

Note that $z+1 < \omega+1 < z+2$.

Exercise 1.1. Use the formula

$$x(\omega+1) = x(2-\frac{1}{\alpha}) + \int_{2-\frac{1}{\alpha}}^{\omega+1} g(x(t-1))\,dt$$

in order to show $x(\omega+1) \leq -\frac{c}{\alpha}$. The last inequality and the monotonicity properties of x on the interval $[\omega, \omega+1] \subset [z, z+2]$ imply

$$x \leq -\frac{c}{\alpha} \quad \text{on } [\omega, \omega+1].$$

Exercise 1.2. Prove that there exists $s \geq \omega+1$ such that

$$\begin{aligned} &x(s) = -\frac{c}{\alpha}, \quad x \leq -\frac{c}{\alpha} \quad \text{on } [\omega+1, s], \\ &x(s+\frac{1}{\alpha}) = 0, \\ &x(t) = c(t-s-\frac{1}{\alpha}) \quad \text{for } \omega+1 \leq t \leq s+1. \end{aligned}$$

Set $\tau = s + \frac{1}{\alpha}$. Observe that x_τ has the same properties as the initial value φ.

Exercise 1.3. Prove that the restriction of x to the interval $[0,\tau]$ defines a periodic solution $y : \mathbb{R} \to \mathbb{R}$ of equation (1.2).

XV.2 Smoothness and oscillation of solutions

Let $C = C([-1,0], \mathbb{R})$. We may cast (1.2) into the form

$$\dot{x}(t) = g(x_t),$$

where g is a map from C into $\mathbb{R}$. Set $g(\varphi) = \alpha f(\varphi(-1))$; then g is continuous and differentiable at $\varphi = 0$ and

$$Dg(0)\varphi = -\alpha\varphi(-1) \quad \text{for } \varphi \in C.$$

These properties of g are not sufficient to apply the existence theory from Section VII.3 since we do not assume local Lipschitz continuity of f away from zero. However, we saw already in Section VII.2 how to construct solutions $x : [-1, \infty) \to \mathbb{R}$ for the initial-value problems

$$\dot{x}(t) = \alpha f(x(t-1)), \qquad x_0 = \varphi \in C$$

by repeated use of the integrated form of equation (1.2):

$$x(t) - x(s) = \alpha \int_s^t f(x(u-1))\,du = \alpha \int_{s-1}^{t-1} f(x(\tau))\,d\tau$$

on the intervals $[0,1], [1,2], \ldots$. These solutions are uniquely determined by φ, α and f; at $t = 0$, there exists a right derivative which extends $\dot{x}$ continuously. We shall write $x^{\varphi}, x^{\varphi,\alpha}$ or $x^{\varphi,\alpha f}$ instead of x when convenient. Slightly more general, we shall also consider solutions $x : [t-1, \infty) \to \mathbb{R}$, where $t \in \mathbb{R}$, and solutions on all of $\mathbb{R}$. We have a continuous dependence on initial data and parameters (Exercise VII.2.11), and the relations $\Sigma_f(t, \varphi, \alpha) = x_t^{\varphi,\alpha f}$, $t \geq 0$, $\varphi \in C$, $\alpha > 0$, define a continuous parameterized semiflow. Recall also the compactness property stated in Exercise VII.2.12.

Proposition 2.1. (*Negative feedback.*) *Let x be a solution of* (1.2).

(i)
$$x > 0 \text{ on } [t,s] \text{ implies } \dot{x} < 0 \text{ on } [t+1, s+1],$$
$$x < 0 \text{ on } [t,s] \text{ implies } \dot{x} > 0 \text{ on } [t+1, s+1].$$

(ii)
$$x \geq 0 \text{ on } [t,\infty) \text{ implies } \dot{x} \leq 0 \quad \text{on } [t+1, \infty),$$
$$x \leq 0 \text{ on } [t,\infty) \text{ implies } \dot{x} \geq 0 \text{ on } [t+1, \infty),$$

and in both cases, $x(s) \to 0$, $x_s \to 0$ as $s \to +\infty$.

Proof. The sign condition in (NF) implies (i). Suppose $x \geq 0$ on $[t, \infty)$. By (i) and $f(0) = 0$,

$$\dot{x} \leq 0 \quad \text{on } [t+1, \infty).$$

Hence

$$x(s) \to x^* \geq 0 \quad \text{as } s \to +\infty.$$

In case $x^* > 0$,

$$\dot{x}(s) \to \alpha f(x^*) < 0,$$

so $x|_{[t,\infty)}$ becomes unbounded, a contradiction to convergence. □

So, monotone solutions converge to an equilibrium, as in the ODE case $\alpha = 0$. However, a sufficiently large delay α in (1.1) forces solutions to oscillate, i.e., to have an unbounded set of zeros in $(0, \infty)$.

Proposition 2.2. (*Oscillation.*) *If $\alpha > 1$ and $x(t) \neq 0$ for some solution x of* (1.2), *then there exists $s > t$ with*

$$\operatorname{sign} x(s) = -\operatorname{sign} x(t).$$

Proof. Let $x(t) > 0$. Choose $\epsilon > 0$ with $\alpha(1-\epsilon) \geq 1$. $f'(0) = -1$ and $f(0) = 0$ imply that there exists $\delta > 0$ with

$$|f(x)| \geq (1-\epsilon)|x| \quad \text{for } |x| \leq \delta.$$

Assume $x \geq 0$ on $[t, \infty)$. Then

$$\dot{x} \leq 0 \quad \text{on } [t+1, \infty), \quad x(s) \to 0 \quad \text{as } s \to \infty.$$

We have $x(t+1) > 0$. Otherwise, $\dot{x}(t+1) = \alpha f(x(t)) < 0$ would yield a contradiction to the assumption. Similarly, $x(t+n) > 0$ for all $n \in \mathbb{N}$. Since $\dot{x} \leq 0$ on $[t+1, \infty)$, we must have $x > 0$ on $[t, \infty)$. There exists $s > t+1$ with $0 < x < \delta$ on $[s, \infty)$. Hence

$$x(s+2) - x(s+1) = \alpha \int_s^{s+1} f \circ x \leq -\alpha(1-\epsilon) \int_s^{s+1} x \leq -x(s+1),$$

a contradiction to $0 < x(s+2)$. □

Using Exercise XI.3.9 on the characteristic equation

$$\lambda + \alpha e^{-\lambda} = 0,$$

one shows that all solutions of the linear equation

$$\dot{x}(t) = -\alpha x(t-1)$$

oscillate if and only if $\alpha > \frac{1}{e}$. The fact that solutions without zeros decay to zero as $t \to \infty$ [Proposition 2.1(ii)] and a comparison technique of Myshkis [199] permit to generalize this to all nonlinear equations (1.2).

Concerning the growth of oscillatory solutions, we have

Proposition 2.3. (*Bounds.*) *Let $m \geq t+1$ be a local extremum for a solution $x : [t-1, \infty) \to \mathbb{R}$ of* (1.2). *Then*

$$|x(m)| \leq \alpha \max_{x([m-2, m-1])} |f|.$$

Proof. Since $0 = \dot{x}(m) = \alpha f(x(m-1))$, we find $x(m-1) = 0$. Hence

$$x(m) = x(m) - x(m-1) = \alpha \int_{m-2}^{m-1} f \circ x.$$

□

XV.3 Slowly oscillating solutions

Definition 3.1. A solution x of (1.2) is called *slowly oscillating* if and only if f there exists $t \in \mathbb{R}$ with

$$|z - z'| > 1$$

for any pair of zeros $z \neq z'$ in $[t, \infty)$.

Slowly oscillating solutions are easily found:

Lemma 3.2. *If a solution $x : [-1, \infty) \to \mathbb{R}$ of equation (1.2) satisfies*

$$x(t) \neq 0 \quad \textit{for } t \in [-1, 0],$$

then x is slowly oscillating. Moreover,

$$\dot{x}(z) \neq 0 \quad \textit{and} \quad |z - z'| > 1 \quad \textit{for all zeros} \quad z \neq z'.$$

Sketch of proof. If there is a first zero z of x, then $z > 0$. By (NF), $\dot{x} \neq 0$ on $[z, z+1)$. Therefore $z' > z + 1$ for any other zero z', and either

$$x > 0 \text{ on some interval } (z, t] \text{ with } t > z + 1$$

or

$$x < 0 \text{ on some interval } (z, t] \text{ with } t > z + 1.$$

Repeat the same argument for the solution $-1 \leq s \mapsto x(t + s)$. □

Corollary 3.3. *The set*

$$\{\varphi \in C \mid x^{\varphi, \alpha f} \textit{ is slowly oscillating } \} \neq \emptyset$$

is open for all f with (NF) and all $\alpha > 0$.

Proof. Suppose $x = x^{\varphi, \alpha f}$ is slowly oscillating. Then, $x(s) \neq 0$ for all s in some interval $[t - 1, t]$, $t \geq 0$. Continuous dependence on initial data yields a neighbourhood U of φ so that for each $\overline{\varphi} \in U$, the state $x_t^{\overline{\varphi}, \alpha f}$ has no zero. Lemma 3.2 implies that

$$-1 \leq s \mapsto x^{\overline{\varphi}, \alpha f}(t + s)$$

is slowly oscillating. So the corollary follows from Definition 3.1. □

Linearizing f at $x = 0$, we obtain the special case

$$\dot{x}(t) = -\alpha x(t-1) \tag{3.1}$$

of (1.2). Condition (NF) is obviously satisfied. In Section XI.3, we described the spectrum of the generator of the semigroup

$$T_\alpha(t), \qquad t \geq 0,$$

defined by solutions of (3.1). We shall now use those results and relate slowly oscillating solutions of (3.1) to the pair of eigenvalues

$$\lambda_{0,\alpha} = u_{0,\alpha} + iv_{0,\alpha}, \qquad \overline{\lambda}_{0,\alpha} = u_{0,\alpha} - iv_{0,\alpha}$$

with a maximal real part. Recall that for every integer $k \geq 0$, the spectral set

$$\Lambda_{k,\alpha} = \{\lambda_{k,\alpha}, \overline{\lambda_{k,\alpha}}\}$$

defines a decomposition

$$\mathcal{C} = \mathcal{C}_{k,\alpha} \oplus \mathcal{Q}_{k,\alpha}$$

into $T_\alpha(t)$-invariant subspaces; $\mathcal{C}_{k,\alpha}$ is the two-dimensional real generalized eigenspace associated with $\Lambda_{k,\alpha}$. Let $P_{k,\alpha}$ denote the projection onto $\mathcal{C}_{k,\alpha}$ associated with this decomposition [Exercises III.7.27, III.7.28 and IV.2.20].

Proposition 3.4. *Let $\alpha > 0$, $\varphi \in \mathcal{C}$. The solution $x^{\varphi,-\alpha I}$ is slowly oscillating if and only if $P_{0,\alpha}\varphi \neq 0$.*

Proof. 1. Let $P_{0,\alpha}\varphi \neq 0$. Recall from Theorem IV.2.16 that there are constants $c \geq 0$ and $\epsilon > 0$ with

$$\|T_\alpha(t)[\varphi - P_{0,\alpha}\varphi]\| \leq ce^{(u_{0,\alpha}-\epsilon)t}\|\varphi - P_{0,\alpha}\varphi\| \quad \text{for all } t \geq 0.$$

Let $y : [-1,\infty) \to \mathbb{R}$ be the solution of (3.1) with $y_0 = P_{0,\alpha}\varphi$:

$$y_t = T_\alpha(t)P_{0,\alpha}\varphi \quad \text{for } t \geq 0.$$

Therefore, as $t \to \infty$,

$$|(x(t) - y(t))e^{-u_{0,\alpha}t}| \to 0 \tag{3.2}$$

and the form of $t \mapsto y(t)e^{-u_{0,\alpha}t}$ (see Section IV.5) implies that for some $t \geq 0$, x has no zero on $[t-1,t]$. Now Lemma 3.2, applied to $-1 \leq s \mapsto x(t+s)$, shows that x is slowly oscillating.
2. Assume $x = x^{\varphi,-\alpha I}$ is slowly oscillating and $P_{0,\alpha}\varphi = 0$. In case $P_{1,\alpha}\varphi \neq 0$, define

$$\hat{\varphi} := \varphi, \quad \hat{x} := x.$$

In case $P_{1,\alpha}\varphi = 0$, define $\hat{\varphi} := \varphi + \chi$ with

$$\chi(t) = \delta \mathrm{Re}\ e^{\lambda_{1,\alpha}t} \quad \text{for } t \in [-1,0].$$

Here, $\delta > 0$ is so small that the solution $\hat{x} := x^{\varphi+\chi,-\alpha I}$ is slowly oscillating (see Corollary 3.3). Note that $\chi \in \mathcal{C}_{1,\alpha}$, so $P_{0,\alpha}\chi = 0$. Therefore, in both cases,

$$P_{0,\alpha}\hat{\varphi} = 0 \neq P_{1,\alpha}\hat{\varphi}$$

and $\hat{x}$ is slowly oscillating. Let $y : [-1,\infty) \to \mathbb{R}$ be the solution of (3.1) with $y_0 = P_{1,\alpha}\hat{\varphi}$. There are constants $c \geq 0$ and $\epsilon > 0$ such that for all $t \geq 0$,

$$\begin{aligned} ce^{(u_{1,\alpha}-\epsilon)t}\|\hat{\varphi} - (P_{0,\alpha} + P_{1,\alpha})\hat{\varphi}\| &\geq \|T_\alpha(t)(\hat{\varphi} - (P_{0,\alpha} + P_{1,\alpha})\hat{\varphi})\| \\ &= \|\hat{x}_t - T_\alpha(t)P_{1,\alpha}\hat{\varphi}\| \\ &= \|\hat{x}_t - y_t\| \\ &\geq |\hat{x}(t) - y(t)|. \end{aligned}$$

Hence

$$(\hat{x}(t) - y(t))e^{-u_{1,\alpha}t} \to 0 \quad \text{as } t \to \infty.$$

According to Section IV.5, the solution $t \mapsto y(t)e^{-u_{1,\alpha}t}$ is periodic with successive zeros at distance $\frac{\pi}{v_{1,\alpha}} < \frac{1}{2}$. This yields a contradiction to the fact that $\hat{x}$ is slowly oscillating. □

The problem whether initial values of slowly oscillating solutions to nonlinear equations (1.2) are dense is still unsolved. A proof for smooth and monotone f is contained in [187]. In any case, Corollary 3.3 and Proposition 3.4 are strong indications that slowly oscillating solutions are important for the parameterized semiflow Σ_f defined by (1.2).

Trajectories

$$0 \leq t \mapsto \Sigma_f(t,\varphi,\alpha) \in \mathcal{C}$$

of slowly oscillating solutions $x^{\varphi,\alpha f}$ are eventually contained in the set

$$\begin{aligned} \mathcal{O} = \{\varphi \in C \mid \varphi \neq 0 \text{ and } &\varphi \geq 0 \text{ or } \varphi \leq 0 \text{ or (there exists} \\ &z \in (-1,0) \text{ with } \varphi \geq 0 \text{ in } [-1,z],\ \varphi \leq 0 \text{ in } [z,0]) \\ &\text{or (there exists } z \in (-1,0) \text{ with} \\ &\varphi \leq 0 \text{ in } [-1,z], \varphi \geq 0 \text{ in } [z,0])\}, \end{aligned}$$

i.e., the set of $\varphi \in \mathcal{C} \setminus \{0\}$ with at most one change of sign. The set $\mathcal{O}$ is a wedge [that is, $(0,\infty) \cdot \mathcal{O} \subset \mathcal{O}$ and $\varphi \in \mathcal{O}$ does not exclude $-\varphi \in \mathcal{O}$], with

$$\overline{\mathcal{O}} = \mathcal{O} \cup \{0\}. \tag{3.3}$$

The formulas for solutions with orbits in eigenspaces imply that for every $\alpha > 0$, $\mathcal{O} \supset C_{0,\alpha} \setminus \{0\}$.

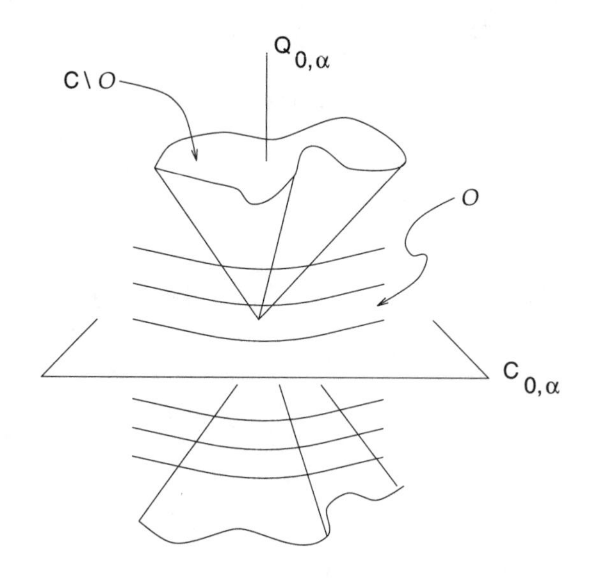

Fig. XV.3.

Corollary 3.5.

(i) *If x is a slowly oscillating solution of (1.2), then there exists $t \in \mathbb{R}$ with*

$$x_s \in \mathcal{O} \quad \textit{for } s \geq t.$$

(ii) *$\varphi \in \mathcal{O}$ implies that $x^{\varphi,\alpha f}$ is slowly oscillating.*

Proof. Assertion (i) follows from Definition 3.1. Let $\varphi \in \mathcal{O}$ and $x = x^{\varphi,\alpha f}$. In case $\varphi \geq 0$,

$$\dot{x} \leq 0 \quad \text{on } (0,1].$$

In case $\varphi \leq 0$ on $[-1,z]$ and $0 \leq \varphi$ on $[z,0]$ for some $z \in (-1,0)$, we have

$$0 \leq \dot{x} \text{ on } (0,z+1] \quad \text{and} \quad \dot{x} \leq 0 \text{ on } [z+1,1].$$

In both cases, we find

$$\dot{x}(t) = \alpha f(\varphi(t-1)) \neq 0 \quad \text{for some } t \in (0,1)$$

since $\varphi \neq 0$. We conclude that for all $s \in (0,1]$,

$$x_s \neq 0,$$

so that $x_s \in \mathcal{O}$. Induction yields

$$x_t \in \mathcal{O} \quad \text{for all } t \geq 0.$$

The same is true for the remaining cases which are possible for $\varphi \in \mathcal{O}$, i.e., $\varphi \leq 0$ or (there exists $z \in [-1,0]$ such that $0 \leq \varphi$ on $[1,z]$ and $\varphi \leq 0$ on $[z,0]$).

In order to show that x is slowly oscillating, it suffices to find $t \geq 0$ with $x(s) \neq 0$ for $t-1 \leq s \leq t$, in view of Lemma 3.2. The arguments given at the beginning of the proof show that there exists $t \geq 0$ with $x_t \geq 0$, or with $x_t \leq 0$. Consider the first case. Then

$$\dot{x} \leq 0 \quad \text{on } [t, t+1].$$

By $x_{t+1} \in \mathcal{O} \subset \mathcal{C} \setminus \{0\}$ and monotonicity,

$$x(t) \neq 0 \quad \text{or} \quad x(t+1) \neq 0.$$

We distinguish the cases $x(t+1) > 0$ (I), $x(t+1) = 0$ (II), $x(t+1) < 0$ and $x(t) = 0$ (III), $x(t+1) < 0$ and $0 < x(t)$ (IV). In case I,

$$x_{t+1} \geq x(t+1) > 0.$$

In case II, there exists $z \in (t, t+1]$ with

$$0 < x \text{ in } [t, z) \quad \text{and} \quad x = 0 \text{ in } [z, t+1],$$

as $x_{t+1} \neq 0$ is decreasing. It follows that

$$\dot{x} < 0 \text{ on } [t+1, z+1) \quad \text{and} \quad \dot{x} = 0 \text{ on } [z+1, t+2].$$

Hence $x < 0$ on $(t+1, t+2]$ and, by continuity, $x_s < 0$ for some $s > t+2$. In case III, there exists $z \in [t, t+1)$ with

$$x = 0 \text{ on } [t, z] \quad \text{and} \quad x < 0 \text{ on } (z, t+1].$$

This implies $\dot{x} = 0$ on $[t+1, z+1]$ so that $x < 0$ on $(z, z+1]$. In case IV, there are z and z' with

$$0 < x \text{ on } [t, z), \quad 0 = x \text{ on } [z, z'], \quad x < 0 \text{ in } (z', t+1].$$

Hence

$$\dot{x} < 0 \text{ on } [t+1, z+1) \quad \text{and} \quad \dot{x} = 0 \text{ on } [z+1, z'+1];$$

and

$$x < 0 \quad \text{on } (z', z'+1].$$

□

From Corollary 3.5(ii) and Proposition 3.4, we infer

Corollary 3.6. *$P_{0,\alpha}\varphi \neq 0$ for all $\alpha > 0$ and all $\varphi \in \mathcal{O}$.*

Slowly oscillating solutions x of the linear equation (3.1) with segments $x_t \in \mathcal{C}_{0,\alpha}$ tend to 0 as $t \to \infty$ if $\alpha < \pi/2$, whereas

$$\|x_t\| \to +\infty \quad \text{as } t \to \infty \text{ for } \alpha > \pi/2.$$

Corollary 3.6 is the first step to a proof that segments of slowly oscillating solutions of nonlinear equations (1.2) increase away from $0 \in \mathcal{C}$ for $\alpha > \pi/2$. This is important for existence and global bifurcation of slowly oscillating periodic solutions.

Remark. It is easy to check that, under additional smoothness conditions on f, there are Hopf bifurcations of periodic solutions at $\alpha = \pi/2$ and at $\alpha = 2k\pi + \frac{\pi}{2}, k \geq 1$. Compare Example X.3.10.

The next aim is to find a wedge $\mathcal{O}' \subset \mathcal{O}$, which absorbs trajectories of slowly oscillating solutions and on which one has an estimate

$$c\|\varphi\| \leq \|P_{0,\alpha}\varphi\| \quad \text{for } \varphi \in \mathcal{O}', \tag{3.4}$$

with a constant $c > 0$. Because this allows us to use the results from Section VIII.5 on unstable behaviour. In order to derive (3.4), we first show the equivalence of this estimate and one where in place of $P_{0,\alpha}$, we have the solution operator $T_\alpha(1)$ of the linear equation (3.1). The latter has the advantage that it is explicitly given by the simple formula

$$T_\alpha(1)\varphi(\theta) = \varphi(0) - \alpha \int_{-1}^{\theta} \varphi(t)\,dt \quad \text{for } \varphi \in \mathcal{C},\ -1 \leq \theta \leq 0. \tag{3.5}$$

Proposition 3.7. *Let $\mathcal{O}' \subset \mathcal{O}$ with $(0,\infty) \cdot \mathcal{O}' \subset \mathcal{O}'$ be given. Then there exists $c > 0$ such that (3.4) holds if and only if there exists $k > 0$ with*

$$k\|\varphi\| \leq \|T_\alpha(1)\varphi\| \quad \textit{for } \varphi \in \mathcal{O}'. \tag{3.6}$$

Proof. 1. Assume (3.6) for some $k > 0$. Exercise VII.2.12 implies that the set

$$\overline{T_\alpha(1)\{\varphi \in \mathcal{O}' \mid \|\varphi\| = 1\}}$$

is compact. Arguments in the proof of Corollary 3.5 (ii) yield

$$\overline{T_\alpha(1)\{\varphi \in \mathcal{O}' \mid \|\varphi\| = 1\}} \subset \overline{\mathcal{O}} = \mathcal{O} \cup \{0\}.$$

By (3.6),

$$\overline{T_\alpha(1)\{\varphi \in \mathcal{O}' \mid \|\varphi\| = 1\}} \subset \mathcal{O}.$$

Corollary 3.6 implies

$$0 \neq P_{0,\alpha}\varphi \quad \text{for } \varphi \in \overline{T_\alpha(1)\{\varphi \in \mathcal{O}' \mid \|\varphi\| = 1\}}.$$

Set

$$c' := \inf\{\|P_{0,\alpha}\varphi\| \mid \varphi \in \overline{T_\alpha(1)\{\varphi \in \mathcal{O}' \mid \|\varphi\| = 1\}}\} > 0.$$

For every $\varphi \in \mathcal{O}'$ with $\|\varphi\| = 1$,

$$c' \leq \|P_{0,\alpha}T_\alpha(1)\varphi\| = \|T_\alpha(1)P_{0,\alpha}\varphi\| \leq \|T_\alpha(1)\|\|P_{0,\alpha}\varphi\|.$$

This implies (3.4), with $c := c'/\|T_\alpha(1)\|$.
2. Let (3.4) hold. Then

$$0 \notin \overline{P_0(\alpha)\{\varphi \in \mathcal{O}' \mid \|\varphi\| = 1\}} \subset \mathcal{C}_{0,\alpha}.$$

The space $\mathcal{C}_{0,\alpha}$ is finite dimensional, and

$$0 \neq T_\alpha(1)\psi \quad \text{for } \psi \in \mathcal{C}_{0,\alpha} \setminus \{0\}.$$

By compactness,

$$0 < \inf\{\|T_\alpha(1)\psi\| \mid \psi \in \overline{P_{0,\alpha}\{\varphi \in \mathcal{O}' \mid \|\varphi\| = 1\}}\},$$

and one can proceed as in Part 1. □

XV.4 The a priori estimate for unstable behaviour

It is not hard to see that for $\alpha > 1$, every slowly oscillating solution x of (1.2) has (for suitable values of t) segments x_t in the convex cone

$$K := \{\varphi \in C \mid \varphi(-1) = 0,\ \varphi \text{ increasing},\ 0 < \varphi(0)\}.$$

Note that $\|\varphi\| = \varphi(0)$ on K, and that $K \subset \mathcal{O}$. How do solutions which start in K look?

Proposition 4.1. *Let $\alpha > 1$, $\varphi \in K$. Set*

$$z_0 := \max\{t \in [-1,0] \mid \varphi = 0 \text{ on } [-1,t]\} \in [-1,0).$$

The zeros of the solution $x := x^{\varphi,\alpha f}$ in (z_0,∞) form a sequence $(z_n)_1^\infty$ with

$$\dot{x}(z_n) \neq 0 \text{ and } z_{n-1} + 1 < z_n \quad \text{for all } n \in \mathbb{N}.$$

In particular,

$$\dot{x} = 0 \quad \text{on } (0, z_0 + 1],$$

and

$$\dot{x} < (>)0 \quad \text{on } (z_{n-1}+1, z_n+1) \quad \text{for } n \text{ odd (even)}.$$

Sketch of proof. By equation (1.2),

$$\dot{x} = 0 \text{ on } (0, z_0 + 1] \quad \text{and} \quad \dot{x} < 0 \text{ on } (z_0 + 1, z_0 + 2].$$

Because of Proposition 2.2,

$$z_1 := \inf\{z > z_0 + 1 : x(z) = 0\}$$

is well defined and $x(z_1) = 0$. Since $x(z_0 + 1) > 0$, $z_1 > z_0 + 1$. If $0 < x$ on (z_0, z_1), then

$$\dot{x} < 0 \quad \text{on } (z_0 + 1, z_1 + 1).$$

Note

$$\hat{\varphi} := x_{z_1+1} \in -K, \quad \text{with } \hat{\varphi} < 0 \text{ on } (-1, 0].$$

Continue as before. □

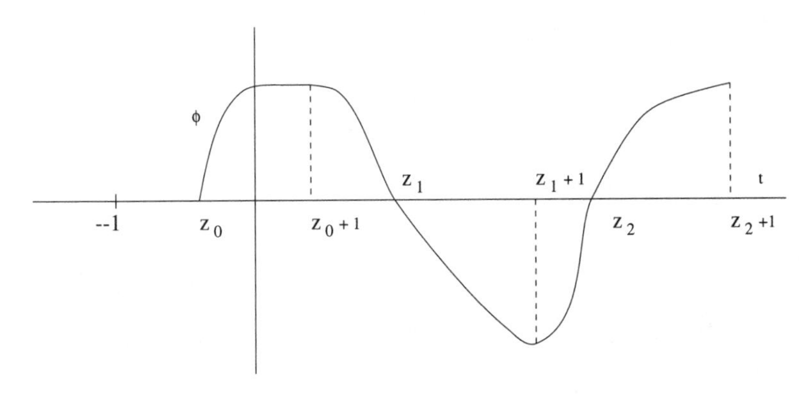

Fig. XV.4.

Later we shall write $z_0(\varphi)$ and $z_n(\varphi, \alpha f)$, $n \geq 1$, when convenient. The cone K is not positively invariant, of course, but the segments $\Sigma_f(t, \varphi, \alpha)$, where $t \geq 0$ and $\varphi \in K$, are contained in the wedge $\mathcal{O}$. The following estimate is crucial. It says that the forward flow corresponding to the initial values $\varphi \in K$ of a *bounded* set of solutions belongs to a subwedge of $\mathcal{O}$ which is determined by an inequality of type (3.6).

Proposition 4.2. *Let $\alpha > 1$ and let $r > 1$ be given. Then there exists $k > 0$ so that*

$$k\|x_t\| \leq \|T_\alpha(1)x_t\|$$

for all $t \geq 0$ and all solutions $x : [-1, \infty) \to \mathbb{R}$ of (1.2) which satisfy

$$x_0 \in K \text{ and } |x(s)| \leq r \quad \text{for all } s \geq -1.$$

Proof. 1. Claim. Let a continuous function $x : [-1, \infty) \to \mathbb{R}$ be given, and let $t \geq 0$. For every pair of reals u and v with

$$t-1 \le u < v \le t,$$

we have

$$\frac{\alpha}{2}\left|\int_u^v x(s)\,ds\right| \le \|T_\alpha(1)x_t\|.$$

Indeed, since

$$\begin{aligned}\alpha\left|\int_u^v x(s)\,ds\right| &= \left|\alpha\int_{u-t}^{v-t} x_t(\theta)\,d\theta\right| \\ &= \left|T_\alpha(1)x_t(v-t) - T_\alpha(1)x_t(u-t)\right|\end{aligned}$$

[see (3.5)]

$$\le 2\|T_\alpha(1)x_t\|,$$

the claim follows.
2. Choose $a \in (0,1)$ and $b > 1$ with

(4.1) $$a|x| < |\alpha f(x)| < b|x| \quad \text{for } 0 < |x| \le r.$$

Define

$$k = \frac{a}{32b}.$$

Let a solution $x : [-1,\infty) \to \mathbb{R}$ of (1.2) be given so that $x_0 \in K$ and $|x(t)| \le r$ for all $t \ge -1$. The zeros of x form a sequence $(z_n)_0^\infty$ as in Proposition 4.1. The points

$$\rho_n = z_n + 1, \qquad n \in \mathbb{N}_0,$$

are local extrema.
3. The desired estimate and formula (3.5) mean that either the value $|x(t)|$ or a piece of the area between the graph of x and the interval $[t-1,t]\times\{0\}$ on the abscissa is not too small, compared to the maximum of the values $|x(s)|$, $t-1 \le s \le t$. In order to estimate such pieces of area, we introduce the lines

$$g_n : \mathbb{R} \to \mathbb{R}, \quad g_n(t) = x(\rho_n) - bx(\rho_n)(t-\rho_n), \quad n \in \mathbb{N}_0.$$

Obviously,

$$g_n(\rho_n) = x(\rho_n) \quad \text{and} \quad g_n(\rho_n + \frac{1}{b}) = 0.$$

Claim. $|g_n(t)| \le |x(t)|$ for $\rho_n \le t \le \rho_n + \frac{1}{b}$, $n \in \mathbb{N}_0$. In particular, $\rho_n + \frac{1}{b} \le z_{n+1}$ for all $n \in \mathbb{N}_0$. To prove the claim observe that for $\rho_n \le t \le \rho_n + \frac{1}{b} < \rho_n + 1$,

$$|\dot{x}(t)| = |\alpha f(x(t-1))| \le b|x(t-1)| \le b|x(\rho_n)| = |\dot{g}_n(t)|.$$

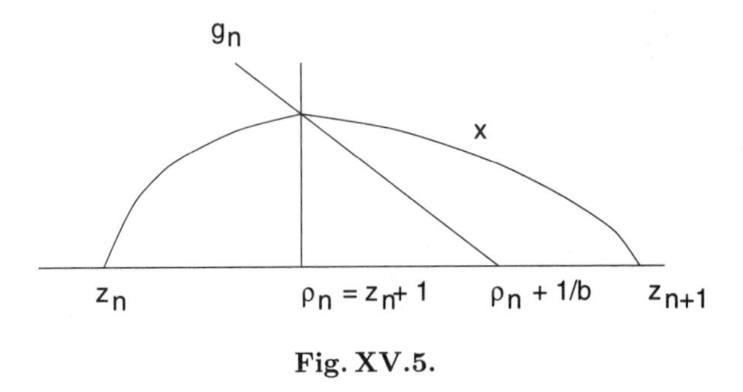

Fig. XV.5.

4. Let $t \geq 0$. In order to estimate $\|T_\alpha(1)x_t\|$, we distinguish the cases
(I) $t-1 \leq \rho_n \leq t$ for some $n \in \mathbb{N}_0$;
(II) $\rho_n \notin [t-1, t]$ for all $n \in \mathbb{N}_0$.
5. Case (I). We consider the subcases

$$\|x_t\| > |x(\rho_n)| \tag{A}$$

and

$$\|x_t\| = |x(\rho_n)| \tag{B}$$

separately.
Case (A). Recall that $|x|$ increases on $[\rho_n - 1, \rho_n] = [z_n, \rho_n]$, decreases on $[\rho_n, z_{n+1}]$ and increases on $[z_{n+1}, z_{n+1}+1]$.

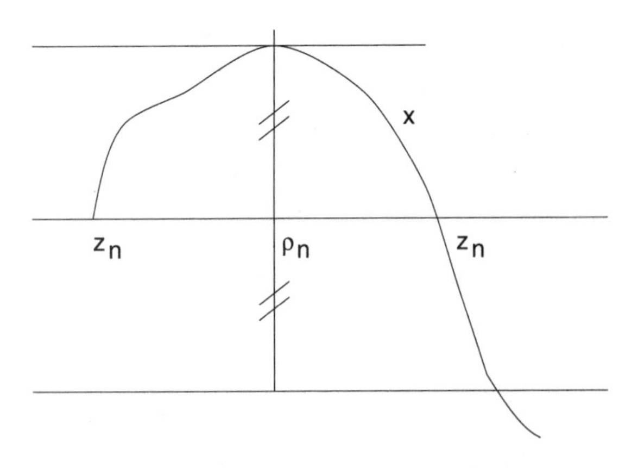

Fig. XV.6.

In the case considered, we obtain from these monotonicity properties that $t \geq z_{n+1}$ and

$$\begin{aligned}\|x_t\| &= |x(t)| \\ &= |T_\alpha(1)x_t(-1)| \le \|T_\alpha(1)x_t\|.\end{aligned}$$

Case (B). We consider the subcases

(i) $$t \le \rho_n + \frac{1}{2b}$$

and

(ii) $$\rho_n + \frac{1}{2b} < t$$

separately.

Subcase (i). Then

$$\begin{aligned}\|T_\alpha(1)x_t\| &\ge |x(t)| \qquad \text{[compare case (A)]} \\ &\ge |g_n(t)| \qquad \text{(see Part 3)} \\ &\ge |g_n(\rho_n + \frac{1}{2b})| = \frac{|x(\rho_n)|}{2} = \frac{\|x_t\|}{2}.\end{aligned}$$

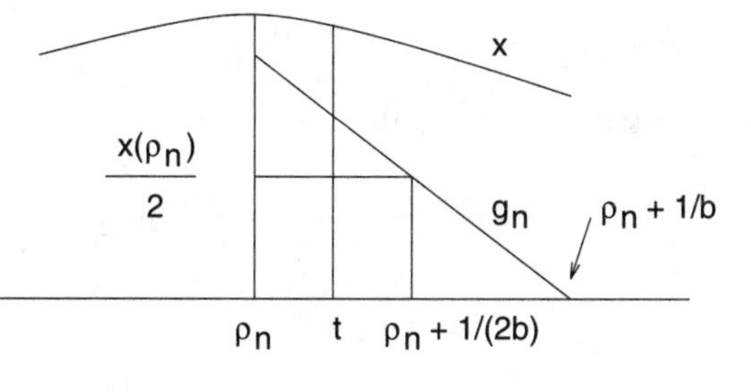

Fig. XV.7.

Subcase (ii). Then $[\rho_n, \rho_n + 1/(2b)] \subset [t-1, t]$. For $\rho_n \le s \le \rho_n + \frac{1}{2b}$, we have

$$\begin{aligned}|x(s)| &\ge |g_n(s)| \qquad \text{(see Part 3)} \\ &\ge \frac{|x(\rho_n)|}{2} = \frac{\|x_t\|}{2}.\end{aligned}$$

Set $u = \rho_n$, $v = \rho_n + \frac{1}{2b}$. Part 1 and the last estimate yield

$$\begin{aligned}\|T_\alpha(1)x_t\| &\ge \frac{\alpha}{2}\Big|\int_u^v x(s)\,ds\Big| \\ &\ge \frac{\alpha}{2}\frac{\|x_t\|}{2}(v-u) = \frac{\alpha}{8b}\|x_t\|.\end{aligned}$$

6. Case (II). Then $t > 1$, and x is monotone on the interval $[t-1, t]$. It follows that either

$$\|x_t\| = |x(t)| \qquad \text{[case (C)]},$$

or

$$\|x_t\| = |x(t-1)| = |x(t)| \qquad \text{[case (D)]},$$

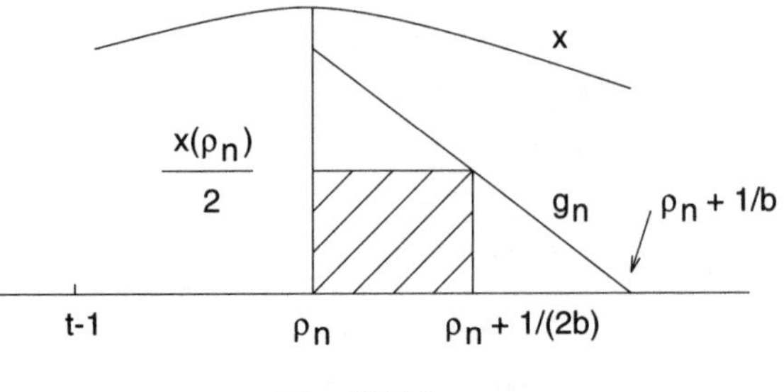

Fig. XV.8.

Case (C). Then

$$\|T_\alpha(1)x_t\| \geq |x(t)| = \|x_t\|.$$

Case (D). We distinguish the subcases

(iii) $\qquad |x(\bar{t})| < |x(t-1)| \qquad$ for some $\bar{t} \in [t-1-\dfrac{1}{2b}, t-1]$

and

(iv) $\qquad |x(s)| \geq |x(t-1)| \qquad$ for some $s \in [t-1-\dfrac{1}{2b}, t-1]$

Subcase (iii). Then

$$|x(\bar{t})| < |x(t-1)| > |x(t)|,$$

and Proposition 4.1 implies that for some $n \in \mathbb{N}_0$,

$$\bar{t} < \rho_n < t. \tag{4.2}$$

As we are in case (II), $\rho_n < t-1$. Using (4.2) and $t-1-\frac{1}{2b} \leq \bar{t}$, we arrive at

$$\begin{aligned} \rho_n < t-1 < \rho_n &+ \frac{1}{2b} \\ &< z_{n+1} \qquad \text{(see Part 3)} \end{aligned}$$

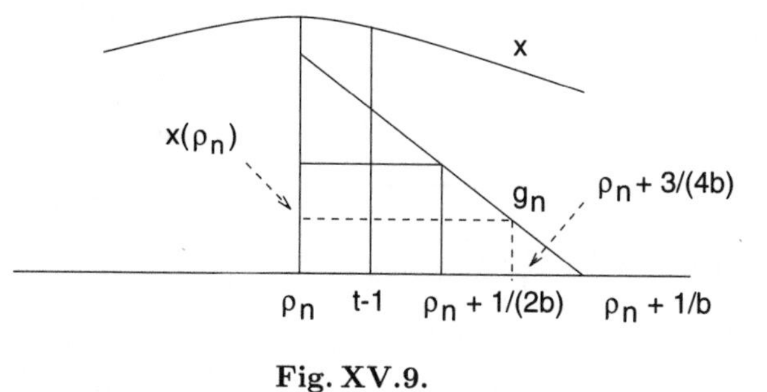

Fig. XV.9.

The fact that $|x|$ decreases on $[\rho_n, z_{n+1}]$ and $t-1 < z_{n+1}$ implies

$$|x(\rho_n)| \geq |x(t-1)|. \tag{4.3}$$

Observe

$$[t-1, t-1+\frac{1}{4b}] \subset [\rho_n, \rho_n + \frac{3}{4b}].$$

For $t-1 \leq s \leq t-1+\frac{1}{4b}$, we obtain

$$\begin{aligned} |x(s)| &\geq |g_n(s)| \qquad \text{[see Part 3]} \\ &\geq |g_n(\rho_n + \frac{3}{4b})| = \frac{1}{4}|x(\rho_n)| \\ &\geq \frac{1}{4}|x(t-1)| \qquad \text{[see (4.3)]} \\ &= \frac{1}{4}\|x_t\| > 0. \end{aligned}$$

Set $u = t-1$, $v = t-1+\frac{1}{4b}$. Part 1 gives

$$\begin{aligned} \|T_\alpha(1)x_t\| &\geq \frac{\alpha}{2}\left|\int_u^v x(s)\, ds\right| \\ &\geq \frac{\alpha}{2}\frac{1}{4}\|x_t\|(v-u) = \frac{\alpha}{2}\frac{1}{4}\|x_t\|\frac{1}{4b}. \end{aligned}$$

Subcase (iv). For all $s \in [t-\frac{1}{2b}, t]$, we have $t-1-\frac{1}{2b} \leq s-1 \leq t-1$; hence

$$|x(s-1)| \geq |x(t-1)| = \|x_t\|,$$

and therefore

$$\begin{aligned} |\dot{x}(s)| &= |\alpha f(x(s-1))| \geq a|x(s-1)| \qquad \text{[see (4.1)]} \\ &\geq a\|x_t\| > 0. \end{aligned}$$

Integration yields

$$|x(t) - x(t - \frac{1}{2b})| \geq a\|x_t\|\frac{1}{2b},$$

so that either

$$|x(t)| \geq \frac{a}{4b}\|x_t\| \tag{4.4}$$

or

$$|x(t - \frac{1}{2b})| \geq \frac{a}{4b}\|x_t\|. \tag{4.5}$$

In case (4.4) holds, we have

$$\|T_\alpha(1)x_t\| \geq |x(t)| \geq \frac{a}{4b}\|x_t\|.$$

It remains to consider the case that (4.5) holds, together with

$$|x(t)| < \frac{a}{4b}\|x_t\|.$$

The combination of these inequalities gives

$$|x(t)| < |x(t - \frac{1}{2b})|. \tag{4.6}$$

We infer that for all $s \in [t-1, t-\frac{1}{2b}]$,

$$\text{sign } x(s) = \text{sign } x(t-1), \tag{4.7}$$

since otherwise x would have a zero $z \in (t-1, t-\frac{1}{2b})$, and the monotonicity of x on $[t-1, t] \supset [z, t]$ would imply $|x(t-\frac{1}{2b})| \leq |x(t)|$, which contradicts (4.6).

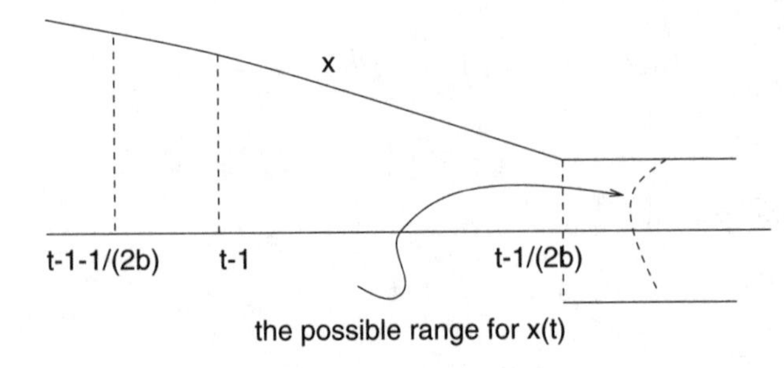

Fig. XV.10.

Equations (4.7), $\|x_t\| = |x(t-1)|$ and the monotonicity of x on $[t-1, t]$ altogether give

$$|x(s)| \geq |x(t - \frac{1}{2b})| \quad \text{for all } s \in [t-1, t - \frac{1}{2b}].$$

Set $u = t - 1$, $v = t - \frac{1}{2b}$. By Part 1,

$$\begin{aligned}
\|T_\alpha(1)x_t\| &\geq \frac{\alpha}{2} |\int_u^v x(s)\, ds| \\
&= \frac{\alpha}{2} \int_u^v |x(s)|\, ds \geq \frac{\alpha}{2} |x(t - \frac{1}{2b})| (1 - \frac{1}{2b}) \\
&\geq \frac{\alpha}{2} \frac{a}{4b} \|x_t\| (1 - \frac{1}{2b}) \qquad \text{[see (4.5)]} \\
&\geq \frac{\alpha a}{16b} \|x_t\|.
\end{aligned}$$

□

The assumption $\alpha > 1$ in Proposition 4.2 can be removed, but this is not of much interest here since we shall use Proposition 4.2 only for $\alpha > \frac{\pi}{2}$.

XV.5 Slowly oscillating solutions which grow away from zero, periodic solutions

From here on, we assume that the function f in equation (1.2) satisfies the boundedness condition (B), in addition to (NF).

One can use arguments as in the proof of the Principle of Linearized Stability (Corollary VII.5.12) in order to show that for $\alpha < \frac{\pi}{2}$, all solutions $x^{\varphi,\alpha f}$, with $\varphi \in \mathcal{C}$ sufficiently close to 0, tend to 0 as $t \to \infty$ and that for $\alpha > \pi/2$, the zero solution of the nonlinear equation (1.2) is unstable.

In the present section we shall prove a result on unstable behaviour in case $\alpha > \pi/2$, which is considerably more difficult: segments x_t of *slowly oscillating* solutions which are initially small leave a neighbourhood of $0 \in \mathcal{C}$ and stay outside. First, we collect some preliminary information. Let $\alpha > \frac{\pi}{2}$ and set

$$r_1 = \alpha \sup f, \qquad r_0 = \alpha \min_{[0,r_1]} f < 0$$

and

$$r = 1 + \max\{-r_0, r_1\}.$$

Choose $k > 0$ so that, with r as just defined, the assertion of Proposition 4.2 holds. Define

$$\mathcal{O}' = \{\varphi \in \mathcal{O} \mid k\|\varphi\| \leq \|T_\alpha(1)\varphi\|\}.$$

According to Proposition 3.7, there exists $c > 0$ such that $\mathcal{O}'$ is contained in the wedge

$$\mathcal{O}_{\alpha f} = \{\varphi \in \mathcal{O} \mid c\|\varphi\| \leq \|P_{0,\alpha}\varphi\|\}.$$

Now consider a slowly oscillating solution $x : [-1, \infty) \to \mathbb{R}$ of (1.2). There exists $s \geq 0$ so that x has no zeros on $[s-1, s]$. An application of Proposition 2.2 yields zeros $z > s$ and $z' > z + 1$ so that

$$x > 0 \quad \text{on } [s-1, z) \quad \text{and} \quad 0 > x \quad \text{on } (z, z'),$$

or vice versa. In either case, there exists $t^* \geq 0$ (namely $t^* = z + 1$ or $t^* = z' + 1$) such that $\varphi = x_{t^*}$ satisfies

$$\varphi \in K \quad \text{and} \quad \varphi(0) \leq r_1.$$

Propositions 4.1 and 2.3 now imply

$$-r \leq r_0 \leq x^{\varphi,\alpha f}(s) \leq r_1 \leq r \quad \text{for all } s \geq 0.$$

Altogether, we infer $|x| \leq r$ on $[t^* - 1, \infty)$, and using Proposition 4.2 and the choice of c above, we find

$$x_s \in \mathcal{O}_{\alpha f} \quad \text{for } s \geq t^*.$$

In case $x_0 \in K$ and $x(0) \leq r_1$, we obtain

$$r_0 \leq x \leq r_1 \quad \text{on } [-1, \infty) \quad \text{and} \quad x_s \in \mathcal{O}_{\alpha f}, \quad \text{for } s \geq c.$$

Theorem 5.1. *Suppose $\alpha > \frac{\pi}{2}$. Then there exist a continuous functional*

$$V_{\alpha f} : C \to [0, \infty)$$

with

$$V_{\alpha f}(s\varphi) = s^2 V_{\alpha f}(\varphi) \quad \text{for all } s \geq 0, \quad \varphi \in C,$$

and a constant $a_{\alpha f} > 0$ with the following properties.

(i) *$\varphi \in K$ and $\varphi(0) = \alpha \sup f$ imply $V_{\alpha f}(\varphi) > a_{\alpha f}$.*

(ii) *Let $x : [-1, \infty) \to \mathbb{R}$ be a slowly oscillating solution of (1.2). There exists $t \geq 0$ such that for all $s \geq t$,*

$$a_{\alpha f} < V_{\alpha f}(x_s) \tag{5.1}$$

and

$$\alpha \min_{[0, \alpha \sup f]} f \leq x(s) \leq \alpha \sup f. \tag{5.2}$$

If $x_0 \in K$ satisfies $a_{\alpha f} \leq V_{\alpha f}(x_0)$ and $x(0) \leq \alpha \sup f$, then (5.1) and (5.2) hold for all $s > 0$.

Proof. 1. Let $V : C \to [0, \infty)$ denote the quadratic Lyapunov functional which is associated with the spectral set $\Lambda = \{\lambda_{0,\alpha}, \overline{\lambda}_{0,\alpha}\}$, according to Section VIII.5. Let $|||\cdot|||$ denote the equivalent norm associated with Λ as in Section VIII.5. There are positive constants γ_1 and γ_2 with

$$\gamma_1\|\varphi\| \le |||\varphi||| \le \gamma_2\|\varphi\| \quad \text{for all } \varphi \in \mathcal{C}.$$

Set

$$q = \frac{\gamma_2}{\gamma_1}\frac{1+c}{c}.$$

Then

$$\begin{aligned}\mathcal{O}_{\alpha f} &\subset \{\varphi \in \mathcal{C} \mid \|(I - P_{0,\alpha})\varphi\| \le \frac{1+c}{c}\|P_{0,\alpha}\varphi\|\} \\ &\subset \{\varphi \in \mathcal{C} \mid |||(I - P_{0,\alpha})\varphi||| \le q|||P_{0,\alpha}\varphi|||\} = K_q.\end{aligned}$$

Theorem VIII.5.8 and Lemma VIII.5.10 yield a constant $\delta > 0$ such that

$$\varphi \in K_q \quad \text{and} \quad 0 < |||\varphi||| \le \delta$$

imply

$$\dot{V}(\varphi) = \lim_{t\downarrow 0} \frac{1}{t}\big(V(x_t^{\varphi,\alpha f}) - V(\varphi)\big) > 0$$

and

$$|||x_t^{\varphi,\alpha f}||| = \delta \qquad \text{for some } t \ge 0.$$

2. Claim. There exists $\epsilon > 0$ such that $\varphi \in \mathcal{O}_{\alpha f}$ and $\|\varphi\| \le \epsilon$ imply $\dot{V}(\varphi) > 0$ and $\|x_t^{\varphi,\alpha f}\| = \epsilon$ for some $t \ge 0$. To prove this, set $\epsilon = \frac{\delta}{\gamma_2}$. Consider $\varphi \in \mathcal{O}_{\alpha f}$ with $\|\varphi\| \le \epsilon$. Then, $\varphi \in K_q$ and $0 < |||\varphi||| \le \delta$. According to Part 1, $\dot{V}(\varphi) > 0$, and there exists $s \ge 0$ with $|||x_s^{\varphi,\alpha f}||| = \delta$. In particular,

$$\|x_s^{\varphi,\alpha f}\| \ge \frac{1}{\gamma_2}|||x_s^{\varphi,\alpha f}||| = \frac{\delta}{\gamma_2} = \epsilon,$$

and, by continuity,

$$\|x_t^{\varphi,\alpha f}\| = \epsilon \qquad \text{for some } t \in [0, s].$$

3. Claim. There exists $a > 0$ such that the open neighbourhood $\{\varphi \in C \mid \|\varphi\| < \epsilon\}$ of $0 \in C$ contains the set

$$\{\varphi \in K_q \mid V(\varphi) \le a\} \supset \{\varphi \in \mathcal{O}_{\alpha f} \mid V(\varphi) \le a\}.$$

In order to prove this, note that Lemma VIII.5.9 yields a constant $c_1 > 0$ such that

$$V(\varphi) \ge c_1\gamma_1^2\|P_{0,\alpha}\varphi\|^2 \quad \text{for all } \varphi \in \mathcal{C}.$$

Choose $a > 0$ so small that

$$\sqrt{\frac{a}{c_1 c^2 \gamma_1^2}} < \epsilon.$$

For $\varphi \in K_q$ and $V(\varphi) \le a$, we obtain

$$\begin{aligned} a \ge V(\varphi) &\ge c_1\gamma_1^2\|P_{0,\alpha}\varphi\|^2 \\ &\ge c_1\gamma_1^2 c^2\|\varphi\|^2;\end{aligned}$$

hence $\|\varphi\| < \epsilon$.
4. Let a slowly oscillating solution $x : [-1, \infty) \to \mathbb{R}$ of (1.2) be given. Recall from the remarks preceding the statement of Theorem 5.1 that $r_0 \leq x \leq r_1$ for $s \geq t^* - 1$ and $x_s \in \mathcal{O}_{\alpha f}$ for $s \geq t^*$. There exists $t \geq t^*$ with $V(x_t) > a$ since, otherwise, $\|x_s\| < \epsilon$ for all $s \geq t^*$ (see Part 3) which leads to a contradiction to Part 2.

Suppose $V(x_s) \leq a$ for some $s > t$. There exists $u \in (t, s]$ such that $V(x_u) = a$ and $V(x_w) > a$ for $t < w < u$. Hence $\dot{V}(x_u) \leq 0$. So Part 3 implies $\|x_u\| < \epsilon$, and we arrive at a contradiction to Part 2.

Suppose now that $x_0 \in K$, $a \leq V(x_0)$ and $x(0) \leq r_1$. Recall $r_0 \leq x \leq r_1$ on $[-1, \infty)$ and $x_s \in \mathcal{O}_{\alpha f}$ for all $s \geq 0$. In case $a < V(x_0)$, the same argument as above yields $a < V(x_s)$ for all $s > 0$. In case $a = V(x_0)$, we have $\|x_0\| < \epsilon$, by Part 3, and $\dot{V}(x_0) > 0$, by Part 2. Hence $a < V(x_s)$ on some interval $(0, \eta]$ where $\eta > 0$. Arguments as above imply $a < V(x_s)$ for all $s \geq \eta$. □

Remark. An analogous result holds if instead of (B), a lower bound for f is assumed. Without any boundedness assumption, one obtains, e.g., that for every $r > 0$, there exists $a_r > 0$ such that for every slowly oscillating solution x bounded by r, there exists $t \geq 0$ with $a_r \leq V(x_s)$ for all $s \geq t$.

The first result on unstable behaviour of slowly oscillating solutions is due to Wright [308]. Stated for (1.2), it says that, under the conditions of Theorem 5.1, there exists $\epsilon > 0$ such that for every slowly oscillating solution x,

$$\lim_{t \to \infty} \sup |x(t)| \geq \epsilon.$$

Theorem 5.1 is stronger. A first application is that ω-limit sets of slowly oscillating solutions are contained in the set

$$\{\varphi \in \mathcal{O} \mid a < V(\varphi), \quad \alpha \min_{[0, \alpha \sup f]} f \leq \varphi \leq \alpha \sup f\}$$

and that the zeros of the solutions $x : \mathbb{R} \to \mathbb{R}$ with segments x_t in such ω-limit sets form a sequence $(z_n)_{-\infty}^{\infty}$ with

$$\dot{x}(z_n) \neq 0 \quad \text{and} \quad z_{n-1} + 1 < z_n \quad \text{for all } n \in \mathbb{Z}.$$

See [287, 302] for a proof. It is also easy to deduce the existence of periodic solutions.

Corollary 5.2. *Let $\alpha > \frac{\pi}{2}$. Then there exist slowly oscillating periodic solutions $x : \mathbb{R} \to \mathbb{R}$ of (1.2) with $x_0 \in K$ and minimal period $z_2(x_0, \alpha f) + 1$.*

Proof. Consider the map $\tau : K \ni \varphi \mapsto z_2(\varphi, \alpha f) \in \mathbb{R}$. Using Exercise VII.2.11 on continuous dependence and the fact that the zeros z_1 and z_2 in Proposition 4.1 are simple, one shows that τ is continuous. Using this and Exercise VII.2.11 once more, one sees that the map

$$\mathcal{P} : K \ni \varphi \mapsto \Sigma_f(\tau(\varphi), \varphi, \alpha) \in K$$

is continuous, too. Set

$$\mathcal{D} = \{\varphi \in K : a \le V(\varphi),\ \varphi(0) \le \alpha \ \sup f\}.$$

The set $\mathcal{D}$ is closed, nonempty, and $\mathcal{P}(\mathcal{D}) \subset \mathcal{D}$, by Theorem 5.1. Furthermore, the set $\overline{\mathcal{P}(\mathcal{D})} \subset \mathcal{D}$ is compact. To prove this, observe

$$0 \le \dot{x}^{\varphi,\alpha f} \le \alpha \sup f \quad \text{on } [z_2(\varphi, \alpha f), z_2(\varphi, \alpha f) + 1] \quad \text{for } \varphi \in K,$$

and apply the Theorem of Ascoli and Arzela.

The affine linear maps from the segments between $V^{-1}(a)$ and the sphere of radius $\alpha \sup f$ on the rays $(0, \infty)\varphi$, $\varphi \in K$, onto the spheres of radii 1 and 2 define a homeomorphism of $\mathcal{D}$ onto the closed convex set

$$K_{12} = \{\varphi \in K \mid 1 \le \varphi(0) \le 2\}.$$

Finally, an application of Schauder's Theorem to the transform $K_{12} \to K_{12}$ of $\mathcal{P}$ yields a fixed point of $\mathcal{P}$ in $\mathcal{D}$, which can be used to define the desired periodic solution. □

We can now graph the pairs (α, φ) of Corollary 5.2, with $x^{\varphi,\alpha f}$, a slowly oscillating periodic solution having period $z_2(\varphi, \alpha f) + 1$, in a schematic diagram:

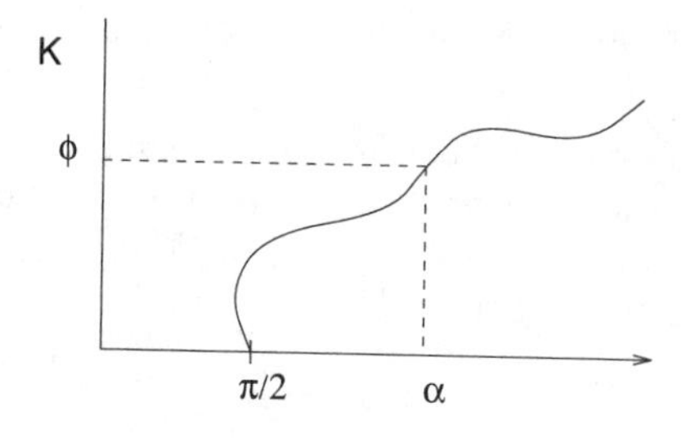

Fig. XV.11.

In case f has additional smoothness properties, one knows, due to the local Hopf Bifurcation Theorem, that a small branch emanating from $(\frac{\pi}{2}, 0)$ exists. More information is contained in a global bifurcation result due to Nussbaum [214], which is stated below as Theorem 5.5.

Let us first introduce operators $\mathcal{P}_f(\cdot, \alpha) : \overline{K} \to \overline{K}$, $\alpha > 0$, such that nonzero fixed points correspond to initial values of slowly oscillating periodic solutions with period $z_2 + 1$. (For $\alpha > \pi/2$, these operators extend the maps from the proof of Corollary 5.2.) Note

$$\overline{K} = K \cup \{0\}.$$

Proposition 5.3. *Let $\varphi \in K$ be given. Set*

$$z_0 = \max\{t \in [-1,0] \mid \varphi(s) = 0 \quad \text{for all } s \in [-1,t]\}.$$

Then $0 < \varphi$ on $(z_0, 0]$, and for $x = x^{\varphi,\alpha f}$,

$$\dot{x} = 0 \quad \text{on } (0, z_0 + 1]$$

and either

$$(5.3) \quad 0 < x \quad \text{on} \quad [0,\infty), \ \dot{x} < 0 \quad \text{on } (z_0+1,\infty), \ x(t) \to 0 \quad \text{as } t \to \infty,$$

or

$$(5.4) \quad \begin{aligned} &\text{there exists a zero } z_1 > z_0 + 1 \text{ of } x \text{ with} \\ &\dot{x} < 0 \quad \text{on } (z_0+1, z_1+1), \ \alpha \min_{[0,\varphi(0)]} f \le x(z_1+1) < 0, \\ &x < 0 \quad \text{on } (z_1, \infty), 0 < \dot{x} \quad \text{on } (z_1+1,\infty), \\ &x(t) \to 0 \quad \text{as } t \to \infty, \end{aligned}$$

or

$$(5.5) \quad \begin{aligned} &\text{there exist zeros } z_1 > z_0 + 1, \ z_2 > z_1 + 1 \text{ of } x \text{ with} \\ &\dot{x} < 0 \quad \text{on } (z_0+1, z_1+1), \quad 0 < \dot{x} \text{ on } (z_1+1, z_2+1), \\ &\alpha \min_{[0,\varphi(0)]} f \le x(z_1+1) < 0, \quad x(z_2+1) < \alpha \max_{[x(z_1+1),0]} f. \end{aligned}$$

Proof. Compare Proposition 2.3 and Proposition 4.1. □

Note that in Proposition 5.3, it is not assumed that $\alpha > 1$, as was in Proposition 4.1, and that the result of Proposition 5.3 holds also if condition (B) is violated. As before, we write $z_1(\varphi, \alpha f)$ and $z_2(\varphi, \alpha f)$ instead of z_1 and z_2 in (5.4), when convenient. Define a map

$$\mathcal{P}_f : (0,\infty) \times \overline{K} \to \overline{K}$$

by

$$\begin{cases} \mathcal{P}_f(\alpha, \varphi) = x^{\varphi,\alpha f}_{z_2(\varphi,\alpha f)+1} & \text{in case of (5.5)} \\ \mathcal{P}(\alpha, \varphi) = 0 & \text{otherwise.} \end{cases}$$

In particular, $\mathcal{P}_f(\alpha, 0) = 0$ for all $\alpha > 0$. If $\varphi \in K$ is a fixed point of $\mathcal{P}_f(\alpha, \cdot)$, then $x^{\varphi,\alpha f}$ extends to a slowly oscillating solution of (1.2) with period $z_2(\varphi, \alpha f) + 1$, and every slowly oscillating periodic solution $x : \mathbb{R} \to \mathbb{R}$ of (1.2) with $x_0 \in K$ and period $z_2(x_0, \alpha f) + 1$ yields a fixed point x_0 of $\mathcal{P}_f(\alpha, \cdot)$.

Corollary 5.4. *The mapping $\mathcal{P}_f$ is continuous and maps sets $(0, \alpha] \times \overline{K}$, $\alpha > 0$, into relatively compact sets.*

Proof. 1. The assertion on compactness follows from an application of the Theorem of Ascoli and Arzela, since in the case of (5.5)

$$0 \le \dot{x}^{\varphi,\tilde{\alpha}f} \le \alpha \sup f \quad \text{on } [z_2(\varphi,\tilde{\alpha}f), z_2(\varphi,\tilde{\alpha}f)+1]$$

for all $\tilde{\alpha} \in (0,\alpha]$ and all $\varphi \in K$.
2. Let $\alpha > 0, \varphi \in \overline{K}, x := x^{\varphi,\alpha f}$.
2.1. In the case of (5.4), one can use Exercise VII.2.11 (continuous dependence) to show that there is a neighbourhood U of (α,φ) in $[0,\alpha+1) \times \overline{K}$ such that for every $(\tilde{\alpha},\tilde{\varphi}) \in U$, there is a first zero $z_1(\tilde{\varphi},\tilde{\alpha}f) > 0$, and that the map

$$U \ni (\tilde{\alpha},\tilde{\varphi}) \mapsto z_1(\tilde{\varphi},\tilde{\alpha}f) \in \mathbb{R}$$

is continuous. It follows that for $(\tilde{\alpha},\tilde{\varphi}) \in U$, either (5.4) or (5.5) holds.
2.2. If (5.5) holds for some (α,φ), then there is a neighbourhood U of (α,φ) in $(0,\alpha+1) \times \overline{K}$ such that for every $(\tilde{\alpha},\tilde{\varphi}) \in U$, (5.5) holds as well and such that the maps

$$U \ni (\tilde{\alpha},\tilde{\varphi}) \mapsto z_1(\tilde{\varphi},\tilde{\alpha}f) \in \mathbb{R}, \qquad U \ni (\tilde{\alpha},\tilde{\varphi}) \mapsto z_2(\tilde{\varphi},\tilde{\alpha}f) \in \mathbb{R}$$

are continuous.
3. In case 2.2, continuity of $\mathcal{P}_f$ at (α,φ) follows from the estimate

$$\begin{aligned}
&\|\Sigma_f(z_2(\tilde{\varphi},\tilde{\alpha}f)+1,\tilde{\varphi},\tilde{\alpha}) - \Sigma_f(z_2(\varphi,\alpha)+1,\varphi,\alpha)\| \\
&\quad \le \|\Sigma_f\big(z_2(\tilde{\varphi},\tilde{\alpha}f)+1,\tilde{\varphi},\tilde{\alpha}\big) - \Sigma_f\big(z_2(\tilde{\varphi},\tilde{\alpha}f)+1,\varphi,\alpha\big)\| \\
&\qquad + \|\Sigma_f\big(z_2(\tilde{\varphi},\tilde{\alpha}f)+1,\varphi,\alpha\big) - \Sigma_f\big(z_2(\varphi,\alpha)+1,\varphi,\alpha\big)\|
\end{aligned}$$

together with the continuity of $U \ni (\tilde{\varphi},\tilde{\alpha}) \mapsto z_2(\tilde{\varphi},\tilde{\alpha}f) \in \mathbb{R}$ and Exercise VII.2.11.
4. In case 2.1, we have $\mathcal{P}_f(\alpha,\varphi) = 0$. Let $\epsilon > 0$ and choose $\delta > 0$ with

$$(\alpha+1) \max_{[-\delta,0]} f < \epsilon.$$

Since $x(t) \to 0$ as $t \to \infty$, there exists

$$t > z_1(\varphi,\alpha f) + 1 \quad \text{with } -\delta < x < 0 \text{ on } [t-1,t].$$

Part 2.1 and Exercise VII.2.11 imply that there is a neighbourhood $U' \subset U$ of (α,φ) in $(0,\alpha+1) \times \overline{K}$ so that for $(\tilde{\alpha},\tilde{\varphi}) \in U'$ and $\tilde{x} := x^{\tilde{\varphi},\tilde{\alpha}f}$,

$$-\delta < \tilde{x} < 0 \text{ on } [t-1,t] \quad \text{and} \quad \tilde{x} < 0 \text{ on } (z_1(\tilde{\varphi},\tilde{\alpha}f),t].$$

If (5.5) holds for $\tilde{x}$, then

$$z_2(\tilde{\varphi},\tilde{\alpha}f) > t \quad \text{and} \quad -\delta < \tilde{x} \le 0 \quad \text{on } [t-1, z_2(\tilde{\varphi},\tilde{\alpha}f)],$$

so that

$$\|\mathcal{P}_f(\tilde{\alpha},\tilde{\varphi}) - \mathcal{P}_f(\alpha,\varphi)\| = \mathcal{P}_f(\tilde{\alpha},\tilde{\varphi})(0) = \tilde{x}(z_2(\tilde{\varphi},\tilde{\alpha}f)+1)$$
$$\leq \alpha \max_{[-\delta,0)} f < \epsilon.$$

In all other cases, we have $\mathcal{P}_f(\tilde{\alpha},\tilde{\varphi}) = 0 = \mathcal{P}_f(\alpha,\varphi)$.
5. In case $x > 0$ on $[0,\infty)$, $\mathcal{P}_f(\alpha,\varphi) = 0$. Let $\epsilon > 0$. Choose $\delta \in (0,\epsilon)$ so small that for

$$\rho(\delta) := (\alpha+1)\min_{[0,\delta]} f < 0,$$

we have

$$(\alpha+1)\max_{[\rho(\delta),0]} f < \epsilon.$$

By (5.3) and Exercise VII.2.11, there exist $t \geq 0$ and a neighbourhood

$$U \subset (0,\alpha+1) \times \overline{K}$$

of (α,φ) so that for all $(\tilde{\alpha},\tilde{\varphi}) \in U$, with $\tilde{x} := x^{\tilde{\varphi},\tilde{\alpha}f}$,

$$0 < \tilde{x} \quad \text{on } [0,t] \quad \text{and} \quad 0 < \tilde{x} < \delta \quad \text{on } [t-1,t].$$

If $\tilde{x}$ satisfies (5.5), we obtain $z_1(\tilde{\varphi},\tilde{\alpha}f) > t$; hence

$$\rho(\delta) \leq \tilde{\alpha}\min_{[0,\delta]} f \leq \tilde{x}\big(z_1(\tilde{\varphi},\tilde{\alpha}f)+1\big) < 0;$$

therefore

$$\|\mathcal{P}_f(\tilde{\alpha},\tilde{\varphi}) - \mathcal{P}_f(\alpha,\varphi)\| = \|\mathcal{P}_f(\tilde{\alpha},\tilde{\varphi})\|$$
$$= \tilde{x}(z_2(\tilde{\varphi},\tilde{\alpha}f)+1)$$
$$\leq (\alpha+1)\max_{[\rho(\delta),0]} f < \epsilon.$$

In all other cases, $\mathcal{P}_f(\tilde{\varphi},\tilde{\alpha}) = 0 = \mathcal{P}_f(\alpha,\varphi)$.
6. If $\varphi = 0$, proceed as in Part 5, with U defined by

$$U = (0,\alpha+1) \times \{\varphi \in \overline{K} \mid \varphi(0) < \delta\}.$$

□

We are now interested in the set

$$\mathcal{F} = \{(\alpha,\varphi) \in (0,\infty) \times \overline{K} \mid \mathcal{P}_f(\alpha,\varphi) = \varphi \neq 0\},$$

i.e., in the nonzero fixed points of the operators $\mathcal{P}_f(\alpha,\cdot)$, $\alpha > 0$.

Theorem 5.5.
(i) *There exists* $\alpha_0 \in (0,\frac{\pi}{2}]$ *such that*

$$\mathcal{F} \subset \big\{(\alpha,\varphi) \in (0,\infty) \times \overline{K} \mid \alpha_0 < \alpha,\ 0 < \varphi(0) \leq \alpha \sup f\big\}.$$

(ii) $(\pi/2,0) \in \overline{\mathcal{F}}$.
(iii) *For* $0 < \alpha \neq \frac{\pi}{2}$, $(\alpha,0) \notin \overline{\mathcal{F}}$.

(iv) *The connected component $\mathcal{F}_0$ of $\overline{\mathcal{F}}$ which contains $(\pi/2, 0)$ is unbounded. For every $\alpha > \pi/2$, there exist $\varphi \in K$ with $(\alpha, \varphi) \in \mathcal{F}_0$.*

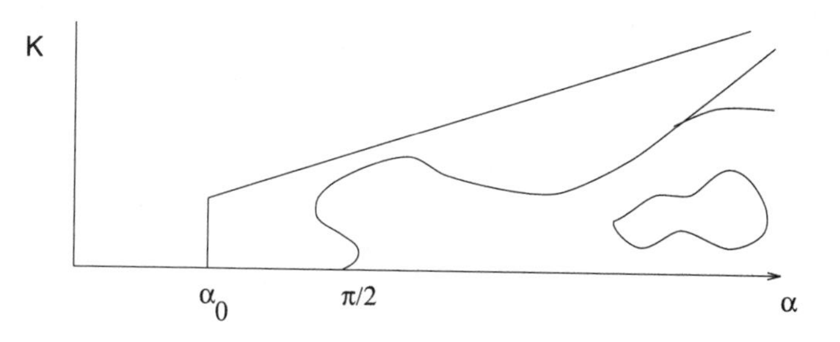

Fig. XV.12.

The proof is contained in Sections 6–8. It makes use of the fixed-point index whose basic properties are recalled in Section 7. In Section 8 we shall see that Theorem 5.1 immediately implies that for $\alpha > \pi/2$, a neighbourhood of $0 \in \overline{K}$ has index 0 with respect to the map $\mathcal{P}_f(\alpha, \cdot)$. This is the heart of the proof. For $\alpha < \pi/2$, a neighbourhood of $0 \in \overline{K}$ has index 1. This is essentially a consequence of linearized stability. The local part of the bifurcation result follows then from the change of index at $\alpha = \frac{\pi}{2}$ (compare Krasnoselskii [152]). The proof of the global part uses topological arguments as in [243].

Finally, let us relate the cone maps $\mathcal{P}_f(\alpha, \cdot)$ to Poincaré maps as in Section XIV.3. Suppose f is C^1 and let $\alpha > 1$. Then the semiflow $\Sigma_f(\cdot, \cdot, \alpha)$ is C^1 on $(1, \infty) \times \mathcal{C}$ and we have

$$D_1 \Sigma_f(t, \varphi, \alpha)1 = \dot{x}_t^{\varphi, \alpha f}, \quad \text{for } t > 1,\ \varphi \in \mathcal{C}.$$

Consider the hyperplane

$$\begin{aligned} H = Y &= \{\varphi \in \mathcal{C} \mid \varphi(-1) = 0\} \\ &= \mathcal{N}(\delta), \end{aligned}$$

where $\delta(\varphi) = \varphi(-1)$ for $\varphi \in \mathcal{C}$. Clearly, $H \supset \overline{K}$. For every $\varphi \in K$, $z_2(\varphi, \alpha f) > 1$ and $\Sigma_f(z_2(\varphi, \alpha f) + 1, \varphi, \alpha) \in H$, we find

$$\delta(D_1 \Sigma_f\big(z_2(\ldots) + 1, \varphi, \alpha)1\big) = \dot{x}^{\varphi, \alpha f}\big(z_2(\varphi, \alpha f)\big) > 0$$

so that the trajectory

$$t \mapsto \Sigma_f(t, \varphi, \alpha)$$

passes transversally through H at $t = z_2(\varphi, \alpha f) + 1$. From Section XIV.3, we infer that there exist an open ϵ-neighbourhood $B_\epsilon(\varphi)$ of φ and a C^1-map

$$\sigma : B_\epsilon(\varphi) \to (0, \infty)$$

such that $\sigma(\varphi) = z_2(\varphi, \alpha f) + 1$ and

$$\Sigma_f(\sigma(\tilde{\varphi}), \tilde{\varphi}, \alpha) \in H$$

for $\tilde{\varphi} \in B_\epsilon(\varphi)$. It is not hard to deduce that for ϵ sufficiently small, $\Sigma_f(\sigma(\cdot), \cdot, \alpha)$ and $\mathcal{P}_f(\alpha, \cdot)$ coincide on $B_\epsilon(\varphi) \cap K$. Incidentally, note that $B_\epsilon(\varphi)$ and $B_\epsilon(\varphi) \cap H$ contain functions in the complement of K. If now φ is a fixed point of $\mathcal{P}_f(\alpha, \cdot)$ in K, then we conclude that restrictions of $\mathcal{P}_f(\alpha, \cdot)$ to sufficiently small neighbourhoods of φ in K are, in fact, given by a Poincaré map of the corresponding periodic orbit, on an open subset $B_\epsilon(\varphi) \cap H$ of the hyperplane $H \supset K$.

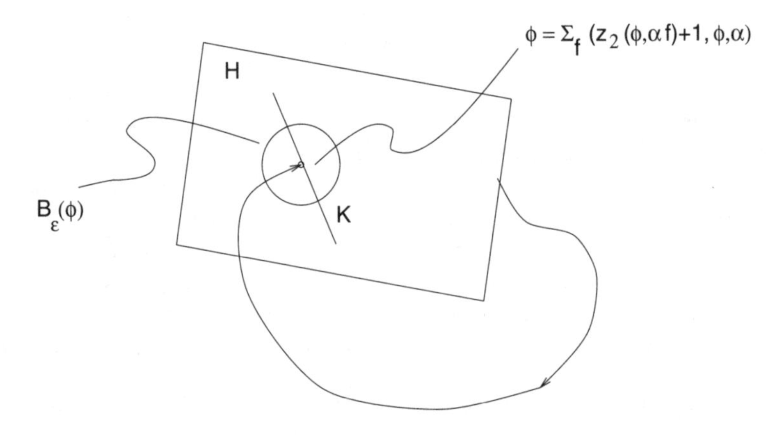

Fig. XV.13.

XV.6 Estimates, proof of Theorem 5.5(i) and (iii)

For $\alpha > 0$, set

$$r(\alpha) := \alpha \sup f.$$

By Proposition 2.3,

$$\|\varphi\| = \varphi(0) \le r(\alpha) \quad \text{for } (\varphi, \alpha) \in \mathcal{F}. \tag{6.1}$$

Proposition 6.1. *There exists $\alpha_0 \in (0, \frac{\pi}{2})$ such that for every $(\alpha, \varphi) \in (0, \alpha_0] \times K$ with (5.5),*

$$x^{\varphi,\alpha f}(z_2(\varphi, \alpha f) + 1) \le \frac{1}{2}\varphi(0).$$

In particular,

$$((0,\alpha_0]\times K)\cap\mathcal{F}=\emptyset.$$

Proof. There exists $\alpha_0\in(0,\frac{\pi}{2})$ so small that for $x\le 2r(\alpha_0)$,

$$|\alpha_0 f(x)|\le\frac{1}{2}|x|.$$

Let $0<\alpha\le\alpha_0$ and $\varphi\in K$ with (5.5) be given; $x:=x^{\varphi,\alpha f}$. In case $\varphi(0)\le 2r(\alpha_0)$,

$$0\le x\le\varphi(0)\le 2r(\alpha_0)\quad\text{on }[-1,z_1].$$

Hence

$$\begin{aligned}0>x(z_1+1)&=\alpha\int_{z_1-1}^{z_1}f\circ x\\&\ge\alpha_0\int_{z_1-1}^{z_1}f\circ x\ge-\frac{1}{2}\varphi(0).\end{aligned}$$

Analogously,

$$x(z_2+1)\le-\frac{1}{2}x(z_1+1)\le\frac{1}{4}\varphi(0).$$

If $2r(\alpha_0)<\varphi(0)$, then Proposition 2.3 yields

$$x(z_2+1)\le r(\alpha)\le r(\alpha_0)<\frac{1}{2}\varphi(0).$$

□

Proposition 6.2. *Let $\alpha_1>0$ and $\alpha_2>\alpha_1$ be given. Then there exists $z_{12}>0$ such that for every $(\alpha,\varphi)\in[\alpha_1,\alpha_2]\times K$ with (5.5) and $\varphi(0)=x^{\varphi,\alpha f}(z_2(\varphi,\alpha f)+1)$,*

$$z_2(\varphi,\alpha f)<z_{12}.$$

Proof. 1. Set

$$r:=\max\{r(\alpha_2),-\alpha_2\min_{[0,r(\alpha_2)]}f\}>0.$$

There exist $a\in(0,1)$ and $b>1$ such that for all $(\alpha,x)\in[\alpha_1,\alpha_2]\times[-r,r]$,

$$a|x|\le|f(x)|\le b|x|.$$

Set

$$z_{12}:=2(2+\frac{\alpha_2 b^2}{a}).$$

Let $\alpha\in[\alpha_1,\alpha_2]$, $\varphi\in K$ with (5.5) be given, and assume that $x:=x^{\varphi,\alpha f}$ satisfies

$$\varphi(0)=x(z_2+1).$$

Using Proposition 5.3, one finds

(6.2) $$|x| \le r \quad \text{on } [-1, z_2 + 1].$$

2. Proof of $x(z_1 + 1) \le -\frac{\varphi(0)}{\alpha b}$: Suppose not, then (6.2) and

$$-\frac{\varphi(0)}{\alpha b} \le x \le 0 \quad \text{on } [z_1, z_2]$$

would imply

$$x(z_2 + 1) = \alpha \int_{z_2 - 1}^{z_2} f \circ x \le -\alpha b x(z_1 + 1) < \varphi(0),$$

a contradiction.

3. Proof of $z_1 \le 2 + \frac{\alpha_2 b^2}{a}$: Suppose $z_1 \ge 2$. We have

$$x(z_1 - 1) > \frac{\varphi(0)}{\alpha^2 b^2},$$

since, otherwise,

$$0 \le x \le \min\{r, \frac{\varphi(0)}{\alpha^2 b^2}\} \quad \text{on } [z_1 - 1, z_1]$$

and

$$x(z_1 + 1) = \alpha \int_{z_1 - 1}^{z_1} f \circ x \ge -\alpha b \frac{\varphi(0)}{\alpha^2 b^2},$$

a contradiction to Part 2. Therefore, it follows that

$$r \ge x > \frac{\varphi(0)}{\alpha^2 b^2} \quad \text{on } [0, z_1 - 1] \supset [0, 1];$$

hence

$$\dot{x} \le -\alpha a \frac{\varphi(0)}{\alpha^2 b^2} \quad \text{on } [2, z_1]$$

and

$$\varphi(0) \ge x(2) = -\int_2^{z_1} \dot{x} \ge \alpha a \frac{\varphi(0)}{\alpha^2 b^2}(z_1 - 2).$$

4. Proof of $z_2 - z_1 \le 2 + \alpha_2 b^2 / a$: Suppose $z_2 \ge z_1 + 2$. Note

(6.3) $$0 > x(z_1 + 1) = \alpha \int_{z_1 - 1}^{z_1} f \circ x \ge -\alpha b \varphi(0).$$

As in Part 2,

$$x(z_2 - 1) \le -\frac{\varphi(0)}{\alpha b}.$$

(Otherwise

$$\max\{-r, -\frac{\varphi(0)}{\alpha b}\} < x \le 0 \quad \text{on } [z_2 - 1, z_2]$$

so that

$$x(z_2+1) = \alpha \int_{z_2-1}^{z_2} f \circ x < \alpha b(-(\frac{\varphi(0)}{\alpha b})) = -\varphi(0),$$

a contradiction.) It follows that

$$-r \le x \le -\frac{\varphi(0)}{\alpha b} \quad \text{on } [z_1+1, z_2-1],$$

and therefore

$$\dot{x} \ge \alpha a \frac{\varphi(0)}{\alpha b} \quad \text{on } [z_1+2, z_2].$$

Consequently,

$$\begin{aligned} -\alpha b \varphi(0) &\le x(z_1+1) \qquad \text{[see (6.3)]} \\ &\le x(z_1+2) = -\int_{z_1+2}^{z_2} \dot{x} \\ &\le -\frac{a\varphi(0)}{b}(z_2 - (z_1+2)). \end{aligned}$$

□

Corollary 6.3. *Let an interval* $[\alpha', \alpha''] \subset (0, \infty)$ *with* $\frac{\pi}{2} \notin [\alpha', \alpha'']$ *be given. Then there exists* $\epsilon > 0$ *such that for all*

$$(\alpha, \varphi) \in [\alpha', \alpha''] \times \{\varphi \in \overline{K} \mid 0 < \|\varphi\| \le \epsilon\}, \qquad \mathcal{P}_f(\alpha, \varphi) \ne \varphi.$$

Proof. 1. By Proposition 6.2, there exists $\tilde{z} > 0$ with

$$z_2(\varphi, \alpha f) + 1 < \tilde{z}$$

for all $(\alpha, \varphi) \in [\alpha', \alpha''] \times \{\varphi \in \overline{K} \mid 0 \ne \varphi = \mathcal{P}_f(\alpha, \varphi)\}$.
2. Suppose the assertion is false. Then there is a sequence of $(\alpha_n, \varphi_n) \in [\alpha', \alpha''] \times \overline{K}$, $n \in \mathbb{N}$, such that

$$0 \ne \mathcal{P}_f(\alpha_n, \varphi_n) = \varphi_n \to 0 \quad \text{as } n \to \infty.$$

Set $z_{2,n} = z_2(\varphi_n, \alpha_n f)$ and $\psi_n := T_{\alpha_n}(z_{2,n}+1)\frac{1}{\|\varphi_n\|}\varphi_n$ for $n \in \mathbb{N}$. Exercise VII.2.12 implies that $\overline{\{\psi_n \mid n \in \mathbb{N}\}}$ is compact. So it follows that there are convergent subsequences

$$\psi_{n_j} \to \psi, \quad \alpha_{n_j} \to \alpha \in [\alpha', \alpha''], \quad z_{2,n_j} \to z \in [1, \tilde{z}-1].$$

Note that $\alpha \ne \frac{\pi}{2}$.
3. Proposition VII.5.6 on uniform differentiability of $\Sigma_f(t, \cdot, \tilde{\alpha})$ for $0 \le t \le \tilde{z}$ and $\alpha' \le \tilde{\alpha} \le \alpha''$ implies

$$\|\frac{1}{\|\varphi_{n_j}\|}\varphi_{n_j} - \psi_{n_j}\| = \frac{\|\Sigma_f(z_{2,n_j}+1, \varphi_{n_j}, \alpha_{n_j}) - T_{\alpha_{n_j}}(z_{2,n_j}+1)\varphi_{n_j})\|}{\|\varphi_{n_j}\|} \to 0$$

as $j \to \infty$. Therefore

$$\frac{1}{||\varphi_{n_j}||}\varphi_{n_j} \to \psi \quad \text{as } j \to \infty$$

and we get $\psi \in \overline{K}$ and $\|\psi\| = 1$. Using Exercise VII.2.11 and continuity of $0 \leq t \mapsto T_\alpha(t)\psi$, we obtain

$$\psi = T_\alpha(z+1)\psi \in K \subset \mathcal{O}.$$

By Corollary 3.6,

$$0 \neq P_{0,\alpha}\psi = T_\alpha(z+1)P_{0,\alpha}\psi,$$

and there exists a nonzero periodic solution of (3.1) with initial value in $\mathcal{C}_{0,\alpha}$. This is a contradiction to $\alpha \neq \frac{\pi}{2}$. □

The first assertion of Theorem 5.5 now follows from Proposition 6.1 and Proposition 5.3. The third assertion is a consequence of Corollary 6.3. Further note that

$$\mathcal{P}_f(\{\alpha\} \times \overline{K}) \subset \{\varphi \in \overline{K} \mid \varphi(0) \leq r(\alpha)\} \quad \text{for all } \alpha > 0, \tag{6.4}$$

by Proposition 5.3. We have the following consequences of Theorem 5.1:

(i) For $\alpha > \frac{\pi}{2}$ and $0 \neq \varphi \in \overline{K}$ with $V_{\alpha f}(\varphi) \leq a_{\alpha f}$,

$$\mathcal{P}_f(\alpha, \varphi) \neq \varphi. \tag{6.5}$$

(ii) For $\alpha > \frac{\pi}{2}$ and $\varphi \in \overline{K}$ with $\varphi(0) = r(\alpha)$,

$$a_{\alpha f} < V_{\alpha f}(\varphi). \tag{6.6}$$

XV.7 The fixed-point index for retracts in Banach spaces, Whyburn's lemma

A subset R of a topological space Y is called a *retract* (of Y) if and only if there exists a continuous map $r : Y \to R$ such that $r|_R = 1$. The map r is called a retraction. Retracts of Y are closed in Y. Examples of retracts are closed convex subsets of Banach spaces.

Let Ω denote the set of triples (W, F, R), where W is an open subset of a retract R of a real Banach space X and $F : D_F \to R$ is a map such that

$\overline{W} \subset D_F$,

$F|_{\overline{W}}$ is continuous,

$\overline{F(W)}$ is compact,

$\varphi \neq F(\varphi)$ for all $\varphi \in \partial W$.

Openness, closure and boundaries refer to the topology on R induced by X. There exists a unique map

$$\text{ind} : \Omega \to \mathbb{Z},$$

called the *fixed-point index*, which has the following properties:

(i) (Normalization.) If $(W, F, R) \in \Omega$ and $F(\overline{W}) = \{x\}$ for some $x \in W$, then

$$\text{ind}\,(W, F, R) = 1.$$

(ii) (Additivity.) Let $(W, F, R) \in \Omega$. If $W_1 \subset W$ and $W_2 \subset W$ are open and disjoint and if $\varphi \neq F(\varphi)$ for all $\varphi \in \overline{W} \setminus (W_1 \cup W_2)$, then

$$\text{ind}\,(W, F, R) = \text{ind}\,(W_1, F, R) + \text{ind}\,(W_2, F, R).$$

(iii) (Homotopy invariance.) Let a compact interval $J \subset \mathbb{R}$, an open subset $W \subset R$ and a map $H : D_H \to R$ be given so that

$$D_H \supset \overline{W} \times J,$$
$$H\big|_{\overline{W} \times J} \text{ is continuous }, \qquad \overline{H(\overline{W} \times J)} \text{ is compact},$$
$$\varphi \neq H(\varphi, t) \quad \text{for all } (\varphi, t) \in \partial U \times J.$$

Then the map

$$J \ni t \mapsto \text{ind}\,(W, H(\cdot, t), R) \in \mathbb{Z}$$

is constant.

(iv) (Permanence.) If $(W, F, R) \in \Omega$ and if $Q \subset R$ is a retract of R such that $F(\overline{W}) \subset Q$, then

$$(W \cap Q, F_Q, Q) \in \Omega$$

[where $F_Q : \overline{W \cap Q} \to Q, F_Q(\varphi) := F(\varphi)$] and

$$\text{ind}\,(W, F, R) = \text{ind}\,(W \cap Q, F_Q, Q).$$

For a proof, see, e.g., [94, 225] and the references given in [225]. An application of the additivity property with $W = \emptyset$ shows

$$\text{ind}\,(\emptyset, F, R) = 0 \tag{7.1}$$

for all maps $F : D_F \to R$. Setting $W_1 = \emptyset = W_2$, one sees that

(7.2) $\text{ind}\,(W, F, R) \neq 0$ implies the existence of a fixed point $\varphi \in W$ of F.

Corollary 7.1. *Let $R \subset X$ be a retract, $W \subset R$ open and contractible (i.e., there are a point $\varphi_0 \in W$, a compact interval $[a, b]$ and a continuous map $h : \overline{W} \times [a, b] \to R$ such that*

$$h(\cdot, a) = I, \quad h(\overline{W} \times [a,b]) \subset \overline{W}, \quad h(\overline{W} \times \{b\}) = \{\varphi_0\}).$$

Then

$$\text{ind}\,(W, F, R) = 1$$

for every map $F : D_F \to R$ *such that* $(W, F, R) \in \Omega$ *and* $F(\overline{W}) \subset W$.

Proof. Set $H = F \circ h$. Then H is continuous, $\overline{H(\overline{W} \times [a,b])}$ is compact and $H(\overline{W} \times [a,b]) \subset W$. Hence

$$H(\varphi, t) \neq \varphi \quad \text{for all } (\varphi, t) \in \partial W \times [a,b].$$

The permanence and normalization properties yield

$$\begin{aligned} \text{ind}\,(W, F, R) &= \text{ind}\,(W, H(\cdot, a), R) \\ &= \text{ind}\,(W, H(\cdot, b), R) = 1 \end{aligned}$$

□

Corollary 7.2. *For* $(R, F, R) \in \Omega$ *with* $R \neq \emptyset$ *closed and convex,*

$$\text{ind}\,(R, F, R) = 1.$$

Proof. Fix $\varphi_0 \in R$. Set $h(\varphi, t) := t\varphi + (1-t)\varphi_0$ for $(\varphi, t) \in R \times [0,1]$. Apply Corollary 7.1. □

Nonlocal arguments in the next section require a more general form of the homotopy invariance property. Let a retract $R \subset X$, a compact interval $J \subset \mathbb{R}$, an open subset $W \subset R \times J$ and a map $H : D_H \to R$ be given so that the following are satisfied:

$$\begin{aligned} &D_H \supset \overline{W}, \\ &H\,|_{\overline{W}} \text{ is continuous and } \overline{H(\overline{W})} \text{ is compact,} \\ &\varphi \neq H(\varphi, t) \quad \text{for all } (\varphi, t) \in \partial W. \end{aligned}$$

Then all fibers

$$W_t := \{\varphi \in R : (\varphi, t) \in W\}, \quad t \in J,$$

are open subsets of R, and we have for the maps

$$H_t = H(\,\cdot\,, t)\,|_{\overline{W_t}}, \quad t \in J.$$

Lemma 7.3. *For all* $t \in J$,

$$(W_t, H_t, R) \in \Omega,$$

and the map $J \ni t \mapsto \text{ind}\,(W_t, H_t, R) \in \mathbb{Z}$ *is constant.*

For a proof, see, e.g., [94, 225]. Finally, we need

Lemma 7.4. (Whyburn [306], Kuratowski [158].) *Let disjoint closed subsets A and B of a compact topological space Y be given. Either there exists a connected closed set $C_{AB} \subset Y$ such that*

$$A \cap C_{AB} \neq \emptyset \neq B \cap C_{AB},$$

or there exist disjoint closed sets $D_A \subset Y$ and $D_B \subset Y$ such that

$$A \subset D_A, B \subset D_B, \qquad D_A \cup D_B = Y.$$

For a proof, see, e.g., [3].

XV.8 Proof of Theorem 5.5(ii) and (iv)

Let f with properties (NF) and (B) be given.

Proposition 8.1. *For every $\alpha > \pi/2$, there exists $\epsilon(\alpha) > 0$ such that*

$$\text{ind}\left(\overline{K} \cap B_\epsilon(0), \mathcal{P}_f(\alpha, \cdot), \overline{K}\right) = 0 \quad \textit{for } 0 < \epsilon < \epsilon(\alpha).$$

Proof. Let $\alpha > \pi/2$. Set $K_\alpha := \{\varphi \in \overline{K} \mid \varphi(0) \leq r(\alpha)\}$. The map

$$\mathcal{P}_\alpha : \overline{K} \to K_\alpha, \qquad \mathcal{P}_f(\varphi) = \mathcal{P}_f(\alpha, \varphi)$$

is continuous, and $\overline{\mathcal{P}_\alpha(\overline{K})}$ is compact (Corollary 5.4). Further, K_α is a retract of $\overline{K}$ [consider the radial retraction which maps $\varphi \in K$ with $\varphi(0) = \|\varphi\| > r(\alpha)$ onto $r(\alpha)\frac{1}{\|\varphi\|}\varphi$]. Recall the definition of $a_{\alpha f}$ and $V_{\alpha f}$ from (6.5) and (6.6) and define $D_\alpha = \{\varphi \in K_\alpha \mid a_{\alpha f} < V_{\alpha f}(\varphi)\}$. By Theorem 5.1, $\mathcal{P}_\alpha$ maps the closure of D_α in K_α, i.e., the set

$$\overline{D}_\alpha = \{\varphi \in K_\alpha \mid a_{\alpha f} \leq V_{\alpha f}(\varphi)\},$$

into D_α. Choose $\varphi_0 \in K_\alpha$ such that $\varphi_0(0) = r(\alpha)$. Define a map $h : \overline{D}_\alpha \times [0,2] \to K_\alpha$ by

$$h(\varphi, t) = \begin{cases} (1-t)\varphi + tr(\alpha)\dfrac{1}{\|\varphi\|}\varphi & \text{on } \overline{D}_\alpha \times [0,1], \\[2ex] (2-t)r(\alpha)\dfrac{1}{\|\varphi\|}\varphi + (t-1)\varphi_0 & \text{on } \overline{D}_\alpha \times [1,2]. \end{cases}$$

Then h is continuous, and we have

$$
\begin{aligned}
h(\varphi,0) &= \varphi && \text{on } \overline{D}_\alpha, \\
h(\varphi,2) &= \varphi_0 && \text{on } \overline{D}_\alpha, \\
h(\varphi,t)(0) &\le r(\alpha) && \text{on } \overline{D}_\alpha \times [0,1], \\
h(\varphi,t)(0) &= r(\alpha) && \text{on } \overline{D}_\alpha \times [1,2], \\
V_{\alpha f}\big(h(\varphi,t)\big) &\ge a_{\alpha f} && \text{on } \overline{D}_\alpha \times [0,1].
\end{aligned}
$$

[Use $V_{\alpha f}\big(h(\varphi,t)\big) = (1 - t + tr(\alpha)\frac{1}{\|\varphi\|})^2 V_{\alpha f}(\varphi) \ge V_{\alpha f}(\varphi) \ge a_{\alpha f}$.]
Using Theorem 5.1(i), we infer

$$V_{\alpha f}\big(h(\varphi,t)\big) > a_{\alpha f} \quad \text{on } \overline{D}_\alpha \times [1,2].$$

An application of Corollary 7.1 yields

$$\operatorname{ind}(D_\alpha, \mathcal{P}_\alpha, K_\alpha) = 1.$$

By Corollary 7.2,

$$\operatorname{ind}(K_\alpha, \mathcal{P}_\alpha, K_\alpha) = 1.$$

Choose $\epsilon(\alpha) > 0$ so small that

$$B_{\epsilon(\alpha)}(0) \subset \{\varphi \in C \mid \quad V_{\alpha f}(\varphi) < a_{\alpha f}\}.$$

Let $\epsilon \in (0, \epsilon(\alpha))$. Property (6.5) excludes fixed points of $\mathcal{P}_\alpha$ with $\epsilon \le \|\varphi\|$ and $V_{\alpha f}(\varphi) \le a_{\alpha f}$. The additivity property of the index yields

$$\operatorname{ind}(K_\alpha, \mathcal{P}_\alpha, K_\alpha) = \operatorname{ind}(K_\alpha \cap B_\epsilon(0), \mathcal{P}_\alpha, K_\alpha) + \operatorname{ind}(D_\alpha, \mathcal{P}_\alpha, K_\alpha).$$

Hence

$$0 = \operatorname{ind}(K_\alpha \cap B_\epsilon(0), \mathcal{P}_\alpha, K_\alpha) = \operatorname{ind}(\overline{K} \cap B_\epsilon(0), \mathcal{P}_\alpha, \overline{K})$$

due to permanence of the index. □

Proposition 8.2. *For every $\alpha \in (0, \frac{\pi}{2})$, there exists $\epsilon(\alpha) > 0$ such that*

$$\operatorname{ind}(\overline{K} \cap B_\epsilon(0), \mathcal{P}_f(\alpha, \cdot), \overline{K}) = 1 \quad \text{for } 0 < \epsilon < \epsilon(\alpha).$$

Proof. 1. In case $\alpha_0 < \alpha < \frac{\pi}{2}$, choose $\epsilon(\alpha)$ in $(0, r(\alpha_0))$ so small that for $\alpha_0 \le \beta \le \alpha$ and for $0 \ne \varphi \in \overline{K} \cap \overline{B_{\epsilon(\alpha)}(0)}$,

$$\mathcal{P}_f(\beta, \varphi) \ne \varphi \qquad \text{(Corollary 6.3)}.$$

For $0 < \epsilon < \epsilon(\alpha)$, we obtain

$$
\begin{aligned}
I &= \operatorname{ind}(\overline{K} \cap B_\epsilon(0), \mathcal{P}_f(\alpha, \cdot), \overline{K}) \\
\text{(homotopy invariance)} \qquad &= \operatorname{ind}(\overline{K} \cap B_\epsilon(0), \mathcal{P}_f(\alpha_0, \cdot), \overline{K}) \\
&= \operatorname{ind}(\overline{K} \cap B_\epsilon(0), \mathcal{P}_{\alpha_0}, K_{\alpha_0})
\end{aligned}
$$

(compare the proof of Proposition 8.1).

Due to Proposition 6.1, $\mathcal{P}_f(\alpha_0, \varphi) \neq \varphi$ for $0 \neq \varphi \in \overline{K}$. Therefore

$$\begin{aligned} I &= \operatorname{ind}(\overline{K} \cap B_\epsilon(0), \mathcal{P}_{\alpha_0}, K_{\alpha_0}) \\ \text{(additivity)} \quad &= \operatorname{ind}(K_{\alpha_0}, \mathcal{P}_{\alpha_0}, K_{\alpha_0}) \\ \text{(Corollary 7.2)} \quad &= 1. \end{aligned}$$

2. For $0 < \alpha \leq \alpha_0$, choose $\epsilon(\alpha) \in (0, r(\alpha))$ and use the last arguments of Part 1. □

Proof of Theorem 5.5 (*ii*). Consider a neighbourhood $[\alpha, \beta] \times \overline{B_\epsilon(0)}$ of $(\frac{\pi}{2}, 0)$ where

$$0 < \alpha, \quad 0 < \epsilon \leq \min\{\epsilon(\alpha), \epsilon(\beta)\}.$$

Suppose

$$\emptyset = \mathcal{F} \cap \{(\gamma, \varphi) \in \mathbb{R} \times C \mid \alpha \leq \gamma \leq \beta,\ \|\varphi\| = \epsilon\},$$

i.e., $\mathcal{P}_f(\gamma, \varphi) \neq \varphi$ for $\alpha \leq \gamma \leq \beta$ and $\varphi \in \overline{K}$ with $\|\varphi\| = \epsilon$. Using homotopy invariance, we conclude

$$\operatorname{ind}\left(\overline{K} \cap B_\epsilon(0), \mathcal{P}_f(\beta, \cdot), \overline{K}\right) = \operatorname{ind}\left(\overline{K} \cap B_\epsilon(0), \mathcal{P}_f(\alpha, \cdot), \overline{K}\right),$$

a contradiction to Propositions 8.1 and 8.2. □

Proof of Theorem 5.5 (*iv*). To prove that $\mathcal{F}_0$ is unbounded, we argue by contradiction and assume that $\mathcal{F}_0$ is bounded.
1. There exist $\alpha_1 \in (0, \alpha_0)$ and $\alpha_2 > \frac{\pi}{2}$ such that

$$\mathcal{F}_0 \subset (\alpha_1, \alpha_2) \times \overline{K}.$$

Indeed, since $\mathcal{F} \subset [\alpha_0, \infty) \times \overline{K}$, we have $\mathcal{F}_0 \subset \overline{\mathcal{F}} \subset [\alpha_0, \infty) \times \overline{K}$ but, by assumption, the set

$$\{\alpha > 0 \mid \text{ There exists } \varphi \in \overline{K} \text{ such that } (\alpha, \varphi) \in \mathcal{F}_0\}$$

is bounded.
2. The set $\mathcal{F}_0$ is compact. Since $\mathcal{F}_0$ is a closed subset of $(0, \infty) \times \overline{K}$ and contained in

$$\begin{aligned} \overline{\mathcal{F}} \cap ([\alpha_1, \alpha_2] \times \overline{K}) &\subset \{(\alpha, \varphi) \in (0, \infty) \times \overline{K} \mid \alpha_1 \leq \alpha \leq \alpha_2,\ \mathcal{P}_f(\alpha, \varphi) = \varphi\} \\ &\subset [\alpha_1, \alpha_2] \times \mathcal{P}_f\left([\alpha_1, \alpha_2] \times \overline{K}\right) \\ &\subset [\alpha_1, \alpha_2] \times \overline{\mathcal{P}_f\left([\alpha_1, \alpha_2] \times \overline{K}\right)}, \end{aligned}$$

where the last set is compact (Corollary 5.4).
3. There exist an open neighbourhood N_0 of $\mathcal{F}_0$ in the metric space $(0, \infty) \times \overline{K}$ and $\epsilon_0 > 0$ such that $N_0 \subset (\alpha_1, \alpha_2) \times \overline{K}$ and

$$\begin{aligned} \overline{N}_0 \cap \left(\{\alpha_1\} \times (\overline{K} \cap \overline{B_{\epsilon_0}(0)})\right) &= \emptyset, \\ \overline{N}_0 \cap \left(\{\alpha_2\} \times (\overline{K} \cap \overline{B_{\epsilon_0}(0)})\right) &= \emptyset. \end{aligned}$$

This follows from Corollary 6.3, which implies $(\alpha_1, 0) \notin \overline{\mathcal{F}}$ and $(\alpha_2, 0) \notin \overline{\mathcal{F}}$.
4. We show that there exists an open neighbourhood $N \subset N_0$ of $\mathcal{F}_0$ so that

$$(\alpha, \varphi) \in \partial N, \quad \mathcal{P}_f(\alpha, \varphi) = \varphi \quad \text{imply} \quad \varphi = 0, \quad \alpha_1 < \alpha < \alpha_2.$$

To prove this, apply Lemma 7.4 to the set

$$M = \overline{N}_0 \cap \overline{\mathcal{F}}.$$

We have $M \subset \big([\alpha_1, \alpha_2] \times \overline{K}\big) \cap \overline{\mathcal{F}}$ and M is compact (compare Part 2). The subsets $A = \mathcal{F}_0$ and $B = (\partial N_0) \cap \overline{\mathcal{F}}$ are disjoint and closed. Suppose there exists a connected subset C_{AB} such that

$$A \cap C_{AB} \neq \emptyset, \qquad B \cap C_{AB} \neq \emptyset.$$

Then $C_{AB} \subset \overline{\mathcal{F}}$ and $C_{AB} \cap \mathcal{F}_0 \neq \emptyset$ imply $C_{AB} \subset \mathcal{F}_0 = A$ (since $\mathcal{F}_0$ is a connected component of $\overline{\mathcal{F}}$), a contradiction to

$$A \cap B = \emptyset, \qquad C_{AB} \cap B \neq \emptyset.$$

Therefore Lemma 7.4 guarantees the existence of disjoint compact subsets K_A and K_B of M such that

$$A \subset K_A, \quad B \subset K_B, \quad K_A \cup K_B = M.$$

We have $K_A \subset N_0$, since

$$\begin{aligned} K_A \subset M &= \overline{N}_0 \cap \overline{\mathcal{F}} = (N_0 \cup \partial N_0) \cap \overline{\mathcal{F}} \\ &= (N_0 \cap \overline{\mathcal{F}}) \cup (\partial N_0 \cap \overline{\mathcal{F}}) \\ &\subset N_0 \cup K_B. \end{aligned}$$

Set $d = \frac{1}{4}\text{dist}\,(K_A, K_B) > 0$ and

$$N = \{(\alpha, \varphi) \in N_0 \mid \text{dist}\,(K_A, (\alpha, \varphi)) < d\}.$$

It is clear that N is open and $\mathcal{F}_0 = A \subset K_A \subset N$. We show that $\alpha, \varphi) \in \partial N$ and $\mathcal{P}_f(\alpha, \varphi)) = \varphi$ imply $\varphi = 0$: Suppose $(\alpha, \varphi) \in \partial N$ and $\mathcal{P}_f(\alpha, \varphi)) = \varphi \neq 0$. Then

$$(\alpha, \varphi) \in \mathcal{F} \quad \text{and} \quad (\alpha, \varphi) \in \overline{N} \subset \overline{N}_0,$$

so that

$$(\alpha, \varphi) \in M = K_A \cup K_B.$$

Since $(\alpha, \varphi) \in \partial N$, we have $\text{dist}\,(K_A, (\alpha, \varphi)) \leq d$. Therefore $(\alpha, \varphi) \notin K_B$, and $(\alpha, \varphi) \in K_A \subset N$, which contradicts $(\alpha, \varphi) \in \partial N$.

Next, we prove

$$\alpha_1 < \alpha < \alpha_2 \quad \text{for } (\alpha, \varphi) \in \partial N \text{ with } \mathcal{P}_f(\alpha, \varphi) = \varphi.$$

We have

$$(\alpha, \varphi) = (\alpha, 0) \in \partial N \subset \overline{N} \subset \overline{N}_0 \subset [\alpha_1, \alpha_2] \times \overline{K},$$

and, by Part 3,

$$(\alpha_1, 0) \notin \overline{N}_0 \not\ni (\alpha_2, 0).$$

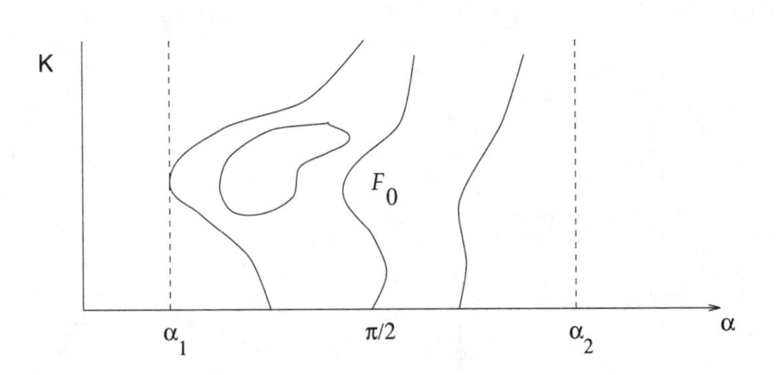

Fig. XV.14.

5. There exist $\epsilon_{00} \in (0, \epsilon_0), \alpha_{11} \in (\alpha_1, \frac{\pi}{2})$ and $\alpha_{22} \in (\frac{\pi}{2}, \alpha_2)$ such that

$$[\alpha_{11}, \alpha_{22}] \times (\overline{K} \cap \overline{B_{\epsilon_{00}}(0)}) \subset N, \tag{8.1}$$

since $(\frac{\pi}{2}, 0) \in \mathcal{F}_0 \subset N$ and N is open. Corollary 6.3 allows one to choose $\epsilon > 0$,

$$\epsilon < \min\{\epsilon_{00}, \epsilon(\alpha_1), \epsilon(\alpha_2)\} \quad \text{(recall Propositions 8.1 and 8.2)} ,$$

such that

$$\mathcal{P}_f(\alpha, \varphi) \neq \varphi$$

for $\alpha \in [\alpha_1, \alpha_{11}] \cup [\alpha_{22}, \alpha_2]$ and $\varphi \in K$ with $0 < \|\varphi\| \leq \epsilon$.
6. For $\alpha_{11} \leq \alpha \leq \alpha_{22}$, set

$$N_\alpha = \{\varphi \in \overline{K} \mid (\alpha, \varphi) \in N\};$$

then N_α is an open subset of $\overline{K}$. Note that $(\alpha, \varphi) \in \partial N$ for $\varphi \in \partial N$. There are no fixed points of $\mathcal{P}_f(\alpha, \cdot)$ on ∂N_α, since, otherwise,

$$\mathcal{P}_f(\alpha, \varphi) = \varphi \in \partial N_\alpha$$

and $(\alpha, \varphi) \in \partial N$; hence $\varphi = 0$ (see Part 4). But $0 \in N_\alpha$ [see (8.1)], a contradiction to $\varphi \in \partial N_\alpha$.

The generalized homotopy property of Lemma 7.3 and Corollary 5.4 now yield

$$\operatorname{ind}(N_{\alpha_{11}}, \mathcal{P}_f(\alpha_{11}, \cdot), \overline{K}) = \operatorname{ind}(N_{\alpha_{22}}, \mathcal{P}_f(\alpha_{22}, \cdot), \overline{K}). \tag{8.2}$$

For $\alpha \in [\alpha_1, \alpha_{22}] \cup [\alpha_{22}, \alpha_2]$, set

$$D_\alpha = N_\alpha \setminus \overline{B_\epsilon(0)}.$$

There is no fixed point of $\mathcal{P}_f(\alpha, \cdot)$ on ∂D_α, since $\varphi \in \partial D_\alpha$ implies

$$\|\varphi\| = \epsilon \quad \text{or} \quad ((\alpha, \varphi) \in \partial N \quad \text{and} \quad \epsilon < \|\varphi\|);$$

in both cases, $\varphi \neq \mathcal{P}_f(\alpha, \varphi)$, due to (8.1) and to Part 4.

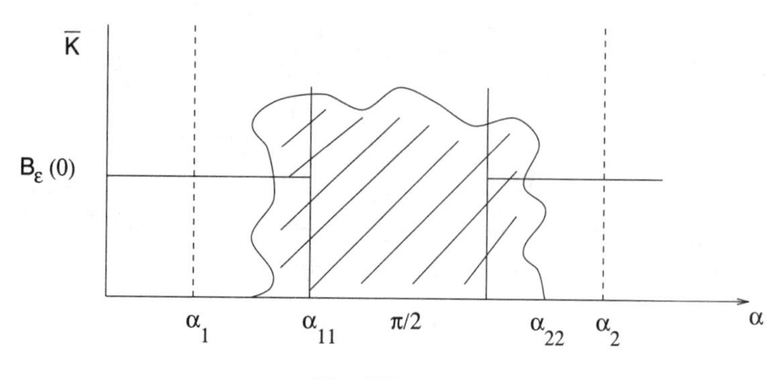

Fig. XV.15.

Note that $D_{\alpha 1} = \emptyset$, as $N \subset N_0 \subset (\alpha_1, \alpha_2) \times \overline{K}$. Lemma 7.3 and (7.1) yield

$$0 = \operatorname{ind}(D_{\alpha_1}, \mathcal{P}_f(\alpha_1, \cdot), \overline{K}) = \operatorname{ind}(D_{\alpha_{11}}, \mathcal{P}_f(\alpha_{11}, \cdot), \overline{K}).$$

The homotopy property of the index and Proposition 8.2 yield

$$\begin{aligned} 1 &= \operatorname{ind}(\overline{K} \cap B_\epsilon(0), \mathcal{P}_f(\alpha_1, \cdot)\overline{K}) \\ &= \operatorname{ind}(\overline{K} \cap B_\epsilon(0), \mathcal{P}_f(\alpha_{11}, \cdot), \overline{K}). \end{aligned}$$

We have

$$N_{\alpha_{11}} \setminus ((\overline{K} \cap B_\epsilon(0)) \cup D_{\alpha_{11}}) \subset \{\varphi \in \overline{K} \mid \|\varphi\| = \epsilon\}.$$

It follows that $\mathcal{P}_f(\alpha_{11}, \cdot)$ has no fixed points on

$$N_{\alpha_{11}} \setminus ((\overline{K} \cap B_\epsilon(0)) \cup D_{\alpha_{11}}).$$

So the additivity property of the index gives

$$1 = \operatorname{ind}(N_{\alpha_{11}}, \mathcal{P}_f(\alpha_{11}, \cdot), \overline{K}).$$

In the same way, we find $D_{\alpha_2} = \emptyset$ and, using Proposition 8.1,

$$0 = \operatorname{ind}(N_{\alpha_{22}}, \mathcal{P}_f(\alpha_{22}, \cdot), \overline{K}),$$

which is a contradiction to (8.2). This proves that $\mathcal{F}_0$ is unbounded.

For every $\alpha > \frac{\pi}{2}$, there exists $\varphi \in K$ (i.e., $\varphi \neq 0$!) such that $(\alpha, \varphi) \in \mathcal{F}_0$: suppose $\mathcal{F}_0 \cap (\{\alpha\} \times K) = \emptyset$ for some $\alpha > \frac{\pi}{2}$. As $(\alpha, 0) \notin \mathcal{F}_0$ [see Theorem 5.5 (iii)],

$$\mathcal{F}_0 \cap (\{\alpha\} \times K) = \emptyset.$$

Connectedness and $(\frac{\pi}{2}, 0) \in \mathcal{F}_0$ exclude points $(\beta, \varphi) \in (\alpha, \infty) \times \overline{K}$ in $\mathcal{F}_0$. Using (6.4), we find

$$\mathcal{F}_0 \subset \{(\beta, \varphi) \in (0, \alpha] \times \overline{K} : \|\varphi\| \leq r(\alpha)\},$$

which contradicts the fact that $\mathcal{F}_0$ is unbounded. □

XV.9 Comments

In 1955 and 1958 there appeared two inspiring studies of nonlinear differential delay equations, Wright's work [308] on

$$\dot{x}(t) = -\alpha x(t-1)[1 + x(t)]$$

and the paper [144] of Kakutani and Marcus on

$$y'(t) = [A - By(t - \tau)]y(t).$$

Both equations are essentially equivalent to equation (1.2) with $f(x) = 1 - e^x$. In years before, similar equations had been investigated in a more heuristic way by biologists who were interested in understanding better regular fluctuations of the size of populations in a constant environment [132, 54, 55, 56].

Wright proved the result mentioned in Section 5 on sustained bounded oscillations: For $\alpha > \frac{\pi}{2}$, i.e., when the equilibrium solution $t \mapsto 0$ is linearly unstable, then there exists $\epsilon > 0$ such that

$$\limsup_{t \to \infty} |x(t)| \geq \epsilon \tag{9.1}$$

for every solution which has no zero in the initial interval $(-1, 0)$. This must have been a strong stimulant to look for periodic solutions. The first paper on existence by Jones [140] already contains the concept which should lead to the most general existence results: consider the map P which assigns to each initial function φ in the convex cone

$$K = \{\varphi \in C([-1, 0], \mathbb{R}) \mid \varphi(-1) = 0, \ \varphi \text{ increasing}, \ 0 < \varphi(0)\}$$

the segment x_{z_2+1} of the corresponding solution $x = x^\varphi$ of (1.2), where z_2 is the second zero of x in $[0, \infty)$. Fixed points of $\mathcal{P}$ define periodic solutions.

The difficulty is the following: $\mathcal{P}$ is continuous and compact as a map into $\mathcal{C}$, and it maps K into itself. But K is not closed and there may be no fixed point at all. Continuation of $\mathcal{P}$ to

$$\overline{K} = K \cup \{0\}$$

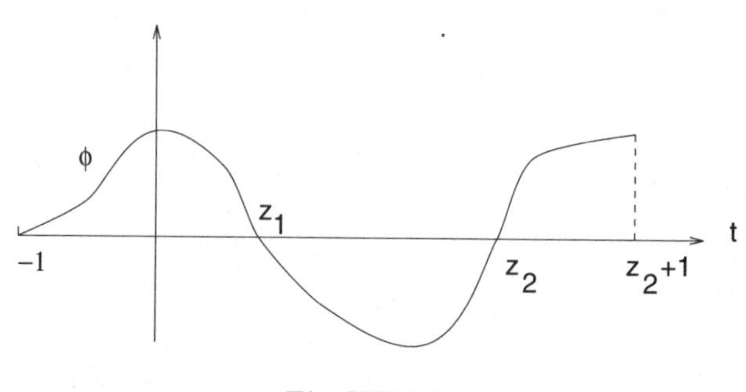

Fig. XV.16.

yields $\mathcal{P}(0) = 0$, a fixed point which corresponds to the equilibrium solution $t \mapsto 0$. So the problem is to convert instability of the zero solution into a sufficient condition for the existence of a second fixed point.

Fixed-point theorems designed for this were obtained by Browder [20]. See also [129]. Ejectivity, a discrete version of (9.1) for a given fixed point, implies the existence of another fixed point. In [211, 214] Nussbaum proved that ejective fixed points in infinite dimensions have index zero, derived estimates like (9.1) for equation (1.2) and verified that they imply ejectivity of the fixed point $0 \in K$. These results led to his theorems on existence and global bifurcation of periodic solutions [214] which are formulated here as Theorem 5.5. See also [215].

The proof of Theorem 5.5 given here goes back to Walther [289]. The difference in the approach in [214] can be recapitulated as follows. Compared to (9.1), we derive more detailed information about unstable behaviour of slowly oscillating solutions. Initially, small trajectories leave a neighbourhood of $0 \in \mathcal{C}$ and cannot reenter. This permits one to find a $\mathcal{P}$-invariant closed subset in K which is bounded away from the vertex 0; one may then use Schauder's theorem or the index to obtain fixed points $\varphi \neq 0$. The result that the fixed point 0 has index 0 follows easily by means of the additivity property.

Other early results on existence of periodic solutions are due to Dunkel [81], Grafton [93], Chow [32] and Pesin [237]. Nussbaum proved a result which is complementary in the sense that the fixed points which define periodic solutions lie between the attractive origin and an expanded region of the invariant cone [212, 213].

A further general result is the global Hopf bifurcation theorem of Nussbaum [217].

Having established the existence of periodic solutions, one asks for more. Many qualitative results have been obtained since 1970. Nussbaum studied analyticity and the range of periods along the continuum of Theorem 5.5 [210, 216, 227].

Kaplan and Yorke [145, 146, 147] introduced two elementary approaches to existence which became useful for the study of stability of periodic solutions. The first one requires the function f in equation (1.2) to be odd. Periodic solutions with period 4 and with the symmetry property

$$x(t) = -x(t-2) \tag{9.2}$$

are obtained from the planar Hamiltonian system

$$\begin{aligned} \dot{x} &= \alpha f(y), \\ \dot{y} &= -\alpha f(x). \end{aligned}$$

The second approach applies to monotone f in equation (1.2) and is based on the investigation of the plane curves given by the evaluations

$$t \mapsto (x(t), x(t-1))$$

of the trajectories $t \mapsto x_t$ in the true phase space $C([-1,0], \mathbb{R})$. These methods were used for results on stability [146, 147, 24], on subcritical bifurcation [286], and on uniqueness (Nussbaum [220]). The uniqueness result holds for nonlinearities f in equation (1.2) which are odd, monotone, bounded, negative and convex on $\mathbb{R}_+$. Nonuniqueness occurs for certain odd, nonmonotone functions f. Nussbaum [222, 223] and Chapin [29, 30, 31] obtained precise estimates of the shape of periodic solutions for large α [also in cases of nonuniqueness].

Numerical results indicated that in cases of non-monotone nonlinearities, the continuum of Theorem 5.5 may be a complicated set of intersecting curves. See, for example the early computations of Hadeler [99] and of Jürgens, Peitgen and Saupe [141].

The method from [145] yields a smooth curve of periodic orbits, all with period 4 and symmetry (9.2), which bifurcates at $\alpha = \frac{\pi}{2}$ from the equilibrium. This "primary branch" is contained in the continuum of Theorem 5.5. In cases of uniqueness, it coincides with the continuum, whereas in cases of nonuniqueness, it may be a proper subset.

For periodic solutions with rational period, as in the primary branch, the Floquet multipliers are given by the zeros of a holomorphic function on $\mathbb{C} \setminus \{0\}$; see the method developed by Walther [291]. This can be used to detect a change of the index along the primary branch, for classes of odd, hump-shaped nonlinearities. It follows that nonsymmetric periodic orbits bifurcate from the primary branch [291].

For further classes of non-monotone nonlinearities, Dormayer [73, 76, 77] obtained results on smooth bifurcations from the primary branch which preserve a generalized version of the symmetry property (9.2). In [75], he determined the location of the Floquet multipliers for large α.

An interesting open question concerning the structure of the continuum of Theorem 5.5 is whether there exist period doubling bifurcations from the primary branch.

The method from [291] and ideas from [34] led to results on exponential stability in cases of uniqueness [42] and to examples with unstable hyperbolic periodic solutions [137].

A priori results on Floquet multipliers of periodic solutions of equation (1.2) with f monotone, but not necessarily odd, were obtained by Walther in [299, 300] and by Mallet-Paret and Sell [185]. These results play a role in investigations of the global dynamics, which is discussed in Chapter XVI.

Another approach to Floquet multipliers is due to Xie [310, 311, 312]. This work is based on asymptotic estimates for α large and yields results on uniqueness and stability, in particular, for Wright's equation (1.4).

Back to the existence of periodic solutions: a rather recent achievement are Poincaré–Bendixson theorems which apply to equation (1.2) with smooth and monotone f and yield periodic solutions, among others. See Smith [258, 259], Walther [298, 300, 302] and Mallet-Paret and Sell [186].

Next we discuss some results on periodic solutions for autonomous differential delay equations that are more general than equation (1.2).

First consider equation (1.2) for a function f with multiple zeros x_j so that

$$\sigma_j(x - x_j)f(x - x_j) > 0 \quad \text{for } 0 < |x - x_j| \text{ small},$$

where $\sigma_j \in \{\pm 1\}$. This is the case of several steady states, with locally positive or negative feedback; it includes periodic functions f. Furumochi [86] proved that in such cases, there exist "periodic solutions of the second kind", analogues of the periodic rotations of the pendulum. In the series of Walther's papers [292, 294, 296] and in the work of Chow and Deng [33], it was shown that periodic solutions of the second kind arise in bifurcations from heteroclinic trajectories.

For bifurcation of periodic solutions with long period from homoclinic solutions, see Walther [295].

Nussbaum and Peitgen [228] studied periodic solutions with the symmetry (9.2) in case f is odd and has multiple zeros.

At this point, before proceeding to equations of a different form, it is worth noting that most dynamical phenomena, which up to now are known to occur in autonomous differential delay equations, are found already in the seemingly narrow class of equations of the form (1.2). This underlines the importance of these simple looking equations as guiding examples. It indicates also that the relations between the shape of the nonlinearity f and the dynamics of equation (1.2) are of a subtle nature.

For scalar equations like

$$\epsilon\dot{x}(t) = -\mu x(t) + f(x(t - \alpha)), \tag{9.3}$$

for equations with several time lags, or with distributed delays, a number of results on periodic solutions have been obtained by Kaplan and

Yorke [147], Chow [32], Pesin [237], Hadeler and Tomiuk [100], Nussbaum [217, 218, 219, 221, 224], Alt [4], Angelstorf [7], Walther [285, 298, 300, 302, 303], Arino and Chérif [10], Cao [25] and others.

An interesting problem concerns the relation between periodic solutions of equation (9.3), with $\mu = \alpha = 1$, and periodic solutions of the *singular limit* equation

$$0 = -x(t) + f(x(t-1))$$

which are determined by the one-dimensional map f.

Results on this problem are due to Nussbaum [226], Mallet-Paret and Nussbaum [177, 178, 179, 180, 183], Ivanov and Sharkovski [138], Chow and Huang [37], Hale and Huang [106], Chow, Hale and Huang [35], Chow, Lin and Mallet-Paret [38] and Chow and Mallet-Paret [40].

Applications motivate the consideration of equation (9.3) with a state-dependent delay

$$\alpha = r(x(t)).$$

See the work of Mallet-Paret and Nussbaum [181, 182], Nussbaum [211] and Kuang and Smith [154, 155]. Existence of periodic solutions to equations with multiple state-dependent time lags is established by Mallet-Paret, Nussbaum and Paraskevopoulos [184].

Periodic solutions of *systems* of first-order autonomous differential delay equations, or of single *higher order equations*, were studied by Furumochi [87], Smith [258, 259], an der Heiden [116], an der Heiden, Longtin, Mackey, Milton and Scholl [117], Chow and Huang [36], Mallet-Paret and Sell [186] Nussbaum [215, 218, 224], Mahaffy [174], Hale and Huang [106], Hale and Ivanov [108] and Wu [309].

The periodicity results mentioned so far have in common that there is always at least one nonlocal aspect in each of them. For *local results* on Hopf bifurcation, see, for example, the work of Chafee [28] Cushing [57], Martelli, Schmitt and Smith [190], Stech [261, 263, 264, 265], Dormayer [72, 74] and Chow and Mallet-Paret [40]. For a case of small delays, see Arino and Hbid [11].

Chapter XVI

On the global dynamics of nonlinear autonomous differential delay equations

XVI.1 Negative feedback

In this section we use the prototype equation

$$\dot{x}(t) = f(x(t-1)), \tag{1.1}$$

with a smooth function $f : \mathbb{R} \to \mathbb{R}$, in order to illustrate basic results on the long-term behaviour of solutions and on the organization of the phase space. We assume that f satisfies the condition

$$xf(x) \neq 0, \quad \text{for } x \neq 0 \tag{NF}$$

for negative feedback and that f is bounded from above or from below.

Most of the results mentioned in the sequel hold and were proved for more general classes of equations such as the decay-delay equation

$$\dot{x}(t) = -\mu x(t) + f(x(t-1)), \qquad \mu > 0, \tag{1.2}$$

or equations of the form

$$\dot{x}(t) = g(x(t), x(t-1)).$$

Let S denote the semiflow of equation (1.1) on the state space $C = C([-1,0], \mathbb{R})$,

$$S(t, \varphi) = x_t \quad \text{for all } t \geq 0,$$

where $x : [-1, \infty) \to \mathbb{R}$ is the solution of equation (1.1) satisfying $x_0 = \varphi$.

The investigation of global properties of S is foremost facilitated by the presence of a compact global attractor $A \subset C$. The subsequent definition is equivalent to the one given by Hale [105]. A *complete trajectory* of a semiflow $\Sigma : \mathbb{R}^+ \times M \to M$ is a curve $\sigma : \mathbb{R} \to M$ such that for all $s \in \mathbb{R}$ and $t \geq 0$,

$$\sigma(t+s) = \Sigma(t, \sigma(s)).$$

A subset $N \subset M$ is called *invariant* if for each $x \in N$ there exists a complete trajectory σ with $\sigma(0) = x$ and $\sigma(\mathbb{R}) \subset N$.

Exercise 1.1. Show that a set $N \subset M$ is invariant if and only if $\Sigma(t, N) = N$ for all $t > 0$.

A *compact global attractor* for Σ is a compact invariant set $A \subset M$ which attracts bounded sets, in the sense that for every bounded set $B \subset M$ and for every open set $U \supset A$, there exists $t \geq 0$ such that for all $s \geq t$,

$$\Sigma(s, B) \subset U.$$

Exercise 1.2. Show that a compact global attractor contains each bounded set $B \subset M$ which satisfies

$$B \subset \Sigma(t, B) \quad \text{for all } t > 0.$$

The result of the last exercise implies the maximality property of compact global attractors, namely, that they contain every compact invariant set. Furthermore, it now becomes obvious that compact global attractors necessarily coincide with the union of all orbits of bounded complete trajectories. In particular, stationary points, periodic points and all ω-limits sets with the properties stated in Proposition VII.2.6, are contained in a compact global attractor.

What else can be said about the compact global attractor A of the semiflow S of equation (1.1)?

Exercise 1.3. Give an example of a function f so that equation (1.1) has a periodic solution with period $\frac{4}{5}$.

According to remarks in Chapter XV, the subset

$$(\overline{\mathcal{O}} \cap A) \subset A$$

is positively invariant. This is a first indication that slowly oscillating solutions are important in the global dynamics. Kaplan and Yorke conjectured in [147] that the initial data for slowly oscillating solutions form an open and dense set in $\mathcal{C}$, provided that f is C^1-smooth, $f(0) = 0$, $f'(x) < 0$ for all $x \in \mathbb{R}$ and $\inf f > -\infty$ or $\sup f < \infty$. In [187] it is shown that under these conditions, all other trajectories constitute a closed graph of codimension 2 in $\mathcal{C}$. In particular, it follows that the conjecture is correct. The result in [187] also holds for equation (1.2). It remains an open question whether the smoothness and monotonicity hypotheses can be replaced by the weaker condition (NF).

The coarse structure of the attractor is clarified by an important result of Mallet-Paret [176]. He obtained a Morse decomposition which is defined in terms of oscillation frequencies.

A Morse decomposition of a compact metric space M, with a flow F on it, is a finite sequence of mutually disjoint compact invariant subsets

$S_1, \ldots, S_k$ with the property that for each $y \in M$, there exist indices $\kappa \geq \kappa'$ such that $\alpha(y) \in S_\kappa$ and $\omega(y) \in S_{\kappa'}$, and in case $\kappa = \kappa'$, $F(t, y) \in S_\kappa$ for all $t \in \mathbb{R}$.

In order to comply with this definition, we need a flow, not a semiflow. So consider instead of the attractor A the set Ψ of bounded solutions $x : \mathbb{R} \to \mathbb{R}$, endowed with the compact-open topology, and instead of the semiflow S the flow F on Ψ which is given by translation, i.e.,

$$(t, x) \mapsto x(t + \cdot) \quad \text{for } t \in \mathbb{R},\ x \in \Psi.$$

The Morse decomposition of Ψ is defined by means of an integer-valued Lyapunov functional V on $\Psi \setminus \{0\}$. Let $0 \neq x \in \Psi$. In case there are no zeros of x on $\mathbb{R}_+$, set $V(x) = 1$; otherwise, define $V(x)$ to be the number of zeros of x in $(\sigma - 1, \sigma]$, where

$$\sigma = \inf\left(x^{-1}(0) \cap \mathbb{R}^+\right).$$

The function V turns out to be bounded and is nonincreasing along the flow on Ψ.

In case the stationary point $0 \in \mathcal{C}$ of S is hyperbolic and unstable, the Morse decomposition of Ψ is as follows. Let N^* denote the number of eigenvalues of the generator for the linearized semiflow

$$t \mapsto D_2 S(t, 0)$$

in the open right half-plane.

Exercise 1.4. Show that $N^* \geq 2$ is even.

Exercise 1.5. Consider equation (1.1). Show that under the above condition, $-\infty = \inf x^{-1}(0)$ and $\sup x^{-1}(0) = \infty$ for all $x \in \Psi$.

Define $S_{N^*} = \{0\}$ and

$$S_N = \left\{x \in \Psi : V\left(x(t + \cdot)\right) = N \text{ for all } t, \text{ and } 0 \notin \alpha(x) \cup \omega(x)\right\}$$

for every odd $N \in \mathbb{N}$. (S_N is left undefined for even integers $N \neq N^*$.) Observe that S_1 consists of slowly oscillating solutions. The periodicity results of Chapter XV imply

$$S_1 \neq \emptyset.$$

It can be shown that there exists an odd integer $N_0 \geq 3$ so that

$$S_N = \emptyset \quad \text{for } N \geq N_0$$

and

$$S_{N_0 - 2} \neq \emptyset.$$

The Morse decomposition of Ψ is then given by the sets S_N where $N = N^*$ or $N \in \{1, \ldots, N_0 - 2\}$ odd.

Note that Ψ is the disjoint union of the sets of the Morse decomposition and of the sets

$$C_{\kappa\kappa'} = \{x \in \Psi : \alpha(x) \in S_\kappa,\ \omega(x) \in S_{\kappa'}\}, \quad \kappa > \kappa',$$

of connecting orbits. The sets $C_{\kappa\kappa'}$ capture transient behaviour between the possible oscillation frequencies and connections to the stationary point. For results on existence of such heteroclinic trajectories, see [84, 198, 298, 300].

Relations between the dynamics on Ψ and flows of vectorfields were established by McCord and Mischaikow [192].

Some insight into the nature of the sets S_κ and $C_{\kappa\kappa'}$ is obtained from the analysis of simple limiting cases. In Section XVI.2 we compute a semiflow for equation (1.1) with the discontinuous nonlinearity $f = -$ sign. This stepfunction may be regarded as the limit of monotone, odd, uniformly bounded smooth functions f_α, $\alpha > 0$, such that $f'_\alpha(0) \to -\infty$ as $\alpha \to \infty$. The computations of Section XVI.2 reveal that in this special case analogues of the sets S_κ are given by single periodic solutions. In general, however, the sets S_κ are larger.

Next we discuss results on planar dynamics in the attractor A. The additional hypothesis on f for these is

$$f'(x) < 0 \quad \text{for all } x.$$

We saw in Chapter XV that the set $\overline{\mathcal{O}}$ is positively invariant under S. It follows that S induces a semiflow F on the complete metric space $\overline{\mathcal{O}}$ and that F has a compact global attractor $A(F) \subset (\overline{\mathcal{O}} \cap A)$. Either $A(F) = \{0\}$, or $A(F)$ consists of all segments of solutions in S_1, together with $\varphi = 0$ and all heteroclinic connections $t \mapsto x_t \in \mathcal{O}$ between $\varphi = 0$ and the orbit of a certain periodic solution in S_1.

Exercise 1.6. Give an example where $A(F) \subset A$, $A(F) \neq A$.

The main result of Walther [302] says that if $A(F)$ is nontrivial, i.e., $A(F) \neq \{0\}$, then $A(F)$ is a Lipschitz continuous graph which is homeomorphic to a closed disk in $\mathbb{R}^2$ and bordered by a periodic orbit. On $A(F)$, the semiflow extends to a complete flow with certain smoothness properties. So, in view of openness and density of the data for slowly oscillating solutions [187], one may say that the typical long-term behaviour of solutions is governed by a smooth vector field in the plane.

All orbits of slowly oscillating periodic solutions lie nested into each other in $A(F)$, with $\varphi = 0$ in the interior.

Every slowly oscillating periodic solution is of the type studied in Chapter XV, i.e., if $z_0 < z_1 < z_2$ are consecutive zeros, then the minimal period is $z_2 - z_0$.

Further information about $A(F)$ is obtained by Walther in [298], for the case that the stationary point $\varphi = 0$ is linearly unstable. Then $A(F)$

contains a neighbourhood W of 0 in $A(F)$ which is a C^1-graph diffeomorphic to an open disk in $\mathbb{R}^2$ and bordered by a periodic orbit. W is formed by the curves $t \mapsto x_t$ in $A(F)$ so that $\alpha(x) = \{0\}$. All these curves, except the stationary one, converge to the periodic orbit $\overline{W} \setminus W$ as $t \to \infty$.

An example where

$$\overline{W} = A(F) = A$$

is worked out in [303]. The conditions in [303] are $f'(0) < \frac{\pi}{2}$ and for all $x \in \mathbb{R}$,

$$f(x) = -f(-x) \qquad \text{and} \qquad -1.9 < f'(x).$$

The main result of Walther in [300] describes the unstable set of a slowly oscillating periodic solution $y : \mathbb{R} \to \mathbb{R}$. If y is hyperbolic and unstable, then the set

$$U = \{\varphi \in C \mid \text{There exists a solution } x : \mathbb{R} \to \mathbb{R} \text{ such that } x_0 = \varphi \\ \text{and } x_t \to |\eta| \quad \text{as } t \to -\infty\},$$

where $|\eta| = \{y_t \mid t \in \mathbb{R}\}$, is a two-dimensional graph which is diffeomorphic to an open annulus. The bordering set $\overline{U} \setminus U$ is either given by two periodic orbits or by one periodic orbit and the stationary state.

Existence of hyperbolic unstable slowly oscillating periodic solutions is shown in [137].

The results on planar dynamics in A, for monotone functions f, rely on properties of the wedge $\mathcal{O}$ as discussed in Chapter XV, on generalizations of the a priori estimate of Proposition XV.4.2 and on studies of Floquet multipliers [42, 299].

Further results on asymptotic behaviour of solutions were obtained by Herz [122]. They involve a Lyapunov functional which is adapted from neural network theory and assert, among others, convergence to a symmetric shape in case the function f in equation (1.1) is odd.

XVI.2 A limiting case

Functions like

$$f_\alpha : x \mapsto -\frac{2}{\pi} \arctan(\alpha x), \qquad \alpha > 0,$$

which are monotone, odd and bounded may be regarded as the simplest nonlinearities in the prototype equation (1.1). They converge pointwise to the step function

$$s : x \mapsto -\text{sign}(x)$$

as α increases to $+\infty$; the slopes $f'_\alpha(0)$ which determine the linearized equation along the zero solution tend to $+\infty$.

In this section we shall see that the equation

$$\dot{x}(t) = s(x(t-1)) \tag{2.1}$$

generates a semiflow on a suitable metric space and that this semiflow can be computed explicitly.

A solution of equation (2.1) is defined to be a continuous function $x : I \to \mathbb{R}$, $I = \mathbb{R}$ or $I = [t_0 - 1, \infty)$ for some $t_0 \in \mathbb{R}$, which satisfies the integrated version of equation (2.1), namely

$$x(t) = x(t') + \int_{t'-1}^{t-1} s \circ x, \tag{2.2}$$

for all $t \geq t'$ with $t' - 1 \in I$. As usual,

$$x_t(\theta) := x(t+\theta)$$

whenever $t - 1 \in I$, $\theta \in [-1, 0]$. Let $\mathcal{C} = C\big([-1,0], \mathbb{R}\big)$.

Exercise 2.1. Show that each $\varphi \in \mathcal{C}$ continues to a solution on $[-1, \infty)$.

Observe that initial data φ and ψ in $\mathcal{C}$ which satisfy

$$\varphi^{-1}(y) = \psi^{-1}(y), \quad \text{for } y \in \{-1, 0, 1\} = s(\mathbb{R})$$

and

$$\varphi(0) = \psi(0)$$

yield solutions which coincide for $t \geq 0$. For every solution $x : [t_0 - 1, \infty) \to \mathbb{R}$, the restriction to $[t_0, \infty)$ is composed of a sequence of straight lines with slopes $-1, 0$ and 1. A comparison of the solutions defined by data $\varphi > 0$, $\varphi = 0$ in $\mathcal{C}$ shows that there is no continuous dependence on initial data if we admit the whole space $\mathcal{C}$. Therefore we restrict attention to the subset

$$X = \{\varphi \in \mathcal{C} \mid \varphi^{-1}(0) \text{ finite}\},$$

equipped with the metric $d : (\varphi, \psi) \mapsto \|\varphi - \psi\|$.

Exercise 2.2. Prove the following assertions. The solutions of equation (2.1) on the interval $[-1, \infty)$ define a continuous semiflow

$$S : [0, \infty) \times X \to X.$$

Solutions x with $x_0 \in X$ are, for $t \geq 0$, composed of straight lines with slopes -1 and 1; flat pieces do not occur. The zeros of x are isolated.

Proposition 2.3. *For every $\varphi \in X$, there is a strictly increasing sequence of zeros $t_n = t_n(\varphi)$, $n \in \mathbb{N}$, of the corresponding solution $x : [-1, \infty) \to \mathbb{R}$ in the interval $(0, \infty)$ such that*

$$\begin{aligned}
&\text{sign } x(t_n-) = -\text{sign } x(t_n+) \quad \text{for all } n,\\
&\text{sign } x(0) = 0 \text{ or sign } x(0) = \text{sign } x(t_1-),\\
&\text{sign } x(t) = \text{sign } x(t_1-) \quad \text{for } 0 < t < t_1,\\
&\text{sign } x(t) = \text{sign } x(t_n-) \quad \text{for } n \geq 2 \text{ and } t_{n-1} < t < t_n,\\
&\textit{and } t_n \to \infty \quad \textit{as } n \to \infty.
\end{aligned}$$

The map $\varphi \mapsto t_1(\varphi)$ is continuous.

Proof. Equation (2.2) and $\varphi \in X$ imply $|x| > 0$ on some interval $(0, \epsilon), \varepsilon > 0$. Assume $x > 0$ on $[0, \infty)$. Equation (2.1) yields $\dot{x} = -1$ on $[1, \infty)$, which leads to a contradiction. □

Periodic solutions are easily found. The most obvious one is the function $x^{(0)} : \mathbb{R} \to \mathbb{R}$ given by

$$x^{(0)}(t) = t \quad \text{for } -1 \leq t \leq 1$$

and

$$x^{(0)}(t) = -x^{(0)}(t-2) \quad \text{on } \mathbb{R};$$

$x^{(0)}$ has minimal period 4 and is slowly oscillating.

Exercise 2.4. Every slowly oscillating solution of equation (2.1) merges into the orbit of $x^{(0)}$ in X in finite time.

The strong stability property expressed in the last exercise does not come unexpectedly since the nonlinearity s in equation (2.1) is a limit of functions f_α for which (1.1) has a slowly oscillating periodic solution $x^{(\alpha)}$ which is unique up to translations in time, satisfies

$$x^{(\alpha)}(t) = -x^{(\alpha)}(t-2) \quad \text{on } \mathbb{R},$$

has an exponentially stable orbit in C and attracts all trajectories of slowly oscillating solutions.

Exercise 2.5. Prove the statements of the last paragraph, using results from [220].

In addition to $x^{(0)}$, $-x^{(0)}$ and translates thereof, there exist a countable number of "rapidly oscillating" periodic solutions $x^{(N)} : \mathbb{R} \to \mathbb{R}$, $N \in 2\mathbb{N}$, given by

$$x^{(N)}(t) = t \quad \text{for } -\frac{1}{2N+1} \leq t \leq \frac{1}{2N+1}$$

and

$$x^{(N)}(t) = -x^{(0)}\left(t - \frac{2}{2N+1}\right) \quad \text{on } \mathbb{R}.$$

The minimal period of $x^{(N)}$ is $4/(2N+1)$.

The existence of these rapidly oscillating periodic solutions should be seen in connection with the fact that, as α increases to ∞, more and more complex conjugate pairs of characteristic values of the linearized equation

$$\dot{x}(t) = f'_\alpha(0)x(t-1)$$

move into the right half-plane, giving rise to Hopf bifurcations. See also the results on rapidly oscillating periodic solutions in [72, 297, 177].

We shall study the semiflow S in terms of a return map which is defined by the simple zeros of the solutions. In view of the preceding remarks, it is clear that a solution starting at some $\varphi \in X$ depends only on $\varphi(0)$ and on the ordered sequence

$$z_N < \cdots < z_1$$

of zeros of φ in $(-1, 0)$ which are simple in the sense that

$$\text{sign } \varphi(\mathrm{z_n}-) = -\text{sign } \varphi(\mathrm{z_n}+), \quad \text{for n} = 1, \ldots, \mathrm{N}.$$

(If no zero of this type exists, we set $N = 0$.) In case $\varphi(0) = 0$, we define, in addition,

$$z_0 = 0$$

and

$$s_n = \text{sign } \varphi(z_n-) \quad \text{for } n = 0, \ldots, N$$

so that

$$s_n = (-1)^n s_0 \quad \text{for } n = 0, \ldots, N.$$

Furthermore, for $\varphi \in X$ with $\varphi(0) = 0$ and $N \geq 1$, we set

$$v_n = z_n - z_{n-1} \qquad \text{if } n = 1, \ldots N,$$

$$v_{N+1} = 1 - \sum_{n=1}^{N} v_n;$$

and in case $N = 0$, $v_1 = 1$. The next result is obvious from equation (2.1).

Proposition 2.6. *Consider $\varphi \in X$ with $\varphi(0) = 0$. The local extrema of the corresponding solution x in the interval $[0, 1]$ are given by*

$$x(v_{N+1}) = -s_N v_{N+1},$$
$$x(v_{N+1} + v_N) = -(s_N v_{N+1} + s_{N-1} v_n),$$
$$\vdots$$
$$x\Big(\sum_{n=1}^{N+1} v_n\Big) = x(1) = -\sum_{n=0}^{N} s_n v_{n+1}.$$

We set

$$w_n = - \sum_{\nu=N-n}^{N} s_\nu v_{\nu+1},$$

for $\varphi \in X$ with $\varphi(0) = 0$ and for $n = 0, \ldots, N$. Note that

$$\text{sign } w_0 = -s_N \neq 0.$$

When convenient, we shall write

$$N(\varphi), z_n(\varphi), s_n(\varphi), v_n(\varphi), w_n(\varphi)$$

instead of N, z_n, s_n, v_n, w_n. Let us compute

$$\psi = x_{t_1}$$

for the solution $x : [-1, \infty) \to \mathbb{R}$ of (2.1) given by $x_0 = \varphi \in X$.

In case

$$\text{sign } \mathrm{w_n} \in \{0, \text{sign } \mathrm{w_0}\} \quad \text{for all n} \in \{0, \ldots, \mathrm{N}\},$$

x does not change sign on $[0, 1]$. This implies

$$t_1 \geq 1 \quad \text{and} \quad \dot{x} = -\text{sign } \mathrm{w_0} \quad \text{on } [1, \mathrm{t_1} + 1),$$

hence

$$\psi(\theta) \in \{0, \text{sign } \mathrm{w_0}\} = \{0, -\mathrm{s_N}\} \quad \text{for all } \theta \in [-1, 0].$$

Set $j(\varphi) := N + 1$ in this case. In the other case, there exists a smallest $n \in \{1, \ldots, N\}$ such that

$$\text{sign } \mathrm{w_n} = -\text{sign } \mathrm{w_0}.$$

Set $j(\varphi) = n$. Necessarily,

$$j = j(\varphi) \quad \text{is an odd number}$$

and we have

$$v_{N+1} + \cdots + v_{N-j+2} < t_1 < v_{N+1} + \cdots + v_{N-j+1}.$$

The relations

$$x(v_{N+1} + \cdots + v_{N-j+2}) = w_{j-1}$$

and

$$|\dot{x}| = 1 \quad \text{on } (v_{N+1} + \cdots + v_{N-j+2},\ v_{N+1} + \cdots + v_{N-j+1})$$

imply

$$t_1 = v_{N+1} + \cdots + v_{N-j+2} + |w_{j-1}|.$$

Corollary 2.7. *Let $\varphi \in X$ be given with $\varphi(0) = 0$. Consider the solution $x : [-1, \infty) \to X$ of equation (2.1) with $x_0 = \varphi$. The segment $\psi := x_{t_1(\varphi)}$ has the following properties.*

(i) *If $j(\varphi) = N(\varphi) + 1$, then $N(\psi) = 0$.*

(ii) *If $j(\varphi) \le N(\varphi)$ and if $N(\varphi)$ is even, then*

$$N(\psi) = N(\varphi) - j(\varphi) + 1 \quad \textit{is even}$$

and

$$v_1(\psi) = t_1(\varphi), \; v_2(\psi) = v_1(\varphi), \; \ldots \; \textit{and } v_{N(\psi)}(\psi) = v_{N-j(\varphi)}(\varphi).$$

(iii) *If $j(\varphi) \le N(\varphi)$ and if $N(\varphi)$ is odd, then*

$$N(\psi) = N(\varphi) - j(\varphi) \quad \textit{is even}$$

and

$$v_1(\psi) = t_1(\varphi) + v_1(\varphi), \; v_2(\psi) = v_2(\varphi), \ldots, v_{N(\psi)}(\psi) = v_{N(\varphi)-j(\varphi)}(\varphi).$$

Proof. To prove assertions (ii) and (iii), set $N = N(\varphi)$, $j = j(\varphi)$ and $t_1 = t_1(\varphi)$. We have

$$-\sum_{n=1}^{N-j+1} v_n = v_{N+1} + \cdots + v_{N-j+2} - 1 < t_1 - 1$$

$$< v_{N+1} + \ldots + v_{N-j+1} - 1 = -\sum_{n=1}^{N-j} v_n.$$

It follows that the zeros of x in $(t_1 - 1, t_1) \cap (-1, 0)$ where a change of sign occurs are given by

$$z_{N-j}, z_{N-j+1}, \ldots, z_1.$$

On $(0, t_1)$, there is no change of sign. At $t = 0$, a change of sign occurs if and only if

$$(-1)^N s_0 = s_N = -\text{sign } \text{w}_0 = -\text{sign } \text{x}(0+) = \text{sign } \text{x}(0-) = \text{s}_0;$$

the latter is equivalent to $N \in 2\mathbb{N}_0$. Recall that j is odd. □

Summarizing, we find that for each solution x which starts in X the segment

$$\widetilde{\varphi} = x_{t_1} \in X$$

satisfies $\widetilde{\varphi}(0) = 0$, and all further segments

$$\widetilde{\varphi} = x_{t_n} \in X, \quad 2 \le n \in \mathbb{N},$$

satisfy $\tilde{\varphi}(0) = 0$ and $N(\tilde{\varphi}) \in 2\mathbb{N}_0$. Each trajectory $S(\cdot, \varphi)$, $\varphi \in X$, passes through the set

$$X_0 = \{\tilde{\varphi} \in X : \tilde{\varphi}(0) = 0,\ N(\tilde{\varphi}) \in 2\mathbb{N}_0\},$$

and the return map

$$R : X_0 \ni \varphi \mapsto S(t_1(\varphi), \varphi) \in X_0$$

is well defined and continuous.

Note that the initial data $x_0^{(N)}$ of the periodic solutions introduced above are fixed points of the map R. We have

$$N(x_0^N) = N \quad \text{for all } N \in 2\mathbb{N}_0;$$

for $N \geq 2$ even, $x_0^{(N)} \in X_0$ has zeros in $(-1, 0)$ at

$$z_n = -\frac{2n}{2N+1}, \quad n = 1, \ldots, N;$$

hence

$$v_n(x_0^{(N)}) = \frac{2}{2N+1} \quad \text{for } n = 1, \ldots, N$$

and

$$v_{N+1}(x_0^{(N)}) = \frac{1}{2N+1}.$$

In order to investigate the map R, we associate with it a transformation on the vectors $(v_N, \ldots, v_1)$ determined by elements $\varphi \in X_0$. Recall that for a given $\varphi \in X_0$, the number $N(\varphi)$, the sign $s_0(\varphi)$ and the distances $v_N(\varphi), \ldots, v_1(\varphi)$ between the successive zeros (with a change of sign) determine the solution $x : [-1, \infty) \to \mathbb{R}$ of (2.1) with $x_0 = \varphi$ completely, for $t \geq 0$. In case $N(\varphi) \geq 2$, we have

$$v_n(\varphi) > 0 \quad \text{for } n = 1, \ldots, N(\varphi)$$

and

$$0 < \sum_{n=1}^{N(\varphi)} v_n(\varphi) < 1.$$

We define

$$\Omega_0 = \{x_0^{(0)},\ -x_0^{(0)}\}$$

and for $N \in 2\mathbb{N}$

$$\Omega_N = \{v \in \mathbb{R}^N \mid 0 < v_n \quad \text{for } n = 1, \ldots, N \text{ and } \sum_1^N v_n < 1\} \times \{-1, 1\}.$$

For $(v, \sigma) \in \Omega_N,\ N \geq 2$, we set

$$v_{N+1} = 1 - \sum_{n=1}^{N} v_n$$

and

$$w_n(v) = \sum_{\nu=N-n}^{N} (-1)^\nu v_{\nu+1} \quad \text{for } n = 0, \dots, N.$$

Let Ω denote the disjoint union of the topological spaces Ω_N, $N \in 2\mathbb{N}_0$. The "coordinate map" $V : X_0 \to \Omega$ given by

$$V(\varphi) = \begin{cases} ((v_1(\varphi), \dots, v_{N(\varphi)}(\varphi)), s_0(\varphi)) & \text{if } N(\varphi) \geq 2, \\ x_0^{(0)} & \text{if } N(\varphi) = 0 \text{ and } s_0(\varphi) < 0, \\ -x_0^{(0)} & \text{if } N(\varphi) = 0 \text{ and } s_0(\varphi) > 0, \end{cases}$$

is surjective, and we have

$$V \circ R = f \circ V,$$

where the map $f : \Omega \to \Omega$ is defined as follows:

(i) $f(x_0^{(0)}) = -x_0^{(0)}, \quad f(-x_0^{(0)}) = x_0^{(0)}$.

(ii) In case $(v, \sigma) \in \Omega_N$, $N \geq 2$, and $w_n(v) \geq 0$ for all $n \in \{0, \dots, N\}$,

$$f(v, \sigma) = \sigma x_0^{(0)}.$$

(iii) In case $(v, \sigma) \in \Omega_N$, $N \geq 2$, and $w_n(v) < 0$ for some $n \in \{0, \dots, N\}$, set

$$j(v) = \min\{n \in \{0, \dots N\} : w_n(v) < 0\}$$

and

$$f(v, \sigma) = \Big(\Big(\sum_{n=N-j+1}^{N} v_{n+1} + w_{j-1}(v), v_1, \dots, v_{N-j} \Big), -\sigma \Big),$$

with $j = j(v)$.

Observe that in the last case, necessarily, $j(v)$ is an odd number. On the subsets

$$\Omega_{N0} = \{(v, \sigma) \in \Omega_N : w_1(v) < 0\}, \quad N \geq 2,$$

we have $j(v) = 1$. Therefore

$$\begin{aligned} f(v, \sigma) &= \big((2v_{N+1}, v_1, \dots v_{N-1}), -\sigma \big) \\ &= \Big(\Big(2 - 2\sum_{n=1}^{N} v_n, v_1, \dots, v_{N-1}\Big), -\sigma \Big) \in \Omega_N; \end{aligned}$$

the first component, $f_1(v, \omega) \in \mathbb{R}^N$, is given by the restriction of the affine linear map

$$A_N : v \mapsto \begin{pmatrix} 2 \\ 0 \\ \vdots \\ 0 \end{pmatrix} + \begin{pmatrix} -2 & \dots & -2 \\ 1 & & \\ & \ddots & \\ & & 1 \end{pmatrix} v$$

to the subset

$$\Omega_{N01} = \{v \in \mathbb{R}^N \mid 0 < v_n \quad \text{for } n = 1, \dots, N, \sum_{n=1}^{N} v_n < 1 and$$

$$1 - \sum_{n=1}^{N} v_n - v_N < 0\}$$

of the open standard simplex in $\mathbb{R}^N$. On the complementary subsets

$$\Omega_{N1} = \Omega_N \setminus \Omega_{N0}, \qquad N \geq 2,$$

we have either $f(v, \sigma) \in \Omega_0$ or

$$3 \leq j(v) \leq N(v) - 1 \quad \text{and} \quad f(v, \sigma) \in \Omega_{N-j(v)+1}.$$

Proposition 2.8. *Let $N \in 2\mathbb{N}$. Each iterate $(A_N)^k$, $k \geq 1$, has exactly one fixed point, namely,*

$$v^{(N)} = (\frac{2}{2N+1}, \dots, \frac{2}{2N+1}).$$

All eigenvalues λ of the linear part of A_N at $v^{(N)}$ satisfy

$$|\lambda| > 1.$$

Proof. The linear map $\tilde{A}_N : v \mapsto A_N(v + v^{(N)}) - v^{(N)}$ is given by multiplication with the matrix

$$M_N = \begin{pmatrix} -2 & \dots & -2 \\ 1 & & \\ & \ddots & \\ & & 1 \end{pmatrix}.$$

The vector $v \in \mathbb{R}^N$ is a fixed point of $(A_N)^k$ if and only if $v - v^{(N)}$ is a fixed point of $(\tilde{A}_N)^k$. It remains to show that all eigenvalues of M_N satisfy $|\lambda| > 1$, i.e., that the unstable space of $\tilde{A}_N$ is $\mathbb{R}^N$. This is the content of the next exercise. □

Exercise 2.9. Consider an eigenvalue $\lambda \in \mathbb{C}$ of the matrix M_N from the last proof, and an eigenvector $z \in \mathbb{C}^N$. Prove $|\lambda| > 1$.
Hints: Show $z_n = \lambda^{N-n} z_N$ for $n = 1, \dots, N$; $\lambda^N = -2\sum_{n=0}^{N-1} \lambda^n$, $\lambda \neq 1$ and $\lambda^N = -2\frac{1-\lambda^N}{1-\lambda}$, $2 = \lambda^N + \lambda^{N+1}$.

Exercise 2.10. Show that every periodic solution of (2.1) is a translate of some $x^{(N)}$, $N \in 2\mathbb{N}_0$.

Now we can describe the structure of the semiflow S. Let $\varphi \in X$ be given. Consider the solution $x : [-1, \infty) \to \mathbb{R}$ of (2.1) with $x_0 = \varphi$. There exists a minimal $t \geq 0$ such that

$$x_t \in X_0 \quad \text{and} \quad V(x_t) \in \Omega_N \quad \text{for some } N \in 2\mathbb{N}_0.$$

In case $V(x_t)_1 = v^{(N)}$, the trajectory $t \mapsto x_t$ merges into the periodic orbit

$$o_N = \{x_t^{(N)} : t \in \mathbb{R}\}.$$

For $N = 0$, there is no further possibility. In case $N \geq 2$ and $V(x_t)_1 \neq v^{(N)}$, either

$$V(x_t) \in \Omega_{N1} \quad \text{or} \quad V(x_t) \in \Omega_{N0}.$$

The latter implies that the trajectory reaches the set Ω_{N1} at some $t' > t$, due to the instability of the fixed point $v^{(N)}$ of the map A_N (Proposition 4.8). It follows that for some $t'' > t'$,

$$x_{t''} \in X_0 \quad \text{and} \quad V(x_{t''}) \in \Omega_J,$$

where $0 \leq J \leq N - 2$.

Observe that the trajectory $t \mapsto x_t$ eventually merges into one of the periodic orbits o_N, $N \in 2\mathbb{N}_0$. These periodic orbits correspond to the sets $S_1, S_3, \ldots$ of the Morse decomposition of the attractor mentioned in Section 1. Analogues of the connecting sets $C_{\kappa\kappa'}$ are the unstable sets of the orbits o_N; they are given by the finite dimensional sets Ω_{N0}.

One can show that each orbit o_N, $N \geq 2$, has a heteroclinic connection to o_0 and that the domain of attraction of o_0 is open and dense in X. In particular, almost every solution becomes slowly oscillating in finite time.

XVI.3 Chaotic dynamics in case of negative feedback

If the function f in (1.1) is not monotone, then the semiflow F on $\overline{\mathcal{O}}$ may be complicated. For a discontinuous f, Peters computed a return map associated with a periodic orbit which is on a one-dimensional set conjugate to a chaotic interval map [238]. Numerical results inspired by the work of Mackey and Glass [173], and Lasota and Wazewska-Czyzewska [167], suggested the existence of smooth, hump-shaped functions f for which the attractor $A(F)$ contains chaotic dynamics. A proof of such a result is given by Lani-Wayda and Walther [165, 166].

It starts with the theorem on existence of an unstable hyperbolic periodic solution y of the slowly oscillating type from [137], for monotone f. Guided by ideas developed in [297] (cf. Section 4 below), f is then deformed outside the numerical range $y(\mathbb{R})$ so that the new equation has a solution $h : \mathbb{R} \to \mathbb{R}$ which is homoclinic to y, i.e.,

$$h_t \to o = \{y_s \mid s \in \mathbb{R}\} \quad \text{as } t \to \pm\infty,$$

and $h_t \notin o$ for some t. The deformed nonlinearity is of class C^1 and satisfies the condition (NF) for negative feedback.

In terms of a Poincaré map P on an open subset of a hyperplane $H \subset C$, with fixed point y_0, one obtains a homoclinic trajectory $(\varphi_n)_{-\infty}^{\infty}$, i.e.,

$$\begin{aligned} \varphi_{n+1} &= P(\varphi_n) && \text{for all } n \in \mathbb{Z}, \\ \varphi_n &\to y_0 && \text{as } n \to \pm\infty, \\ \varphi_n &\neq y_0 && \text{for some } n. \end{aligned}$$

P has no continuous inverse defined on an open subset of H, i.e., is certainly not a diffeomorphism.

Let $\mathcal{W}^u$ and $\mathcal{W}^s$ denote local unstable and local stable manifolds of the map P at y_0. For integers $m < 0$ and $n > 0$ sufficiently large, the points on the homoclinic trajectory satisfy the transversality condition

$$DP^{n-m}(\varphi_m) T_{\varphi_m} \mathcal{W}^u \oplus T_{\varphi_n} \mathcal{W}^s = H. \tag{3.1}$$

The verification of condition (3.1) is based on a priori results about the Floquet multipliers of the periodic solution y. Condition (3.1) permits one to apply Theorems 5.1 and 5.2 of [267]. Theorem 5.2 of [267] provides a description of all complete trajectories $(\psi_n)_{-\infty}^{\infty}$ of P in a neighbourhood U of the homoclinic loop

$$L = \{y_0\} \cup \{\varphi_n \mid n \in \mathbb{Z}\}$$

in terms of symbol sequences $(a_n) \in \{0, 1, \ldots, J\}^{\mathbb{Z}}$ of the form

$$\ldots 012 \ldots J \underbrace{0 \ldots 0}_{\geq M \text{ times}} 12 \ldots J0 \ldots,$$

for certain integers $J \geq 1$ and $M \geq 1$. The index shift induced by P on its complete trajectories in U is a homeomorphism (with respect to the product topology on $U^{\mathbb{Z}}$) which is conjugate to the index shift on the space of symbol sequences just described. (In terms of P, one gets nontrivial equivariance with the symbol shift.)

A result of Lani-Wayda in [162, 163] shows that the points ψ_n on the complete trajectories in U form a hyperbolic set for the map P which contains L and is maximal in U.

The solutions $x : \mathbb{R} \to \mathbb{R}$ of (1.1) corresponding to the complete trajectories of P in U are all slowly oscillating and have segments x_t in $A(F)$.

Remark **3.1**. In case of a periodic orbit of an ODE with a transversal homoclinic solution, the problem describing all complete solutions $x : \mathbb{R} \to \mathbb{R}^n$ in a suitable neighbourhood of the homoclinic loop is known as the Poincaré–Birkhoff problem. Its complete solution is due to Shilnikov [250]. Closely related are Smale's result on the embedding of a symbol shift into an iterate of a diffeomorphism with a transversal homoclinic point, and the introduction of hyperbolic sets for diffeomorphisms [252, 253].

The first results on existence of chaotic dynamics generated by RFDE (cf. Section 4 below) involve Poincaré maps which are not one-to-one and can be reduced to chaotic interval maps, or which have no continuous inverse defined on an open set. This motivated generalizations of the work of Shilnikov and Smale to the case of C^1-maps in Banach spaces. Hale and Lin [109] obtained a version of the Shilnikov result which is applicable to the Poincaré maps arising in RFDE. Their proof is based on inclination lemmas which are of independent interest. In [266, 267], Steinlein and Walther presented another proof. First the definition of a hyperbolic set is widened so that it becomes useful also for C^1-maps without inverse. Then it is shown that the Shadowing Lemma remains valid for hyperbolic sets of C^1-maps. [This permits one to find, close to a hyperbolic set, complete trajectories with a predesigned behaviour, by shadowing suitable "pseudo-trajectories" in the hyperbolic set.] Theorem 5.1 in [267] shows that under a transversality condition, homoclinic loops are hyperbolic sets. Then the Shadowing Lemma is used to derive Theorem 5.2 in [267], which is a variant of the Hale-Lin-Shilnikov result which is somewhat easier to apply. Further results on hyperbolic sets and shadowing for C^1-maps are contained in [162, 163].

XVI.4 Mixed feedback

Many equations of the form (1.1) or (1.2) which arise in applications involve nonlinearities f with several zeros, so that the condition (NF) for global negative feedback is violated, and f is not monotone. For some classes of such functions, existence of chaotic behavior has been proved.

One method of proof relies on the explicit computation of solutions. It applies to smooth functions f close to step functions. There exist examples with periodic solutions (not of the slowly oscillatory type) and homoclinic solutions so that associated Poincaré maps on one-dimensional submanifolds are conjugate to chaotic interval maps [290, 118]. For transversality in these cases, see [110, 164].

More general ideas of how to find hyperbolic unstable periodic orbits and transversal homoclinic and heteroclinic connections are developed by Walther in [297]. They apply to situations which cannot be reduced to the study of interval maps and require the results in [109, 266, 267] in order to establish the existence of chaotic dynamics.

Results on homoclinic and heteroclinic trajectories which are associated with other dynamic phenomena were obtained by Arino and Seguier [9, 112, 24] and others.

XVI.5 Some global results for general autonomous RFDE

A version of the Kupka–Smale Theorem on genericity, now in the class of equations

$$\dot{x}(t) = f(x_t) \tag{5.1}$$

with smooth $f : C([-h,0], \mathbb{R}^n) \to \mathbb{R}^n$, was proved by Mallet-Paret [175]. A generic result on connecting trajectories is due to Hale and Lin [111]. Exponential dichotomies were studied by Lin [170].

Work on monotone semiflows of RFDE is due to Smith [254] and Smith and Thieme [256, 257].

Appendix I

Bounded variation, measure and integration

I.1 Functions of bounded variation

Let $f : [a, b] \to \mathbb{R}$ be a given function. The *total variation function* $V_a(f)$ is defined by

$$V_a(f)(t) = \sup_{P(a,t)} \sum_{j=1}^{N} \left|f(\sigma_j) - f(\sigma_{j-1})\right|,$$

where $P(a, t)$ denotes a *partition* $a = \sigma_0 < \sigma_1 < \cdots < \sigma_N = t$ of $[a, t]$. When $V(f)(b)$ is bounded, we say that f is of *bounded variation.*

Suppose that f is a nondecreasing function on $[a, b]$; then

$$\begin{aligned} V_a(f)(t) &= \sup_{P(a,t)} \sum_{j=1}^{N} \left|f(\sigma_j) - f(\sigma_{j-1})\right| = \sup_{P(a,t)} \sum_{j=1}^{N} f(\sigma_j) - f(\sigma_{j-1}) \\ &= \sup_{P(a,t)} f(t) - f(a) = f(t) - f(a). \end{aligned}$$

So for nondecreasing and, likewise, for nonincreasing functions, the variation is easily computed and given by

$$V_a(f)(t) = \left|f(t) - f(a)\right|. \tag{1.1}$$

The following exercise shows that for piecewise monotone functions, the variation can also be computed explicitly.

Exercise 1.1. Let $f : [a, b] \to \mathbb{R}$ and suppose that $a \le c \le b$. Then

$$V_a(f)(b) = V_a(f)(c) + V_c(f)(b). \tag{1.2}$$

(The variation is additive on intervals.)

Exercise 1.2. Prove that

$$V_a(f + g)(t) \le V_a(f)(t) + V_a(g)(t)$$

and conclude that the sum of two functions of bounded variation is of bounded variation.

Exercise 1.3. Let $f : [a, b] \to \mathbb{R}$ be of bounded variation. Prove that

(i) if f is left-continuous at b, then $V_a(f)$ is left-continuous at b;
(ii) if f is right-continuous at a, then $V_a(f)$ is right-continuous at a;
(iii) if f is continuous at t, $a < t < b$, then $V_a(f)$ is continuous at t.

Theorem 1.4. *Let $f : [a, b] \to \mathbb{R}$ be given. The function f is of bounded variation on $[a, b]$ if and only if there are nondecreasing functions $g, h : [a, b] \to \mathbb{R}$ such that $f = g - h$. Moreover, if f is (left) right-continuous, then so are g and h.*

Proof. If $f = g - h$ with g and h nondecreasing, then Exercise 1.2 implies that f is of bounded variation. On the other hand, if f is of bounded variation, then choose

$$g(t) = \frac{1}{2}[f(a) + V_a(f)(t) + f(t)], \qquad h(t) = \frac{1}{2}[f(a) + V_a(f)(t) - f(t)].$$

It is not difficult to see that g and h are nondecreasing. The last assertion follows from Exercise 1.3. □

Corollary 1.5. *The product of two functions of bounded variation is of bounded variation.*

Corollary 1.6. *If $f : [a, b] \to \mathbb{R}$ is of bounded variation, then the points of discontinuity form at most a countable set.*

Proof. Without loss of generality, assume that f is nondecreasing. For $a < c < b$, $f(c+) = \lim_{t \downarrow c} f(t)$ and $f(c-) = \lim_{t \uparrow c} f(t)$ both exist. Since f is nondecreasing and bounded, $f(c+) \geq f(c-)$ and the set

$$D_\epsilon = \{t \in [a, b] \mid f(t+) - f(t-) \geq \epsilon\}$$

is finite. Therefore the set $\bigcup_{j=1}^{\infty} D_{2^{-j}}$ is countable. Obviously, this set contains all points where f is discontinuous. □

We shall call a function $f : [a, b] \to \mathbb{R}$ of bounded variation *normalized* if $f(a) = 0$ and f is continuous from the right on the **open** interval (a, b) [that is, $f(\tau) = f(\tau+)$ at every point $\tau \in (a, b)$]. We shall write $f \in \mathrm{NBV}$ to express that f is a normalized bounded variation function. A complex-valued function f is of (normalized) bounded variation if and only if $\operatorname{Re} f$ and $\operatorname{Im} f$ are of (normalized) bounded variation. When f is defined on $\mathbb{R}_+$, we use the same terminology and f is called of bounded variation if $\lim_{t \to \infty} V_0(f)(t)$ is bounded.

Let $\eta : [a, b] \to \mathbb{C}$ and $f : [a, b] \to \mathbb{C}$ be given. For any partition $P = \{\sigma_0, \ldots, \sigma_N\}$ of $[a, b]$ and any choice of $\tau_j \in [\sigma_{j-1}, \sigma_j]$, we introduce the sum

$$S(f, P, \eta) = \sum_{j=1}^{N} f(\tau_j)(\eta(\sigma_j) - \eta(\sigma_{j-1})).$$

Suppose $A \in \mathbb{C}$ exists such that for all $\epsilon > 0$, there exists a $\delta > 0$ such that

$$|A - S(f, P, \eta)| < \epsilon$$

for all partitions P with width

$$\mu(P) = \max_{1 \leq j \leq N} (\sigma_j - \sigma_{j-1}) < \delta$$

and any choice of "intermediate" points τ_j. We then say that f is *Riemann-Stieltjes integrable* with respect to η over $[a, b]$ [or, in short, $f \in S(\eta)$] and we shall write

$$A = \int_a^b f(\tau)\, d\eta(\tau) = \int_a^b f\, d\eta.$$

Theorem 1.7. *Let $f : [a, b] \to \mathbb{R}$ be continuous. If η is of bounded variation, then $f \in S(\eta)$ and*

$$\Big|\int_a^b f\, d\eta\Big| \leq \big(\max_{[a,b]} |f|\big)\, V_a(\eta)(b). \tag{1.3}$$

Proof. Let $\epsilon > 0$. Since f is uniformly continuous on $[a, b]$, there exists a $\delta > 0$ such that $|f(t_1) - f(t_2)| < \epsilon$ whenever $|t_1 - t_2| < \delta$. Choose two partitions P_1 and P_2 of width less than δ and let $P^* = P_1 \cup P_2$ denote the common refinement of P_1 and P_2. If $P_1 = \{\sigma_0, \ldots, \sigma_N\}$ and $P^* = \{\sigma_0^*, \ldots, \sigma_{N^*}^*\}$, then for any i there are integers k and l such that $[\sigma_{j-1}^*, \sigma_j^*] \subseteq [\sigma_{i-1}, \sigma_i]$ for $l \leq j \leq k$. So for $s_j \in [\sigma_{j-1}^*, \sigma_j^*]$ and $t_i \in [\sigma_{i-1}, \sigma_i]$, we find

$$\begin{aligned} \Big|f(t_i)\big(\eta(\sigma_i) - \eta(\sigma_{i-1})\big) - \sum_{j=k}^{l} f(s_j)(\eta(\sigma_j^*) - \eta(\sigma_{j-1}^*))\Big| \\ < \epsilon \sum_{j=k}^{l} |\eta(\sigma_j^*) - \eta(\sigma_{j-1}^*)|. \end{aligned}$$

Hence

$$|S(f, P_1, \eta) - S(f, P^*, \eta)| < \epsilon V_a(\eta)(b).$$

Similarly, one proves $|S(f, P_2, \eta) - S(f, P^*, \eta)| < \epsilon V_a(\eta)(b)$ and so

$$|S(f, P_1, \eta) - S(f, P_2, \eta)| < 2\epsilon V_a(\eta)(b).$$

We may now invoke the Cauchy criterium to conclude that the limit

$$\lim_{\mu(P) \downarrow 0} S(f, P, \eta) \quad \text{exists}$$

and this proves $f \in S(\eta)$. The proof of (1.3) is left as an exercise. □

Theorem 1.8. (*Partial integration.*) *Let f and η be functions defined on $[a,b]$. If $f \in S(\eta)$, then $\eta \in S(f)$ and*

$$\int_a^b f\,d\eta = f(b)\eta(b) - f(a)\eta(a) - \int_a^b \eta\,df.$$

Proof. The proof is an immediate consequence of the identity

$$\sum_{j=1}^{N} f(t_j)(\eta(\sigma_j) - \eta(\sigma_{j-1})) = f(b)\eta(b) - f(a)\eta(a) - \sum_{j=1}^{N-1} \eta(\sigma_j)(f(t_{j+1}) - f(t_j)).$$

The details are left to the reader. □

The following theorem follows from the intermediate value theorem.

Theorem 1.9. *Let f be an integrable function on $[a,b]$. If η is a function such that η' exists and is integrable on $[a,b]$, then $f \in S(\eta)$ and*

$$\int_a^b f\,d\eta = \int_a^b f(t)\eta'(t)\,dt.$$

I.2 Abstract integration

We start with some definitions and basic results. The main reference will be Rudin [245].

A collection Σ of subsets of a topological space X is said to be a *σ-algebra* in X if Σ has the following properties:

(i) X belongs to Σ.
(ii) If A belongs to Σ, then the complement of A belongs to Σ.
(iii) If A_j, $j = 1, 2, \ldots$, belongs to Σ, then $\bigcap_{j=1}^{\infty} A_j$ belongs to Σ.

A mapping f from X into a topological space Y is said to be *measurable* if $f^{-1}(V)$ belongs to Σ for every open set in Y.

There exists a smallest σ-algebra $\mathcal{B}$ in X such that every open set in X belongs to $\mathcal{B}$. The members of $\mathcal{B}$ are called *Borel sets.*

A (positive) *measure* is a function $\mu : \Sigma \to [0, \infty]$ which is countably additive, that is,

$$\mu(\bigcap_{j=1}^{\infty} A_n) = \sum_{j=1}^{\infty} \mu(A_j)$$

whenever $A_n \in \Sigma$ and $A_i \cap A_j = \emptyset$ for $i \neq j$. When $\Sigma = \mathcal{B}$ we say that μ is a *Borel measure.* If $f : X \to [0, \infty]$ is measurable and $A \in \Sigma$, we define the integral $\int_A f \, d\mu$ by

$$\int_A f \, d\mu = \sup \sum_{j=1}^{N} c_j \mu(A_j \cap A),$$

where the supremum is taken over all measurable step functions $s = \sum_{j=1}^{N} c_j \chi_{A_j}$ such that $0 \leq s \leq f$.

If $X = \mathbb{R}$, there exist a positive measure m (the *Lebesgue measure*) defined on a σ-algebra Σ in $\mathbb{R}$ such that

(i) $m([a, b]) = b - a$;

(ii) Σ contains all Borel sets;

(iii) the measure m is translation invariant, i.e., $m\big([a+x, b+x]\big) = m\big([a, b]\big)$ for any $x \in \mathbb{R}$;

(iv) the integral with respect to m, which is called the Lebesgue integral, extends the Riemann integral in $\mathbb{R}$.

The Lebesgue measure is unique (modulo normalization) in the sense that whenever μ is a positive translation invariant Borel measure on $\mathbb{R}$ such that $\mu(K) < \infty$ for every compact set K, there must be a constant c such that $\mu(A) = cm(A)$ for all Borel sets $A \subset \mathbb{R}$. The set of functions g such that

$$\int |g| \, dm < \infty$$

is denoted by L^1 and a function $g \in L^1$ is called Lebesgue integrable.

Let $X = I \subseteq \mathbb{R}$, $\mathcal{B}$ the Borel σ-algebra on I and m the Lebesgue measure on I obtained by restriction from $\mathbb{R}$. A countable collection $\{A_i\}$ of members of $\mathcal{B}$ is called a *partition* of A if $A_i \cap A_j = \emptyset$ whenever $i \neq j$, and

$$A = \bigcup_{j=1}^{\infty} A_j.$$

A *complex Borel measure* is a complex function $\mu : \mathcal{B} \to \mathbb{C}$ such that for all partitions of $A \in \mathcal{B}$

$$\mu(A) = \sum_{j=1}^{\infty} \mu(A_j).$$

The series converges absolutely, since every rearrangement of it must also converge. The function $|\mu| : \mathcal{B} \to [0, \infty)$ defined by

$$|\mu|(A) = \sup \sum_{j=1}^{\infty} |\mu(A_j)|, \qquad A \in \mathcal{B},$$

where the supremum is taken over all partitions $\{A_i\}$ of A, defines a measure on $\mathcal{B}$ which dominates μ:

$$|\mu(A)| \leq |\mu|(A), \qquad A \in \mathcal{B}.$$

The measure $|\mu|$ is called the *total variation measure* and $|\mu|(I)$ is called the *total variation* of μ on I. For example, if $g \in L^1$, then

$$\lambda(A) = \int_A g\, dm, \qquad A \in \mathcal{B},$$

defines a complex Borel measure on I and $|\lambda|$ is given by

$$|\lambda|(A) = \int_A |g|\, dm, \qquad A \in \mathcal{B}.$$

The space of complex Borel measures on I provided with the norm

$$\|\lambda\| = |\lambda|(I)$$

is a complex Banach space. As a consequence of the Radon-Nikodym Theorem (see [245, Theorem 6.12]), there exists a measurable function $h : I \to \mathbb{C}$ such that $|h(t)| = 1$ for all $t \in I$ and such that

$$\mu(A) = \int_A h\, d|\mu| \quad \text{for all } A \in \mathcal{B} \tag{2.1}$$

Therefore we can define integration of a measurable function f with respect to a complex Borel measure μ by the formula

$$\int_I f\, d\mu = \int_I fh\, d|\mu|. \tag{2.2}$$

In particular,

$$\mu(A) = \int_I \chi_A\, d\mu.$$

Define $L^1(\mu)$ to be the collection of all complex measurable functions g on I for which

$$\int_I |g|\, d\mu < \infty.$$

The elements of $L^1(\mu)$ are called Lebesgue integrable functions with respect to the Borel measure μ. The definition of the integral (2.2) by means of a Lebesgue integral makes it possible to apply results from abstract integration theory.

Theorem 2.1. (*Dominated convergence.*) *Let λ and μ be given complex Borel measures on I. Suppose $\{f_j\}$ is a sequence of complex measurable functions on I such that*

$$f(t) = \lim_{j\to\infty} f_j(t)$$

exists pointwise for every $t \in I$. If there exists a function $\chi \in L^1(\mu)$ such that for every j

$$|f_j(t)| \leq \chi(t) \quad a.e.$$

with respect to μ, then $f \in L^1(\mu)$ and

$$\lim_{j\to\infty} \int_I |f - f_j| \, d\mu = 0.$$

Let $f : I \to \mathbb{C}$ be continuous. It is clear from the definition that the mapping

$$f \mapsto \int_I f \, d\mu$$

is a bounded linear functional on $C(I)$, whose norm is no larger than $|\mu|(I)$. That all bounded linear functionals on $C(I)$ are obtained in this way is the content of the Riesz theorem [245]:

Theorem 2.2. *To each bounded linear functional Λ on $C(I)$, there corresponds a unique Borel measure μ such that*

$$\Lambda(f) = \int_I f \, d\mu.$$

Moreover, $\|\Lambda\| = |\mu|(I)$.

Let $I_1, I_2 \subseteq \mathbb{R}$. For $A \subseteq I_1 \times I_2$ and $f : A \to \mathbb{C}$, we define

$$\begin{aligned} A_t &= \{s \mid (t,s) \in A\}, \qquad & f_t(s) &= f(t,s), \\ A^s &= \{t \mid (t,s) \in A\}, \qquad & f^s(t) &= f(t,s). \end{aligned}$$

Suppose λ and μ are complex Borel measures on I_1 and I_2, respectively. If $A \subset I_1 \times I_2$ is a Borel set, then

$$\int_{I_1} \lambda(A_t) \, d\mu(t) = \int_{I_2} \mu(A^s) \, d\lambda(s)$$

and one defines the product $\mu \times \lambda$ of μ and λ by one of the integrals above,

$$(\mu \times \lambda)(A) = \int_{I_1} \lambda(A_t) \, d\mu(t).$$

This product is a complex measure and one has the Fubini theorem [245]:

Theorem 2.3. *Let $f : I_1 \times I_2 \to \mathbb{C}$ be a Borel measurable function. If*

$$\int_{I_1 \times I_2} |f| \, d|\mu \times \lambda| < \infty,$$

then

$$\int_{I_1} \Big(\int_{I_2} f_t \, d\lambda \Big) \, d\mu(t) = \int_{I_1 \times I_2} f \, d(\mu \times \lambda) = \int_{I_2} \Big(\int_{I_1} f^s \, d\mu \Big) \, d\lambda(s).$$

Let λ and μ be complex Borel measures on $\mathbb{R}$. Associate to each Borel set $A \subset \mathbb{R}$, the set

$$A_2 = \{(t, s) \in \mathbb{R}^2 \mid t + s \in A\}$$

and define the convolution $\mu * \lambda : \mathcal{B} \to \mathbb{C}$

$$(\mu * \lambda)(A) = (\mu \times \lambda)(A_2), \qquad A \in \mathcal{B}.$$

Then $\mu * \lambda$ is a complex Borel measure on $\mathbb{R}$ and

$$\begin{aligned} \big|(\mu * \lambda)(A)\big| &= \big|(\mu \times \lambda)(A_2)\big| \\ &\le \int_{\mathbb{R}} \Big(\int_{\mathbb{R}} d|\lambda| \Big) \, d|\mu| \\ &\le \|\lambda\| \|\mu\|. \end{aligned}$$

So $||\mu * \lambda|| \le ||\lambda|| \, ||\mu||$. Furthermore, if $f : \mathbb{R} \to \mathbb{R}$ is Borel measurable,

$$\int_{\mathbb{R}} f \, d(\mu * \lambda) = \int_{\mathbb{R}} \int_{\mathbb{R}} f(t + s) \, d\lambda(s) \, d\mu(t).$$

So, by taking $f = \chi_A$ we find

$$\mu * \lambda(A) = \int_{\mathbb{R}} \mu(A - s) \, d\lambda(s), \tag{2.3}$$

where $A - s = \{t - s \mid t \in A\}$ denotes the translate of A. The convolution of complex Borel measures extends the convolution of functions in the following sense. Let f and g be L^1-functions. When

$$\mu(A) = \int_A f \, dm, \qquad \lambda(A) = \int_A g \, dm, \quad A \in \mathcal{B},$$

denote the complex Borel measures associated with f and g, respectively, then

$$(\mu * \lambda)(A) = \int_A (f * g) \, dm,$$

where

$$(f * g)(t) = \int_{-\infty}^{\infty} f(t - s) g(s) \, ds.$$

Since $\|\mu\| = \int |f|\, dm$, it follows that $\|f * g\|_1 \leq \|f\|_1 \|g\|_1$ and $f * g$ belongs to L^1. The next theorem states that complex Borel measures and functions of bounded variation are "one and the same".

Theorem 2.4. *There exists a one-to-one correspondence between elements of NBV and complex Borel measures on* $\mathbb{R}_+$ *expressed by*

$$\eta(t) = \mu_\eta([0,t]),$$

where $\eta \in NBV$ *and* μ_η *is a complex Borel measure. The above correspondence is one-to-one because of the normalization of* η*, i.e.,* $\eta(0) = 0$ *and* η *is continuous from the right.*

The theorem implies that we can extend the Riemann-Stieltjes integral using (2.2) and this has a pleasant consequence that the abstract integration theory becomes available for the (extended) Riemann-Stieltjes integral. (See also Hino, Murakami and Naito [125].)

Corollary 2.5. *If* α *and* β *belong to NBV, then the Riemann-Stieltjes convolution*

$$(\alpha * \beta)(t) = \int_{[0,t]} \alpha(t-s)\, d\beta(s)$$

exists and belongs to NBV.

Exercise 2.6. Let μ and λ be complex Borel measures on $\mathbb{R}$. Define the Laplace transform

$$\overline{\mu}(z) = \int_{\mathbb{R}_+} e^{-zt}\, d\mu(t)$$

for those values of z for which the integral exists. Prove that there exists a σ_μ such that the integral converges for $\operatorname{Re} z > \sigma_\mu$ and derive the identity

$$\overline{\mu}(z)\overline{\lambda}(z) = \int_{\mathbb{R}_+} e^{-zt} d(\mu * \lambda), \qquad \operatorname{Re} z \geq \max\{\sigma_\mu, \sigma_\lambda\}. \tag{2.4}$$

Appendix II

Introduction to the theory of strongly continuous semigroups of bounded linear operators and their adjoints

This appendix consists of four parts. First, we present some basic material concerning $\mathcal{C}_0$-semigroups. Here we do not give proofs, as these may be found in many good textbooks, such as [233], but rather suggest (series of) exercises which together yield a complete proof of a certain result. We adopt more or less the same strategy in the second part, which is an interlude on absolutely continuous functions and their relation with the operator of differentiation and the semigroup of translation in L^1. The third part deals with adjoint semigroups. Here we do give detailed proofs, as we expect that not all our readers can easily lay a hold on the basic references [124], [23] and [48]. The fourth part deals with spectral theory and asymptotic behaviour for large time. Here we suggest exercises related to the proof of some results, whereas for others we simply refer to the appropriate literature.

II.1 Strongly continuous semigroups

Let X be a complex Banach space and let, for each $t \geq 0$, $T(t) : X \to X$ be a bounded linear operator. Then the family $\{T(t)\}_{t\geq 0}$ is called a *strongly continuous semigroup*, or a $\mathcal{C}_0$*-semigroup*, whenever the following three properties hold:

(i) $T(0) = I$;

(ii) $T(t)T(s) = T(t+s)$, for $t, s \geq 0$;

(iii) for all $x \in X$, $\|T(t)x - x\| \to 0$ as $t \downarrow 0$.

The *infinitesimal generator* A of $\{T(t)\}_{t\geq 0}$ is defined by

$$\mathcal{D}(A) = \{x \mid \lim_{h\downarrow 0} \frac{1}{h}(T(h)x - x) \quad \text{exists}\},$$

$$Ax = \lim_{h\downarrow 0} \frac{1}{h}(T(h)x - x).$$

So A is the derivative of $T(t)$ at $t = 0$ and is, in general, unbounded. In the following, we shall formulate a number of basic properties of such a semigroup $\{T(t)\}_{t \geq 0}$, its infinitesimal generator A and their interrelationship.

The first result is a direct consequence of the semigroup property [(ii)] and the strong continuity at $t = 0$ [(iii)].

Proposition 1.1. *$t \mapsto T(t)x$ is continuous from $\mathbb{R}_+$ into X.*

Exercise 1.2. Use the Banach–Steinhaus theorem (also called the uniform boundedness principle) to show that there must exist $a > 0$ and $M \geq 1$ such that

$$\|T(t)\| \leq M \quad \text{for } 0 \leq t \leq a.$$

Next, use the semigroup property to establish the following result.

Proposition 1.3. *There exist $\omega \in \mathbb{R}$ and $M \geq 1$ such that*

$$\|T(t)\| \leq Me^{\omega t}.$$

In principle, $\mathcal{D}(A)$ could contain only $x = 0$. As a first step to proving that $\mathcal{D}(A)$ actually is "large", we note

Proposition 1.4. *For arbitrary $t > 0$,*

(i) *for any $x \in X$, $\int_0^t T(s)x\,ds \in \mathcal{D}(A)$ and*

$$A\Big(\int_0^t T(s)x\,ds\Big) = T(t)x - x;$$

(ii) *$\mathcal{D}(A)$ is $T(t)$-invariant and for $x \in \mathcal{D}(A)$*

$$\frac{d}{dt}T(t)x = AT(t)x = T(t)Ax.$$

Here the integral is a Riemann-integral of a continuous function and so

$$\frac{1}{h}\int_t^{t+h} T(s)x\,ds \to T(t)x \quad \text{as } h \downarrow 0.$$

This observation and some elementary formula manipulation, exploiting the semigroup property, are all that one needs to verify the assertions [concerning (ii), also use that a function having a *continuous* right derivative is differentiable]. As a direct consequence of (i) and the above observation, we have

Proposition 1.5. *$\mathcal{D}(A)$ is dense in X.*

We recall that an unbounded operator L is *closed* if and only if for every sequence $x_n \in \mathcal{D}(L)$ such that $x_n \to x$ and $Lx_n \to y$ as $n \to \infty$, it necessarily follows that $x \in \mathcal{D}(L)$ and $Lx = y$ (in other words, if and only if the graph of L is closed in $X \times X$). (See Section II.4 for more information about closed operators.)

Proposition 1.6. *A is closed.*

Exercise 1.7. Verify this.
Hint: use the identity in Proposition 1.4(ii) to rewrite

$$T(t)x_n - x_n = \int_0^t \frac{d}{ds} T(s)x_n \, ds.$$

Can one operator be the infinitesimal generator of two different semigroups? The answer is no!

Exercise 1.8. Let both $S = \{S(t)\}_{t\geq 0}$ and $T = \{T(t)\}_{t\geq 0}$ have infinitesimal generator A. Prove that $S = T$.
Hint: Choose $t > 0$ and $x \in \mathcal{D}(A)$ and consider the function $f : [0,t] \to X$ defined by $f(s) = T(t-s)S(s)x$.

Exercise 1.9. Let $u : [0, t_e] \to \mathcal{D}(A)$ satisfy

$$\begin{cases} \dfrac{du}{dt} = Au, & 0 \leq t \leq t_e, \\ u(0) = x. \end{cases}$$

Prove that necessarily $u(t) = T(t)x$.

In view of the exponential estimate of Proposition 1.3, the Laplace transform

$$R(z)x = \int_0^\infty e^{-zs} T(s)x \, ds$$

exists (as an improper Riemann integral) for all $z \in \mathbb{C}$ with $\operatorname{Re} z$ sufficiently large. The following definition allows us to be more precise about "sufficiently large".

Definition 1.10. The *growth bound* (also called the *type*) of $\{T(t)\}_{t\geq 0}$ is the real number ω_0 defined by

$$\omega_0 = \inf\{\omega \mid \exists M \geq 1 \quad \text{such that } \|T(t)\| \leq Me^{\omega t}\}.$$

If A were bounded, we could write $T(t) = e^{tA}$ and then it would follow at once that $R(z) = (zI - A)^{-1}$ for $z \in \rho(A)$, with $\rho(A)$ denoting the resolvent set of A (cf. II.4). Actually the identity holds for unbounded A as well.

Proposition 1.11. *The set $\{z \mid \operatorname{Re} z > \omega_0\}$ belongs to $\rho(A)$, and for z in this set, the identity*

$$(zI - A)^{-1} = R(z)$$

holds.

Exercise 1.12. Prove the last proposition.
Hint: Consider

$$\frac{1}{h}(T(h) - I) \int_0^t e^{-zs} T(s)x \, ds.$$

Let, after some formula manipulation, h tend to zero and then t to infinity (exploit that A is closed!).

We are now ready to formulate the central result of the theory, the famous theorem of Hille and Yosida (and Phillips, Feller and Miyadera), which gives necessary and sufficient conditions for an operator A to be the infinitesimal generator of a $\mathcal{C}_0$-semigroup.

Theorem 1.13. *A linear operator A is the infinitesimal generator of a $\mathcal{C}_0$-semigroup $T(t)$ satisfying $\|T(t)\| \leq Me^{\omega t}$ if and only if*

(i) *A is closed and $\mathcal{D}(A)$ is dense in X;*

(ii) *the resolvent set $\rho(A)$ contains the ray (ω, ∞) and*

$$\|(zI - A)^{-k}\| \leq \frac{M}{(z-\omega)^k} \quad \text{for } z > \omega, \ k = 1, 2, \ldots.$$

The following exercises together yield a proof of this theorem. The first supplements Propositions 1.3 and 1.11 to give a proof of the "only if" part.

Exercise 1.14. Let A be the infinitesimal generator of a $\mathcal{C}_0$-semigroup $T(t)$.

(i) Derive the estimate

$$\|(zI - A)^{-1}\| \leq \frac{M}{(z-\omega)}$$

from the identity $(zI - A)^{-1} = R(z)$.

(ii) Show that

$$\frac{d^k}{dz^k} R(z) = \int_0^\infty (-s)^k e^{-zs} T(s) \, ds$$

and

$$\frac{d^k}{dz^k}(zI - A)^{-1} = (-1)^k k! (zI - A)^{-k-1}.$$

Hint: Use the resolvent equation

$$(zI - A)^{-1} - (\lambda I - A)^{-1} = (\lambda - z)(zI - A)^{-1}(\lambda I - A)^{-1}.$$

(iii) Conclude that

$$(zI - A)^{-k-1} = \frac{1}{k!}\int_0^\infty s^k e^{-zs} T(s)\, ds$$

and derive from this identity, the estimate

$$\|(zI - A)^{-k-1}\| \le M(z-\omega)^{-k-1}.$$

In the next five exercises, the starting point is a closed operator A with a dense domain $\mathcal{D}(A)$, satisfying the estimates of the theorem. The idea is to construct a semigroup $T(t)$ by a limiting procedure and then to verify that A is its infinitesimal generator.

Exercise 1.15. Show that $\lim_{z\to\infty} z(zI - A)^{-1}x = x$ for all $x \in X$.
Hint: First, check for $x \in \mathcal{D}(A)$. Next, use that $\mathcal{D}(A)$ is dense and that we have a uniform bound for $\|z(zI - A)^{-1}\|$.

Definition 1.16. Given A, we define its so-called *Yosida approximation* A_z by

$$A_z = zA(zI - A)^{-1} = z^2(zI - A)^{-1} - zI.$$

Note that A_z is a *bounded* operator and that for $x \in \mathcal{D}(A)$

$$\lim_{z\to\infty} A_z x = \lim_{z\to\infty} zA(zI - A)^{-1}x = \lim_{z\to\infty} z(zI - A)^{-1}Ax = Ax.$$

Since A_z is bounded, its exponential

$$e^{tA_z} = \sum_{n=0}^{\infty} \frac{t^n}{n!} A_z^n$$

is well defined.

Exercise 1.17. Verify that for $z > 2|\omega|$ and $t \ge 0$

$$\|e^{tA_z}\| \le Me^{\frac{z\omega t}{z-\omega}} \le Me^{\theta(\omega)\omega t},$$

where $\theta(\omega) = 2$ if $\omega > 0$ and $\theta(\omega) = 1$ if $\omega \le 0$.

Exercise 1.18. Verify that for $z, \lambda > 2|\omega|$ and $t \ge 0$,

$$\|e^{tA_z}x - e^{tA_\lambda}x\| \le M^2 e^{\theta(\omega)\omega t} t\|(A_z - A_\lambda)x\|.$$

Hint:

$$e^{tA_z}x - e^{tA_\lambda}x = \int_0^1 \frac{d}{ds}(e^{tsA_z}e^{t(1-s)A_\lambda}x)\, ds.$$

Exercise 1.19. Show that $T(t)x = \lim_{z\to\infty} e^{tA_z}x$ exists and defines a C_0-semigroup satisfying $\|T(t)\| \le Me^{\omega t}$.
Hint: Again consider first $x \in \mathcal{D}(A)$ and then use the density of $\mathcal{D}(A)$ in combination with the uniform estimate of Exercise 1.17. Note that the convergence is uniform for t in a compact set.

Exercise 1.20. Verify that $T(t)$ as defined in the last exercise has infinitesimal generator A.
Hint: Derive the identity $T(t)x - x = \int_0^t T(s)Ax\,ds$ from the corresponding one for A_z. Conclude that the infinitesimal generator has to be an extension of A. But $(zI - A)^{-1}$ exists for large z!

This concludes the "proof" of the Hille-Yosida theorem. As a kind of dessert we have one more exercise.

Exercise 1.21. Show that $\mathcal{D}(A^k)$ is dense in X for any $k \ge 1$.

In conclusion of this first part, we note that one can pose additional conditions on A such that one can represent the semigroup by an inverse Laplace transform, i.e.,

$$T(t) = \frac{1}{2\pi i}\int_\Gamma (zI - A)^{-1} e^{zt}\,dz,$$

where the contour Γ is contained in some sector of the complex plane as indicated in the following picture:

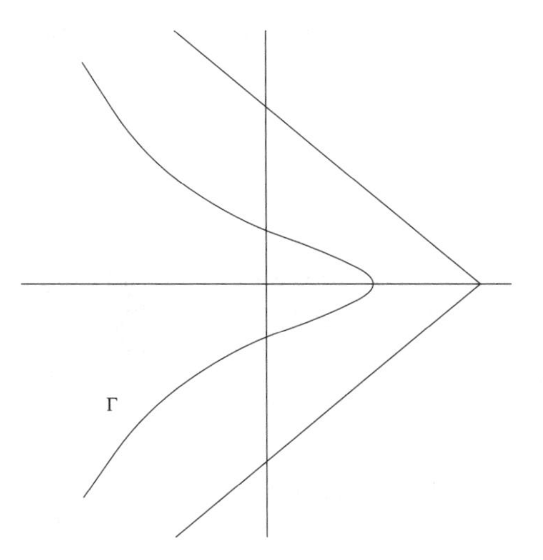

We then enter the realm of *analytic semigroups*, i.e., semigroups which can be extended from the positive reals into a sector of the complex plane.

This theory is very important when one wants to deal with parabolic partial differential equations (using the spectral theory of elliptic partial differential equations). For the kind of differential equations studied in this book, the additional conditions do not hold and the semigroups are not analytic. Hence we refrain from stating results about analytic semigroups and refer instead to [121], [85], [270] and [233].

II.2 Interlude: absolute continuity

We now show how the abstract theory can be used to great advantage when proving certain results in real function theory (see [23] for an extensive elaboration of such ideas).

Definition 2.1. A function $f : \mathbb{R} \to \mathbb{C}$ is called *absolutely continuous* if and only if for every $\epsilon > 0$ there exists a $\delta > 0$ such that

$$\sum_{j=1}^{N} (\beta_j - \alpha_j) < \delta \qquad \text{implies} \quad \sum_{j=1}^{N} \left| f(\beta_j) - f(\alpha_j) \right| < \epsilon,$$

whenever $(\alpha_1, \beta_1), \ldots, (\alpha_N, \beta_N)$ are disjoint intervals.

We now give two equivalent characterisations of absolute continuity. For the proof of the first we refer to [245]

Theorem 2.2. *A function $f : \mathbb{R} \to \mathbb{C}$ is absolutely continuous if and only if there exists $g \in L^1_{loc}(\mathbb{R})$ such that*

$$f(a) = f(0) + \int_0^a g(\alpha)\, d\alpha.$$

In this characterisation, we see that f should have an almost everywhere defined pointwise derivative which belongs to L^1, at least locally. There are other ways to think about differentiation in an L^1-context. Indeed, if we define

$$\big(T(t)f\big)(a) = f(a+t),$$

then we may ask whether $t^{-1}\big(T(t)f - f\big)$ has a limit in $L^1(\mathbb{R})$ as $t \downarrow 0$. Necessarily, this then is a "global" concept and since $\mathbb{R}$ is noncompact, we have to pay special attention to the behaviour at infinity. But apart from this aspect, the two ways of defining a derivative in an L^1-sense actually amount to the same, as we show now.

Theorem 2.3. *A function $f : \mathbb{R} \to \mathbb{C}$ is absolutely continuous with $g = f' \in L^1(\mathbb{R})$ if and only if*

$$\|\frac{1}{t}(T(t)f - f) - g\| \to 0 \quad \text{as } t \downarrow 0,$$

where the norm is the L^1-norm.

Exercise 2.4. Show that $T(t)$ defines a $\mathcal{C}_0$-semigroup (in fact a group!) on $L^1(\mathbb{R})$, i.e., show that translation is continuous in L^1.
Hint: Use that continuous functions with compact support are dense.

Exercise 2.5. Choose z with $\operatorname{Re} z > 0$ and put $h = (zI - A)^{-1}f$. Use Proposition 1.11 to show that

$$h(\alpha) = h(0) + \int_0^a (zh(\alpha) - f(\alpha))\, d\alpha.$$

Let us define the operator D on $L^1(\mathbb{R})$ by

$$\mathcal{D}(D) = \{f \mid f \in \mathrm{AC}(\mathbb{R}),\ \text{i.e.,}\ f(a) = f(0) + \int_0^a g(\alpha)\, d\alpha,\ g \in L^1(\mathbb{R})\},$$

$$Df = g.$$

Then it follows from the last exercise that, with A the infinitesimal generator of the semigroup $T(t)$ of translations on $L^1(\mathbb{R})$, we have

$$\mathcal{D}(A) \subset \mathcal{D}(D) \quad \text{and} \quad (zI - D)(zI - A)^{-1} = I.$$

To prove that actually $A = D$ we have to verify one more thing.

Exercise 2.6. Prove that for $f \in \mathcal{D}(D)$

$$(zI - A)^{-1}(zI - D)f = f.$$

Hint: Use Proposition 1.11 and partial integration.

From this characterization of the generator of translation acting on L^1-functions defined on all of $\mathbb{R}$, one can easily derive results for translation acting on L^1-functions defined on subsets of $\mathbb{R}$. We state a typical example.

Theorem 2.7. *Consider on $L^1(\mathbb{R}_-)$ the $\mathcal{C}_0$-semigroup*

$$\big(S(t)f\big)(a) = \begin{cases} f(a+t), & a \le -t, \\ 0, & a > -t \end{cases}$$

and let B denote its infinitesimal generator. Then

$$\mathcal{D}(B) = \{f \mid f \in \mathrm{AC}(\mathbb{R}_-),\ \textit{i.e.},\ f(a) = \int_0^a g(\alpha)\, d\alpha \ \textit{with}\ g \in L^1(\mathbb{R}_-)\},$$

$$Bf = g.$$

Idea of the proof: The subspace $L^1_-(\mathbb{R}) = \{f \in L^1(\mathbb{R}) \mid f(a) = 0 \text{ for a.e. } a \geq 0\}$ is a closed left-translation invariant subspace. So the restriction of the left-translations to this subspace yields a $\mathcal{C}_0$-semigroup and, as one easily verifies, its generator is the restriction, i.e.,

$$(\mathcal{D}(A) \cap L^1_-(\mathbb{R}), A).$$

Finally, use the isometric isomorphism between $L^1_-(\mathbb{R})$ and $L^1(\mathbb{R}_-)$.

II.3 Adjoint semigroups

Exercise 3.1. Consider the group of translations on $L^\infty(\mathbb{R})$. Show that it is *not* strongly continuous.

In the definition of a $\mathcal{C}_0$-semigroup, the algebraic conditions (i) $T(0) = I$ and (ii) $T(t)T(s) = T(t+s)$ for $t, s \geq 0$ are directly related to the interpretation of $T(t)$ as the operator giving the state of a time-invariant system at time t in terms of the state at time zero. The third condition, the strong continuity, is of another category. It says that we want orbits to be continuous with respect to the norm topology on X. It is a remarkable fact that we would *not* get another class of semigroups if we replace the norm topology by the weak topology here:

Theorem 3.2. (*Weak equals strong.*) *If $T(t)$ is a semigroup of bounded linear operators on a Banach space X such that for any $x \in X$*

$$\text{weak-}\lim_{t \downarrow 0} T(t)x = x,$$

then $T(t)$ is strongly continuous. Moreover, if we define

$$\widetilde{A}x = \text{weak-}\lim_{h \downarrow 0} \frac{1}{h}(T(h)x - x), \quad \mathcal{D}(\widetilde{A}) = \{x \mid \textit{this limit exists}\},$$

then $\widetilde{A} = A$.

For a proof see [59, 124].

The matter becomes different if we consider (*nonreflexive*) dual spaces and equip them with the *weak* topology* (which is by definition the weakest topology for which elements of X define, by the given pairing, continuous linear functionals on X^*). First, we recall some definitions and facts.

Let A be a densely defined operator, then the *adjoint operator* A^* is defined as follows:

$$\mathcal{D}(A^*) = \{x^* \in X^* \mid \exists y^* \in X^* \text{ such that } \langle x^*, Ax\rangle = \langle y^*, x\rangle, \text{ forall } x \in \mathcal{D}(A)\}$$

and then

$$A^* x^* = y^*.$$

Note that we need A to be densely defined in order to have uniqueness of y^* (otherwise A^* is multivalued).

Proposition 3.3.

(i) *A^* is weak* closed, i.e., the graph of A^* is closed in $X^* \times X^*$ equipped with the weak* topology;*

(ii) *if A is closed, $\mathcal{D}(A^*)$ is weak* dense;*

(iii) *if A is bounded, so is A^* and $\|A\| = \|A^*\|$;*

(iv) *if $z \in \rho(A)$, then $z \in \rho(A^*)$ and*

$$\left((zI - A)^{-1}\right)^* = (zI - A^*)^{-1}.$$

Exercise 3.4. Prove these statements.

We now return to semigroups. Let $T(t)$ be a $\mathcal{C}_0$-semigroup on X. Let $T^*(t) := \left(T(t)\right)^*$ be the adjoint operators on the dual space X^*.

Theorem 3.5.

(i) *$T^*(t)$ is a weak* continuous semigroup, i.e., $T^*(0) = I$, $T^*(t)T^*(s) = T^*(t+s)$ for $t, s \geq 0$ and for given $x^* \in X^*$, $t \mapsto T^*(t)x^*$ is continuous from $\mathbb{R}_+$ into X^* equipped with the weak* topology.*

(ii) *A^* is the weak* generator of $T^*(t)$, i.e.,*

$$\frac{1}{h}\left(T^*(h)x^* - x^*\right) \overset{*}{\rightarrow} y^* \quad \text{as } h \downarrow 0$$

if and only if $x^* \in \mathcal{D}(A^*)$ and $y^* = A^* x^*$.

(iii) *$\mathcal{D}(A^*)$ is $T^*(t)$ invariant and $A^*T^*(t) = T^*(t)A^*$ on $\mathcal{D}(A^*)$.*

Proof. (i) is straightforward.
(ii) Assume the first assertion holds. Consider $x \in \mathcal{D}(A)$; then

$$\begin{aligned}\langle y^*, x\rangle &= \lim_{h\downarrow 0} \frac{1}{h}\langle T^*(h)x^* - x^*, x\rangle \\ &= \lim_{h\downarrow 0} \frac{1}{h}\langle x^*, T(h)x - x\rangle = \langle x^*, Ax\rangle\end{aligned}$$

from which we conclude that $x^* \in \mathcal{D}(A^*)$ and $y^* = A^*x^*$. Conversely, assume that the second assertion holds. Then for every $x \in X$

$$\begin{aligned}\frac{1}{h}\langle T^*(h)x^* - x^*, x\rangle &= \frac{1}{h}\langle x^*, T(h)x - x\rangle \\ &= \frac{1}{h}\langle x^*, A\int_0^h T(s)x\,ds\rangle \\ &= \frac{1}{h}\langle y^*, \int_0^h T(s)x\,ds\rangle = \langle y^*, x\rangle \quad \text{as } h \downarrow 0.\end{aligned}$$

(iii) Take $x \in \mathcal{D}(A)$ and $x^* \in \mathcal{D}(A^*)$; then

$$\begin{aligned}\langle T^*(t)A^*x^*, x\rangle &= \langle A^*x^*, T(t)x\rangle = \langle x^*, AT(t)x\rangle \\ &= \langle x^*, T(t)Ax\rangle = \langle T^*(t)x^*, Ax\rangle\end{aligned}$$

from which we conclude that $T^*(t)x^* \in \mathcal{D}(A^*)$ and

$$A^*T^*(t)x^* = T^*(t)A^*x^*.$$

□

Our next step is to show that we can obtain a $\mathcal{C}_0$-semigroup again by a suitable restriction procedure. In the process, we need the following analogue of Proposition 1.4 (also see Lemma 3.15 below).

Lemma 3.6. *For $x^* \in \mathcal{D}(A^*)$ and $x \in X$, we have*

$$\langle T^*(t)x^* - x^*, x\rangle = \int_0^t \langle T^*(s)A^*x^*, x\rangle\,ds.$$

Proof.

$$\begin{aligned}\int_0^t \langle T^*(s)A^*x^*, x\rangle\,ds &= \int_0^t \langle A^*x^*, T(s)x\rangle\,ds \\ &= \langle A^*x^*, \int_0^t T(s)x\,ds\rangle \\ &= \langle x^*, A\int_0^t T(s)x\,ds\rangle \\ &= \langle x^*, T(t)x - x\rangle = \langle T^*(t)x^* - x^*, x\rangle.\end{aligned}$$

□

A^*, being the adjoint of A, does satisfy the estimates of the Hille-Yosida theorem but is not necessarily densely defined. We now show that the part of A^* in $\overline{\mathcal{D}(A^*)}$ generates a $\mathcal{C}_0$-semigroup.

Definition 3.7. $X^{\odot} = \{x^* \in X^* \mid \lim_{h\downarrow 0} \|T^*(h)x^* - x^*\| = 0\}$.

Proposition 3.8.

(i) $X^{\odot}$ *is a norm-closed,* $T^*(t)$*-invariant subspace of* X^*;

(ii) $X^{\odot} = \overline{\mathcal{D}(A^*)}$.

Proof. (i) The invariance follows from

$$\|T^*(t+h)x^* - T^*(t)x^*\| \le \|T^*(t)\|\,\|T^*(h)x^* - x^*\|.$$

Suppose $x_n^* \in X^{\odot}$ and $x^* \in X^*$ are such that $\|x_n^* - x^*\| \to 0$; then

$$\|T^*(h)x^* - x^*\| \le \|T^*(h)x^* - T^*(h)x_n^*\| + \|T^*(h)x_n^* - x^*\|$$

shows that $x^* \in X^{\odot}$.

(ii) Assume $x^* \in \mathcal{D}(A^*)$; then, by Lemma 3.6,

$$\left|\langle T^*(t)x^* - x^*, x\rangle\right| \le \big(\sup_{0\le s\le t} \|T^*(s)\|\big)\|A^*x^*\|\,\|x\|t$$

[so note that this is a Lipschitz estimate for the norm of $T^*(t)x^* - x^*$ when $x^* \in \mathcal{D}(A^*)$] and therefore $x^* \in X^{\odot}$.

Next, let $x^* \in X^{\odot}$ and define $x_t^* = t^{-1}\int_0^t T^*(s)x^*\,ds$; then $\|x_t^* - x^*\| \to 0$ as $t \downarrow 0$ [since $t \mapsto T^*(t)x^*$ is (norm) continuous for $x^* \in X^{\odot}$!]. We shall show that $x_t^* \in \mathcal{D}(A^*)$ and that $A^*x_t^* = t^{-1}(T^*(t)x^* - x^*)$. Again we use something like Lemma 3.6, but now with $x \in \mathcal{D}(A)$.

$$\begin{aligned}\langle x_t^*, Ax\rangle &= \frac{1}{t}\int_0^t \langle T^*(s)x^*, Ax\rangle\,ds = \frac{1}{t}\int_0^t \langle x^*, T(s)Ax\rangle\,ds \\ &= \frac{1}{t}\langle x^*, \int_0^t T(s)Ax\,ds\rangle = \frac{1}{t}\langle x^*, T(t)x - x\rangle \\ &= \frac{1}{t}\langle T^*(t)x^* - x^*, x\rangle\end{aligned}$$

which shows what was stated above. □

Notation 3.9. By $T^{\odot}(t)$ we denote the $\mathcal{C}_0$-semigroup on $X^{\odot}$ obtained by restricting $T^*(t)$ to the invariant closed subspace $X^{\odot}$.

Theorem 3.10. $A^{\odot}$ *is the part of* A^* *in* $X^{\odot}$, *i.e.,*

$$\mathcal{D}(A^{\odot}) = \{x^* \in \mathcal{D}(A^*) \mid A^*x^* \in X^{\odot}\}, \qquad A^{\odot}x^* = A^*x^*.$$

Proof. $\mathcal{D}(A^{\odot}) \subset \mathcal{D}(A^*)$ since norm convergence implies weak* convergence. Moreover, $A^{\odot}x^* = A^*x^*$ for $x^* \in \mathcal{D}(A^{\odot})$. Now consider $x^* \in \mathcal{D}(A^*)$ such that $A^*x^* = y^* \in X^{\odot}$. By Lemma 3.6,

$$\langle T^*(t)x^* - x^*, x\rangle = \int_0^t \langle T^*(s)y^*, x\rangle\, ds, \quad \text{for all } x \in X.$$

Since $s \mapsto T^*(s)y^*$ is norm-continuous, we may rewrite this as

$$T^*(t)x^* - x^* = \int_0^t T^*(s)y^*\, ds.$$

Hence, using continuity once more,

$$\frac{1}{t}\big(T^*(t)x^* - x^*\big) \to y^* \qquad \text{in norm as } t \downarrow 0.$$

So $x^* \in \mathcal{D}\big(A^{\odot}\big)$ and $A^{\odot}x^* = y^* = A^*x^*$. □

Starting from the $\mathcal{C}_0$-semigroup $T^{\odot}(t)$ defined on $X^{\odot}$, we can repeat our construction and introduce semigroups $T^{\odot *}(t)$ on $X^{\odot *}$ and $T^{\odot\odot}(t)$ on $X^{\odot\odot}$. A linear mapping $j : X \to X^{\odot *}$ is defined by

$$\langle jx, x^{\odot}\rangle = \langle x^{\odot}, x\rangle$$

and since $X^{\odot}$ is weak* dense in X^*, this mapping is injective. In order to study the continuity of j, we introduce another norm on X.

Definition 3.11.

$$\|x\|' = \sup\big\{\big|\langle x^{\odot}, x\rangle\big| \mid x^{\odot} \in X^{\odot},\ \|x^{\odot}\| \le 1\big\}.$$

Lemma 3.12.

$$\|x\|' \le \|x\| \le M\|x\|',$$

where M is such that $\|T(t)\| \le Me^{\omega t}$ for some ω.

Note that, whenever $M = 1$, e.g., contraction semigroups, one has $\|\cdot\|' = \|\cdot\|$. For the proof of this lemma, as well as for later use, we need some auxiliary results.

Interlude 3.13. (The weak* integral.) Let a and $b > a$ be real numbers and let $q : [a, b] \to X^*$ be such that

$$\langle q(\cdot), x\rangle \in L^1([a, b], \mathbb{C}) \qquad \text{for all } x \in X.$$

We claim that there exists $Q \in X^*$ for which

$$\langle Q, x\rangle = \int_a^b \langle q(\sigma), x\rangle\, d\sigma \qquad \text{for all } x \in X$$

and we shall call Q the *weak* integral* of q over $[a, b]$.

To substantiate our claim, we introduce a linear mapping

$$\mathcal{L} : X \to L^1([a,b], \mathbb{C})$$

by defining

$$(\mathcal{L}x)(t) = \langle q(t), x \rangle, \qquad a \le t \le b, \quad x \in X.$$

Then $\mathcal{L}$ is closed since

(i) if $x_n \to x$ in X, then $(\mathcal{L}x_n)(t) \to (\mathcal{L}x)(t)$ in $\mathbb{C}$ for all $t \in [a,b]$,

(ii) so if $\mathcal{L}x_n \to g$ in L^1, then necessarily $g(t) = (\mathcal{L}x)(t)$ for almost all t.

By the Closed Graph Theorem [245], we can conclude that $\mathcal{L}$ is bounded, i.e.,

$$\int_a^b |\langle q(\sigma), x \rangle| \, d\sigma \le K \|x\|$$

for some constant K. Hence $x \mapsto \int_a^b \langle q(\sigma), x \rangle \, d\sigma$ defines a continuous linear functional and $Q \in X^*$ must exist such that

$$\langle Q, x \rangle = \int_a^b \langle q(\sigma), x \rangle \, d\sigma.$$

We shall simply write $Q = \int_a^b q(\sigma) \, d\sigma$. Note that the argument does not rely on $[a,b]$ being a finite interval.

A drawback of the weak* integral is that it does not commute with general bounded linear operators on X^*. However,

Lemma 3.14. *Let $L : X \to X$ be a bounded linear operator. Then*

$$L^* \int_a^b q(\sigma) \, d\sigma = \int_a^b L^* q(\sigma) \, d\sigma.$$

The proof is a trivial verification. This ends the interlude.

We can now state a much stronger version of Lemma 3.6.

Lemma 3.15.

(i) *For every $x^* \in X^*$ and $h > 0$,*

$$\int_0^h T^*(s) x^* \, ds \in \mathcal{D}(A^*)$$

and

$$A^* \int_0^h T^*(s) x^* \, ds = T^*(h) x^* - x^*;$$

(ii)

$$\| \int_0^h T^*(s) x^* \, ds \| \le \frac{M}{\omega} (e^{\omega h} - 1) \|x^*\|.$$

Proof. Some of the following arguments were already used for the special case $x^* \in X^\odot$ in the proof of Proposition 3.8(ii). We have

$$\begin{aligned}\left(T^*(t) - I\right) \int_0^h T^*(s)x^*\,ds &= \int_t^{t+h} T^*(s)x^*\,ds - \int_0^h T^*(s)x^*\,ds \\ &= \int_h^{h+t} T^*(s)x^*\,ds - \int_0^t T^*(s)x^*\,ds \\ &= \left(T^*(h) - I\right) \int_0^t T^*(s)x^*\,ds.\end{aligned}$$

So t^{-1} times this expression converges as $t \downarrow 0$ in the weak* topology with limit $T^*(h)x^* - x^*$. □

Proof of Lemma 3.12. Note that $\|x\|' \leq \|x\|$ since the supremum is taken over a smaller set. There exists $x^* \in X^*$ such that $\|x^*\| = 1$ and $\langle x^*, x\rangle = \|x\|$. Define

$$x_h^* = \frac{1}{h} \int_0^h T^*(s)x^*\,ds \in X^\odot;$$

then $x_h^* \xrightarrow{*} x^*$ as $h \downarrow 0$ and

$$\|x_h^*\| \leq M \frac{e^{\omega h} - 1}{\omega h}$$

by Lemma 3.15. Hence

$$\|x\|' \geq \limsup_{h\downarrow 0} \frac{|\langle x_h^*, x\rangle|}{\|x_h^*\|} = \frac{|\langle x^*, x\rangle|}{M} = \frac{\|x\|}{M}.$$

□

Corollary 3.16. *$j : X \to X^{\odot *}$ defined by $\langle jx, x^\odot\rangle = \langle x^\odot, x\rangle$ is a continuous embedding (not necessarily onto). When we equip X with the prime norm, j is norm-preserving.*

Proposition 3.17.

(i) $jT(t) = T^{\odot *}(t)j$;

(ii) $j(X) \subset X^{\odot\odot}$.

The proof consists of trivial formula manipulation. Finally, we show that there is no need to introduce a prime norm on $X^\odot$.

Proposition 3.18.

$$\|x^\odot\| = \sup\left\{ \left|\langle x^\odot, x\rangle\right| \mid x \in X,\ \|x\|' \leq 1 \right\}.$$

Proof.

$$\|x^{\odot}\| = \sup\{|\langle x^{\odot}, x\rangle| \mid \|x\| \le 1\} \le \sup\{|\langle x^{\odot}, x\rangle| \mid \|x\|' \le 1\}$$

since $\|x\|' \le \|x\|$; hence $\|x\| \le 1$ guarantees that $\|x\|' \le 1$. On the other hand, the definition of the prime norm is such that

$$\|x\|' \ge \frac{|\langle x^{\odot}, x\rangle|}{\|x^{\odot}\|}; \qquad \text{hence} \quad \|x^{\odot}\| \ge \frac{|\langle x^{\odot}, x\rangle|}{\|x\|'}$$

from which it follows that $\|x^{\odot}\| \ge \sup\{|\langle x^{\odot}, x\rangle| \mid \|x\|' \le 1\}$. □

Theorem 3.19. *Define the Favard class*

$$\text{Fav}\,(T^*) = \{x^* : \limsup_{h\downarrow 0} \frac{1}{h}\|T^*(h)x^* - x^*\| < \infty\}.$$

Then $\text{Fav}\,(T^*) = \mathcal{D}(A^*)$.

Proof. 1. ($\supseteq$) If $x \in \mathcal{D}(A^*)$, then (Lemma 3.6)

$$\limsup_{h\downarrow 0} \frac{1}{h}\|T^*(h)x^* - x^*\| < \infty.$$

2. ($\subseteq$) Take $x \in \mathcal{D}(A)$, arbitrary, then

$$\begin{aligned} |\langle x^*, Ax\rangle| &= |\lim_{h\downarrow 0}\langle T^*(h)x^* - x^*, x\rangle| \\ &\le \Big(\limsup_{h\downarrow 0} \frac{1}{h}\|T^*(h)x^* - x^*\|\Big)\,\|x\|. \end{aligned}$$

Hence $x \mapsto \langle x^*, Ax\rangle$ extends to a bounded linear map from X into $\mathbb{C}$ and consequently there exists $y^* \in X^*$ such that

$$\langle x^*, Ax\rangle = \langle y^*, x\rangle, \qquad \text{for all} \quad x \in \mathcal{D}(A),$$

from which we conclude that $x^* \in \mathcal{D}(A^*)$. □

Similarly one proves that $\text{Fav}\,(T) = j^{-1}(\mathcal{D}(A^{\odot *}))$ (see [23]).

Definition 3.20. X is called $\odot$-*reflexive* with respect to $T(t)$ when $j(X) = X^{\odot\odot}$.

Theorem 3.21. (*Phillips - de Pagter* [124, 231, 204].) *X is $\odot$-reflexive with respect to $T(t)$ if and only if $(zI - A)^{-1}$ is weakly compact.*

For a demonstration that the $\odot *$-framework is very well suited for treating perturbation problems, we refer to Chapter III. In this context, the embedding operator j is often omitted in the notation. For instance, when the bounded operator $B : X \to X^{\odot *}$ is considered as a perturbation of the semigroup generator $A_0^{\odot *}$, we should, assuming $\odot$-reflexivity, define $A^{\odot *} = A_0^{\odot *} + Bj^{-1}$ with domain $\mathcal{D}(A^{\odot *}) = \mathcal{D}(A_0^{\odot *})$. Frequently, we will identify X with $X^{\odot\odot}$ and omit the symbol j^{-1} from such formulas.

II.4 Spectral theory and asymptotic behaviour

As we have stated before, a linear operator $L : \mathcal{D}(L) \to X$ with domain $\mathcal{D}(L)$ in X is called *closed* if and only if its *graph*

$$\{(x, Lx) \mid x \in \mathcal{D}(L)\}$$

is a closed subset of $X \times X$ equipped with the norm $\|(x,y)\| = \|x\| + \|y\|$; or, equivalently, if $\mathcal{D}(L)$ endowed with the graph norm

$$\|x\|_L = \|x\| + \|Lx\|$$

is a Banach space, which we then denote by X_L. Bounded operators defined on the whole space are closed and so are the generators of a $\mathcal{C}_0$-semigroup.

Exercise 4.1. Let $L : \mathcal{D}(L) \to X$ be a closed operator. Show that L^{-1} is, if it exists, a closed operator.

Exercise 4.2. Let $L : \mathcal{D}(L) \to X$ be a linear operator. Show that L is closed if the set

$$\{z \in \mathbb{C} \mid (zI - L)^{-1} \quad \text{exists and is bounded on } X\} \neq \emptyset.$$

A complex number z belongs to the *resolvent set* $\rho(L)$ of a linear operator L on a complex Banach space X if and only if the *resolvent* $(zI - L)^{-1}$ exists and is bounded, i.e.,

(i) $zI - L$ is one-to-one (injective);

(ii) $\overline{\mathcal{R}(zI - L)} = X$;

(iii) $(zI - L)^{-1}$ is bounded.

Note that for closed operators, (iii) is superfluous, since it follows automatically from the other assumptions by the closed graph theorem. The *spectrum* $\sigma(L)$ is by definition the complement of $\rho(L)$ in $\mathbb{C}$. By $\mathcal{L}(X)$ we denote the Banach algebra of bounded linear operators on X endowed with the operator norm.

Exercise 4.3. Let $L : \mathcal{D}(L) \to X$ be a closed operator. Prove that, for any $z, \lambda \in \rho(L)$,

(i) $(zI - L)^{-1}(\lambda I - L)^{-1} = (\lambda I - L)^{-1}(zI - L)^{-1}$,

(ii) $(zI - L)^{-1} - (\lambda I - L)^{-1} = (\lambda - z)(zI - L)^{-1}(\lambda I - L)^{-1}$.

The equation in (ii) is called the *resolvent equation.*

Exercise 4.4. Let $L \in \mathcal{L}(X)$ be such that $\|L\| < 1$. Prove that

$$(I - L)^{-1} = \sum_{n=0}^{\infty} L^n \in \mathcal{L}(X)$$

and $\|(I - L)^{-1}\| \leq (1 - \|L\|)^{-1}$.

Exercise 4.5. Let L be a closed operator and let $\lambda \in \rho(L)$. Prove that $z \in \rho(L)$ for $z \in \mathbb{C}$ which satisfy

$$|z - \lambda| \|(\lambda I - L)^{-1}\| < 1$$

and that then

$$(zI - L)^{-1} = \sum_{n=0}^{\infty} (-1)^n (z - \lambda)^n (\lambda I - L)^{-(n+1)}.$$

Conclude that the resolvent set is open and that $z \mapsto (zI - L)^{-1}$ is holomorphic from $\rho(L)$ into $\mathcal{L}(X)$ (see Appendix III.1).
Hint: Use both of the foregoing exercises.

For a bounded operator $L : X \to X$, we define the *spectral radius* $r_\sigma(L)$ by

$$r_\sigma(L) = \sup\{|z| \mid z \in \sigma(L)\}.$$

Proposition 4.6.

$$r_\sigma(L) = \inf_{n \geq 1} \|L^n\|^{\frac{1}{n}} = \lim_{n \to \infty} \|L^n\|^{\frac{1}{n}}.$$

The proof of this proposition combines two ideas. The first is that the series expansion

$$(zI - L)^{-1} = \sum_{n=1}^{\infty} z^{-n} L^{n-1}$$

which holds for $|z| > \|L\|$ by Exercise 4.4 can actually be extended to $|z| > r_\sigma(L)$ since $z \mapsto (zI - L)^{-1}$ is holomorphic for $|z| > r_\sigma(L)$ (Exercise 4.5). The second is that $\|L^n\|^{\frac{1}{n}}$ converges for $n \to \infty$. The next exercise makes this second idea precise, though in a continuous "time" setting.

Exercise 4.7. A function $\alpha : \mathbb{R}_+ \to \mathbb{R} \cup \{-\infty\}$ is called *subadditive* if

$$\alpha(t + s) \leq \alpha(t) + \alpha(s).$$

Let α be subadditive and bounded from above on $[0, 1]$. Show that

$$-\infty \leq \inf_{t>0} \frac{1}{t} \alpha(t) = \lim_{t \to \infty} \frac{1}{t} \alpha(t) < +\infty.$$

Exercise 4.8. Let $L : X \to X$ be a bounded operator and let p be a polynomial. Prove the spectral mapping theorem for polynomials:

$$p(\sigma(L)) = \sigma(p(L)).$$

Exercise 4.9. Let $L : X \to X$ be a bounded operator such that L^n is compact for some $n \in \mathbb{N}$. Describe the spectrum of L.

In Definition 1.10, the *growth bound* ω_0 of a semigroup $T(t)$ was defined by

$$\omega_0 = \inf\{\omega \mid \exists M \geq 1 \quad \text{such that } \|T(t)\| \leq Me^{\omega t}\}.$$

So, clearly, ω_0 is related to the large time behaviour of $T(t)$ and, conversely, if we know ω_0, we can deduce conclusions about the large time behaviour. The aim of this part of the appendix is to investigate how much information we can obtain about ω_0 and in which manner. In particular, we want to know what we can learn about ω_0 from the spectral analysis of the infinitesimal generator A.

The first proposition gives some more information about the relation between $T(t)$ and ω_0.

Proposition 4.10.
(i) $\omega_0 = \inf_{t>0} \frac{1}{t} \log \|T(t)\| = \lim_{t\to\infty} \frac{1}{t} \log \|T(t)\|$;
(ii) $r_\sigma\big(T(t)\big) = e^{\omega_0 t}$.

Exercise 4.11. Prove the last proposition.
Hint: Use Exercise 4.7 with $\alpha(t) = \log \|T(t)\|$.

Definition 4.12. The *spectral bound* $s(A)$ is defined by

$$s(A) = \sup\{\operatorname{Re}\lambda \mid \lambda \in \sigma(A)\}.$$

Exercise 4.13. Prove the inequality $\omega_0 \geq s(A)$.

Exercise 4.14. Show that the obvious conjecture $\omega_0 = s(A)$ is false.
Hint: Follow [95] or [202, pp. 61–62] in considering $X = C_0(\mathbb{R}_+) \cap L^1(\mathbb{R}_+, e^a da)$ with norm

$$\|f\| = \|f\|_\infty + \|f\|_1 = \sup\{|f(a)| \mid a \in \mathbb{R}_+\} + \int_0^\infty |f(a)|e^a \, da$$

and

$$(T(t)f)(a) = f(a+t).$$

Deduce that $\omega_0 = 0$ and $s(A) = -1$.

Note that this exercise also shows that the spectral mapping theorem $\sigma(T(t)) = e^{t\sigma(A)}$ cannot be true.

In order to proceed, we need another auxiliary result in the spirit of Proposition 1.4 (i). The proof is left as an exercise to the reader.

Lemma 4.15.

(i) $(e^{zt}I - T(t))x = (zI - A) \int_0^t e^{z(t-s)}T(s)x\,ds;$

(ii) *for* $x \in \mathcal{D}(A)$, $(e^{zt}I - T(t))x = \int_0^t e^{z(t-s)}T(s)(zI - A)x\,ds.$

Exercise 4.16. Use Lemma 4.15 to prove the inclusion

$$e^{t\sigma(A)} \subset \sigma(T(t)), \quad \text{for } t \geq 0.$$

Hint: Define $Q = (e^{zt}I - T(t))^{-1}$ and $B = \int_0^t e^{z(t-s)}T(s)\,ds$ and show that (i) $BQ = QB$, (ii) $(zI - A)BQ = I$ and (iii) $QB(zI - A)x = x$ for $x \in \mathcal{D}(A)$.

The *point spectrum* $\sigma_p(L)$ is the set of those $\lambda \in \mathbb{C}$ for which $\lambda I - L$ is not one-to-one, i.e., $L\varphi = \lambda\varphi$ for some $\varphi \neq 0$. One then calls λ an *eigenvalue* and φ an *eigenvector* corresponding to λ. We are now ready to state a major result, which says that for the point spectrum, we do have a spectral mapping principle.

Theorem 4.17. $e^{t\sigma_p(A)} = \sigma_p(T(t)) \setminus \{0\}$.

Exercise 4.18. Use Lemma 4.15 (ii) to prove that $e^{t\sigma_p(A)} \subset \sigma_p(T(t))$.

Exercise 4.19. Fix $t > 0$. Let $\lambda \in \mathbb{C}$ and $x \neq 0$ be such that $(e^{\lambda t}I - T(t))x = 0$. Let f_k denote the Fourier coefficients of the t-periodic function $s \mapsto e^{-\lambda s}T(s)x$, i.e.,

$$f_k = \frac{1}{t}\int_0^t e^{-\frac{2\pi ik}{t}s}(e^{-\lambda s}T(s)x)\,ds.$$

Note that at least one of the f_k is different from zero (otherwise $x = 0$). Prove that $\lambda_k = \lambda + \frac{2\pi ik}{t}$ is an eigenvalue of A with corresponding eigenvector f_k when $f_k \neq 0$.

The definitions and results of the rest of this appendix will not be used in the book. Since they are, however, quite important in other applications, we add them for completeness.

Definition 4.20. The (Kuratowski) *measure of noncompactness* $\alpha(V)$ of a subset V of a Banach space X is defined by

$$\alpha(V) = \inf\{d > 0 \mid \text{there exists a finite number of sets } V_1, \dots, V_n \text{ with}$$
$$\operatorname{diam} V_i \leq d \text{ such that } V = \bigcup_{j=1}^n V_j\}.$$

For a bounded linear operator L, define

$$|L|_\alpha = \inf\{k > 0 \mid \alpha(L(V)) \leq k\,\alpha(V) \quad \text{for all bounded sets } V\}.$$

To put things into perspective we also note that one can define the ball-measure of noncompactness $\gamma(V)$ by

$$\gamma(V) = \inf\{d > 0 \mid V \text{ can be covered by a finite number of balls having radii } \leq d\}$$

and

$$|L|_\gamma = \inf\{k > 0 \mid \gamma(L(V)) \leq k\,\gamma(V) \quad \text{for all bounded sets } V\}.$$

Proposition 4.21. $\gamma(V) \leq \alpha(V) \leq 2\gamma(V)$.

Definition 4.22. For a closed, densely defined operator L, we define the (Browder) *essential spectrum* by $\lambda \in \sigma_{\text{ess}}(L)$ if and only if at least one of the following three conditions holds:

(i) λ is a limit point of $\sigma(L)$;

(ii) $\mathcal{R}(\lambda I - L)$ is not closed;

(iii) $\bigcup_{k \geq 0} \mathcal{N}((\lambda I - L)^k)$ is infinite dimensional.

The *essential spectral radius* $r_{\text{ess}}(L)$ is defined by

$$r_{\text{ess}}(L) = \sup\{|z| \mid z \in \sigma_{\text{ess}}(L)\}.$$

Theorem 4.23. *Let $\lambda \in \sigma(L) \setminus \sigma_{\text{ess}}(L)$; then λ is a pole of $(zI - L)^{-1}$ and the residue is an operator of finite rank. So, in particular, λ is an eigenvalue of finite algebraic multiplicity.*

Theorem 4.24. [(i) *is due to R. Nussbaum.*] *Let L be bounded.*

(i)

$$r_{\text{ess}}(L) = \lim_{n\to\infty} \left(|L^n|_\alpha\right)^{\frac{1}{n}} = \lim_{n\to\infty} \left(|L^n|_\gamma\right)^{\frac{1}{n}}.$$

(ii) *For all z with $|z| > r_{\text{ess}}(L)$, $zI - L$ is a Fredholm operator with index zero.*

(iii)

$$r_{\text{ess}}(L) = \lim_{n\to\infty} \left(\|L^n\|_{\text{ess}}\right)^{\frac{1}{n}},$$

where $\|L\|_{\text{ess}} = \operatorname{dist}(L, \mathcal{K}) = \inf\{\|L - K\| \mid K \in \mathcal{K}\}$ with $\mathcal{K}$ the set of all compact operators on X.

Definition 4.25. For a C_0-semigroup, we now define the *essential growth bound* ω_{ess} by

$$\omega_{\text{ess}} = \lim_{t\to\infty} \frac{1}{t} \log \bigl|T(t)\bigr|_{\alpha}.$$

Proposition 4.26. *For $t \geq 0$, $r_{\text{ess}}\bigl(T(t)\bigr) = e^{\omega_{\text{ess}} t}$.*

[The proof of Proposition 4.10 (ii) carries over with appropriate modification.]

We are now ready to state the main result of this part of the section.

Theorem 4.27. $\omega_0 = \max\bigl\{s(A), \omega_{\text{ess}}\bigr\}$.

Proof. Since $|L|_\alpha \leq \|L\|$, we know that $\omega_{\text{ess}} \leq \omega_0$. If $w_{\text{ess}} < \omega_0$, there exists an eigenvalue λ of $T(t)$ such that $|\lambda| = r\bigl(T(t)\bigr)$. So, by Theorem 4.17, there must exist $\mu \in \sigma_p(A)$ such that $e^{\mu t} = \lambda$ and therefore $|e^{\mu t}| = e^{\operatorname{Re}\mu t} = |\lambda| = r\bigl(T(t)\bigr) = e^{\omega_0 t}$. We conclude that in this case $\omega_0 = \sup\bigl\{\operatorname{Re}\mu \mid \mu \text{ is an eigenvalue of } A\bigr\}$ which is a stronger conclusion than stated in the theorem. □

Exercise 4.28. Assume that, for some semigroup $T(t)$, $\lambda_d \in \mathbb{R}$ exists such that, for some $\alpha \in \mathbb{R}$,

(i) $s(A) = \lambda_d$,

(ii) $\lambda_d + i\alpha k \in \sigma_p(A)$ for all $k \in \mathbb{Z}$.

Prove that $\omega_{\text{ess}} \geq \lambda_d$.

Hint: Choose t such that $\alpha t \pi^{-1}$ is irrational and use that the spectrum is a closed subset of $\mathbb{C}$.

Appendix III

The operational calculus

Let $L : \mathcal{D}(L) \to X$ be a closed operator. In the previous appendix we have seen that $\rho(L)$ is open and $z \mapsto (zI - L)^{-1}$ is holomorphic from $\rho(L)$ into $\mathcal{L}(X)$. This enables us to use contour integration to obtain spectral decompositions for isolated points in the spectrum of L.

We start with some information about vector-valued functions.

III.1 Vector-valued functions

In this section we shall be concerned with functions F defined on some abstract set Ω with values in a complex Banach space X (such functions are called "vector valued", to distinguish them from the subclass of $\mathbb{R}$ or $\mathbb{C}$-valued functions). So, to each point $a \in \Omega$, there corresponds a vector $F(a) \in X$. In the case where $X = \mathcal{L}(Y, Z)$, the space of linear bounded operators from a Banach space Y into a Banach space Z, we sometimes call F an operator-valued function. The general reference for this section is the book by Hille and Phillips [124].

Let $\Omega \subseteq \mathbb{C}$ and $f : \Omega \to X$. The function F is *weakly continuous* at $z = z_0$ if

$$\lim_{z \longrightarrow z_0} \left| \langle x^*, F(z) - F(z_0) \rangle \right| = 0$$

for each $x^* \in X^*$, and *norm-continuous* if

$$\lim_{z \longrightarrow z_0} \| F(z) - F(z_0) \| = 0.$$

The function F is called weakly (in norm) differentiable at $z = z_0$ if there is an element $F'(z_0) \in X$ such that for $\delta \in \mathbb{C}$ with $z_0 + \delta \in \Omega$, the difference quotient tends weakly (norm) to $F'(z_0)$ as $|\delta| \to 0$. Note that if F is weakly differentiable at z_0, then

$$z \mapsto \langle x^*, F(z) \rangle$$

is a differentiable scalar-valued function. The definition of a holomorphic vector-valued function now becomes

Definition 1.1. Let $\Omega \subseteq \mathbb{C}$ be open. If $F : \Omega \to X$, then F is said to be holomorphic in Ω if

$$z \mapsto \langle x^*, F(z) \rangle$$

is a holomorphic function for every choice of $x^* \in X^*$.

From the Uniform Boundedness Principle, it follows that holomorphic vector-valued functions are norm continuous and norm differentiable.

Let $I = [0,1] \subseteq \mathbb{R}$ and let $\alpha : I \to \mathbb{C}$ be a complex function which is continuous and of bounded variation on I. If $\Gamma = \{\alpha(t) \mid t \in I\}$, then Γ is called a *rectifiable* curve in the complex plane. If $\alpha(t_1) \neq \alpha(t_2)$ unless $t_1 = t_2$, or $\{t_1, t_2\} = \{0, 1\}$, Γ is called a *simple Jordan* curve. If Γ is a simple Jordan curve such that $\alpha(0) = \alpha(1)$, then Γ is said to be closed. If Γ is a closed rectifiable curve in $\mathbb{C}$ and $a \notin \Gamma$, then the *winding number* of Γ about a is defined by

$$n(\Gamma; a) = \frac{1}{2\pi i} \int_\Gamma \frac{dz}{z - a}.$$

The number $n(\Gamma; a)$ is an integer and is constant on each component of $\mathbb{C} \setminus \Gamma$ and vanishes on the unbounded component of $\mathbb{C} \setminus \Gamma$. Let $\Omega \subseteq \mathbb{C}$ be open. If Γ is a rectifiable curve in Ω and $F : \Omega \to X$ is a continuous vector-valued function, then

$$\int_\Gamma F(z)\, dz = \int_0^1 F(\alpha(t))\, d\alpha(t)$$

is well defined.

Theorem 1.2. *Let X be a complex Banach space and Ω an open subset of $\mathbb{C}$. If $F : \Omega \to X$ is holomorphic and $\Gamma = \{\gamma_1, \ldots, \gamma_m\}$ is a finite collection of closed rectifiable curves in Ω such that $\sum_{j=1}^m n(\gamma_j; a) = 0$ for all $a \in \mathbb{C} \setminus \Omega$, then*

$$\int_\Gamma F(z)\, dz = \sum_{j=1}^m \int_{\gamma_j} F(z)\, dz = 0.$$

Proof. For any $x^* \in X^*$

$$\begin{aligned} \langle x^*, \int_\Gamma F(z)\, dz \rangle &= \int_\Gamma \langle x^*, F(z) \rangle\, dz \\ &= \sum_{j=1}^m \int_{\gamma_j} \langle x^*, F(z) \rangle\, dz = 0 \end{aligned}$$

by the scalar version of the Cauchy theorem. □

The fact that linear operators commute with the integral turns out to be a useful tool to extend classical function theory to vector-valued functions. But one has to be careful. The Plancherel theorem, which states that the Laplace transform of an L^2-function is an L^2-function along the imaginary axis, is known not to be true in general Banach spaces (the Hilbert space structure is essential).

III.2 Bounded operators

First, we consider the case that $L \in \mathcal{L}(X)$. In this case we know that $(zI - L)^{-1}$ has a power series expansion when $|z| > r_\sigma(L)$. A closed rectifiable curve Γ is positively oriented if for every $a \notin \Gamma$, the winding number $n(\Gamma; a)$ is either 0 or 1 (see Section 1). In this case the inside of Γ is defined by $\{z \in \mathbb{C} \setminus \Gamma \mid n(\Gamma; z) = 1\}$ and the outside of Γ is defined by $\{z \in \mathbb{C} \setminus \Gamma \mid n(\Gamma; z) = 0\}$. If $\Gamma = \{\gamma_1, \ldots, \gamma_m\}$ is a collection of closed rectifiable curves, then Γ is positively oriented if $\gamma_i \cap \gamma_j = \emptyset$ for $i \neq j$ and if $n(\Gamma; z) = \sum_{j=1}^m n(\gamma_j; z)$ is either 0 or 1.

Let $L : X \to X$ be a bounded operator. The spectrum of L is a compact subset of $\mathbb{C}$. If Ω is an open neighbourhood of $\sigma(L)$ (not necessarily connected), then there exists a positively oriented system of infinitely differentiable curves $\Gamma = \{\gamma_1, \ldots, \gamma_m\}$ in $\Omega \setminus \sigma(L)$ such that $\sigma(L)$ is contained inside and $\mathbb{C} \setminus \Omega$ is outside Γ. (See Conway [52].) Let $f : \Omega \to \mathbb{C}$ be analytic and define

$$f(L) = \frac{1}{2\pi i} \int_\Gamma f(z)(zI - L)^{-1}\, dz. \tag{2.1}$$

To show that $f(L)$ is well defined, i.e., independent of the choice of Γ, let $\Gamma' = \{\gamma_1', \ldots, \gamma_k'\}$ denote another positively oriented contour such that $\sigma(L)$ is inside Γ'. For $j = 1, \ldots, k$, define $\gamma_{m+j}(t) = \gamma_j'(1-t)$, $0 \le t \le 1$. Set $\Gamma^* = \{\gamma_j \mid 1 \le j \le m + k\}$. Then

$$\begin{aligned} n(\Gamma^*; z) &= \sum_{j=1}^{m+k} n(\gamma_j; z) \\ &= n(\Gamma; z) - n(\Gamma'; z) \\ &= \begin{cases} 0 - 0 & \text{for } z \in \mathbb{C} \setminus \Omega, \\ 1 - 1 & \text{for } z \in \sigma(L). \end{cases} \end{aligned} \tag{2.2}$$

So, Γ^* is a system of closed curves in $\Omega \setminus \sigma(L) = \Omega'$. Since $z \mapsto f(z)(zI - L)^{-1}$ is analytic on Ω', equation (2.2) and Cauchy's theorem (see Theorem 1.2) imply

$$\begin{aligned} 0 &= \int_{\Gamma^*} f(z)(zI - L)^{-1}\, dz \\ &= \int_\Gamma f(z)(zI - L)^{-1}\, dz - \int_{\Gamma'} f(z)(zI - L)^{-1}\, dz. \end{aligned}$$

This shows that $f(L)$ is well defined. Given a linear operator L, Ω always denotes a neighbourhood of $\sigma(L)$ and Γ a positively oriented system of infinitely differentiable curves chosen as above.

Exercise 2.1. (Operational calculus). Prove that if $f, g : \Omega \to \mathbb{C}$ are analytic functions, then

(i) $(f+g)(L) = f(L) + g(L)$;

(ii) $(\alpha f)(L) = \alpha f(L), \quad \alpha \in \mathbb{C}$;

(iii) $(fg)(L) = f(L)g(L)$.

(For the last identity one has to use the resolvent equation.)

A subset of $\sigma(L)$ which is both open and closed *in* $\sigma(L)$ is called a *spectral set.* If σ is a spectral set, then σ and $\sigma(L)\setminus\sigma$ are compact sets such that $\sigma \cap (\sigma(L) \setminus \sigma) = \emptyset$. Therefore, there exist $\Omega_1 \supset \sigma$, $\Omega_2 \supset (\sigma(L) \setminus \sigma)$ such that $\Omega_1 \cap \Omega_2 = \emptyset$. Set $\Omega = \Omega_1 \cup \Omega_2$ and define an analytic function $g : \Omega \to \mathbb{C}$ by

$$g(z) = \begin{cases} 1 & \text{for } z \in \Omega_1, \\ 0 & \text{for } z \in \Omega_2. \end{cases} \tag{2.3}$$

According to (2.1), the operator $g(L)$ is defined and we will denote this operator by P_σ, i.e.,

$$P_\sigma = g(L).$$

Since $g(z)g(z) = g(z)$ for $z \in \Omega$, the operational calculus shows immediately that

$$P_\sigma P_\sigma = P_\sigma,$$

so that P_σ is a bounded projection. The definition of P_σ does not depend on the particular choice of Ω_1 and Ω_2 (see Dunford-Schwartz [80, VII.3.18]). The projection P_σ is called *the spectral projection associated with* σ.

Exercise 2.2. Let τ and σ be spectral sets of a bounded operator $L : X \to X$. Verify the following properties:

(i) $P_\sigma = 0$ if $\sigma = \emptyset$;

(ii) $P_\sigma = I$ if $\sigma = \sigma(L)$;

(iii) $P_{\sigma\cap\tau} = P_\sigma P_\tau = P_\tau P_\sigma$;

(iv) $P_{\sigma\cup\tau} = P_\sigma + P_\tau - P_\sigma P_\tau$.

III.3 Unbounded operators

There exists an extension of the operational calculus to closed unbounded operators, which we shall discuss next. Let $L : \mathcal{D}(L) \to X$ be a closed unbounded operator. The spectrum of L need no longer be compact. We will assume that $\rho(L)$ is not empty, so that $\sigma(L)$ is not the whole plane. Fix a $\lambda \in \rho(L)$; then $(\lambda I - L)^{-1}$ is a bounded operator and one defines an operator calculus for L using $(\lambda I - L)^{-1}$. Let $S = \mathbb{C} \cup \{\infty\}$ denote the one-point compactification of $\mathbb{C}$ and define a homeomorphism $\Phi : S \to S$ by

$$w = \Phi(z) = (\lambda - z)^{-1}, \qquad \Phi(\infty) = 0, \quad \Phi(\lambda) = \infty.$$

Exercise 3.1. Let $L : \mathcal{D}(L) \to X$ be a closed unbounded operator. Show that

$$\Phi(\sigma(L) \cup \{\infty\}) = \sigma((\lambda I - L)^{-1})$$

and that the relation

$$\varphi(w) = f(\Phi^{-1}(w))$$

determines a one-to-one correspondence between the class of functions f analytic in a neighbourhood of $\sigma(L)$ and $\{\infty\}$ and the class of functions φ analytic in a neighbourhood of $\sigma((\lambda I - L)^{-1})$.
Hint: Note that $0 \in \sigma((\lambda I - L)^{-1})$ since the inverse is unbounded.

Given an analytic function f on a neighbourhood of $\sigma(L)$ and $\{\infty\}$, one defines

$$f(L) = \varphi((\lambda I - L)^{-1}), \tag{3.1}$$

where $\varphi(w) = f(\Phi^{-1}(w))$. Let Ω be an open set containing $\sigma(L)$; then $\Omega' = \Phi(\Omega)$ is an open set containing the compact set $\sigma((\lambda I - L)^{-1})$. As above, there exists a positively oriented system of infinitely differentiable curves $\Gamma = \{\gamma_1, \ldots, \gamma_m\}$ in $\Omega \setminus \sigma(L)$ such that $\sigma(L)$ is contained inside Γ and $\mathbb{C} \setminus \Omega$ is outside Γ. Since $w = (\lambda - z)^{-1}$, we have

$$z = \lambda - \frac{1}{w} \quad \text{and} \quad dz = \frac{dw}{w^2}.$$

Therefore, if $\Gamma' = \Phi(\Gamma)$ and $B = (\lambda I - L)^{-1}$, then

$$\begin{aligned}
\frac{1}{2\pi i}\int_\Gamma f(z)(zI - L)^{-1}\,dz &= \frac{1}{2\pi i}\int_{\Gamma'} \varphi(w)\Big(\lambda I - L - \frac{1}{w}I\Big)^{-1}\frac{dw}{w^2}\\
&= \frac{1}{2\pi i}\int_{\Gamma'} \varphi(w)B(wI - B)^{-1}\frac{dw}{w}\\
&= \frac{1}{2\pi i}\int_{\Gamma'} \varphi(w)\big[-w^{-1}I + (wI - B)^{-1}\big]\,dw\\
&= \varphi(B) - \varphi(0)I\\
&= f(L) - f(\infty)I.
\end{aligned}$$

So we find the following formula for $f(L)$:

$$f(L) = f(\infty)I + \frac{1}{2\pi i}\int_\Gamma f(z)(zI - L)^{-1}\,dz, \tag{3.2}$$

where Γ is as defined above and $f(\infty)$ exists since $f(z)$ is bounded as $z \to \infty$. We have the following result.

Exercise 3.2. The identities (i)–(iii) in Exercise 2.1 remain true for (3.2). Show this.

Remark. For a polynomial p, the operator $p(L)$ cannot be defined using (3.2). It is possible, however, to extend the operational calculus to polynomials as well. (See [80], [272].)

Let $L : \mathcal{D}(L) \to X$ be a closed operator with nonempty resolvent set. In addition to the concept of spectrum, we need the concept of the *extended spectrum* $\sigma_e(L)$ of L. This is defined to be the same as $\sigma(L)$ if $L \in \mathcal{L}(X)$, but consists of $\sigma(L)$ and the point ∞ in the one-point compactification of the complex plane if L is not in $\mathcal{L}(X)$. Observe that $\sigma_e(L)$ is always compact and nonempty. A subset σ of $\sigma_e(L)$ is called a *spectral set* of L if it is both open and closed *in* $\sigma_e(L)$. If σ is a spectral set, then σ and $\sigma_e(L) \setminus \sigma$ are disjunct compact sets in S and there exist open sets $\Omega_1 \supset \sigma$ and $\Omega_2 \supset \big(\sigma_e(L) \setminus \sigma\big)$ such that $\Omega_1 \cap \Omega_2 = \emptyset$. Set $\Omega = \Omega_1 \cup \Omega_2$ and define $g : \Omega \to \mathbb{C}$ by

$$g(z) = \begin{cases} 1 & \text{for } z \in \Omega_1, \\ 0 & \text{for } z \in \Omega_2; \end{cases}$$

then g is analytic on this neighbourhood Ω of $\sigma(L)$. According to (3.1), the operator $g(L)$ is defined and we will denote this operator by

$$P_\sigma = g(L).$$

Using Exercises 3.1 and 3.2, one shows, as in the bounded case, that P_σ is a bounded projection and that the definition of P_σ does not depend on the particular choice of Ω_1 and Ω_2. As in the bounded case, the projection P_σ is called *the spectral projection associated with* σ.

If $\sigma = \{\lambda\}$ is an isolated point of $\sigma(L)$, then σ is a spectral set and the resolvent can be expanded in a Laurent series around λ

$$(zI - L)^{-1} = \sum_{l=0}^{\infty} (z-\lambda)^l\, C_l + \sum_{l=1}^{\infty} (z-\lambda)^{-l}\, D_l \tag{3.3}$$

for $0 < |z-\lambda| < \delta$ and δ sufficiently small. (See, for example, [143, III.5,A].) The coefficients C_l and D_l are elements of $\mathcal{L}(X)$ and are given by the standard formulas

$$C_l = \frac{1}{2\pi i} \int_{\Gamma_\lambda} (z-\lambda)^{-l-1} (zI - L)^{-1}\, dz, \tag{3.4}$$

$$D_l = \frac{1}{2\pi i} \int_{\Gamma_\lambda} (z-\lambda)^{l-1} (zI - L)^{-1}\, dz, \tag{3.5}$$

where $\Gamma_\lambda = \{z \mid |z - \lambda| = \eta < \delta\}$. Define the following neighbourhood of $\sigma(L)$:

$$\Omega = B_{\eta/2}(\lambda) \cup \left(\mathbb{C} \setminus \overline{B_\eta(\lambda)}\right),$$

where $B_r(y) = \{x \in \mathbb{C} \mid |x - y| < r\}$. The functions $f_l : \Omega \to \mathbb{C}$ defined by

$$f_l(z) = \begin{cases} 0 & \text{if } |z-\lambda| < \eta/2 \\ (z-\lambda)^{-l-1} & \text{if } |z-\lambda| > \eta \end{cases} \quad \text{for } l \geq 0,$$

$$f_l(z) = \begin{cases} (z-\lambda)^{-l-1} & \text{if } |z-\lambda| < \eta/2 \\ 0 & \text{if } |z-\lambda| > \eta \end{cases} \quad \text{for } l < 0$$

are analytic on Ω and $f_l(\infty) = 0$ for all integers l. Observe that

$$C_l = -f_l(L), \quad l \geq 0,$$
$$D_l = f_{-l}(L), \quad l \geq 1.$$

In particular, D_1, the residue of $(zI - L)^{-1}$ at λ, is the projection P_σ, where σ is the spectral set consisting of the single point λ. In this case, we set $P_\lambda = P_\sigma$ and find the Dunford integral for the spectral projection

$$P_\lambda = \frac{1}{2\pi i} \int_{\Gamma_\lambda} (zI - L)^{-1}\, dz, \tag{3.6}$$

where Γ_λ is a simple closed rectifiable curve enclosing λ, but no other points in the spectrum of L.

Exercise 3.3. Derive the following formulas:

(i) $(L - \lambda I)C_0 = D_1 - I$;

(ii) $(L - \lambda I)^l C_l = C_0, \quad l = 0, 1, \ldots;$

(iii) $D_{l+1} = (L - \lambda I)^l D_1, \quad l = 0, 1, \ldots.$

We recall the following definitions. If λ is an isolated point in the spectrum of L, then the smallest number q_λ such that

$$\mathcal{R}(P_\lambda) = \mathcal{N}((\lambda I - L)^{q_\lambda})$$

is called the *ascent* of $\lambda I - L$. Furthermore, a point λ is called a pole of $z \mapsto (zI - L)^{-1}$ of order q if and only if $q \geq 1$, $D_q \neq 0$ and $D_l = 0$ when $l > q$. From Exercise 3.3(iii) we see that $D_{l+1} = 0$ if $D_l = 0$. Hence λ is a pole of order q if $D_q \neq 0$ and $D_{q+1} = 0$.

We end this section with the following theorem (see, for example, Taylor and Lay [272, Theorem V.10.1]).

Theorem 3.4. *Let $L : \mathcal{D}(L) \to X$ be a closed operator. If λ is a pole of $(zI - L)^{-1}$ of order q, then λ is an eigenvalue of L. The ascent of $\lambda I - L$ is equal to q. The range of the projection $P_\lambda = D_1$ is the nullspace of $(\lambda I - L)^q$,* *and the range of* $I - P_\lambda$ *is the range of* $(\lambda I - L)^q$. *Further, the following decomposition of* X *holds:*

$$X = \mathcal{R}\big((\lambda I - L)^q\big) \oplus \mathcal{N}\big((\lambda I - L)^q\big), \tag{3.7}$$

where both subspaces are closed and L*-invariant.*

In addition to Theorem 3.4, one can formulate sufficient conditions for an isolated $\lambda \in \sigma(L)$ to be a pole of the resolvent: if $\lambda \in \sigma(L)$ and $\lambda I - L$ has finite ascent, then λ is a pole of $(zI - L)^{-1}$. So, in particular, if $L \in \mathcal{L}(X)$ is compact, then each nonzero point of $\sigma(L)$ is a pole of $(zI - L)^{-1}$.

Appendix IV

Smoothness of the substitution operator

Let E and F be Banach spaces and let f be a mapping from E into F. If h is a mapping from $\mathbb{R}$ into E, then we define the mapping $\tilde{f}(h)$ from $\mathbb{R}$ into F by

$$\tilde{f}(h)(s) = f(h(s)).$$

In case f is C^k we define, for $1 \leq l \leq k$, multilinear mappings $\Phi^l(h)$ as follows. If $g_1, \ldots, g_l$ are mappings from $\mathbb{R}$ into E, then $\Phi^l(h)(g_1, \ldots, g_l)$ is the mapping from $\mathbb{R}$ into F defined by

$$\Phi^l(h)(g_1, \ldots, g_l)(s) = D^l f(h(s))(g_1(s), \ldots, g_l(s)),$$

and we define $\Phi^0(h) = \tilde{f}(h)$.

Recall that $BC^\eta(\mathbb{R}, E)$ consists of the continuous functions f such that the norm

$$||f||_\eta = \sup_{s \in \mathbb{R}} e^{-\eta|s|} ||f(s)|| < \infty.$$

The space $C_b^k(\mathbb{R}, E)$ consists of the functions f which are k times differentiable, the function itself and all its k derivatives being bounded and continuous.

Lemma 1.1. *Let f be a mapping from E into F.*

(i) *If f is continuous, then $\tilde{f} : C(\mathbb{R}, E) \to C(\mathbb{R}, F)$ is continuous*

(ii) *If f is globally Lipschitz continuous, then $\tilde{f} : BC(\mathbb{R}, E) \to BC(\mathbb{R}, F)$ is globally Lipschitz continuous with the same Lipschitz constant.*

(iii) *If $f \in C_b^k$, $k \geq 0$ and $f(0) = 0$, then $\tilde{f} : BC_0(\mathbb{R}, E) \to BC_0(\mathbb{R}, F)$ is C^k.*

(iv) *Let η be positive and $k \geq 1$. If $f \in C_b^k$ and $f(0) = 0$, then $\tilde{f} : BC^{-\eta}(\mathbb{R}, E) \to BC^{-\eta}(\mathbb{R}, F)$ is C^k.*

(v) *Let η_1 and η_2 be positive constants such that $k\eta_1 < \eta_2$. If $f \in C_b^k$, then $\tilde{f} : BC^{\eta_1}(\mathbb{R}, E) \to BC^{\eta_2}(\mathbb{R}, F)$ is C^k.*

(vi) *In* (iii), (iv) *and* (v) *and for* $1 \leq l \leq k$, *the identity*

$$D_l \tilde{f} = \Phi^l$$

holds.

Remark. In (i), we cannot replace the Fréchet spaces of continuous functions by the corresponding Banach spaces of *bounded* continuous functions. We have added (i) to illustrate that the regularity of the substitution operator is a little bit subtle, but actually we will not use (i). In (iv), the statement is false (in general) if $k = 0$.

Proof. (i) In $C(\mathbb{R}, E)$ we choose the metric d_E, compatible with the topology of this space, as follows:

$$d_E(g, h) = \sum_{i=1}^{\infty} \frac{2^{-i} p_i(g-h)}{1 + p_i(g-h)}$$

where p_i is the seminorm $p_i(f) = \max_{[-i,i]} ||f(t)||$. Similarly, we define d_F. For given ϵ positive, we need to show that δ positive exists such that $d_E(g, h) \leq \delta$ implies that $d_F(\tilde{f}(g), \tilde{f}(h)) \leq \epsilon$. Choose N such that $\sum_N^\infty 2^{-i} \leq \frac{\epsilon}{2}$. There exists $\delta > 0$ such that $p_N(g-h) \leq 2^{N+1}\delta$ implies that $p_N(\tilde{f}(g) - \tilde{f}(h)) \leq \frac{\epsilon}{2}$. If $d_E(g, h) \leq \delta$ and $\delta < 2^{-(N+1)}$, then $p_N(g-h) \leq 2^{N+1}\delta$. Hence

$$\begin{aligned} d_F(\tilde{f}(g), \tilde{f}(h)) &= \sum_{i=1}^{\infty} \frac{2^{-i} p_i(\tilde{f}(g), \tilde{f}(h))}{1 + p_i(\tilde{f}(g), \tilde{f}(h))} \\ &\leq \sum_{i=1}^{N-1} \frac{2^{-i} p_i(\tilde{f}(g), \tilde{f}(h))}{1 + p_i(\tilde{f}(g), \tilde{f}(h))} + \sum_{i=N}^{\infty} \frac{2^{-i} p_i(\tilde{f}(g), \tilde{f}(h))}{1 + p_i(\tilde{f}(g), \tilde{f}(h))} \\ &\leq \sum_{i=1}^{N-1} 2^{-i} p_N(\tilde{f}(g), \tilde{f}(h)) + \sum_{i=N}^{\infty} 2^{-i} \\ &\leq \frac{\epsilon}{2} + \frac{\epsilon}{2} = \epsilon. \end{aligned}$$

(ii) First, note that $\tilde{f}(h)$ is bounded and continuous and that, consequently, the mapping is well defined. If κ is the Lipschitz constant for f, then

$$\begin{aligned} ||\tilde{f}(h) - \tilde{f}(g)||_{BC(\mathbb{R},F)} &= \sup_{s \in \mathbb{R}} ||\tilde{f}(h)(s) - \tilde{f}(g)(s)|| \\ &= \sup_{s \in \mathbb{R}} ||f(h(s)) - f(g(s))|| \\ &\leq \kappa \sup_{s \in \mathbb{R}} ||h(s) - g(s)|| \\ &= \kappa ||h - g||_{BC(\mathbb{R},E)}. \end{aligned}$$

(iii) $\tilde{f}(h)$ is continuous and

$$\lim_{|s|\to\infty} ||\tilde{f}(h)(s)|| = \lim_{|s|\to\infty} ||f(h(s))|| = 0.$$

So, $\tilde{f}(h) \in BC_0(\mathbb{R}, F)$. Next we show that $\tilde{f}$ is continuous. Let $h \in BC_0(\mathbb{R}, E)$ and choose $\xi > 0$. For $\epsilon \in \mathbb{R}$,

$$\begin{aligned}
&\sup_{||g||=1} ||\tilde{f}(h+\epsilon g) - \tilde{f}(h)||_{BC_0(\mathbb{R},F)} \\
&\qquad = \sup_{||g||=1} \sup_{s\in\mathbb{R}} ||\tilde{f}(h+\epsilon g)(s) - \tilde{f}(h)(s)|| \\
&\qquad = \sup_{||g||=1} \sup_{s\in\mathbb{R}} ||f(h(s)+\epsilon g(s)) - f(h(s))||.
\end{aligned}$$

As $||h(s)|| \to 0$ for $|s| \to \infty$ and $||g|| = 1$, we can find A and ϵ_1 positive such that for $|\epsilon| < \epsilon_1$

$$\sup_{||g||=1} \sup_{|s|\geq A} ||f(h(s)+\epsilon g(s)) - f(h(s))|| \leq \xi.$$

Given the compact interval $[-A, A]$, we can find ϵ_2 positive such that for $||y|| < \epsilon_2$

$$\sup_{[-A,A]} ||f(h(s)+y) - f(h(s))|| \leq \xi.$$

This shows that for $|\epsilon| \leq \min\{\epsilon_1, \epsilon_2\}$

$$\sup_{||g||=1} \sup_{|s|\in\mathbb{R}} ||f(h(s)+\epsilon g(s)) - f(h(s))|| \leq \xi,$$

or, in words, that $\tilde{f}$ is continuous at h.

Let $U = BC_0(\mathbb{R}, E)$, $\Phi^0 = \tilde{f}$ and for $1 \leq l \leq k$, let

$$\Phi^l : BC_0(\mathbb{R}, E) \to \mathcal{L}(\underbrace{U \times \ldots \times U}_{l \text{ times}}, BC_0(\mathbb{R}, F))$$

be as defined above. Similar arguments as before show that all maps Φ^m, $1 \leq m \leq k$, are continuous. It remains to prove that for $0 \leq m < k$, Φ^m is differentiable, and that for $h, g_1, , \ldots, g_{m+1}$ in $BC_0(\mathbb{R}, E)$ we have

$$D\Phi^m(h)(g_{m+1})(g_1, \ldots, g_m)(s) = \Phi^{m+1}(h)(g_1(s), \ldots, g_{m+1}(s)). \tag{1.1}$$

If $m = 0$, $h \in BC_0(\mathbb{R}, E)$ and $\epsilon > 0$ then

$$\begin{aligned}
&\sup_{||g||=1} \frac{1}{\epsilon} ||\tilde{f}(h+\epsilon g) - \tilde{f}(h) - \epsilon\Phi^1(h)(g)|| \\
&\qquad = \sup_{||g||=1} \sup_{s\in\mathbb{R}} \frac{1}{\epsilon} ||f(h(s)+\epsilon g(s)) - f(h(s)) - \epsilon Df(h(s))g(s)|| \\
&\qquad \leq \sup_{||y||\leq\epsilon} \sup_{s\in\mathbb{R}} ||Df(h(s)+y) - Df(h(s))||.
\end{aligned}$$

The last term approaches zero as $\epsilon \to 0$. This proves the claim for $m = 0$. Now consider the case $1 \leq m < k$. For $h \in BC_0(\mathbb{R}, E)$ and $\epsilon > 0$,

$$\begin{aligned}
&\sup_{||g_{m+1}||=1} \frac{1}{\epsilon}||\Phi^m(h + \epsilon g_{m+1}) - \Phi^m(h) - \epsilon\Phi^{m+1}(h)g_{m+1}|| \\
&= \sup_{s\in\mathbb{R}} \sup_{||g_1||=1} \dots \sup_{||g_{m+1}||=1} \frac{1}{\epsilon}||D^m f(h(s) + \epsilon g_{m+1}(s))(g_1(s), \dots g_m(s)) \\
&\quad - D^m f(h(s))(g_1(s), \dots, g_m(s)) - \epsilon D^{m+1} f(h(s))(g_1(s), \dots, g_{m+1}(s))|| \\
&\leq \sup_{s\in\mathbb{R}} \sup_{||y||\leq\epsilon} ||D^{m+1} f(h(s) + y) - D^{m+1} f(h(s))||.
\end{aligned}$$

The continuity of $D^{m+1}f$ at 0 and $\lim_{|s|\to\infty} h(s) = 0$ imply that the last term tends to zero as $\epsilon \to 0$. So the claim is true for $m = l$.

(iv) First, we observe that $s \mapsto \tilde{f}(h)(s)$ is continuous. Since

$$\begin{aligned}
||f(h(s))|| &\leq \int_0^1 ||Df(\tau h(s))h(s)||d\tau \\
&\leq \{\int_0^1 ||Df(0) - Df(\tau h(s))||d\tau + ||Df(0)||\}\,||h(s)||,
\end{aligned}$$

we have the estimate

$$e^{\eta|s|}||f(h(s))|| \leq ||h||_{-\eta}(\int_0^1 ||Df(0) - Df(\tau h(s))||d\tau + ||Df(0)||).$$

As $h(s) \to 0$ for $|s| \to \pm\infty$ and Df is continuous at 0, it follows that $f \circ h \in BC^{-\eta}(\mathbb{R}, F)$.

Continuity of $\tilde{f}$ at $h \in BC^{-\eta}(\mathbb{R}, E)$ follows from the estimate

$$\begin{aligned}
&\sup_{||g||_{-\eta}=1} ||\tilde{f}(h + \epsilon g) - \tilde{f}(h)||_{-\eta} \\
&\quad\leq \sup_{||g||_{-\eta}=1} \sup_{s\in\mathbb{R}} e^{\eta|s|}||f(h(s) + \epsilon g(s)) - f(h(s))|| \\
&\quad\leq \sup_{||g||_{-\eta}=1} \sup_{s\in\mathbb{R}} \sup_{\tau\in(0,1)} e^{\eta|s|}||g(s)||\;||Df(h(s) + \epsilon\tau g(s))|| \\
&\quad\leq \sup_{||y||\leq\epsilon} \sup_{s\in\mathbb{R}} ||Df(h(s) + y)||,
\end{aligned}$$

where the last term tends to 0 as $\epsilon \to 0$, due to the the continuity of Df at 0 and

$$\lim_{|s|\to\infty} ||h(s)|| = 0.$$

Let $U = BC^{-\eta}(\mathbb{R}, E)$, $\Phi^0 = f$ and for $1 \leq m \leq k$, define

$$\Phi^m : BC^{-\eta}(\mathbb{R}, E) \to \mathcal{L}(\underbrace{U \times \dots \times U}_{m \text{ times}}, BC^{-\eta}(\mathbb{R}, F))$$

as above. All maps Φ^m are continuous. We show that for $0 \le m < k$, Φ^m is differentiable and that (1.1) holds in this case. For $m = 0$ this is a consequence of the estimate

$$\sup_{||g||_{-\eta}=1} \frac{1}{\epsilon}||\tilde{f}(h+\epsilon g) - \tilde{f}(h) - \epsilon\Phi^1(h)(g)||_{-\eta}$$

$$= \sup_{||g||_{-\eta}=1} \sup_{s\in\mathbb{R}} e^{\eta|s|}\frac{1}{\epsilon}||f(h(s)+\epsilon g(s)) - f(h(s)) - \epsilon Df(h(s))g(s)||$$

$$\le \sup_{||y||\le\epsilon} \sup_{s\in\mathbb{R}} ||Df(h(s)+y) - Df(h(s))||,$$

for $h \in BC^{-\eta}(\mathbb{R}, E)$ and $\epsilon > 0$. The last term tends to 0 as $\epsilon \to 0$. In case $1 \le m < k$ we have

$$\sup_{||g_{m+1}||_{-\eta}=1} \frac{1}{\epsilon}||\Phi^m(h+\epsilon g_{m+1}) - \Phi^m(h) - \epsilon D\Phi^m(h)(g_{m+1})||_{-\eta}$$

$$= \sup_{||g_1||_{-\eta}=1} \dots \sup_{||g_{m+1}||_{-\eta}=1} \sup_{s\in\mathbb{R}} \frac{1}{\epsilon}e^{\eta|s|}||D^m f(h(s)+\epsilon g_{m+1}(s))$$
$$(g_1(s),\dots,g_m(s)) - D^m f(h(s))(g_1(s),\dots,g_m(s))$$
$$- \epsilon D^{m+1} f(h(s))(g_1(s),\dots g_{m+1}(s))||$$
$$\le \sup_{||y||\le\epsilon} \sup_{s\in\mathbb{R}} ||D^{m+1}f(h(s)+y) - D^{m+1}f(h(s))||,$$

for $h \in BC^{-\eta}(\mathbb{R}, E)$ and $\epsilon > 0$. The last term tends to 0 as $\epsilon \to 0$. This shows that , Φ^m is differentiable and $D\Phi^m = \Phi^{m+1}$ also.

(v) First we prove the continuity of the mapping $h \to \tilde{f}(h)$ from $BC^{\eta_1}(\mathbb{R}, E)$ into $BC^{\eta_2}(\mathbb{R}, F)$. For $h \in BC^{\eta_1}(\mathbb{R}, E)$ and $\epsilon \in \mathbb{R}$

$$\sup_{||g||_{\eta_1}=1} ||\tilde{f}(h+\epsilon g) - \tilde{f}(h)||_{\eta_2}$$

$$= \sup_{||g||_{\eta_1}=1} \sup_{s\in\mathbb{R}} e^{-\eta_2|s|}||f(h(s)+\epsilon g(s)) - f(h(s))||.$$

Let $\xi > 0$ be given. Choose A positive such that

$$2e^{-\eta_2 A} \sup_{x\in E} ||f(x)|| \le \xi. \tag{A.2}$$

Then

$$\sup_{||g||_{\eta_1}=1} \sup_{|s|\ge A} e^{-\eta_2|s|}||f(h(s)+\epsilon g(s)) - f(h(s))|| \le \xi.$$

On the interval $[-A, A]$, $||g(s)||$ is bounded by $||g||_{\eta_1}e^{\eta_1 A}$. We can choose $\hat{\epsilon} = \hat{\epsilon}(\xi)$ such that for $|\epsilon| \le \hat{\epsilon}$

$$\sup_{||y||\le||g||_{\eta_1}e^{\eta_1 A}} \sup_{s\in[-A,A]} ||f(h(s)+\epsilon y) - f(h(s))|| \le \xi.$$

This shows that

$$\sup_{||g||_{\eta_1}=1} \sup_{s\in\mathbb{R}} e^{-\eta_2|s|}||f(h(s)+\epsilon g(s)) - f(h(s))|| \to 0,$$

as $\epsilon \to 0$. Next, we let $U = BC^{\eta_1}(\mathbb{R}, E)$, $\Phi^0 = \tilde{f}$ and we define

$$\Phi^m : BC^{\eta_1}(\mathbb{R}, E) \to \mathcal{L}(\underbrace{U \times \ldots \times U}_{m \text{ times}}, BC^{\eta_2}(\mathbb{R}, F))$$

for $1 \leq m \leq k$ as above. In order to obtain the continuity of Φ^1 at $h \in BC^{\eta_1}(\mathbb{R}, E)$, observe that for every $\epsilon > 0$ we have the estimates

$$\begin{aligned}
&\sup_{||g||_{\eta_1}=1} ||\Phi^1(h+\epsilon g) - \Phi^1(h)|| \\
&\quad = \sup_{||g||_{\eta_1}=1} \sup_{||\hat{g}||_{\eta_1}\leq 1} ||(\Phi^1(h+\epsilon g) - \Phi^1(h))(\hat{g})||_{\eta_2} \\
&\quad = \sup_{||g||_{\eta_1}=1} \sup_{||\hat{g}||_{\eta_1}\leq 1} \sup_{s\in\mathbb{R}} e^{-\eta_2|s|}||Df(h(s)+\epsilon g(s)) - Df(h(s))\hat{g}(s)|| \\
&\quad \leq \sup_{||g||_{\eta_1}=1} \sup_{s\in\mathbb{R}} e^{(\eta_1-\eta_2)|s|}||Df(h(s)+\epsilon g(s)) - Df(h(s))|| \\
&\quad \leq \sup_{s\in\mathbb{R}} \sup_{||y||\leq e^{\eta_1|s|}} e^{(\eta_1-\eta_2)|s|}||Df(h(s)+\epsilon y) - Df(h(s))||,
\end{aligned}$$

and proceed as in the proof that $\Phi^0 = \tilde{f}$ is continuous. Similarly one can show that all maps Φ^m, $1 \leq m \leq k$, are continuous. It remains to prove that for $0 \leq m < k$, Φ^m is differentiable and $D\Phi^m = \Phi^{m+1}$. If $m = 0$, $\epsilon > 0$ and $h \in BC^{\eta_1}(\mathbb{R}, E)$, then

$$\begin{aligned}
&\sup_{||g||_{\eta_1}=1} \frac{1}{\epsilon}||\tilde{f}(h+\epsilon g) - \tilde{f}(h) - \epsilon\Phi^1(h)(g)||_{\eta_2} \\
&\quad = \sup_{||g||_{\eta_1}=1} \sup_{s\in\mathbb{R}} e^{-\eta_2|s|}\frac{1}{\epsilon}||f(h(s)+\epsilon g(s)) - f(h(s)) - \epsilon Df(h(s))g(s)|| \\
&\quad \leq \sup_{s\in\mathbb{R}} \sup_{||y||\leq e^{\eta_1|s|}} e^{(-\eta_2+\eta_1)|s|}||Df(h(s)+\epsilon y) - Df(h(s))||.
\end{aligned}$$

In order to see that the last term tends to 0 as $\epsilon \to 0$, let $\delta > 0$ be given, choose $A > 0$ with

$$2e^{(-\eta_2+\eta_1)A} \sup_{x\in E} ||Df(x)|| < \delta,$$

and find $\epsilon(\delta) > 0$ so that for

$$|s| \leq A, \quad ||y|| \leq e^{\eta_1 A}, \quad 0 < \epsilon < \epsilon(\delta)$$

we have

$$||Df(h(s)+\epsilon y) - Df(h(s))|| < \delta.$$

This yields that $\Phi^0 = \tilde{f}$ is differentiable at h, with

$$D\Phi^0(h) = \Phi^1(h).$$

Finally, in the case $1 \leq m < k$, we show that Φ^m is differentiable and that $D\Phi^m = \Phi^{m+1}$. For $h \in BC^{\eta_1}(\mathbb{R}, E)$ and $\epsilon > 0$ we have

$$\begin{aligned}
&\sup_{||g_{m+1}||_{\eta_1}=1} \frac{1}{\epsilon}||\Phi^m(h + \epsilon g_{m+1}) - \Phi^m(h) - \epsilon D\Phi^m(h) g_{m+1}|| \\
&= \sup_{s\in\mathbb{R}} \sup_{||g_1||_{\eta_1}=1} \cdots \sup_{||g_{m+1}||_{\eta_1}=1} \frac{1}{\epsilon}||D^m f(h(s) + \epsilon g_{m+1}(s))(g_1(s), \ldots, g_m(s)) \\
&\quad - D^m f(h(s))(g_1(s), \ldots, g_m(s)) - \epsilon D^{m+1} f(h(s))(g_1(s), \ldots, g_{m+1}(s))||_{\eta_2} \\
&\leq \sup_{s\in\mathbb{R}} \sup_{||y||\leq e^{\eta_1|s|}} e^{(-\eta_2+(m+1)\eta_1)|s|}||D^{m+1} f(h(s) + \epsilon y) - D^{m+1} f(h(s))||.
\end{aligned}$$

We have chosen η_2 such that the exponential has a negative exponent. This allows us, using the same reasoning as before, to conclude that the last term approaches zero as $\epsilon \to 0$. □

In the next lemma we address the smoothness of the substitution operator in the case that f is only differentiable on a open subset of E.

Lemma 1.2. *Let Λ, E and F be Banach spaces and $f \in C_b^0(E, F)$. Let V be an open subset of E and suppose that the restriction of f to V is of class C^1 and $\sup_{x\in V} ||Df(x)|| = M < \infty$. Let $\tilde{f}$ be the substitution operator associated with f from $BC^\zeta(\mathbb{R}, E)$ into $BC^\eta(\mathbb{R}, F)$ with $\eta > \zeta > 0$. Let H be a mapping of class C^1 from Λ into $BC^\zeta(\mathbb{R}, E)$, with range contained in V. Then the mapping $\lambda \mapsto \tilde{f}(H(\lambda))$ is of class C^1 and $D(\tilde{f} \circ H)(\lambda) = \Phi^1(H(\lambda)) \cdot DH(\lambda)$, where Φ^1 is defined by $(\Phi^1(u)v)(s) = Df(u(s))v(s)$ for all $v \in BC^\zeta(\mathbb{R}, E)$ and $u \in BC^\zeta(\mathbb{R}, E)$ with $u(\mathbb{R}) \subset V$.*

Proof. First, we observe that

$$\begin{aligned}
\tilde{f}(H(\lambda))(s) - \tilde{f}(H(\mu))(s) - (\Phi^1(H(\mu))DH(\mu)(\lambda - \mu))(s) \\
= R_1(\lambda, \mu)(s) + R_2(\lambda, \mu)(s),
\end{aligned}$$

where

$$\begin{aligned}
R_1(\lambda, \mu)(s) = \int_0^1 &(Df(\theta\, H(\lambda)(s) + (1 - \theta)\, H(\mu)(s)) - Df(H(\mu)(s))) \\
&\times DH(\mu)(\lambda - \mu)(s)\, d\theta,
\end{aligned}$$

$$\begin{aligned}
R_2(\lambda, \mu)(s) = \int_0^1 &(Df(\theta\, H(\lambda)(s) + (1 - \theta)\, H(\mu)(s)) \\
&\times (H(\lambda)(s) - H(\mu)(s) - (DH(\mu)(\lambda - \mu))(s)))\, d\theta.
\end{aligned}$$

It follows that

$$
\begin{aligned}
&||\tilde{f} \circ H(\lambda) - \tilde{f} \circ H(\mu) - \Phi^1(H(\mu))DH(\mu)(\lambda - \mu)||_\eta \\
&\quad \leq |\lambda - \mu| ||DH(\mu)||_\zeta \sup_{s \in \mathbb{R}} \Big[e^{-(\eta-\zeta)|s|} \\
&\qquad \int_0^1 ||Df(\theta\, H(\lambda)(s) + (1-\theta)\, H(\mu)(s)) - Df(H(\mu)(s))||\, d\theta \Big] \\
&\qquad + M \sup_{s \in \mathbb{R}} e^{-(\eta-\zeta)|s|} e^{-\zeta|s|} ||H(\lambda)(s) - H(\mu)(s) - (DH(\mu)(\lambda-\mu))(s)|| \\
&\quad \leq |\lambda - \mu|\, ||DH(\mu)||_\zeta \max\{2Me^{-(\eta-\zeta)A}, \\
&\qquad \sup_{s \in [-A,A]} \int_0^1 ||Df(\theta\, H(\lambda)(s) + (1-\theta)\, H(\mu)(s)) - Df(H(\mu)(s))||\, d\theta\} \\
&\qquad + M||H(\lambda) - H(\mu) - DH(\mu)(\lambda - \mu)||_\zeta.
\end{aligned}
$$

Fix some $\epsilon > 0$. Let $A > 0$ be such that $2Me^{-(\eta-\zeta)A} < \epsilon$. Let

$$\Omega = \{H(\mu)(s) : \ s \in [-A, A]\}.$$

Since Ω is compact there exists $\delta_1 > 0$ such that

$$||Df(x + \tilde{x}) - Df(x)|| < \epsilon \text{ if } x \in \Omega \text{ and } ||\tilde{x}|| < \delta_1.$$

As $\sup_{s\in[-A,A]} ||H(\lambda)(s) - H(\mu)(s)|| \to 0$ as $\lambda \to \mu$, there exists $\delta_2 > 0$ such that $|\lambda - \mu| < \delta_2$ implies $||H(\lambda)(s) - H(\mu)(s)|| < \delta_1$ for all $s \in [-A, A]$. As H is differentiable at μ, there exists $\delta_3 > 0$ such that $|\lambda - \mu| < \delta_3$ implies $||H(\lambda) - H(\mu) - DH(\mu)(\lambda - \mu)||_\zeta \leq |\lambda - \mu|\epsilon$. We conclude that $|\lambda - \mu| < \min\{\delta_2, \delta_3\}$ implies that

$$||\tilde{f} \circ H(\lambda) - \tilde{f} \circ H(\mu) - \Phi^1(H(\mu))DH(\mu)(\lambda - \mu)||_\eta \leq |\lambda - \mu| \cdot (1 + ||DH(\mu)||)\epsilon,$$

which proves that $\tilde{f} \circ H$ is differentiable at μ.

Next we show that the derivative is continuous. Indeed

$$
\begin{aligned}
&||\Phi^1(H(\lambda))DH(\lambda) - \Phi^1(H(\mu))DH(\mu)||_\eta \\
&\leq ||\Phi^1(H(\lambda))(DH(\lambda) - DH(\mu))||_\eta \\
&\qquad + ||(\Phi^1(H(\lambda)) - \Phi^1(H(\mu)))(DH(\mu))||_\eta \\
&= ||Df(H(\lambda)(\cdot))(DH(\lambda)(\cdot) - DH(\mu)(\cdot))||_\eta \\
&\qquad + ||(Df(H(\lambda)(\cdot)) - Df(H(\mu)(\cdot)))DH(\mu)(\cdot)||_\eta \\
&\leq M||DH(\lambda) - DH(\mu)||_\zeta + \max\{2Me^{-(\zeta-\eta)A}||DH(\mu)||_\zeta, \\
&\qquad ||DH(\mu)||_\zeta \sup_{s\in[-A,A]} ||Df(H(\lambda)(s)) - Df(H(\mu)(s)))||\}
\end{aligned}
$$

Similar arguments as above show that the second term approaches zero if $\lambda \to \mu$. As DH is continuous, we conclude that the derivative is continuous. This completes the proof. □

In Sections VII.3, VII.4 and VII.6, we need a simpler version of Lemma 1.1, with functions defined on a compact interval instead of the whole real line. So let a compact interval $I \subset \mathbb{R}$, Banach spaces E and F, an open subset $U \subset E$ and a map $g : U \to F$ be given. The spaces $C(I,E)$ and $C(I,F)$ are equipped with the supremum norm, i.e.,

$$\|h\| = \max_{s \in I} \|h(s)\|.$$

Set $C(I,U) = \{h \in C(I,E) \mid h(I) \subset U\}$ and consider the substitution operator

$$G : C(I,U) \to C(I,F)$$

defined by

$$G(h) = g \circ h.$$

Exercise 1.3. Show that $C(I,U)$ is an open subset of $C(I,E)$.

Exercise 1.4. Suppose g is continuous and $K \subset U$ is a compact set. Show that g is uniformly continuous on K, in the sense that for every $\epsilon > 0$, there exists $\delta > 0$ such that for all $x \in K$ and $\overline{x} \in E$ with $\|x - \overline{x}\| \leq \delta$, we have $\overline{x} \in U$ and

$$\|g(\overline{x}) - g(x)\| \leq \epsilon.$$

Lemma 1.5. *If g is C^k, $0 \leq k < \infty$, then G is C^k.*

Exercise 1.6. Prove the assertion of Lemma 1.5 for $k = 0$.
Hint: In order to show continuity at $h \in C(I,U)$, set $K = h(I)$ and use Exercise 1.4.

Sketch of the proof of Lemma 1.5 *for* $k = 1$. For $h \in C(I,U)$, $\overline{h} \in C(I,E)$ and $s \in I$, set

$$(A(h)\overline{h})(s) = Dg(h(s))\overline{h}(s).$$

The maps $A(h) : C(I,E) \to C(I,F)$, $h \in C(I,U)$, are linear and continuous.

Exercise 1.7. The map $A : C(I,U) \to \mathcal{L}(C(I,E), C(I,F))$ is continuous.
Hint: Let $h \in C(I,U)$ and $\epsilon > 0$ be given; then the set $K = h(I)$ is compact. Apply Exercise 1.4 to the map Dg to conclude that there exists a $\delta > 0$ such that for $\overline{h} \in C(I,U)$ with $\|h - \overline{h}\| < \delta$,

$$\|A(\overline{h}) - A(h)\| = \sup_{\|\tilde{h}\| \leq 1} \|(A(\overline{h}) - A(h))(\tilde{h})\| \leq \epsilon.$$

In order to verify that $A = DG$, let $h \in C(I,U)$ and $\epsilon > 0$ be given. Set $K = h(I)$ and choose δ according to Exercise 1.4 [with Dg instead of

$g]$. Then, for every $\overline{h} \in C(I, E)$ with $\|h - \overline{h}\| < \delta$ and for every $s \in I$ and $t \in [0,1]$, we obtain

$$h(s) + t(\overline{h}(s) - h(s)) \in U$$

and

$$\|Dg(h(s) + t(\overline{h}(s) - h(s))) - Dg(h(s))\| \leq \epsilon.$$

Therefore, for every $s \in I$

$$\begin{aligned}
&\|(G(\overline{h}) - G(h) - A(h)(\overline{h} - h))(s)\| \\
&\quad = \|g(\overline{h}(s)) - g(h(s)) - Dg(h(s))(\overline{h}(s) - h(s))\| \\
&\quad = \|\int_0^1 [Dg(h(s) + t(\overline{h}(s) - h(s))) - Dg(h(s))]\,(\overline{h}(s) - h(s))\,dt\| \\
&\quad \leq \epsilon\|\overline{h}(s) - h(s)\|,
\end{aligned}$$

which yields

$$\|G(\overline{h}) - G(h) - A(h)(\overline{h} - h)\| \leq \epsilon\|\overline{h} - h\|.$$

□

For the cases $k \geq 2$ in Lemma 1.5, we refer to [121, p. 64].

Appendix V

Tangent vectors, Banach manifolds and transversality

We collect the basic facts about Banach manifolds which are used in Chapter XIV and which put the local invariant manifolds constructed in Chapters VIII and IX into an appropriate framework.

V.1 Tangent vectors of subsets of Banach spaces

Let a subset M of a real Banach space X be given. Let $x \in M$. A *tangent vector* of M at x is an element $v \in X$ so that there is a differentiable curve $c : I \to X$, $I \subset \mathbb{R}$ an open interval containing 0, with the properties

$$c(0) = x, \quad c(I) \subset M, \quad Dc(0)1 = v.$$

The set of all tangent vectors of M at x is denoted by T_xM.

Exercise 1.1. Show that $\mathbb{R} \cdot T_xM \subset T_xM$.

Exercise 1.2. Give an example where T_xM is not a linear space.

Exercise 1.3. Let $U \subset X$ be an open set. Let Y be a real Banach space. Let a differentiable map $f : U \to Y$, subsets $M \subset U$ and $N \subset Y$ and a point $x \in M$ be given. Assume $f(M) \subset N$. Show that

$$Df(x)T_xM \subset T_{f(x)}N.$$

V.2 Banach manifolds

Here and in the next section we largely follow Abraham and Robbin [1]. Compare also Lang [161]. We define C^r-manifolds for integers $r \geq 0$, C^r maps between C^r-manifolds and tangent vectors of such manifolds.

Let a Hausdorff space X be given. A *chart* is a homeomorphism from an open subset of X onto an open subset of a real Banach space.

Charts $\alpha : U \to A$ and $\beta : V \to B$ are called *C^r-compatible* if and only if the map

$$\alpha(U \cap V) \ni \xi \mapsto \beta(\alpha^{-1}(\xi)) \in \beta(U \cap V)$$

is a C^r-diffeomorphism between open subsets of Banach spaces. A set of pairwise C^r-compatible charts is called a *C^r-atlas* of X. A C^r-atlas is called *maximal* if and only if it contains every chart which is C^r-compatible with all of its elements.

Exercise 2.1. Show that every C^r-atlas extends to a maximal atlas.

A pair $(X, \mathcal{A})$ where $\mathcal{A}$ is a maximal C^r-atlas is called a *C^r-manifold.* In the notation, the atlas $\mathcal{A}$ is usually suppressed; one writes X instead of $(X, \mathcal{A})$.

It should be obvious that a maximal atlas of a C^r-manifold X generates a maximal C^s-atlas whenever $0 \le s \le r$, i.e., X becomes a C^s-manifold in a natural way. Taking restrictions of charts, one sees that every open subset of a C^r-manifold becomes a C^r-manifold, too. Open subsets of Banach spaces are considered as C^r-manifolds for any integer $r \ge 0$, with the maximal C^r-atlas determined by the identity map.

Let X and Y be C^r-manifolds. A map $f : X \to Y$ is called a *C^r-map* if and only if for every chart $\alpha : U \to A$ of X and for every chart $\beta : V \to B$ of Y with $f(U) \subset V$, the induced map

$$A \ni \xi \mapsto \beta(f(\alpha^{-1}(\xi)))$$

is a C^r-map from the open subset A of a Banach space into the Banach space containing the range B of β.

A *C^r-curve* in a C^r-manifold X is a C^r-map c from an open interval $I \subset \mathbb{R}$ into X. Let a point x in a C^r-manifold X with $r \ge 1$ be given. Consider C^r-curves $c : I \to X$ and $d : J \to X$ so that $0 \in I \cap J$ and $c(0) = x = d(0)$. The curves c and d are called *tangent at* 0 if and only if there exists a chart $\alpha : U \to A$ with $x \in U$ and $\epsilon > 0$ so that $c((-\epsilon, \epsilon)) \cap d((-\epsilon, \epsilon)) \subset U$ and

$$D(\alpha \circ c|_{(-\epsilon,\epsilon)})(0)1 = D(\alpha \circ d|_{(-\epsilon,\epsilon)})(0)1,$$

where D denotes the usual derivative of a map from the interval $(-\epsilon, \epsilon)$ into the Banach space containing the range A of α.

Exercise 2.2. Show that tangency is an equivalence relation.

A *tangent vector* of X at x is an equivalence class of C^r-curves $c : I \to X$ with $c(0) = x$, with respect to the tangency relation. The set of all tangent vectors of X at x is denoted by $(TX)_x$.

Exercise 2.3. Let $\alpha : U \to A$ be a chart of a C^r-manifold X with $r \geq 1$. Let E be the Banach space containing A. Let $x \in U$. Show that the map

$$c \mapsto D(\alpha \circ c)(0)1$$

on the C^r-curves $c : I \to X$ with $c(I) \subset U$, $0 \in I$ and $c(0) = x$ induces a bijection from $(TX)_x$ onto E.

The bijection of Exercise 2.3 provides a vector space structure and a norm $|\cdot|_\alpha$ on $(TX)_x$ so that $(TX)_x$ becomes a Banach space.

Exercise 2.4. Show that all charts $\alpha : U \to A$ with $x \in U$ define the same vector space structure on $(TX)_x$ and that all norms $|\cdot|_\alpha$ are equivalent.

From now on, we consider tangent sets $(TX)_x$ of C^r-manifolds X, with $r \geq 1$, at points $x \in X$ equipped with the unique vector space structure and the unique topology which are generated by the charts $\alpha : U \to A$ with $x \in U$.

Exercise 2.5. Let X and Y be C^r-manifolds, $r \geq 1$. Let a C^r-map $f : X \to Y$ be given. Let $x \in X$, and set $y = f(x)$. Show that composition with curves induces a linear continuous map

$$(Tf)_x : (TX)_x \to (TY)_y.$$

The map $(Tf)_x$ in Exercise 2.5 is called the *linearization* of f at x, or the differential of f at x.

V.3 Submanifolds and transversality

A subset M of a C^r-manifold, equipped with the relative topology, is called a *submanifold* if and only if for every $x \in M$, there exist real Banach spaces E, F, open neighbourhoods A of 0 in E and B of 0 in F and a chart $\alpha : U \to A \times B$ so that

$$x \in U, \quad \alpha(x) = (0,0), \quad \alpha(U \cap M) = A \times \{0\}.$$

The chart α is called a submanifold chart for M.

Exercise 3.1. Show that open subsets of C^r-manifolds are submanifolds.

Exercise 3.2. Let M be a submanifold. Show that the restriction of submanifold charts $\alpha : U \to A \times B$ for M to the sets $U \cap M$ constitute a C^r-atlas of charts α_M on M.

We consider submanifolds as C^r-manifolds, with the maximal C^r-atlas given by the C^r-atlas of Exercise 3.2.

Exercise 3.3. Let a C^r-submanifold X and a submanifold M of X be given. Assume $r \geq 1$. Show that the inclusion map $i : M \ni x \mapsto x \in X$ is a C^r-map and that the linearizations $(Ti)_x$ at points $x \in M$ are injective maps onto closed subspaces with closed complementary spaces.
Hint: Let a submanifold chart $\alpha : U \to A \times B$ for M be given. Let $E \supset A$ and $F \supset B$ denote real Banach spaces as in the definition of a submanifold. Let $x \in M \cap U$. Consider the norms

$$(e, f) \mapsto |e| + |f|$$

on $E \times F$, $|\cdot|_\alpha$ on $(TX)_x$, and $|\cdot|_{\alpha_M}$ on $(TM)_x$. With respect to these norms, the map $(Ti)_x$ is an isometry.

It is convenient to omit from here on the differential $(Ti)_x$ from Exercise 3.3 and to consider tangent spaces $(TM)_x$ of a submanifold M of a C^r-manifold X as subspaces of the tangent spaces $(TX)_x$.

Exercise 3.4. Let a submanifold M of a C^r-manifold X, a submanifold N of a C^r-manifold Y and a map $f : M \to N$ be given. Show that f is a C^r-map if and only if the map

$$M \ni x \mapsto f(x) \in Y$$

is a C^r-map.

Next, we define transversality of maps and submanifolds. Let C^1-manifolds X and Y, a C^1-map $f : X \to Y$, a submanifold N of Y, and a point $x \in X$ be given. Set $y = f(x)$. The map f is called *transversal* to N at y if and only if either $y \notin N$ or the following conditions are satisfied:

(i) $y \in N$;
(ii) the closed subspace $(Tf)_x^{-1}(TN)_y$ has a closed complementary subspace in $(TX)_x$;
(iii) the space $(Tf)_x(TX)_x$ contains a closed complementary subspace for $(TN)_y$ in $(TY)_y$.

Finally, we briefly discuss submanifolds of a real Banach space X, considered as a C^r-manifold for some integer $r \geq 1$.

Exercise 3.5. Assume that there exist closed subspaces E and F of X so that $X = E \oplus F$. Let an open subset U of the Banach space E and a C^r-map $f : U \to F$ be given. Show that the "graph" $M = \{e + f(e) \mid e \in U\}$ is a submanifold of X.

Exercise 3.6. Let E and F be given as in Exercise 3.5. Let a vector $x \in X$ be given. Show that $M = x + E$ is a submanifold of X with $T_yM = E$ for all $y \in M$. (See Section 1 for the definition of T_yM!)

The next exercises show that in case of a submanifold M of a Banach space X it is not necessary to work with tangent vectors as defined in Section 2. Instead, one can use tangent vectors introduced in Section 1. This is often more convenient. Let $x \in M$ be given.

Exercise 3.7. The set $T_x M$ is a closed linear subspace of X with a complementary closed subspace.

Exercise 3.8. For every $v \in T_x M$, there exists a C^r-curve $c : I \to X$ so that

$$0 \in I, \quad c(0) = x, \quad c(I) \subset M, \quad Dc(0)1 = v. \tag{3.1}$$

Exercise 3.9. The map $c \mapsto Dc(0)1$ defined on the set of C^r-curves $c : I \to X$ satisfying (3.1) induces a topological isomorphism

$$i_x : (TM)_x \to T_x M.$$

The map i_x from Exercise 3.9 permits one to compute the linearization of a C^1-map $f : M \to N$, where N is a submanifold of a real Banach space Y, by the usual differentiation of maps from open intervals $I \subset \mathbb{R}$ into the Banach space Y, as follows. Set $y = f(x)$, and define the derivative of f at x by

$$Df(x) = i_y \circ (Tf)_x \circ i_x^{-1}.$$

Let $v \in T_x M$ be given. Choose a C^r-curve $c : I \to X$ so that (3.1) holds. Then we have

$$Df(x)v = D\big(f_y \circ c\big)(0)1,$$

where $f_y : M \ni x \mapsto f(x) \in Y$.

Exercise 3.10. (Local representation of submanifolds of a Banach space.) Show that there exist open neighbourhoods U of x in X and V of 0 in the Banach space $T_x M$, a closed complementary subspace Q for $T_x M$ in X and a C^r-map $f : V \to Q$ so that

$$f(0) = 0, \qquad Df(0) = 0$$

and

$$M \cap U = x + \{v + f(v) \mid v \in V\}.$$

Appendix VI

Fixed points of parameterized contractions

Let X and P denote Banach spaces over the field $\mathbb{K}$, where $\mathbb{K} = \mathbb{R}$ or $\mathbb{K} = \mathbb{C}$.

Proposition 1.1. (*Smooth dependence of fixed points on parameters*). *Let U_0 and U be open subsets of X and let V be an open subset of P. Suppose*

$$\overline{U}_0 \subset U,$$

and let a map $f : U \times V \to X$ be given which satisfies

$$f(\overline{U}_0 \times V) \subset \overline{U}_0.$$

Suppose there is a constant $q \in [0, 1)$ such that for all (φ, p) and (ψ, p) in $\overline{U}_0 \times V$, we have

$$\|f(\psi, p) - f(\varphi, p)\| \leq q\|\psi - \varphi\|.$$

Assume f is C^k, where $0 \leq k < \infty$. Then the map $F : V \to X$, which is defined by the relations

$$F(p) = x \in \overline{U}_0 \quad \textit{and} \quad f(x, p) = x,$$

is C^k.

Exercise 1.2. Prove the case $k = 0$ of Proposition 1.1.
Hint:

$$\begin{aligned} \|F(p) - F(\overline{p})\| &\leq \|f(F(p), p) - f(F(p), \overline{p})\| + \|f(F(p), \overline{p}) - f(F(\overline{p}), \overline{p})\| \\ &\leq \|f(F(p), p) - f(F(p), \overline{p})\| + q\|F(p) - F(\overline{p})\|. \end{aligned}$$

We need the assertion of Proposition 1.1 for $k = 0$ and $k = 1$ in Section VII.3.

Exercise 1.3. Suppose the map f in Proposition 1.1 is C^1, and the map F is C^1. Let $p \in V$. Show that $Y = DF(p)$ satisfies

$$(I - D_1 f(F(p), p)) \circ Y = D_2 f(F(p), p).$$

Proof of Proposition 1.1. We shall give the proof for the case that f is C^1. (Compare pp. 13ff in [121].)
1. The hypotheses imply that for all $(x,p) \in \overline{U}_0 \times V$, we have

$$\|D_1 f(x,p)\| \leq q,$$

so it follows that the maps $I - D_1 f\big(F(p),p\big) \in \mathcal{L}(X,X)$, $p \in V$, are bijective with continuous inverse and that

$$\|I - D_1 f\big(F(p),p\big)\| \leq \frac{1}{1-q} \quad \text{for all } p \in V.$$

The map $Y : V \to \mathcal{L}\big(P,X\big)$ defined by

$$Y(p) = \big(I - D_1 f\big(F(p),p\big)\big)^{-1} \circ D_2 f\big(F(p),p\big)$$

is continuous. Let $p \in V$ be given. We shall show that $Y(p)$ is the derivative of F at p.
2. There exists a $\delta_1 \in (0,1)$ such that for $x \in X$ and $\overline{p} \in P$ with $\|x\| + \|\overline{p}\| < \delta_1$, we have $F(p) + x \in U$ and $p + \overline{p} \in V$. For such x and $\overline{p}$, set

$$\begin{aligned} \Delta(x,\overline{p}) = {} & f\big(F(p)+x, p+\overline{p}\big) - f\big(F(p),p\big) \\ & - D_1 f\big(F(p),p\big)x - D_2 f\big(F(p),p\big)\overline{p}. \end{aligned}$$

The differentiability of f implies that for every $\epsilon > 0$, there exists $\delta(\epsilon) \in (0,\delta_1)$ with

$$\|\Delta(x,\overline{p})\| \leq \epsilon\big(\|x\| + \|\overline{p}\|\big)$$

for $(x,\overline{p}) \in X \times P$ with $\|x\| + \|\overline{p}\| < \delta(\epsilon)$.
3. It is convenient to set

$$\gamma(\overline{p}) = F(p+\overline{p}) - F(p) \quad \text{for } \overline{p} \in P \text{ with } \|\overline{p}\| \leq \delta_1.$$

The continuity of F (see Exercise 1.2) implies there exists $\delta_2 \in (0,\delta_1)$ such that for all $\overline{p} \in P$ with $\|\overline{p}\| \leq \delta_2$, we have

$$\|\gamma(\overline{p})\| \leq \delta_1,$$

so that $\Delta(\gamma(\overline{p}),\overline{p})$ is defined.
Claim. For all $\overline{p} \in P$ with $\|\overline{p}\| \leq \delta_2$, we have

$$F(p+\overline{p}) - F(p) - Y(p)\overline{p} = \big(I - D_1 f\big(F(p),p\big)\big)^{-1}\Delta(\gamma(\overline{p}),\overline{p}).$$

Proof. We have

$$\begin{aligned} & \big(I - D_1 f\big(F(p),p\big)\big)\big(F(p+\overline{p}) - F(p) - Y(p)\overline{p}\big) \\ & \quad = \big(I - D_1 f\big(F(p),p\big)\big)\gamma(\overline{p}) - D_2 f\big(F(p),p\big) \end{aligned}$$

and

$$
\begin{aligned}
\big(I - D_1 f(F(p),p)\big)\gamma(\overline{p}) &= \gamma(\overline{p}) - D_1 f\big(F(p),p\big)\gamma(\overline{p}) \\
&= F(p+\overline{p}) - F(p) - D_1 f\big(F(p),p\big)\gamma(\overline{p}) \\
&= f(F(p)+\gamma(\overline{p}), p+\overline{p}) - f(F(p),p) \qquad (1.1)\\
&\qquad - D_1 f\big(F(p),p\big)\gamma(\overline{p}) \\
&= D_2 f\big(F(p),p\big) + \Delta(\gamma(\overline{p}),\overline{p}).
\end{aligned}
$$

4. Claim. There exists $\delta_3 \in (0,\delta_2)$ such that for all $\overline{p} \in P$ with $\|\overline{p}\| \le \delta_3$, we have

$$\|\gamma(\overline{p})\| \le \big(2(1-q)^{-1}\|D_2 f\big(F(p),p\big)\| + 1\big)\|\overline{p}\|.$$

Proof. Set $\epsilon = (1-q)/2$, and choose $\delta(\epsilon)$ according to Part 2. There exists $\delta_3 \in (0,\delta_2)$ such that $\|\overline{p}\| \le \delta_3$ implies $\|\gamma(\overline{p})\| + \|\overline{p}\| \le \delta(\epsilon)$, and thereby

$$\|\Delta(\gamma(\overline{p}),\overline{p})\| \le \epsilon\big(\|\gamma(\overline{p})\| + \|\overline{p}\|\big).$$

Using (1.1), we obtain

$$
\begin{aligned}
\|\gamma(\overline{p})\| &= \|\big(I - D_1 f\big(F(p),p\big)\big)^{-1}\big(D_2 f\big(F(p),p\big)\overline{p} + \Delta(\gamma(\overline{p}),\overline{p})\big)\| \\
&\le \frac{1}{1-q}\big(\|D_2 f\big(F(p),p\big)\|\,\|\overline{p}\| + \epsilon\big(\|\gamma(\overline{p})\| + \|\overline{p}\|\big)\big),
\end{aligned}
$$

therefore

$$\frac{1}{2}\|\gamma(\overline{p})\| \le \frac{1}{1-q}\big(\|D_2 f\big(F(p),p\big)\| + (1-q)/2\big)\|\overline{p}\|.$$

5. Now let $\epsilon > 0$ be given. Define

$$\overline{\epsilon} = (1-q)\epsilon\big(2(1-q)^{-1}\|D_2 f\big(F(p),p\big)\| + 2\big)^{-1}.$$

Consider $\delta(\overline{\epsilon}) \in (0,\delta_1)$ according to Part 2. The continuity of F and the definition of γ imply that there exists $\delta \in (0,\delta_3)$ such that for $\|\overline{p}\| \le \delta$, we have

$$\|\gamma(\overline{p})\| + \|\overline{p}\| \le \delta(\overline{\epsilon}).$$

Then

$$
\begin{aligned}
\|\Delta(\gamma(\overline{p}),\overline{p})\| &\le \overline{\epsilon}\big(\|\gamma(\overline{p})\| + \|\overline{p}\|\big) \qquad \text{(see Part 2)} \\
&\le \overline{\epsilon}\big(2(1-q)^{-1}\|D_2 f\big(F(p),p\big)\| + 1 + 1\big)\|\overline{p}\| \quad \text{(see Part 4)} \\
&= (1-q)\epsilon\|\overline{p}\|.
\end{aligned}
$$

Using the claim in Part 3, we infer

$$\|F(p+\overline{p}) - F(p) - Y(p)\overline{p}\| \le \epsilon\|\overline{p}\|$$

for $\|\overline{p}\| \le \delta$. □

The spectral theory of linear age-dependent population dynamics is presented in Exercises IV.5.14–5.25. As these exercises are quite detailed already, we refrain from elaborating them here.

Appendix VII

Linear age-dependent population growth: elaboration of some of the exercises

Exercise II.2.2.

(i) In Exercise I.2.2(ii) we noted that translation is continuous in L_1. Hence T_0 defined by (2.9) is strongly continuous. Clearly, T_0 is nilpotent, i.e., $T_0(t)$ is the zero operator for t sufficiently large (in fact for $t \geq h$). So the Laplace transform exists for $\lambda = 0$. The formula

$$-A_0^{-1} = \int_0^\infty T_0(t)\,dt$$

becomes, in the present situation,

$$-(A_0^{-1}\varphi)(a) = \int_0^a \varphi(a-t)\,dt = \int_0^a \varphi(\sigma)\,d\sigma,$$

and from this, (2.10) follows directly.

(ii) For $\varphi \in \mathcal{D}(A_0)$, the differential equation

$$\frac{d}{dt}T_0(t)\varphi = A_0 T_0(t)\varphi$$

holds. If we put $u(t,a) = (T_0(t)\varphi)(a)$, we can rewrite this as

$$\begin{aligned} \frac{\partial u}{\partial t}(t,a) &= -\frac{\partial u}{\partial a}(t,a), \\ u(t,0) &= 0, \end{aligned}$$

where the second equation stems from the requirement that $u(t,\cdot) \in \mathcal{D}(A_0)$.

Exercise II.5.6. From Exercise II.2.2 we know that

$$(A_0^{-1}\varphi)(a) = \int_0^a \varphi(\alpha)d\alpha.$$

A straightforward integration by parts then shows that

$$((A_0^{-1})^*\psi)(a) = \int_h^a \psi(\alpha)\,d\alpha.$$

Since $(A_0^{-1})^* = (A_0^*)^{-1}$ and the class of Lipschitz functions coincides with the class of functions which are integrals of L^∞-functions, we conclude that

$$\mathcal{D}(A_0^*) = \{f \in \mathrm{Lip} \mid f(a) = 0 \text{ for } a \geq h\}, \quad A_0^* f = \dot{f}.$$

When taking the closure with respect to the sup-norm, continuity and being zero for $a \geq 0$ “survive”, but the Lipschitz condition gets lost, so

$$X^\odot = \{f \in C(\mathbb{R}_+, \mathbb{C}) \mid f(a) = 0 \text{ for } a \geq h\}.$$

Since $A_0^\odot$ is the part of A_0^* in $X^\odot$, it follows at once that

$$\mathcal{D}(A_0^\odot) = \{f \in C^1(\mathbb{R}_+, \mathbb{C}) \mid f(a) = 0 \text{ for } a \geq h\}, \quad A_0^\odot f = \dot{f}.$$

In order to determine the $\odot *$-operators, we go through the same set of calculations once more. First, note that

$$((A_0^\odot)^{-1} f)(a) = \int_h^a f(\alpha) d\alpha$$

and that, by integration by parts,

$$\int_0^h \int_h^a f(\alpha)\,d\alpha\,d\psi(a) = -\int_0^h f(a)\,d\Big(\int_0^a \psi(\alpha)\,d\alpha\Big).$$

We conclude that

$$\mathcal{D}(A_0^{\odot *}) = \{\varphi \mid \varphi(a) = \int_0^a \psi(\alpha)\,d\alpha \text{ for } a \geq 0, \text{ for some } \psi \in \mathrm{NBV}\},$$

$$A_0^{\odot *}\varphi = -\psi.$$

Finally, since $(T_0^\odot(t)f)(a) = f(a+t)$ and, for f with support in $[0,h)$,

$$\int_0^h f(a+t)\,d\varphi(a) = \int_0^{h-t} f(a+t)\,d\varphi(a) = \int_t^h f(\alpha)\,d_\alpha\varphi(\alpha - t),$$

we find that

$$(T_0^{\odot *}(t)\varphi)(a) = \begin{cases} \varphi(a-t), & a \geq t, \\ \varphi(0) = 0, & a < t\,. \end{cases}$$

In order that for all $f \in X^\odot$,

$$\langle f, \varphi\rangle = \langle j(\varphi), f\rangle,$$

we must have that

$$\int_0^h f(a)\varphi(a)\,da = \int_0^h f(a)\,dj(\varphi)(a),$$

and hence

$$j(\varphi)(a) = \int_0^a \varphi(\alpha)\,d\alpha.$$

Using that $X^{\odot\odot} = \overline{\mathcal{D}(A_0^{\odot *})}$ and the arguments of the proof of Theorem II.5.2, we deduce that $j(X) = X^{\odot\odot}$, i.e., X is $\odot$-reflexive with respect to T_0. (Alternatively, we can deduce this from the compactness of A_0^{-1} and the Phillips-de Pagter Theorem A.II.3.21.)

Exercise III.4.11. In terms of our general notation, we have

$$r_i^* = \beta_i \quad \text{and} \quad r_i^{\odot *} = \delta_i.$$

Since $(T_0^*(\sigma)\beta_i)(a) = \beta_i(a+\sigma)$, we have $\int_0^t (T_0^*(\sigma)\beta_i)(a)\,d\sigma = \int_0^t \beta_i(a+\sigma)\,d\sigma$ and so

$$\langle \delta_j, \int_0^t T_0^*(\sigma)\beta_i\,d\sigma \rangle = \int_0^t \beta_{ij}(\sigma)\,d\sigma.$$

According to the definition (see Corollary 3.2 and Lemma 2.15), this equals $\int_0^t R_{0ij}(\sigma)\,d\sigma$ and therefore $R_0 = \beta$.

Since

$$(T_0^{\odot *}(t-\tau)\delta_i)(a) = \begin{cases} 0, & \tau \le t-a, \\ e_i, & \tau > t-a, \end{cases}$$

we find that

$$\int_0^t (T_0^{\odot *}(t-\tau)\delta_i)(a)\,\eta(\tau)d\tau = e_i \int_{\max(0,t-a)}^t \eta(\tau)\,d\tau,$$

where the right hand side is an element of $X^{\odot *} = \mathrm{NBV}$. If we conceive of the right hand side as an element of $L^1 = X \cong X^{\odot\odot}$, we should write it as

$$\begin{cases} 0, & a \ge t, \\ e_i\,\eta(t-a), & a < t, \end{cases}$$

[Recall the embedding $j(\varphi)(a) = \int_0^a \varphi(\alpha)d\alpha$ from Exercise II.5.6.] Define $y_{0i}(t) = \langle r_i^*, T_0(t)\varphi \rangle = \langle \beta_i, T_0(t)\varphi \rangle = \int_0^t \beta_i(\alpha)\varphi(\alpha - t)\,d\alpha$ and let y be the solution of the RE

$$y = \beta * y + y_0.$$

According to Theorem III.3.6,

$$T(t)\varphi = T_0(t)\varphi + \sum_{i=1}^n \int_0^t T_0^{\odot *}(t-\tau)\,\delta_i\,y_i(\tau)\,d\tau,$$

and consequently,

$$(T(t)\varphi)(a) = \begin{cases} \varphi(a-t), & a \ge t, \\ y(t-a), & a < t. \end{cases}$$

Starting from the first order PDE, we define $y(t) = u(t,0)$. Solving the PDE along characteristics, we obtain

$$u(t,a) = \begin{cases} \varphi(a-t), & a \geq t, \\ y(t-a), & a < t, \end{cases}$$

and substitution of this expression into the boundary condition yields

$$y(t) = \int_0^t \beta(\alpha) y(t-\alpha)\, d\alpha + \int_t^h \beta(\alpha)\varphi(\alpha - t)\, d\alpha,$$

which is exactly the RE $y = \beta * y + y_0$.

Exercise III.4.12. For the present example, $B^* : X^\odot \to X^*$ is given by

$$B^* f = \sum_{j=1}^n f_j(0)\beta_j.$$

Since $A^* = A_0^* + B^*$ and $A^\odot$ is the part of A^* in $X^\odot$, we have

$$\mathcal{D}(A^\odot) = \{ f \in \text{Lip} \mid f(a) = 0,\ a \geq h,\ \dot{f} + \sum_{j=1}^n f_j(0)\beta_j \text{ continuous}\},$$

$$A^\odot f = \dot{f} + \sum_{j=1}^n f_j(0)\beta_j.$$

In order to determine A, we must take the embedding j into account. First, observe that

$$j^{-1}(\mathcal{D}(A_0^{\odot *})) = \{\varphi \mid \varphi \in \text{NBV}\}$$

and that for $\varphi \in j^{-1}(\mathcal{D}(A_0^{\odot *}))$, we have

$$A_0^{\odot *} j(\varphi) = -\varphi.$$

Hence

$$A_0^{\odot *} j(\varphi) + B\varphi = -\varphi + \sum_{i=1}^n \langle \beta_i, \varphi \rangle\, \delta_i.$$

The right hand side is absolutely continuous with the value zero at zero [in other words, the right hand side belongs to $j(X) = X^{\odot\odot}$] if and only if φ is absolutely continuous and

$$\varphi_i(0) = \langle \beta_i, \varphi \rangle,$$

and in that case, j^{-1} applied to the right hand side gives $-\dot{\varphi}$. We conclude that

$$\mathcal{D}(A) = \{\varphi \mid \varphi \in \text{AC},\ \varphi(0) = \sum_{i=1}^n \langle \beta_i, \varphi \rangle = \langle \beta, \varphi \rangle_n \}, \quad A\varphi = -\dot{\varphi}.$$

Exercise III.5.6. Exactly as in the beginning of Exercise III.4.7, we find that

$$\int_0^t z_0(\tau)\,d\tau = \langle \delta, \int_0^t T_0^*(\sigma) f\,d\sigma \rangle = \int_0^t f(\sigma)\,d\sigma,$$

and, consequently, $z_0(t) = f(t)$. By Theorem 3.9 we now know that

$$\begin{aligned}(T^*(t)f)(a) &= (T_0^*(t)f)(a) + \sum_{j=1}^{n} \Big(\int_0^t T_0^*(t-\tau)\, r_j^*\, z_j(\tau)\,d\tau \Big)(a) \\ &= f(t+a) + \sum_{j=1}^{n} \int_0^t \beta_j(t-\tau+a)\, z_j(\tau)\,d\tau,\end{aligned}$$

where z is a solution of the RE

$$z = z_0 + \beta^T * z.$$

Appendix VIII
The Hopf bifurcation theorem

In this appendix we will state and prove the Hopf bifurcation theorem for a finite dimensional system of ODE. The presentation here is mainly based on a paper of Crandall and Rabinowitz [53], and employs the so-called Lyapunov-Schmidt reduction.

Consider the system of ODE's

$$\dot{x} = f(x, \mu), \tag{1.1}$$

where $x \in \mathbb{R}^n$ and $\mu \in \mathbb{R}$. We assume that

H1 $f \in C^k(\mathbb{R}^n \times \mathbb{R}, \mathbb{R})$, $k \geq 2$, $f(0, \mu) = 0$,

H2 $A = D_1 f(0, \mu_0)$ has simple (i.e. of algebraic multiplicity one) eigenvalues at $\pm i\omega_0$, $\omega_0 > 0$, and no other eigenvalue of A belongs to $i\omega_0\mathbb{Z}$,

H3 $\mathrm{Re}\,(D\sigma(\mu_0)) \neq 0$, where $\sigma(\mu)$ is the branch of eigenvalues of $D_1 f(0, \mu)$ through $i\omega_0$ at $\mu = \mu_0$.

Let p be an eigenvector of A at $i\omega_0$. We choose q in $\mathbb{C}^n$ such that $A^T q = i\omega_0\, q$ and

$$q \cdot p = \sum_{i=1}^{n} q_i\, p_i = 1. \tag{1.2}$$

This can be done as we assume that the eigenvalue of A at $i\omega_o$ is simple, see Exercise III.7.9(ii) and Theorem IV.2.5(vi)

Theorem 1.1. (*Hopf bifurcation for ODE's.*) *Let the above hypotheses be satisfied. Then there exist C^{k-1} functions $\epsilon \mapsto \mu^*(\epsilon)$, $\epsilon \mapsto \omega^*(\epsilon)$ and $\epsilon \mapsto x^*(\epsilon)$, defined for ϵ sufficiently small, taking values in $\mathbb{R}$, $\mathbb{R}$ and $C(\mathbb{R}, \mathbb{R}^n)$ respectively, such that at $\mu = \mu^*(\epsilon)$, $x^*(\epsilon)(t)$ is a $\frac{2\pi}{\omega^*(\epsilon)}$ periodic solution of (1.1). Moreover, μ^* and ω^* are even, $\mu(0) = \mu_0$, $\omega(0) = \omega_0$, $x^*(-\epsilon)(t) = x^*(\epsilon)(t + \frac{\pi}{\omega^*(\epsilon)})$ and $x^*(\epsilon)(t) = \epsilon\mathrm{Re}\,(e^{i\omega_0 t}p) + o(\epsilon)$, uniformly on compacta in $\mathbb{R}$, for $\epsilon \downarrow 0$. In addition, if x is a small periodic solution of this equation with μ close to μ_0 and minimal period close to $\frac{2\pi}{\omega_0}$, then $x(t) = x^*(\epsilon)(t + \theta)$ and $\mu = \mu^*(\epsilon)$ for some ϵ and some $\theta \in [0, 2\pi/\omega^*(\epsilon))$ (with ϵ unique modulo its sign).*

Notation. We let Z be the space of continuous, 2π-periodic functions from $\mathbb{R}$ into $\mathbb{R}^n$, provided with the supremum norm: for u in Z, $||u||_Z = max_{s\in\mathbb{R}}|u(s)|$. X is the space of one time continuously differentiable, 2π-periodic functions from $\mathbb{R}$ into $\mathbb{R}^n$, provided with the norm: $||u||_X = ||u||_Z + ||Du||_Z$. We shall use that X is embedded into Z.

For $u \in Z$ we denote by u_k its kth Fourier coefficient:

$$u_k = \tfrac{1}{2\pi}\int_0^{2\pi} e^{-ikt}u(t)\,dt, \tag{1.3}$$

and by $P_N u$ the Nth order Fourier approximation

$$(P_N u)(t) = \sum_{-N}^{N} u_k e^{ikt}. \tag{1.4}$$

As u is real valued we have $u_k = \overline{u_{-k}}$. If $u \in Z$ then certainly $P_N u \to^{L_2} u$ and $\sum |u_k|^2 < \infty$.

We identify the complexification $Z_{\mathbb{C}}$ with the space of continuous, 2π-periodic functions from $\mathbb{R}$ into $\mathbb{C}^n$ (see Exercise III.7.9) and we consider Z as the subspace of $Z_{\mathbb{C}}$ consisting of real-valued functions. The complexification of X is handled similarly. On $Z_{\mathbb{C}}$ we define the bilinear form

$$[\,u, v\,] = \frac{1}{2\pi}\int_0^{2\pi} u(s)\cdot v(s)\,ds. \tag{1.5}$$

We let J be the bounded linear operator from X into Z given by

$$J = -\omega_0\frac{d}{ds} + A. \tag{1.6}$$

We define the elements ϕ and ϕ^* of $X_{\mathbb{C}}$ by

$$\phi(s) = e^{is}\,p, \quad \phi^*(s) = e^{-is}\,q. \tag{1.7}$$

For $\theta \in \mathbb{R}$ the translation operator $S(\theta)\colon Z \to Z$ is defined by $(S(\theta)u)(s) = u(s+\theta)$.

Lemma 1.2. $\mathcal{N}(J) = \{\, z\phi + \overline{z\phi} \mid z \in \mathbb{C}\,\}$.

Proof. Clearly $J_{\mathbb{C}}\phi = 0$, where $J_{\mathbb{C}}$ denotes the complexification of J. So the set at the right hand side belongs to $\mathcal{N}(J)$. Conversely, consider $u \in \mathcal{N}(J)$. From the identity

$$0 = \tfrac{1}{2\pi}\int_0^{2\pi} e^{-ikt}(Ju)(t)\,dt$$

we obtain by integration by parts the identity

$$(-i\omega_0 k + A)\, u_k = 0$$

from which it follows that $u_k = 0$ for $k \neq \pm 1$ and that u_1 is a multiple of p and u_{-1} is a multiple of $\overline{p}$. Fourier theory and the fact that u is real valued now imply that $u = z\phi + \overline{z}\overline{\phi}$ for some $z \in \mathbb{C}$. □

Exercise 1.3. Show that $\mathcal{N}(J)$ is translation invariant. Prove that modulo translation $\mathcal{N}(J)$ is one-dimensional. More precisely, show that

$$\mathcal{N}(J) = \{\, \epsilon S(\theta)(\phi + \overline{\phi}) \mid \epsilon \in \mathbb{R}_+,\ \theta \in [0, 2\pi)\,\}$$

Exercise 1.4. Show that $S(\pi)(\phi + \overline{\phi}) = -(\phi + \overline{\phi})$.

Lemma 1.5. $\mathcal{R}(J) = \{\, u \in Z : \ [\phi^*, u] = 0\,\}$.

Proof. That $[\phi^*, h] = 0$ when $h = Ju$ for some $u \in X$, follows from partial integration.

Now assume that $h \in Z$ is such that $[\phi^*, h] = 0$. We define

$$u_k = (-i\omega_0 k + A)^{-1}\, h_k,$$

noting that even for $k = \pm 1$ this definition makes sense as $[\phi^*, h] = 0$ guarantees that $q \cdot h_1 = 0$ which in turn implies that $\overline{q} \cdot h_{-1} = 0$. We have $\sum_{-\infty}^{\infty} |k u_k|^2 < \infty$. This gives that

$$\begin{aligned}\Big(\sum_{k\in\mathbb{Z}} |u_k|\Big)^2 &= \Big(\sum_{k\in\mathbb{Z}} (1+k)^{-1}(1+k)|u_k|\Big)^2 \\ &\overset{\text{CauchySchwarz}}{\leq} \sum_{k\in\mathbb{Z}} (1+k)^{-2} \sum_{k\in\mathbb{Z}} (1+k)^2 |u_k|^2 \\ &< \infty.\end{aligned}$$

The absolute convergence of the sequence $\{u_k\}$ implies that $u(s) = \sum u_k e^{iks}$ defines a continuous function. As P_N and J commute, it follows from (1.4) that

$$A P_N u - \omega_0 \frac{d}{ds}(P_N u) = P_N h.$$

Integration yields

$$\omega_0\, (P_N u)(s) - \omega_0\, (P_N u)(0) = \int_0^s \big(A P_N u(\tau) - P_N h(\tau)\big)\, d\tau.$$

Taking the limit $N \to \infty$ we find

$$\omega_0 \, u(s) - \omega_0 \, u(0) = \int_0^s \left(Au(\tau) - h(\tau)\right) d\tau.$$

[Here we use that if $h_n \to h$ in L^2 then for any $g \in L^2$ we have

$$\lim_{n\to\infty} \int_0^s g(\tau)h_n(\tau)\, d\tau = \int_0^s g(\tau)h(\tau)\, d\tau.$$

In particular we have that $\lim_{n\to\infty} \int_0^s h_n(\tau)\, d\tau = \int_0^s h(\tau)\, d\tau$.] Hence u is of class C^1 and satisfies

$$\omega_0 \frac{d}{ds} u(s) = Au(s) - h(s).$$

□

Corollary 1.6. *J is a Fredholm operator of index* 0.

Exercise 1.7. In this exercise we give an alternative for the proof of the previous lemma that does not use Fourier theory.

(i) Show that finding a 2π-periodic solution of the equation $Ju = h$ is equivalent to finding $u(0)$ such that

$$(I - exp(\tfrac{-2\pi}{\omega_0} A))u(0) = \tfrac{1}{\omega_0} \int_0^{2\pi} exp(\tfrac{-\tau}{\omega_0} A)\, h(\tau)\, d\tau.$$

(ii) Show that H2 implies that $\mathcal{N}(exp(\frac{-2\pi}{\omega_0} A) - I) = \mathrm{span}\{\mathrm{Re}\, p, \mathrm{Im}\, p\}$.
(iii) Derive the analogue of (ii) for A^T.
(iv) Show that $\int_0^{2\pi} exp(\frac{-\tau}{\omega_0} A)\, h(\tau)\, d\tau \in \mathcal{R}(I - exp(\frac{-2\pi}{\omega_0} A))$ is equivalent to $[\phi^*, h] = 0$.

On Z we define the operator

$$Pu = [\phi^*, u]\, \phi + \overline{[\phi^*, u]\phi}. \tag{1.8}$$

Exercise 1.8. Prove the following four statements:

(i) P is a projection operator.
(ii) $X \cap \mathcal{R}(P) = \mathcal{N}(J)$.
(iii) $\mathcal{R}(I - P) = \mathcal{R}(J)$.
(iv) $Ju = h \Leftrightarrow \begin{cases} Ph = 0 \\ Ju = (I - P)h \end{cases}$
(v) P commutes with $S(\theta)$.

Proof of Theorem 1.1, *partly in the form of exercises.* We introduce the unknown period as a parameter by applying the rescaling of the time $s = \omega t$.

Periodic solutions of (1.1) are in one to one correspondence with zeros of the mapping $G: X \times \mathbb{R}^2 \to Z$ defined by

$$G(u, \mu, \omega)(s) = -\omega \frac{du}{ds}(s) + f(u(s), \mu). \tag{1.9}$$

We write G as the sum of its linearization about $(u, \mu, \omega) = (0, \mu_0, \omega_0)$ and the remainder: $G = J + N$, with J given in (1.6) and $N : X \times \mathbb{R}^2 \to Z$ defined by

$$N(u, \mu, \omega)(s) = (\omega_0 - \omega)\frac{du}{ds}(s) + f(u(s), \mu) - A\,u(s). \tag{1.10}$$

In X we write

$$u = \epsilon S(\theta)(\phi + \overline{\phi}) + v, \quad v \in \mathcal{R}(J) \cap X. \tag{1.11}$$

Solving the equation $G = 0$ is equivalent to solving the pair of equations (see Exercise 1.8.(iv))

$$\begin{cases} Jv + (I - P)N(\epsilon S(\theta)(\phi + \overline{\phi}) + v, \mu, \omega) = 0 \\ PN(\epsilon S(\theta)(\phi + \overline{\phi}) + v, \mu, \omega) = 0. \end{cases} \tag{1.12}$$

It is a consequence of Lemma 1.2 that J is an isomorphism of $\mathcal{R}(J) \cap X$ onto $\mathcal{R}(J)$. By the implicit function theorem we can solve from the first equation $v = v^*(\epsilon, \theta, \mu, \omega)$ such that $v^*(0, \theta, \mu_0, \omega_0) = 0$. For further purpose we observe that

$$v^*(\epsilon, \theta, \mu, \omega) = \mathcal{O}\big(\epsilon^2 + |\epsilon|(|\mu - \mu_0| + |\omega - \omega_0|)\big). \tag{1.13}$$

Also observe that J, P and N all commute with $S(\theta)$. By uniqueness this has the consequence that

$$S(\psi)v^*(\epsilon, \theta, \mu, \omega) = v^*(\epsilon, \theta + \psi, \mu, \omega). \tag{1.14}$$

Writing $z = \epsilon e^{i\theta}$ the bifurcation equations, obtained by plugging v^* in the second equation of (1.12), read

$$0 = \tilde{g}(z, \overline{z}, \mu, \omega) = [\phi^*, N(z\phi + \overline{z\phi} + v^*(\epsilon, \theta, \mu, \omega), \mu, \omega)]. \tag{1.15}$$

Exercise 1.9. Show that it is a consequence of (1.5), (1.14) and the commutativity properties of S mentioned above that

$$e^{i\theta}\tilde{g}(z, \overline{z}, \mu, \omega) = \tilde{g}(e^{i\theta}z, e^{-i\theta}\overline{z}, \mu, \omega), \qquad \text{for all } \theta \in S^1. \tag{1.16}$$

We will in the next two exercises investigate some of the consequences of (1.16).

Exercise 1.10. Let g be a polynomial of degree $2k+1$ in the variables z and $\overline{z}$ such that $e^{-i\theta}g(e^{i\theta}z, e^{-i\theta}\overline{z}) = g(z,\overline{z})$ for all $\theta \in [0, 2\pi)$. Show that g is necessarily of the form

$$g(z,\overline{z}) = z\sum_{j=0}^{k} a_j|z|^{2j},$$

with $a_j \in \mathbb{C}$, $j = 0, \dots, k$.

Exercise 1.11. Let g be a function of the variables z, $\overline{z}$ with values in $\mathbb{C}$, which is k-times continuously differentiable, $k \geq 1$. By this we mean that with $z = x + iy$ the mapping $(x, y) \mapsto g(z, \overline{z})$ from $\mathbb{R}^2$ into $\mathbb{C}$ is k-times continuously differentiable. Let g satisfy $e^{-i\theta}g(e^{i\theta}z, e^{-i\theta}\overline{z}) = g(z, \overline{z})$ for all $\theta \in [0, 2\pi)$.

(i) Show that the mapping

$$h(z,\overline{z}) = \begin{cases} \dfrac{1}{|z|}g(z,\overline{z}) & \text{for } z \neq 0 \\ \lim\limits_{|z|\to 0} \dfrac{g(z,\overline{z})}{|z|} & \text{for } z = 0 \end{cases}$$

is $k-1$ times continuously differentiable (in the above sense).

(ii) Show that $h(-\epsilon, -\epsilon) = h(\epsilon, \epsilon)$ for all $\epsilon \in \mathbb{R}$.

The mapping $\tilde{g}$ is of class C^k. It follows from Exercise 1.10 that we can write its Taylor series approximation as the product of z and a polynomial of $|z|^2$. That this is also true for the remainder term is not trivial to prove and can be found in [241, 305]. For the proof of the Hopf bifurcation result this is not crucial, as Exercise 1.11 shows that we can divide out the trivial solution and end up with a function that is even in ϵ. But for the presentation here we invoke the deeper result.

We write $\tilde{g}$ as

$$\tilde{g}(z,\overline{z},\mu,\omega) = zg(|z|^2,\mu,\omega) = zg(\epsilon^2,\mu,\omega), \tag{1.17}$$

for some function g, which is of class C^m, $m = \text{entier}(\frac{k}{2})$. Our achievement is twofold: we can now divide out the trivial solution ($z = 0$) and we get rid of the phase shift. We look for solutions of the equation

$$g(\epsilon^2,\mu,\omega) = 0. \tag{1.18}$$

The next step is to look at the mapping

$$(\hat{\mu},\hat{\omega}) \longrightarrow \frac{\partial g}{\partial \mu}\hat{\mu} + \frac{\partial g}{\partial \omega}\hat{\omega} \tag{1.19}$$

at the bifurcation point ($\mu = \mu_0, \omega = \omega_0, \epsilon = 0$).

Exercise 1.12. Show that it follows from (1.10), (1.11), (1.13) and (1.15) that

$$(1.20)\qquad \begin{aligned}(\hat{\mu},\hat{\omega}) \mapsto &-\hat{\omega}[e^{-is}q, \frac{d}{ds}e^{is}p] + \hat{\mu}[e^{-is}q, D_{1,2}f(0,\mu_0)e^{is}p] \\ &= -i\hat{\omega} + \hat{\mu}\, q \cdot D_{1,2}f(0,\mu_0)p.\end{aligned}$$

Exercise 1.13. Show that the same mapping is an isomorphism of $\mathbb{R}^2$ onto $\mathbb{C}$ iff $\operatorname{Re} q \cdot D_{1,2}f(0,\mu_0)p \neq 0$.

This is precisely the transversality condition H2, as we will show in Lemma 1.15 below. The implicit function theorem in this case states that (1.18) determines $\mu = \mu^*(\epsilon^2)$ and $\omega = \omega^*(\epsilon^2)$ such that $\mu^*(0) = \mu_0$ and $\omega^*(0) = \omega_0$. Then we obtain via (1.11) the solution

$$u^*(\epsilon) = \epsilon(\phi + \overline{\phi}) + v^*(\epsilon, 0, \mu^*(\epsilon^2), \omega^*(\epsilon^2)).$$

Rescaling the time again yields

$$(1.21)\qquad x^*(\epsilon)(t) = \epsilon(e^{i\omega^*(\epsilon^2)t}p + e^{-i\omega^*(\epsilon^2)t}\overline{p}) + v^*(\epsilon, 0, \mu^*(\epsilon^2), \omega^*(\epsilon^2))(\omega^*(\epsilon^2)t).$$

From (1.13) and (1.21) we conclude that, uniformly on compacta in $\mathbb{R}$

$$(1.22)\qquad x^*(\epsilon)(t) = \epsilon(e^{i\omega_0 t}p + e^{-i\omega_0 t}\overline{p}) + o(\epsilon).$$

Exercise 1.14. Show that shifting the time by half of the period and simultaneously changing ϵ to $-\epsilon$ leaves $x^*(\epsilon)(\cdot)$ invariant

This, in combination with the proof of the next lemma completes the proof of the theorem. □

Lemma 1.15. $\operatorname{Re} \frac{d\sigma}{d\mu}(\mu_0) = \operatorname{Re} q \cdot D_{1,2}f(0,\mu_0)p$.

Proof. Since

$$\mathbb{C}^n = \mathcal{R}(A - i\omega_0 I) \oplus \operatorname{span}\{p\},$$

we define a mapping $\mathcal{F} : \mathbb{R} \times \mathcal{R}(A - i\omega_0 I) \times \mathbb{C} \to \mathbb{C}^n$ by

$$\mathcal{F}(\mu, r, \lambda) = D_1 f(0,\mu)(p + r) - \lambda(p + r).$$

Observe that $\mathcal{F}(\mu_0, 0, i\omega_0) = 0$. The derivative of $\mathcal{F}$ at this point with respect to r and λ is given by

$$(\hat{r}, \hat{\lambda}) \mapsto (A - i\omega_0)\hat{r} - \hat{\lambda}p,$$

which is an isomorphism. Hence there exist functions $r(\mu)$ and $\lambda(\mu)$ such that $\lambda(\mu_0) = i\omega_0$, $r(\mu_0) = 0$ and $\mathcal{F}(\mu, r(\mu), \lambda(\mu)) = 0$. Differentiation of this identity with respect to μ and evaluation at $\mu = \mu_0$ yields the result. □

Exercise 1.16. The purpose of this exercise is to find the formula for the direction of Hopf bifurcation. Here we need that f is of class C^k, $k \geq 3$. Starting point is the identity

$$0 = Ju + (\omega_0 - \omega)\frac{du}{ds} + f(u, \mu) - Au.$$

Based on the proof of Theorem 1.1 we substitute the series expansion (dropping all the stars)

$$\begin{aligned}
\omega &= \omega_0 + \omega_2\epsilon^2 + o(\epsilon^2) \\
\mu &= \mu_0 + \mu_2\epsilon^2 + o(\epsilon^2) \\
u &= \epsilon(\phi + \overline{\phi}) + \epsilon^2\psi_2 + o(\epsilon^2), \quad \psi_2 \in \mathcal{R}(J) \cap X.
\end{aligned}$$

(i) Show that

$$\begin{aligned}
J\psi_2 + \frac{1}{2}D_1^2 f(0, \mu_0)(\phi, \phi) &+ D_1^2 f(0, \mu_0)(\phi, \overline{\phi}) \\
&+ \frac{1}{2}D_1^2 f(0, \mu_0)(\overline{\phi}, \overline{\phi}) = 0.
\end{aligned}$$

(ii) Determine ψ_2 from this equation, keeping in mind that $[\phi^*, \psi_2] = 0$.
(iii) Show that

$$\begin{aligned}
J\psi_3 - \omega_2\frac{d}{ds}(\phi + \overline{\phi}) &+ \mu_2 D_{1,2} f(0, \mu_0)(\phi + \overline{\phi}) \\
&+ D_1^2 f(0, \mu_0)((\phi + \overline{\phi}), \psi_2) + \frac{1}{6}D_1^3 f(0, \mu_0)(\phi + \overline{\phi})^3 = 0.
\end{aligned}$$

Introduce the complex number

$$\begin{aligned}
c = \frac{1}{2}\ & q \cdot D_1^3 f(0, \mu_0)(p^2, \overline{p}) + \\
& q \cdot D_1^2 f(0, \mu_0)(-A^{-1} D_1^2 f(0, \mu_0)(p, \overline{p}), p) + \\
\frac{1}{2}\ & q \cdot D_1^2 f(0, \mu_0)((2i\omega_0 - A)^{-1}(D_1^2 f(0, \mu_0)(p, p), \overline{p})).
\end{aligned}$$

(iv) Derive that

$$\begin{aligned}
\mu_2 &= -\frac{\operatorname{Re}(c)}{\operatorname{Re} q \cdot D_{1,2} f(0, \mu_0)p} \\
\omega_2 &= \operatorname{Im}(c) - \frac{\operatorname{Re}(c) \operatorname{Im} q \cdot D_{1,2} f(0, \mu_0)p}{\operatorname{Re} q \cdot D_{1,2} f(0, \mu_0)p}.
\end{aligned}$$

Exercise 1.17. In this exercise we will show that the direction of bifurcation is related to the direction of movement of the critical Floquet exponent. The Floquet exponents are the eigenvalues of

$$J(\epsilon) = -\omega(\epsilon)\frac{d}{ds} + D_1 f(u(\epsilon), \mu(\epsilon)). \tag{1.23}$$

First we note that $J(\epsilon)\dot{u}(\epsilon) = 0$, where $\dot{}$ denotes differentiation with respect to time. This is a consequence of the translation invariance of the periodic solution. Hence, $J(\epsilon)$ has an eigenvalue 0 for all values of ϵ.

(i) Show that $J(0)$ has a double eigenvalue 0, with one eigenvector given by $\lim_{\epsilon\to 0} \frac{\dot{u}(\epsilon)}{\epsilon}$ and a second (independent) eigenvector given by $\lim_{\epsilon\to 0} u'(\epsilon)$, where $'$ denotes differentiation with respect to ϵ.

The idea now is that $J(\epsilon)$ has a two dimensional invariant subspace spanned by $\frac{\dot{u}(\epsilon)}{\epsilon}$ and $u'(0) + o(\epsilon)$. We will consider the equation

(1.24) $$J(\epsilon)y(\epsilon) = \lambda(\epsilon)y(\epsilon) + \gamma(\epsilon)\dot{u}(\epsilon),$$

where λ and γ are real valued.

(ii) Show that if we can solve (1.24) it follows that $\lambda(\epsilon)$ is a Floquet exponent.

At this point we could solve (1.24) by means of the implicit function theorem, just as we did in the proof of the Hopf bifurcation theorem. We will however proceed in the spirit of the previous exercise and compute the first terms in the expansion of the solution. The justification is part (viii) below (see also [53, 135]). We substitute the series expansion

$$\begin{aligned} y &= \phi + \overline{\phi} + \epsilon y_1 + \epsilon^2 y_2 + o(\epsilon^2), \quad [\phi^*, y_k] = 0, \quad k = 1, 2 \\ \lambda &= \lambda_1 \epsilon + \lambda_2 \epsilon^2 + o(\epsilon^2) \\ \gamma &= \gamma_1 \epsilon + \gamma_2 \epsilon^2 + o(\epsilon^2) \end{aligned}$$

and we use the expansion of u, ω and μ that we have computed in the previous exercise.

(iii) Show that at order ϵ we find the equation

$$Jy_1 + D_1^2 f(0, \mu_0)(\phi + \overline{\phi})^2 = \lambda_1(\phi + \overline{\phi}) + i\gamma_1(\phi - \overline{\phi}).$$

Conclude that $\lambda_1 = 0$, $\gamma_1 = 0$ and $y_1 = 2\psi_2$.

(iv) Show that at order ϵ^2 we must solve the equation

$$\begin{aligned} Jy_2 &+ \tfrac{1}{2} D_1^3 f(0, \mu_0)(\phi + \overline{\phi})^3 + \mu_2 D_{1,2} f(0, \mu_0)(\phi + \overline{\phi}) \\ &+ 3D_1^2 f(0, \mu_0)(\psi_2, \phi + \overline{\phi}) = \lambda_2(\phi + \overline{\phi}) + i(\gamma_2 + \omega_2)(\phi - \overline{\phi}) \end{aligned}$$

(v) Apply the solvability condition to this equation and conclude that

$$\lambda_2 = -2\mu_2\, q \cdot D_{1,2} f(0, \mu_0) p.$$

(vi) Conclude that the loss of stability of an equilibrium through a supercritital Hopf bifurcation leads to the birth of stable periodic solutions (assuming the nondegeneracy condition $\mu_2 \neq 0$).

(vii) Formulate the complementary case of (vi) and next formulate the "Principle of Exchange of Stability".

(viii) Justify the expansion by means of the implicit function theorem (see also [53, 135]).

References

[1] Abraham, R. and J. Robbin, *Transversal mappings and flows*, Benjamin, New York, 1967.

[2] Adimy, M., "Integrated semigroups and delay differential equations", *J. Math. Anal. Appl.* **177**, (1993), 125–134.

[3] Alexander, J.C., "A primer on connectivity", pp. 455–483 in *Fixed point theory* (E. Fadell and G. Fournier, eds.), Springer-Verlag, Berlin, 1981.

[4] Alt, W., "Some periodicity criteria for functional differential equations", *Manuscripta Math.* **23**, (1978), 295–318.

[5] Amann, H., *Gewöhnliche Differentialgleichungen*, de Gruyter, Berlin, 1983.

[6] Andronov, A.A. and C.E. Chaikin, *Theory of Oscillations*, Princeton University Press, Princeton, NJ, 1949.

[7] Angelstorf, N., *Spezielle periodische Lösungen einiger autonomer zeitverzögerter Differentialgleichungen mit Symmetrien*, Ph.D. thesis, Bremen, 1980.

[8] Arendt, W. and H. Kellerman, "Integrated solutions of Volterra integro-differential equations and applications", in *Volterra Integro-Differential Equations* (G. Da Prato and M. Iannelli, eds.), Pitman, Boston, 1989.

[9] Arino, O. and P. Seguier, "Existence of oscillating solutions for certain differential equations with delay", pp. 46–64 in *Functional Differential Equations and Approximations of Fixed Points* (H.O. Peitgen and H.-O. Walther, eds.), Springer-Verlag, Berlin, 1979.

[10] Arino, O. and A.A. Chérif, *More on ordinary differential equations which yield periodic solutions of delay differential equations*, Preprint, Université de Pau et du Pays de l'Adour, 1990.

[11] Arino, O. and M.L. Hbid, "Periodic solutions for retarded differential equations close to ordinary ones", *Nonlinear Analysis* **14**, (1990), 23–34.

[12] Arnol'd, V.I., *Ordinary Differential Equations, 3rd ed.*, Springer-Verlag, Berlin, 1992.

[13] Ball, J.M., "Saddle point analysis for an ordinary differential equation in a Banach space", pp. in *Proceedings of a symposium on Nonlinear Elasticity* (R.W. Dickey, ed.), Academic Press, New York, 1973.

[14] Banks, H.T. and A. Manitius, "Projection series for retarded functional differential equations with applications to optimal control problems", *J. Diff. Eqn.* **18**, (1975), 296–332.

[15] Bart, H., I. Gohberg and M.A. Kaashoek, *Minimal Factorization of Matrix and Operator Functions*, Birkhäuser Verlag, Basel, 1979.

[16] Bellman, R. and K.L. Cooke, *Differential-Difference Equations*, Academic Press, New York, 1963.

[17] Boas, R., *Entire Functions*, Academic Press, New York, 1954.

[18] Bogdanov, R.I., "Versal deformation of a singularity of a vector field on the plane in case of zero eigenvalues", *Functional Analysis and its Applications* **9**, (1975), 144–145.

[19] Bogdanov, R.I., "Versal deformation of a singularity of a vector field on the plane in case of zero eigenvalues", *Sel. Math. Sov.* **1**, (1981), 389–421.

[20] Browder F.E., "A further generalization of the Schauder fixed point theorem", *Duke Math. J.* **32**, (1965), 575–578.

[21] Burkhardt, H.J. and U. Halbach, "Sind einfache Zeitverzögerungen die Ursachen für periodische Populationsschwankungen?", *Oecologia* **9**, (1972), 215–222.

[22] Burns, J.A. and T.L. Herdman, "Adjoint semigroup theory for a class of functional differential equations", *SIAM J. Math. Anal.* **7**, (1976), 729–745.

[23] Butzer, P.L. and H. Berens, *Semi-Groups of Operators and Approximation*, Springer-Verlag, New York, 1967.

[24] Cao, Y., "The discrete Lyapunov function for scalar differential delay equations", *J. Diff. Eqn.* **87**, (1990), 365–390.

[25] Cao, Y., *Uniqueness of slowly oscillating periodic solutions*, Georgia Institute of Technology, 1989.

[26] Carr, J., *Applications of the Centre Manifold Theory*, Springer-Verlag, New York, 1981.

[27] van Casteren, J., *Generators of Strongly Continuous Semigroups*, Pitman, Boston, 1985.

[28] Chafee, N.N., "A bifurcation problem for a functional differential equation of finitely retarded type", *J. Math. Anal. Appl.* **35**, (1971), 312–348.

[29] Chapin, S., *Periodic solutions of some nonlinear differential-delay equations and nonuniqueness of periodic solutions*, Ph.D. dissertation, Rutgers University, 1983.

[30] Chapin, S., "Asymptotic analysis of differential-delay equations and nonuniqueness of periodic solutions", *Math. Meth. Applied Sciences* **7**, (1985), 223–237.

[31] Chapin, S. and R.D. Nussbaum, "Asymptotic estimates of the periods of periodic solutions of a differential delay equation", *Michigan Math. J.* **31**, (1984), 215–229.

[32] Chow, S.-N., "Existence of periodic solutions of autonomous functional differential equations", *J. Diff. Eqn.* **15**, (1974), 350–378.

[33] Chow, S.-N. and B. Deng, "Homoclinic and heteroclinic bifurcation in Banach spaces", *Trans. Amer. Math. Soc.* **312**, (1989), 539–587.

[34] Chow, S.-N., O. Diekmann and J. Mallet-Paret, "Stability, multiplicity and global continuation of symmetric periodic solutions of a nonlinear Volterra integral equation", *Japan J. Appl. Math* **2**, (1985), 433–469.

[35] Chow, S.-N., J.K. Hale and W. Huang, "From sine waves to square waves in delay equations", *Proc. Roy. Soc. Edinburgh Sect. A* **120A**, (1992), 223–229.

[36] Chow, S.-N. and W. Huang, *Heteroclinic orbits for singularly perturbed differential-difference equations*, Preprint, Georgia Institute of Technology, 1992.

[37] Chow, S.-N. and W. Huang, *Singular perturbation problems for a system of differential-difference equations*, Preprint, Georgia Institute of Technology, 1991.

[38] Chow, S.N., X.B. Lin, and J. Mallet-Paret, "Transition layers for singularly perturbed delay differential equations with monotone nonlinearities", *J Dyn. Diff. Eqn.* **1**, (1989), 3–44.

[39] Chow, S.N. and K. Lu, "Invariant manifolds for flows in Banach spaces", *J. Diff. Eqn.* **74**, (1988), 285–317.

[40] Chow, S.N. and J. Mallet-Paret, "Singularly perturbed delay-differential equation", in *Coupled Nonlinear Oscillators* (J. Chandra and A.C. Scott, eds.), North-Holland, Amsterdam, 1983.

[41] Chow, S.N. and J. Mallet-Paret, "Integral averaging and bifurcation", *J. Diff. Eqn.* **26**, (1977), 112–159.

[42] Chow, S.N. and H.O. Walther, "Characteristic multipliers and stability of symmetric periodic solutions of $\dot{x}(t) = g(x(t-1))$", *Trans. Amer. Math. Soc.* **307**, (1988), 127–142.

[43] Claeyssen, J.R., "The integral-averaging bifurcation method and the general one-delay equation", *J. Math. Anal. Appl.* **78**, (1980), 429–439.

[44] Clément, Ph., Diekmann, O., Gyllenberg, M., Heijmans, H.J.A.M. and H.R. Thieme, "Perturbation theory for dual semigroups. I. The sun-reflexive case", *Math. Ann.* **277**, (1987), 709–725.

[45] Clément, Ph., Diekmann, O., Gyllenberg, M., Heijmans, H.J.A.M. and H.R. Thieme, "Time dependent perturbations in the sun-reflexive case", *Proc. Roy. Soc. Edinburgh Sect. A* **109**, (1988), 145–172.

[46] Clément, Ph., Diekmann, O., Gyllenberg, M., Heijmans, H.J.A.M. and H.R. Thieme, "Nonlinear Lipschitz continuous perturbations in the sun-reflexive case", pp. 67–89 in *Volterra Integro-Differential Equations in Banach Spaces and Applications* (G. Da Prato and M. Iannelli, eds.), Pitman, Boston, 1989.

[47] Clément, Ph., Diekmann, O., Gyllenberg, M., Heijmans, H.J.A.M. and H.R. Thieme, "The intertwining formula and the canonical pairing", pp. 95–116 in *Semigroup Theory and Applications* (E. Mitidieri and I.I. Vrabie, eds.), Marcel Dekker, New York, 1989.

[48] Clément, Ph., Heijmans, H.J.A.M., et al., *One-parameter semigroups*, North-Holland, Amsterdam, 1987.

[49] Coddington, E.A. and N. Levinson, *Theory of Ordinary Differential Equations*, McGraw-Hill, New York, 1955.

[50] Conway, J.B., *Functions of One Complex Variable* (*2nd ed.*), Springer-Verlag, New York, 1978..

[51] Colonius, F, A. Manitius and D. Salamon, "Structure theory and duality for time varying RFDE", *J. Diff. Eqn.* **78**, (1989), 320–353.

[52] Conway, J.B., *Introduction to Functional Analysis*, Springer-Verlag, New York, 1989.

[53] Crandall, M.C. and P.H. Rabinowitz, "The Hopf bifurcation theorem in infinite dimensions", *Arch. Rat. Mech. Anal.* **67**, (1978), 53–72. [See also, The Hopf bifurcation theorem, MRC Technical Summary Report 1604, April 1976].

[54] Cunningham, W.J., "A nonlinear differential difference equation of growth", *Proc. Nat. Acad. Sci. U.S.A.* **40**, (1954), 709–713.

[55] Cunningham, W.J. and P.J. Wangersky, "On time lags in equations of growth", *Proc. Nat. Acad. Sci. U.S.A.* **42**, (1956), 699–702.

[56] Cunningham, W.J. and P.J. Wangersky, "Time lag in population models", *Cold Spring Harbor Symposia on Quantitative Biology* **22**, (1957), 329–338.

[57] Cushing, J.M., "Periodic solutions of Volterra's population equation with hereditary effects", *SIAM J. Appl. Math.* **31**, (1976), 251–261.

[58] Da Prato, G. and E. Sinestrari, "Differential operators with non dense domain,", *Ann. Sc. Norm. Sup. Pisa* **14**, (1987), 285–344.

[59] E.B. Davis, *One-parameter Semigroups*, Academic Press, New York, 1980.

[60] Delfour, M.C. and A. Manitius, "The structural operator F and its role in the theory of retarded systems I,", *J. Math. Anal. Appl.* **73**, (1980), 466–490.

[61] Delfour, M.C. and A. Manitius, "The structural operator F and its role in the theory of retarded systems II,", *J. Math. Anal. Appl.* **74**, (1980), 359–381.

[62] Desch, W. and W. Schappacher, "On relatively bounded perturbations of linear $\mathcal{C}_0$-semigroups", *Ann. Sc. Norm. Sup. Pisa* **11**, (1984), 327–341.

[63] Diekmann, O., *Volterra integral equations and semigroups of operators*, MC Report TW 197 Centre for Mathematics and Computer Science, Amsterdam, 1982.

[64] Diekmann, O., "A duality principle for delay equations", pp. 84–86 in *Equadiff 5* (M. Gregas, ed.), Teubner Texte zur Math., Leipzig, 1982.

[65] Diekmann, O., "Perturbed dual semigroups and delay equations", pp. 67–74 in *Dynamics of Infinite Dimensional Systems* (S.-N. Chow and J. K. Hale, eds.), Springer-Verlag, New York, 1987.

[66] Diekmann, O and S.A. van Gils, "Invariant manifolds for Volterra integral equations of convolution type", *J. Diff. Eqn.* **54**, (1989), 139–180.

[67] Diekmann, O, M. Gyllenberg and H.J.A.M. Heijmans, "When are two C0-semigroups related by a bounded perturbation?", pp. 153–162 in *Semigroup Theory and Applications* (Ph.Cle'ment, S.Invernizzi,E.Mitidieri and I.I.Vrabie, eds.), Lect.Notes in Pure and Appl.Math., Vol. 116, Marcel Dekker, New York, 1989.

[68] Diekmann, O., J.A.J. Metz and M.W. Sabelis, "Mathematical models of predator-prey-plant interactions in a patchy environment", *Exp. Appl. Acarology* **5**, (1988), 319–342.

[69] Diekmann, O. and R. Montijn, "Prelude to Hopf bifurcation in an epidemic model: analysis of a characteristic equation associated with a nonlinear Volterra integral equation", *J. Math. Biol.* **14**, (1982), 117–127.

[70] Dieudonné, *Foundations of Modern Analysis*, Academic Press, New York, 1960.

[71] Doetsch, G., *Handbuch der Laplace-Transformation Band I*, Birkhäuser, Basel, 1950.

[72] Dormayer, P., "Exact formulae for periodic solutions of $\dot{x}(t+1) = \alpha\big(-x(t)+bx^3(t)\big)$", *J. Appl. Math. Phys.* **37**, (1986), 765–775.

[73] Dormayer, P., "Smooth bifurcation of symmetric periodic solutions of functional differential equations", *J. Diff. Eqn.* **82**, (1989), 109–155.

[74] Dormayer, P., "The stability of special symmetric solutions of $\dot{x}(t) = \alpha f(x(t-1))$ with small amplitudes", *Nonlinear Anal.* **14**, (1990), 701–715.

[75] Dormayer, P., "An attractivity region for characteristic multipliers of special symmetric solutions of $\dot{x}(t) = \alpha f(x(t-1))$", *J. Math. Anal. Appl.* **168**, (1992), 70–91.

[76] Dormayer, P., "Smooth symmetry breaking bifurcation for functional differential equations", *Diff. Int. Eqn.* **5**, (1992), 831–854.

[77] Dormayer, P., "Examples for smooth bifurcation for $\dot{x}(t) = \alpha f(x(t-1))$", *Appl. Anal.*, (to appear).

[78] Driver, R.D., *Ordinary and Delay Differential Equations*, Springer-Verlag, New York, 1977.

[79] Duistermaat, J.J., *Stable Manifolds*, University of Utrecht, 1976.

[80] Dunford, N. and J.T. Schwartz, *Linear Operators, Part II: Spectral Theory*, Wiley, New York, 1963.

[81] Dunkel, G., "Single species model for population growth depending on past history", in *Seminar on Differential Equations and Dynamical Systems* (G.S. Jones, ed.), Springer-Verlag, Berlin, 1968.

[82] El'sgol'ts, L.E., *Introduction to the Theory of Differential Equations with Deviating Argument*, Holden Day Inc., San Francisco, 1966 (Russian edition 1964).

[83] El'sgol'ts, L.E. and S.B. Norkin, *Introduction to the Theory of Differential Equations with Deviating Argument* (revised edition), Academic Press, New York, 1973 (Russian edition 1971).

[84] Fiedler, B. and J. Mallet-Paret, "Connections between Morse sets for delay differential equations", *J. Reine Angew. Math.* **397**, (1989), 23–41.

[85] Friedman, A., *Partial differential equations*, Holt, Rinehart and Winston, New York, 1969.

[86] Furumochi, T., "Existence of periodic solutions of one-dimensional difference-delay equations", *Tôhoku Math. J.* **30**, (1978), 13–35.

[87] Furumochi, T., "Existence of periodic solutions of two-dimensional differential delay equations", *Applicable Analysis* **9**, (1979), 279–289.

[88] Gils, S.A. van, *On a formula for the direction of Hopf bifurcation*, Centre for Mathematics and Computer Science, Twente, TW 225, 1982.

[89] Gohberg, I., S. Goldberg and M.A. Kaashoek, *Classes of Linear Operators, Vol. I*, Birkhäuser, Basel, 1990.

[90] Gohberg, I.C. and E.I. Sigal, "An operator generalization of the logarithmic residue theorem and the theorem of Rouché", *Mat. Sb.* **84**, (1971), 609–629 [Russian: translation: *Math. USSR Sb.* **13**, (1971), 603–625].

[91] Goldstein, J.A., *Semigroups of Linear Operators and Applications*, Oxford Univ. Press, Oxford, 1985.

[92] Golubitsky, M. and D. Schaeffer, *Singularities and Groups in Bifurcation Theory*, Springer-Verlag, Berlin, 1985.

[93] Grafton, R.B., "A periodicity theorem for autonomous functional differential equations", *J. Diff. Eqn.* **6**, (1969), 87–109.

[94] Granas, A., "The Leray-Schrauder index and the fixed point theory for arbitrary ANR's", *Bull. Soc. Math. France* **100**, (1972), 209–228.

[95] Greiner, G., J. Voigt and M. Wolff, "On the spectral bound of the generator of semigroups of positive operators", *J. Operator Th.* **5**, (1981), 245–256.

[96] Greiner, G., "Perturbing the boundary conditions of a generator", *Houston J. Math.* **13**, (1987), 213–229.

[97] Gripenberg, G., S-O. Londen and O. Staffans, *Volterra Integral and Functional Equations*, Cambridge University Press, Cambridge, 1990.

[98] Guckenheimer, J. and P. Holmes, *Nonlinear Oscillations, Dynamical Systems, and Bifurcation of Vector Fields*, Springer-Verlag, New York, 1983.

[99] Hadeler, K.P., "Effective computation of periodic orbits and bifurcation diagrams in delay equations", *Numer. Math.* **34**, (1980), 457–467.

[100] Hadeler, K.P. and F. Tomiuk, "Periodic solutions of difference-differential equations", *Arch. Rat. Mech. Anal.* **65**, (1977), 87–95.

[101] Hale, J.K., *Ordinary Differential Equations, 2nd ed.*, Krieger, Malabar, 1980.

[102] Hale, J.K., *Theory of Functional Differential Equations*, Springer-Verlag, New York, 1977.

[103] Hale, J.K., "Nonlinear oscillations in equations with delay", pp. 157–185 in *Nonlinear Oscillations in Biology* (F.C. Hoppensteadt, ed.), Lectures in Appl. Math. 17, American Mathematical Society, Providence, RI, 1979.

[104] Hale, J.K. and S.M. Verduyn Lunel, *Introduction to Functional Differential Equations*, Springer-Verlag, New York, 1993.

[105] Hale, J.K., *Asymptotic Behaviour of Dissipative Systems*, American Mathematical Society, Providence, RI, 1988.

[106] Hale, J.K. and W. Huang, "Period doubling in singularly perturbed delay equations.", *J. Diff. Eqn.*, (to appear).

[107] Hale, J.K. and W. Huang, "Square and pulse waves in matrix delay differential equations.", *Dynamic Systems Applications* **1**, (1992), 51–70.

[108] Hale J.K. and A.F. Ivanov, "On a high order differential delay equation", *J. Math. Anal. Appl.*, (to appear).

[109] Hale, J.K. and X.-B. Lin, "Symbolic dynamics and nonlinear semiflows", *Annali Mat. Pura Appl.* **144**, (1986), 229–259.

[110] Hale, J.K. and X.-B. Lin, "Examples of transverse homoclinic orbits in delay equations", *Nonlinear Anal.* **10**, (1986), 693–709.

[111] Hale, J.K. and X.-B. Lin, "Heteroclinic orbits for retarded functional differential equations", *J. Diff. Eqn.* **65**, (1986), 175–202.

[112] Hale, J.K. and K. Rybakowski, "On a gradient-like integro-differential equation", *Proc. Roy. Soc. Edinburgh Sect. A* **92**, (1982), 77–85.

[113] Hartman, R., *Ordinary Differential Equations*, Wiley, New York, 1964.

[114] Hassard, B.D., N.D. Kazarinoff and Y.H. Wan, *Theory and Applications of the Hopf Bifurcation*, Cambridge University Press, Cambridge, 1981.

[115] Hastings, A., "Spatial heterogeneity and the stability of predator-prey systems", *Theor. Pop. Biol.* **12**, (1977), 37–48.

[116] an der Heiden, U., "Periodic solutions of a nonlinear second order differential equation with delay", *J. Math. Anal. Appl.* **70**, (1979), 599–609.

[117] an der Heiden, U., A. Longtin, M.C. Mackey, J.G. Milton and R. Scholl, "Oscillatory models in a nonlinear second-order differential equation with delay", *J. Dyn. Diff. Eqn.* **2**, (1990), 423–449.

[118] an der Heiden, U. and H.O. Walther, "Existence of chaos in control systems with delayed feedback", *J. Diff. Eqn.* **47**, (1983), 273–295.

[119] Henry, D., "Small solutions of linear autonomous functional differential equations,", *J. Diff. Eqn.* **8**, (1970), 494–501.

[120] Henry, D., "The adjoint of a linear functional differential equation and boundary-value problems", *J. Diff. Eqn.* **9**, (1971), 55–66.

[121] Henry, D., *Geometric theory of semilinear parabolic equations*, Springer-Verlag, New York, 1981.

[122] Herz, A., "Solutions of $\dot{x}(t) = -g(x(t-1))$ approach the Kaplan-Yorke orbits for odd sigmoid g", *J. Diff. Eqn.*, (to appear).

[123] Hewitt, E. and K. Stromberg, *Real and Abstract Analysis*, Springer-Verlag, Berlin, 1965.

[124] Hille, E. and R. Phillips, *Functional Analysis and Semigroups*, American Mathematical Society, Providence, RI, 1957.

[125] Hino, Y., S. Murakami and T. Naito, *Functional Differential Equations with Infinite Delay*, Springer-Verlag, New York, 1991.

[126] Hirsch, M., C. Pugh and M. Shub, *Invariant Manifolds*, Springer-Verlag, New York, 1977.

[127] Hirsch, M.W. and S. Smale, *Differential Equations, Dynamical Systems, and Linear Algebra*, Academic Press, New York, 1974.

[128] Hopf, E., "Abzweigung einer periodischer Losung von einer Stationären Lösung eines Differential Systems", *Ber. Math. Phys.* **94**, (1942), 1–22.

[129] Horn, W.A., "Some fixed point theorems for compact mappings and flows on a Banach space", *Trans. Amer. Math. Soc.* **149**, (1970), 391–404.

[130] Huang, Y.S. and J. Mallet-Paret, *A homotopy method for locating the Floquet exponents for linear periodic differential delay equations*, Preprint, University of Toledo (Ohio), 1994.

[131] Huang, Y.S. and J. Mallet-Paret, *An infinite-dimensional version of the Floquet theorem*, Preprint, University of Toledo (Ohio), 1994.

[132] Hutchinson, G.E., "Circular causal systems in ecology", *Annals of the New York Academy of Sciences* **50**, (1948), 221–246.

[133] Inaba, H., *Functional Analytic Approach to Age-Structured Population Dynamics*, Ph.D. Dissertation, University of Leiden, 1989.

[134] Iooss, G., *Bifurcation of Maps and Applications*, North-Holland, Amsterdam, 1979.

[135] Iooss, G. and D.D. Joseph, *Elementary Stability and Bifurcation Theory*, Springer-Verlag, Berlin, 1980.

[136] Irwin, M.C., *Smooth Dynamical Systems*, Academic Press, London, 1980.

[137] Ivanov, A.F., B. Lani-Wayda and H.O. Walther, "Unstable hyperbolic periodic solutions of differential delay equations", pp. 301–316 in *Recent Trends in Differential Equations* (R.P. Agarwal, ed.), World Scientific, Singapore, 1992.

[138] Ivanov, A.F. and A.N. Sharkovsky, "Oscillations in singularly perturbed delay equations", pp. 164–224 in *Dynamics Reported Vol. 1 (New Series)* (C.K.R.T. Jones, U. Kirchgraber and H.O. Walther, eds.), Springer-Verlag, New York, 1992.

[139] Janse, V.A.A. and M.W. Sabelis, "Prey dispersal and predator presistence", *Exp. Appl. Acarology* **14**, (1992), 215–231.

[140] Jones G.S., "The existence of periodic solutions of $f'(x) = -\alpha f(x-1)\{1+f(x)\}$", *J. Math. Anal. Appl.* **5**, (1962), 435–450.

[141] Jürgens, H., H.O. Peitgen and D. Saupe, "Topological perturbations in the numerical study of nonlinear eigenvalue and bifurcations problems", pp. 139–181 in *Proc. Symp. Analysis and Computation of Fixed Points* (S.M. Robinson, ed.), Academic Press, New York, 1980.

[142] Kaashoek, M.A. and S.M. Verduyn Lunel, "Characteristic matrices and spectral properties of evolutionary systems", *Trans. AMS* **334**, (1992), 479–517.

[143] Kato, T., *Perturbation Theory for Linear Operators (2nd edn.)*, Springer-Verlag, Berlin, 1976.

[144] Kakutani, S. and L. Markus, "On the non-linear difference-differential equation $y'(t) = [A - By(t-\tau)]y(t)$", pp. 1–18 in *Contributions to the Theory of Nonlinear Oscillations IV* Princeton University Press, Princeton, NJ,, 1958.

[145] Kaplan, J.L. and J.A. Yorke, "Ordinary differential equations which yield periodic solutions of differential delay equations", *J. Math. Anal. Appl.* **48**, (1974), 317–324.

[146] Kaplan, J.L. and J.A. Yorke, "On the stability of a periodic solution of a differential delay equation", *SIAM J. Math. Anal.* **6**, (1975), 268–282.

[147] Kaplan, J.L. and J.A. Yorke, "On the nonlinear differential delay equation $x'(t) = -f(x(t), x(t-1))$", *J. Diff. Eqn.* **23**, (1977), 293–314.

[148] Kellerman, H. and M. Hieber, "Integrated semigroups", *J. Funct. Anal.* **84**, (1989), 160–180.

[149] Kelley, A., "The stable, center-stable, center-unstable and unstable manifolds", *J. Diff. Eqn.* **3**, (1967), 546–570.

[150] Kirchgässner, K. and J. Scheurle, "Bifurcation", pp. 115–129 in *Dynamical Systems : an International Symposium, Volume I* (L. Cesari, J.K. Hale and J.P. Lasalle, eds.), Academic Press, London, 1976.

[151] Kolmanovskii, V.B. and V.R. Nosov, *Stability of Functional Differential Equations*, Academic Press, London, 1986.

[152] Krasnoselskii, M.A., *Topological Methods in the Theory of Nonlinear Integral Equations*, MacMillan, New York, 1964.

[153] Kuang, Y., *Delay Differential Equations with Applications in Population Dynamics*, Academic Press, Boston, 1993.

[154] Kuang, Y. and H. L. Smith, "Slowly oscillating periodic solutions of autonomous state-dependent delay equations", *Nonlinear Anal.* **18**, (1992), 153–176.

[155] Kuang, Y. and H. L. Smith, "Periodic solutions of differential delay differential equations with threshold-type delays", pp. 153–176 in *Oscillation and Dynamics of Delay Equations* (J.R. Graef and J.K. Hale, eds.), Contemp. Math. 129, American Mathematical Society, Providence, RI, 1992.

[156] Kuchment, P., *Floquet Theory for Partial Differential Equations*, Birkhäuser, Basel, 1993.

[157] Kunisch, K. and M. Mastinšek, "Dual semigroups and structural operators for partial functional differential equations with unbounded operators acting on the delays", *Diff. Int. Eqn.* **3**, (1990), 733–756.

[158] Kuratowski, K., *Topology, Vol. 2*, Academic Press, New York, 1968.

[159] Kuznetsov, Yu.A., *Elements of Applied Bifurcation Theory*, Springer-Verlag, New York, 1994.

[160] Lancaster, P. and M. Tismenetsky, *Theory of Matrices (2nd edn.)*, Academic Press, New York, 1985.

[161] Lang, S., *Introduction to Differentiable Manifolds*, John Wiley & Sons, New York and London, 1969.

[162] Lani-Wayda, B., *Hyperbolische Mengen für C^1–Abbildungen in Banachräumen*, Ph.D. dissertation, Universität München, 1991.

[163] Lani-Wayda, B., *Hyperbolic sets, shadowing and persistence for noninvertible mappings in Banach spaces*, Preprint, Universität München, 1992.

[164] Lani-Wayda, B., "Persistence of Poincaré mappings in functional differential equations (with applications to structural stability of chaotic behaviour)", *J. Dyn. Diff. Eqn.*, (to appear).

[165] Lani-Wayda, B. and H.O. Walther, "Chaotic motion generated by delayed negative feedback I: A transversality criterion", *Diff. Int. Eqn.*, (to appear).

[166] Lani-Wayda, B. and H.O. Walther, *Chaotic motion generated by delayed negative feedback II: Construction of nonlinearities*, Preprint, Universität München, 1994.

[167] Lasota, A. and M. Wazewska-Czyzewska, "Matematyczne problemy dynamiki ukladu krwinek czerwonych", *Mat. Stosowana* **6**, (1976), 23–40.

[168] Lefschetz, S., *Differential Equations: Geometric Theory*, Interscience, New York (Dover reprint available), 1962.

[169] Levinson, N. and C. McCalla, "Completeness and independence of the exponential solutions of some functional differential equations", *Studies in Appl. Math.* **53**, (1974), 1–15.

[170] Lin, X.B., "Exponential dichotomies and homoclinic orbits in functional differential equations", *J. Diff. Eqn.* **63**, (1986), 227–254.

[171] Lumer, G., "Solutions généralisées et semi-groupes intégrés", *C.R. Acad. Sci. Paris* **310**, (1990), 577–582.

[172] MacDonald, N., *Biological Delay Systems: Linear Stability Theory*, Cambridge University Press, 1989.

[173] Mackey, M.C. and L. Glass, "Oscillation and chaos in physiological control systems", *Science* **197**, (1977), 287–295.

[174] J. Mahaffy, "Periodic solution for certain protein synthesis models", *J. Math Anal. Appl.* **74**, (1980), 72–105.

[175] Mallet-Paret, J., "Generic periodic solutions of functional differential equations", *J. Diff. Eqn.* **25**, (1977), 163–183.

[176] Mallet-Paret, J., "Morse decompositions for differential delay equations", *J. Diff. Eqn.* **72**, (1988), 270–315.

[177] Mallet-Paret, J. and R.D. Nussbaum, "Global continuation and asymptotic behaviour of periodic solutions of a differential delay equation", *Annali Mat. Pura Appl.* **145**, (1986), 33–128.

[178] Mallet-Paret, J. and R.D. Nussbaum, "A bifurcation gap for a singularly perturbed delay equation", pp. 263–287 in *Chaotic Dynamics and Fractals* (M. Barnsley and S. Demko, eds.), Academic Press, New York, 1986.

[179] Mallet-Paret, J. and R.D. Nussbaum, "Global continuation and complicated trajectories of periodic solutions for a differential-delay equation", pp. 155–167 in *Proc. of Symposia in Pure Math. Vol 45, Part 2* Amer. Math. Soc, Providence, RI, 1986.

[180] Mallet-Paret, J. and R.D. Nussbaum, "A differential-delay equation arising in optics and physiology", *SIAM J. Math. Anal.* **20**, (1989), 249–292.

[181] Mallet-Paret, J. and R.D. Nussbaum, "Boundary layer phenomena for differential delay equations with state dependent time lags I", *Arch. Rat. Mech. Anal.* **120**, (1992), 99–146.

[182] Mallet-Paret, J. and R.D. Nussbaum, "Boundary layer phenomena for differential delay equations with state dependent time lags II", *J. Reine Angew. Math.*, (to appear).

[183] Mallet-Paret, J. and R.D. Nussbaum, "Multiple transition layers in a singularly perturbed differential-delay equation", *Proc. Roy. Soc. Edinburgh Sect. A* **123**, (1993), 1119–1134..

[184] Mallet-Paret, J., R.D. Nussbaum and P. Paraskevopoulos, "Periodic solutions for functional differential equations with multiple state-dependent time lags", *Top. Meth. Nonlinear Anal.* **3**, (1994).

[185] Mallet-Paret, J. and G. Sell, *Systems of differential delay equations I: Floquet multipliers and discrete Lyapunov functions*, Preprint, Brown University, Providence, RI, 1994.

[186] Mallet-Paret, J. and G. Sell, *The Poincaré–Bendixson theorem for monotone cyclic feedback systems with delay*, Preprint, Brown University, Providence, RI, 1994.

[187] Mallet-Paret, J. and H.O. Walther, *Rapid oscillations are rare in scalar systems governed by monotone negative feedback with a time lag*, Preprint, 1994.

[188] Manitius, A., "Completeness and F-completeness of eigenfunctions associated with retarded functional differential equations", *J. Differential Eqn.* **35**, (1980), 1–29.

[189] Marshall, J.E., H. Górecki, K. Walton and A. Korytowski, *Time-Delay Systems: stability and performance criteria with applications*, Ellis Horwood, New York, 1992.

[190] Martelli, M., K. Schmitt and H. Smith, "Periodic solutions of some nonlinear delay-differential equations", *J. Math. Anal. Appl.* **74**, (1980), 494–503.

[191] Massera, J.L., "On Liapounoff's conditions of stability", *Ann. Math.* **50**, (1949), 705–721.

[192] McCord, C. and K. Mischaikow, *On the global dynamics of attractors for scalar delay equations*, Preprint, Georgia Institute of Technology, 1992.

[193] Metz, H. and O. Diekmann, *The Dynamics of Physiologically Structured Populations*, Springer-Verlag, Berlin, 1986.

[194] Mikusin'ski, J.G. and Cz. Ryll-Nardzewski, "Sur le produit de composition", *Studia Math.* **XII**, (1951), 51–57.

[195] Miller, R.K., *Nonlinear Volterra Integral Equations*, Benjamin, Menlo Park, CA, 1971.

[196] Miller, R.K., "Linear Volterra integro-differential equations as semigroups", *Funk. Ekv.* **17**, (1974), 39–55.

[197] Miller, R.K. and G.R. Sell,, *Volterra Integral Equations and Topological Dynamics*, Memoirs of the A.M.S, Vol. 102, American Mathematical Society, Providence, RI, 1970.

[198] Mischaikow, K., *On the existence of connecting orbits for scalar delay equations*, Preprint, Georgia Institute of Technology, 1992.

[199] Myshkis, A.D., *Lineare Differentialgleichungen mit nacheilendem Argument*, Deutsche Verlag der Wissenschaften, Berlin, 1955.

[200] Myshkis, A.D., *General Theory of Differential Equations with Retarded Argument*, A.M.S. Translations, Series I, 4, American Mathematical Society, Providence, RI, 1962 (Russian original 1949).

[201] Myshkis, A.D., *Linear Differential Equations with Retarded Argument*, (Russian), 2nd edition (revised, enlarged from a 1951 original), Izdat. Nauka, Moscow, 1971.

[202] Nagel, R. (ed.), *One-parameter Semigroups of Positive Operators*, Springer-Verlag, New York, 1986.

[203] Natanson, I.P., *Theorie der Funktionen einer Reellen Veränderlichen*, Akademie-Verlag, Berlin, 1954 (Russian edition 1949).

[204] van Neerven, J., *The Adjoint of a Semigroup of Linear Operators*, Springer-Verlag, New York, 1992.

[205] Negrini, P. and A. Tesei, "Attractivity and Hopf bifurcation in Banach spaces", *J. Math. Anal. Appl.* **78**, (1980), 204–221.

[206] Neubrander, F., "Integrated semigroups and their applications to the abstract Cauchy problem", *Pacific J. Math.* **135**, (1988), 111–155.

[207] Neugebauer, A., *Invariante Mannigfaltigkeiten und Neigungslemmata für Abbildungen in Banachräumen*, Diploma thesis, Universität München, 1988.

[208] Nisbett, R.M. and W.S.C. Gurney, *Modelling fluctuating populations*, Wiley, New York, 1982.

[209] Norkin, S.B., *Differential Equations of the 2nd Order with Retarded Argument*, Translations of Math. Monographs 31, American Mathematical Society, Providence, RI, 1972 (Russian original 1965).

[210] Nussbaum, R.D., "Periodic solutions of analytic functional differential equations are analytic", *Michigan Math. J.* **20**, (1973), 249–255.

[211] Nussbaum, R.D., "Periodic solutions of some nonlinear autonomous functional differential equations", *Ann. Mat. Pura Appl.* **101**, (1974), 263–306.

[212] Nussbaum, R.D., "Periodic solutions of some nonlinear autonomous functional differential equations II", *J. Diff. Eqn.* **14**, (1973), 368–394.

[213] Nussbaum, R.D., "A correction of "Periodic solutions of some nonlinear autonomous functional differential equations II"", *J. Diff. Eqn.* **16**, (1974), 548–549.

[214] Nussbaum, R.D., "A global bifurcation theorem with applications to functional differential equations", *J. Funct. Anal.* **19**, (1975), 319–339.

[215] Nussbaum, R.D., "Global bifurcation of periodic solutions of some autonomous functional differential equations", *J. Math. Anal. Appl.* **55**, (1976), 699–725.

[216] Nussbaum, R.D., "The range of periods of periodic solutions of $x'(t) = -\alpha f(x(t-1))$", *J. Math. Anal. Appl.* **58**, (1977), 280–292.

[217] Nussbaum, R.D., "A Hopf global bifurcation theorem for retarded functional differential equations", *Trans. Amer. Math. Soc.* **238**, (1978), 139–163.

[218] Nussbaum, R.D., "Periodic solutions of special differential delay equations: an example in nonlinear functional analysis", *Proc. Roy. Soc. Edinburgh Sect. A* **81**, (1978), 131–151.

[219] Nussbaum, R.D., *Diffential-delay equations with two time lags*, Memoirs of the A.M.S, Vol. 205, American Mathematical Society, Providence, RI, 1978.

[220] Nussbaum, R.D., "Uniqueness and nonuniqueness for periodic solutions of $x'(t) = -g(x(t-1))$", *J. Diff. Eqn.* **34**, (1979), 25–54.

[221] Nussbaum, R.D., "Periodic solutions of nonlinear autonomous functional differential equations", pp. 283–325 in *Functional Differential Equations and Approximations of Fixed Points, Bonn 1978* (H.O. Peitgen and H.O. Walther, eds.), Springer-Verlag, Berlin, 1979.

[222] Nussbaum, R.D., "Asymptotic analysis of some functional differential equations", pp. 277–301 in *Dynamical Systems II: Proc. of a Univ. of Florida International Symposium* (A.R. Bednarek and L. Cesari, eds.), Academic Press, New York, 1982.

[223] Nussbaum, R.D., "Asymptotic analysis of functional differential equations and solutions of long period", *Arch. Rat. Mech. Anal.* **81**, (1983), 373–397.

[224] Nussbaum, R.D., "Circulant matrices and differential-delay equations", *J. Diff. Eqn.* **60**, (1985), 201–217.

[225] Nussbaum, R.D., *The fixed point index and some applications*, Report, Université de Montrèal, 1985.

[226] Nussbaum, R.D., "Boundary layer phenomena for a differential-delay equation", pp. 579–599 in *Oscillation, Bifurcation and Chaos* (F.V. Atkinson, W.F. Langford and A.B. Mingarelli, eds.), Canadian Math. Soc. Conf. Proc., Vol 8, 1987.

[227] Nussbaum, R.D., "Wright's equation has no solutions of period four", *Proc. Roy. Soc. Edinburgh Sect. A* **113**, (1989), 281–288.

[228] Nussbaum, R.D. and H.O. Peitgen, *Special and Spurious Solutions of* $\dot{x}(t) = -\alpha f(x(t-1))$, Memoirs of the A.M.S, Vol. 251, American Mathematical Society, Providence, RI, 1984.

[229] Nussbaum, R.D. and A.J.B. Potter, "Cyclic differential equations and period three solutions of differential delay equations", *J. Diff. Eqn.* **46**, (1982), 379–408.

[230] Oliveira, J.C.F., "Hopf bifurcation for functional differential equations", *Nonlinear Anal.* **4**, (1980), 217–229.

[231] de Pagter, "A characterization of sun-reflexivity", *Math. Ann.* **283**, (1989), 511–518.

[232] Palis, J. and W. de Melo, *Geometric Theory of Dynamical Systems*, Springer-Verlag, New York, 1982.

[233] Pazy, A., *Semigroups of Linear Operators and Applications to Partial Differential Equations*, Springer-Verlag, New York, 1983.

[234] Packham, B.R., *The closing of resonance horns for periodically forced osclillators*, Report, University of Minnesota, 1988.

[235] S. Perlis, *Theory of Matrices*, Addison Wesley, Reading, MA, 1958.

[236] Perron, O., "Die Stabilitätsfrage bei Differentialgleichungen", *Math. Z.* **32**, (1930), 703–728.

[237] Pesin, Ya.B., "On the behavior of a strongly nonlinear differential equation with retarded argument", *Differentsialnye Uravnenija* **10**, (1974), 1025–1036.

[238] Peters, H., *Globales Lösungsverhalten zeitverzögerter Differentialgleichungen am Beispiel von Modellfunktionen*, Ph.D. thesis, Bremen, 1980.

[239] Phillips, R.S., "The adjoint semigroup", *Pacific J. Math.* **5**, (1955), 269–283.

[240] Pliss, V., "Principal reduction in the theory of stability of motion", *Izv. Akad. Nauk. SSSR Mat. Ser.* **28**, (1964), 1297–1324.

[241] Poenaru, V., *Singularites C en presence de symetrie : en particulier en presence de la symetrie d'un groupe de Lie compact*, Lecture Notes in Mathematics, Vol. 510, Springer-Verlag, Berlin, 1976.

[242] Prüss, J., "Linear evolutionary integral equations on the line", pp. 485–513 in *Evolution Equations, Control Theory and Biomathematics* (Ph. Clément and G. Lumer, eds.), Marcel Dekker, New York, 1994.

[243] Rabinowitz, P.H., "Some global results for nonlinear eigenvalue problems", *J. Funct. Anal.* **7**, (1971), 487–513.

[244] Rabier, P.J., "Generalized Jordan chains and two bifurcation theorems of Krasnoselskii,", *Nonlinear Anal. Th. Meth. Appl.* **13**, (1989), 903–934.

[245] Rudin, W., *Real and Complex Analysis (2nd edition)*, McGraw-Hill, New York, 1974.

[246] Ruston, A.F., *Fredholm theory in Banach spaces*, Cambridge University Press, Cambridge, 1986.

[247] Sabelis, M.W. and O. Diekmann, "Overall population stability despite local extinction: the stabilizing influence of prey dispersal from predator-invaded patches", *Theor. Pop. Biol.* **34**, (1988), 169–176.

[248] Schmitt, E., "Über eine Klasse linearer funktionaler Differentialgleichungen", *Math. Ann.* **70**, (1911), 499–524.

[249] Sharpe, F.R. and A.J. Lotka, "A problem in age-distribution", *Philosophical Magazine,* **21**, (1911), 435–438.

[250] Shilnikov, L.P., "On a Poincaré-Birkhoff problem", *Math. U.S.S.R. Sbornik* **3**, (1967), 353–371.

[251] Sijbrand, J., "Properties of center manifolds", *Trans. Amer. Math. Soc.* **289**, (1985), 431–469.

[252] Smale, S., "Diffeomorphisms with many periodic points", pp. 63–80 in *Differential and Combinatorial Topology* Princeton University Press, Princeton, NJ, 1965.

[253] Smale, S., "Differentiable dynamical systems", *Bull. Amer. Math. Soc.* **73**, (1967), 747–817.

[254] Smith, H.L., "Monotone semiflows generated by functional differential equations", *J. Diff. Eqn.* **66**, (1987), 420–442.

[255] Smith, H.L., "A structured population model and a related functional differential equation: Global attractors and uniform persistence", *J. Dyn. Diff. Eqn.* **6**, (1994), 71–99.

[256] Smith, H.L. and H.R. Thieme, "Monotone semiflows in scalar non-quasimonotone functional differential equations", *J. Math. Anal. Appl.* **150**, (1990), 289–306.

[257] Smith, H.L. and H.R. Thieme, "Strongly order preserving semiflows generated by functional differential equations", *J. Diff. Eqn.* **93**, (1991), 332–363.

[258] Smith, R.A., "Existence of periodic orbits of autonomous retarded functional differential equations", *Math. Proc. Camb. Phil. Soc.* **88**, (1980), 89–109.

[259] Smith, R.A., "Poincaré-Bendixson theory for certain retarded functional differential equations", *Diff. Int. Eqn.* **5**, (1992), 213–240.

[260] Staffans, O., "Hopf bifurcation for functional and functional differential equations with infinite delay", *J. Diff. Eqn.* **70**, (1987), 114–151.

[261] Stech, H.W., "The Hopf bifurcation: a stability result and application", *J. Math. Anal. Appl.* **71**, (1979), 525–546.

[262] Stech, H.W., *On the computation of the stability of the Hopf bifurcation*, Virginia Polytechnics Institute, Blacksburg, 1984.

[263] Stech, H.W., "Nongeneric Hopf bifurcation in functional differential equations", *SIAM J. Math. Anal.* **16**, (1985), 1134–1151.

[264] Stech, H.W., "Hopf bifurcation calculation for functional differential equations", *J. Math. Anal. Appl.* **109**, (1985), 472–491.

[265] Stech, H.W., "Nongeneric Hopf bifurcation in a functional differential equation in population biology", *J. Integral Eqn.* **10**, (1985), 343–350.

[266] Steinlein, H. and H.O. Walther, "Hyperbolic sets and shadowing for noninvertible maps", pp. 219–234 in *Advanced Topics in the Theory of Dynamical Systems* (G. Fusco, M. Iannelli and L. Salvadori, eds.), Academic Press, New York, 1989.

[267] Steinlein, H. and H.O. Walther, "Hyperbolic sets, transversal homoclinic trajectories, and symbolic dynamics for C^1-maps in Banach spaces", *J. Dyn. Diff. Eqn.* **2**, (1990), 325–365.

[268] Stépán, G., *Retarded dynamical systems: stability and characteristic functions*, Pitman, Boston, 1989.

[269] Swick, K.E., "Stability and bifurcation in age-dependent population dynamics", *Theor. Pop. Biol.* **20**, (1981), 80–100.

[270] Treves, F., *Basic linear partial differential equations*, Academic Press, New York, 1975.

[271] Takens, F., "Singularities of vector fields", *Publ. Math. IHES* **43**, (1974), 47–100.

[272] Taylor, A.E. and D.C. Lay, *Introduction to Functional Analysis*, Wiley, New York, 1980.

[273] Thieme, H., "Semiflows generated by Lipschitz perturbations of non-densely defined operators", *Diff. Int. Eqn.* **3**, (1990), 1035–1066.

[274] Thieme, H., "Integrated semigroups and integrated solutions to abstract Cauchy problems", *J. Math. Anal. Appl.* **152**, (1990), 416–447.

[275] Titchmarsh, E.C., *The Theory of Functions (2nd edition)*, Oxford University Press, London, 1979.

[276] Vanderbauwhede, A. and S.A. van Gils, "Center manifolds and contractions on a scale of Banach spaces", *J. Funct. Anal.* **71**, (1987), 209–224.

[277] Vanderbauwhede, A., "Centre manifolds, normal forms and elementary bifurcations", pp. 89–169 in *Dynamics Reported Vol. 2* (U. Kirchgraber and H.O. Walther, eds.), Teubner and Wiley, Stuttgart, 1989.

[278] Vanderbauwhede, A. and G. Iooss, "Center manifold Theory in Infinite Dimensions", in *Dynamics Reported (New Series) Vol. 1* (C.K.R.T. Jones and U. Kirchgraber and H.O. Walther, eds.), Springer-Verlag, New York, 1992.

[279] Verduyn Lunel, S.M., "A sharp version of Henry's theorem on small solutions", *J. Diff. Eqn.* **62**, (1986), 266–274.

[280] Verduyn Lunel, S.M., *Exponential type calculus for linear delay equations*, Centre for Mathematics and Computer Science, Tract No. 57 Amsterdam, 1989.

[281] Verduyn Lunel, S.M., "Series expansions and small solutions for Volterra equations of convolution type", *J. Diff. Eqn.* **85**, (1990), 17–53.

[282] Verduyn Lunel, S.M., "The closure of the generalized eigenspace of a class of infinitesimal generators", *Proc. Roy. Soc. Edinburgh Sect. A* **117A**, (1991), 171–192.

[283] Verduyn Lunel, S.M., "About completeness for a class of unbounded operators", *J. Diff. Eqn.*, (to appear).

[284] Verduyn Lunel, S.M., "Series expansions for functional differential equations", *Int. Eqn. Operator Th.*, (to appear).

[285] Walther, H.O., *Über Ejektivität und periodische Lösungen bei Funktionaldifferentialgleichungen mit verteilter Verzögerung*, Habilitationsschrift, Universität München, 1977.

[286] Walther, H.O., "A theorem on the amplitudes of periodic solutions of delay equations, with an application to bifurcation", *J. Diff. Eqn.* **29**, (1978), 396–404.

[287] Walther, H.O., "On instability, ω–limit sets and periodic solutions of nonlinear autonomous differential delay equations", pp. 489–503 in *Functional Differential Equations and Approximation of Fixed Points* (H.O. Peitgen and H.O. Walther, eds.), Springer-Verlag, Berlin, 1979.

[288] Walther, H.O., "Density of slowly oscillating solutions of $\dot{x}(t) = -f(x(t-1))$", *J. Math. Anal. Appl.* **79**, (1981), 127–140.

[289] Walther, H.O., "Delay equations: Instability and the trivial fixed point's index", pp. 231–238 in *Abstract Cauchy Problems and Functional Differential Equations* (F. Kappel and W. Schappacher, eds.), Pitman, London, 1981.

[290] Walther, H.O., "Homoclinic solution and chaos in $\dot{x}(t) = f(x(t-1))$", *Nonlinear Anal.* **5**, (1981), 775–788.

[291] Walther, H.O., "Bifurcation from periodic solutions in functional differential equations", *Math. Z.* **182**, (1983), 269–289.

[292] Walther, H.O., "Bifurcation from a heteroclinic solution in differential delay equations", *Trans. Amer. Math. Soc.* **290**, (1985), 213–233.

[293] Walther, H.O., "Inclination lemmas with dominated convergence", *J. Appl. Math. Phys.* **32**, (1987), 327–337.

[294] Walther, H.O., "Bifurcation from homoclinic to periodic solutions by an inclination lemma with pointwise estimate", in *Dynamics of Infinite Dimensional Systems* (S.N. Chow and J.K. Hale, eds.), Springer-Verlag, New York, 1987.

[295] Walther, H.O., "Homoclinic and periodic solutions of scalar differential delay equations", pp. 243–263 in *Dynamical Systems and Ergodic Theory* Banach Center Publ. 23,, PWN Polish Scientific Publ., Warsaw, 1989.

[296] Walther, H.O., "Bifurcation from a saddle connection in functional differential equations: An approach with inclination lemmas", *Dissertationes Mathematicae* **CCXCI**, (1990).

[297] Walther, H.O., *Hyperbolic periodic solutions, heteroclinic connections and transversal homoclinic points in autonomous differential delay equations*, Memoirs of the A.M.S, Vol. 402, American Mathematical Society, Providence, RI, 1989.

[298] Walther, H.O., "An invariant manifold of slowly oscillating solutions for $\dot{x}(t) = -\mu x(t) + f(x(t-1))$", *J. Reine Angew. Math.* **414**, (1991), 67–112.

[299] Walther, H.O., "On Floquet multipliers of periodic solutions of delay equations with monotone nonlinearities", pp. 349–356 in *Proc. Int. Symp. on Functional Differential Equations, Kyoto* (J. Kato and T. Yoshizawa, eds.), World Scientific, Singapore, 1991.

[300] Walther, H.O., "Unstable manifolds of periodic orbits of a differential delay equation", pp. 177–240 in *Oscillation and Dynamics of Delay Equations* (J.R. Graef and J.K. Hale, eds.), Contemp. Math. 129,, American Mathemathical Society, Providence, RI, 1992.

[301] Walther, H.O., "Corrections to "Bifurcation from a saddle connection in functional differential equations: An approach with inclination lemmas"", *Annales Polonici Mathematici* **LVIII**, (1993), 105–106.

[302] Walther, H.O., *The 2–dimensional attractor of* $x' = -\mu x(t) + f(x(t-1))$, Memoirs of the A.M.S, American Mathemathical Society, Providence, RI, to appear.

[303] Walther, H.O., "A differential delay equations with a planar attractor", in *Proc. of the Int. Conf. on Diff. Eqs Marrakech 1991,* to appear.

[304] Webb, G.F., *Theory of Nonlinear Age-dependent Population Dynamics*, Marcel Dekker, New York, 1985.

[305] Whitney, H., "Differentiable even functions", *Duke Math. J.* **10**, (1943), 159–160.

[306] Whyburn, G.T., *Topological Analysis*, Princeton Univ. Press, Princeton, NJ, 1958.

[307] Widder, D.V., *The Laplace Transform*, Princeton Univ. Press, Princeton, NJ, 1946.

[308] Wright, E.M., "A non-linear differential-difference equation", *J. Reine Angew. Math.* **194**, (1955), 66–87.

[309] Wu, J., *Delay-induced discrete waves of large amplitudes in neural networks with circulant connection matrices*, Preprint, The Fields Institute for Research in the Mathematical Sciences, 1993.

[310] Xie, X., "Uniqueness and stability of slowly oscillating periodic solutions of delay equations with bounded nonlinearity", *J. Dyn. Diff. Eqn.* **3**, (1991), 515–540.

[311] Xie, X., "The multiplier equation and its application to S-solution of differential delay equations", *J. Diff. Eqn.*, (to appear).

[312] Xie, X., "Uniqueness and stability of slowly oscillating periodic solutions of differential delay equations with unbounded nonlinearity", *J. Diff. Eqn.*, to appear.

[313] Yosida, K., *Functional Analysis, 6th edn.*, Springer-Verlag, New York, 1980.

Index

List of symbols

List of notation

AIE	abstract integral equation
$\mathrm{AC}(I,E)$	space of absolutely continuous functions from an interval I into a space E
$\operatorname{adj} M$	matrix of cofactors of a matrix M
$\mathbb{C}^n$	complex n-dimensional vectorspace
$\mathbb{C}^{n\times n}$	space of complex $n \times n$ matrices
$X_{\mathbb{C}}$	complexification of X
$C_b^k(E,F)$	Banach space of k-times continuously differentiable mappings from a Banach space E into a Banach space F with bounded derivatives up to order k
$\mathrm{E}(F)$	exponential type of F
$\operatorname{Im} z$	imaginary part of a complex number z
$L^1(I,E)$	Banach space of L^1-integrable functions from an interval I into a space E equipped with the L^1-norm
$L^\infty(I,E)$	Banach space of L^∞-integrable functions from an interval I into a space E equipped with the L^∞-norm
$\mathcal{N}(L)$	null space of an operator L
$\langle x^*, x\rangle$	pairing between an element $x^* \in X^*$ and $x \in X$

$\mathcal{R}(L)$	range of an operator
$\operatorname{Re} z$	real part of a complex number z
$\mathbb{R}^n$	real n-dimensional vectorspace
$\mathbb{R}_-$	set of negative real numbers
$\mathbb{R}_+$	set of positive real numbers
RE	renewal equation
$\operatorname{Res}_{z=\lambda} F$	residue of F at $z = \lambda$
RFDE	retarded functional differential equation
$\operatorname{tr} L$	trace of a finite rank operator L
M^T	transpose of a matrix M

Applied Mathematical Sciences

(continued from page ii)

61. *Sattinger/Weaver:* Lie Groups and Algebras with Applications to Physics, Geometry, and Mechanics.
62. *LaSalle:* The Stability and Control of Discrete Processes.
63. *Grasman:* Asymptotic Methods of Relaxation Oscillations and Applications.
64. *Hsu:* Cell-to-Cell Mapping: A Method of Global Analysis for Nonlinear Systems.
65. *Rand/Armbruster:* Perturbation Methods, Bifurcation Theory and Computer Algebra.
66. *Hlavácek/Haslinger/Necasl/Lovísek:* Solution of Variational Inequalities in Mechanics.
67. *Cercignani:* The Boltzmann Equation and Its Applications.
68. *Temam:* Infinite Dimensional Dynamical Systems in Mechanics and Physics.
69. *Golubitsky/Stewart/Schaeffer:* Singularities and Groups in Bifurcation Theory, Vol. II.
70. *Constantin/Foias/Nicolaenko/Temam:* Integral Manifolds and Inertial Manifolds for Dissipative Partial Differential Equations.
71. *Catlin:* Estimation, Control, and the Discrete Kalman Filter.
72. *Lochak/Meunier:* Multiphase Averaging for Classical Systems.
73. *Wiggins:* Global Bifurcations and Chaos.
74. *Mawhin/Willem:* Critical Point Theory and Hamiltonian Systems.
75. *Abraham/Marsden/Ratiu:* Manifolds, Tensor Analysis, and Applications, 2nd ed.
76. *Lagerstrom:* Matched Asymptotic Expansions: Ideas and Techniques.
77. *Aldous:* Probability Approximations via the Poisson Clumping Heuristic.
78. *Dacorogna:* Direct Methods in the Calculus of Variations.
79. *Hernández-Lerma:* Adaptive Markov Processes.
80. *Lawden:* Elliptic Functions and Applications.
81. *Bluman/Kumei:* Symmetries and Differential Equations.
82. *Kress:* Linear Integral Equations.
83. *Bebernes/Eberly:* Mathematical Problems from Combustion Theory.
84. *Joseph:* Fluid Dynamics of Viscoelastic Fluids.
85. *Yang:* Wave Packets and Their Bifurcations in Geophysical Fluid Dynamics.
86. *Dendrinos/Sonis:* Chaos and Socio-Spatial Dynamics.
87. *Weder:* Spectral and Scattering Theory for Wave Propagation in Perturbed Stratified Media.
88. *Bogaevski/Povzner:* Algebraic Methods in Nonlinear Perturbation Theory.
89. *O'Malley:* Singular Perturbation Methods for Ordinary Differential Equations.
90. *Meyer/Hall:* Introduction to Hamiltonian Dynamical Systems and the N-body Problem.
91. *Straughan:* The Energy Method, Stability, and Nonlinear Convection.
92. *Naber:* The Geometry of Minkowski Spacetime.
93. *Colton/Kress:* Inverse Acoustic and Electromagnetic Scattering Theory.
94. *Hoppensteadt:* Analysis and Simulation of Chaotic Systems.
95. *Hackbusch:* Iterative Solution of Large Sparse Systems of Equations.
96. *Marchioro/Pulvirenti:* Mathematical Theory of Incompressible Nonviscous Fluids.
97. *Lasota/Mackey:* Chaos, Fractals, and Noise: Stochastic Aspects of Dynamics, 2nd ed.
98. *de Boor/Höllig/Riemenschneider:* Box Splines.
99. *Hale/Lunel:* Introduction to Functional Differential Equations.
100. *Sirovich (ed):* Trends and Perspectives in Applied Mathematics.
101. *Nusse/Yorke:* Dynamics: Numerical Explorations.
102. *Chossat/Iooss:* The Couette-Taylor Problem.
103. *Chorin:* Vorticity and Turbulence.
104. *Farkas:* Periodic Motions.
105. *Wiggins:* Normally Hyperbolic Invariant Manifolds in Dynamical Systems.
106. *Cercignani/Illner/Pulvirenti:* The Mathematical Theory of Dilute Gases.
107. *Antman:* Nonlinear Problems of Elasticity.
108. *Zeidler:* Applied Functional Analysis: Applications of Mathematical Physics.
109. *Zeidler:* Applied Functional Analysis: Main Principles and Their Applications.
110. *Diekmann/van Gils/Verduyn Lunel/Walther:* Delay Equations: Functional-, Complex-, and Nonlinear Analysis.
111. *Visintin:* Differential Models of Hysteresis.
112. *Kuznetsov:* Elements of Applied Bifurcation Theory.
113. *Hislop/Sigal:* Introduction to Spectral Theory: With Applications to Schrödinger Operators.